Kyu-Young Whang   Jongwoo Jeon
Kyuseok Shim   Jaideep Srivastava (Eds.)

Advances in Knowledge Discovery and Data Mining

7th Pacific-Asia Conference, PAKDD 2003
Seoul, Korea, April 30 – May 2, 2003
Proceedings

Springer

Volume Editors

Kyu-Young Wang
Korea Advanced Institute of Science and Technology
Computer Science Department
373-1 Koo-Sung Dong, Yoo-Sung Ku, Daejeon 305-701, Korea
E-mail: kywhang@cs.kaist.ac.kr

Jongwoo Jeon
Seoul National University
Department of Statistics
Sillimdong Kwanakgu, Seoul 151-742, Korea
E-mail: jwjeon@plaza.snu.ac.kr

Kyuseok Shim
Seoul National University
School of Electrical Engineering and Computer Science
Kwanak P.O. Box 34, Seoul 151-742, Korea
E-mail: shim@ee.snu.ac.kr

Jaideep Srivastava
University of Minnesota
Department of Computer Science and Engineering
200 Union St SE, Minneapolis, MN 55455, USA
E-mail: srivasta@umn.edu

Cataloging-in-Publication Data applied for

A catalog record for this book is available from the Library of Congress.

Bibliographic information published by Die Deutsche Bibliothek.
Die Deutsche Bibliothek lists this publication in the Deutsche Nationalbibliografie;
detailed bibliographic data is available in the Internet at <http://dnb.ddb.de>.

CR Subject Classification (1998): I.2, H.2.8, H.3, H.5.1, G.3, J.1, K.4

ISSN 0302-9743
ISBN 3-540-04760-3 Springer-Verlag Berlin Heidelberg New York

Springer-Verlag Berlin Heidelberg New York,
a member of BertelsmannSpringer Science+Business Media GmbH

http://www.springer.de

© Springer-Verlag Berlin Heidelberg 2003
Printed in Germany

Typesetting: Camera-ready by author, data conversion by PTP-Berlin, Stefan Sossna
Printed on acid-free paper   SPIN: 10927366   06/3142   5 4 3 2 1 0

Preface

The 7th Pacific Asia Conference on Knowledge Discovery and Data Mining (PAKDD) was held from April 30 to May 2, 2003 in the Convention and Exhibition Center (COEX), Seoul, Korea. The PAKDD conference is a major forum for academic researchers and industry practitioners in the Pacific Asia region to share original research results and development experiences from different KDD-related areas such as data mining, data warehousing, machine learning, databases, statistics, knowledge acquisition and discovery, data visualization, and knowledge-based systems. The conference was organized by the Advanced Information Technology Research Center (AITrc) at KAIST and the Statistical Research Center for Complex Systems (SRCCS) at Seoul National University. It was sponsored by the Korean Datamining Society (KDMS), the Korea Information Science Society (KISS), the United States Air Force Office of Scientific Research, the Asian Office of Aerospace Research & Development, and KAIST. It was held with cooperation from ACM's Special Group on Knowledge Discovery and Data Mining (SIGKDD).

In its seventh year, PAKDD has reached new heights of recognition as a quality forum for the dissemination and discussion of KDD research. This is evident from the high number of submissions and the highly selective nature of the review process. Out of 215 submissions, only 38 were selected as full papers, and 20 as short papers for inclusion in the proceedings. Given this, we were unable to accommodate many high-quality papers. We thank the authors of all submitted papers for their efforts, which provided the basic ingredient for developing an exceptionally solid technical program. PAKDD has become a truly international forum, as is evident from the fact that papers were submitted from 22 countries across six continents. Similarly, the program committee consisted of 70 members from five continents.

The PAKDD 2003 technical program was enhanced by two keynote speeches, delivered by Dr. Rakesh Agrawal of IBM's Almaden Research Center and Prof. Andrea Buja of the University of Pennsylvania. In addition to the main technical program, PAKDD 2003 also had two tutorials, namely Data Mining for Intrusion Detection (by Aleksandar Lazarevic, Jaideep Srivastava and Vipin Kumar), and Analyzing and Mining Data Streams (by Sudipto Guha and Nick Koudas). Finally, two international workshops, titled "Mining Data for Actionable Knowledge," and "Biological Data Mining," were also organized in conjunction with this meeting. Articles in these workshops are being published separately.

The conference would not have been a success without help from many people, and our thanks go to all of them. We would like to express our special thanks to our honorary chairs, Won Kim at Cyber Database Solutions and Sukho Lee at Seoul National University for providing valuable advice on all aspects of the conference's organization. We also extend our sincere thanks to the international advisory committee members, Dr. Rakesh Agrawal, Prof. Jiawei Han, and Dr.

Kuan-Tsae Huang for their moral support. We owe a great debt to the program committee members for doing their reviews in a timely manner. This is especially commendable, given that most of the reviews were done during the holidays and at short notice. Sang Kyun Cha and Rajeev Rastogi solicited and selected excellent tutorials. Doheon Lee and Bongki Moon arranged several workshops on interesting topics. Ramakrishnan Srikant ensured excellent talks from experienced leaders from industry. Jaehee Cho solicited valuable industry support and coordinated industry demonstrations. Myoung Kim and Byung S. Lee did a tremendous job in distributing the CFP to solicit papers. Byung Yeon Hwang ensured a smooth and successful registration process. Yong Chul Oh, Sungzoon Cho and Jinho Kim handled the local arrangements exceptionally well. As publications chair, Byoung-Kee Yi ensured that all authors submitted the final versions of their papers in a timely and error-free manner. The high-quality proceedings and tutorial notes are a result of this effort. Eui Kyeong Hong and Won Chul Jhee ensured the smooth operation of the conference. Needless to say, all members of the Steering Committee provided their full support in making PAKDD 2003 a success.

April 2003   Kyu-Young Whang
                                                             Jongwoo Jeon
                                                             Kyuseok Shim
                                                         Jaideep Srivastava

PAKDD 2003 Conference Committee

Honorary Chairs:
Won Kim   Cyber Database Solutions, USA,
Ewha Womans University, Korea
Sukho Lee   Seoul National University, Korea

International Advisory Committee:
Rakesh Agrawal   IBM Almaden Research Center, USA
Jiawei Han   University of Illinois, Urbana-Champaign, USA
Kuan-Tsae Huang   Taskco Corporation, Taiwan
Jongwoo Jeon   Seoul National University/SRCCS, Korea

Conference Chair:
Kyu-Young Whang   KAIST/AITrc, Korea

Program Chairs:
Kyuseok Shim   Seoul National University/AITrc, Korea
Jaideep Srivastava   University of Minnesota, USA

Workshop Chairs:
Doheon Lee   KAIST, Korea
Bongki Moon   University of Arizona, USA

Tutorial Chairs:
Sang Kyun Cha   Seoul National University, Korea
Rajeev Rastogi   Lucent Bell Labs, USA

Industrial Program Chairs:
Jaehee Cho   Kwangwoon University, Korea
Ramakrishnan Srikant   IBM Almaden Research Center, USA

Publication Chair:
Byoung-Kee Yi   POSTECH, Korea

Registration Chair:
Byung Yeon Hwang   Catholic University of Korea, Korea

Publicity Chairs:
Myung Kim   Ewha Womans University, Korea
Byung S. Lee   University of Vermont, USA

Local Arrangements:
Yong Chul Oh   Korea Polytechnic University, Korea (Chair)

Sungzoon Cho   Seoul National University, Korea
Jinho Kim   Kangwon National University/AITrc, Korea

Organization Committee:
Eui Kyeong Hong   University of Seoul/AITrc, Korea
Won Chul Jhee   Hongik University, Korea

PAKDD Steering Committee:
Hongjun Lu   Hong Kong University of Science
                                and Technology,
                                Hong Kong (Chair)

Hiroshi Motoda   Osaka University, Japan (Co-chair)
Arbee L.P. Chen   National Tsing Hua University, Taiwan
Ming-Syan Chen   National Taiwan University, Taiwan
David W. Cheung   University of Hong Kong, Hong Kong
Masaru Kitsuregawa   University of Tokyo, Japan
Rao Kotagiri   University of Melbourne, Australia
Huan Liu   Arizona State University, USA
Takao Terano   University of Tsukuba, Japan
Graham Williams   CSIRO, Australia
Ning Zhong   Maebashi Institute of Technology, Japan
Lizhu Zhou   Tsinghua University, China

PAKDD 2003 Program Committee

Raj Acharya	Pennsylvania State University, USA
Roberto Bayardo	IBM Almaden Research Center, USA
Elisa Bertino	University of Milano, Italy
Bharat Bhasker	Indian Institute of Management Lucknow, India
Paul Bradley	digiMine, USA
Arbee L.P. Chen	National Tsing Hua University, Taiwan
Ming-Syan Chen	National Taiwan University, Taiwan
David Cheung	University of Hong Kong, Hong Kong
Daewoo Choi	Hankuk University of Foreign Studies, Korea
Gautam Das	Microsoft Research, USA
Umesh Dayal	HP Labs, USA
Usama Fayyad	digiMine, USA
Venkatesh Ganti	Microsoft Research, USA
Minos Garofalakis	Lucent Bell Labs, USA
Johannes Gehrke	Cornell University, USA
Joydeep Ghosh	University of Texas, Austin, USA
Dimitrios Gunopoulos	University of California, Riverside, USA
Shyam Kumar Gupta	Indian Institute of Technology Delhi, India
Sudipto Guha	University of Pennsylvania, USA
Jayant R. Haritsa	Indian Institute of Science, India
San-Yih Hwang	National Sun-Yat Sen University, Taiwan
Soon Joo Hyun	Information and Communication University, Korea
H.V. Jagadish	University of Michigan, USA
Myoung Ho Kim	KAIST, Korea
Masaru Kitsuregawa	University of Tokyo, Japan
Rao Kotagiri	University of Melbourne, Australia
Nick Koudas	AT&T Research, USA
Vipin Kumar	University of Minnesota, USA
Laks V.S. Lakshmanan	University of British Columbia, Canada
Aleks Lazarevic	University of Minnesota, USA
Kyuchul Lee	Chungnam National University, Korea
Sangho Lee	Soongsil University, Korea
Sang-goo Lee	Seoul National University, Korea
Yoon-Joon Lee	KAIST, Korea
Yugi Lee	University of Missouri, Kansas City, USA
Jianzhong Li	Harbin Institute of Technology, China
Ee-Peng Lim	Nanyang Technological University, Singapore
Tok Wang Ling	National University of Singapore, Singapore
Bing Liu	University of Chicago, USA
Huan Liu	Arizona State University, USA
Hongjun Lu	Hong Kong University of Science and Technology, Hong Kong

Akifumi Makinouch	Kyushu University, Japan
Heikki Mannila	Nokia, Finland
Yoshifumi Masunaga	Ochanomizu University, Japan
Hiroshi Motoda	Osaka University, Japan
Rajeev Motwani	Stanford University, USA
Sham Navathe	Georgia Institute of Technology, USA
Raymond Ng	University of British Columbia, Canada
Beng Chin Ooi	National University of Singapore, Singapore
T.V. Prabhakar	Indian Institute of Technology Kanpur, India
Raghu Ramakrishnan	University of Wisconsin, USA
Keun Ho Ryu	Chungbuk National University, Korea
Shashi Shekhar	University of Minnesota, USA
Kian-Lee Tan	National University of Singapore, Singapore
Pang-Ning Tan	University of Minnesota, USA
Takao Terano	University of Tsukuba, Japan
Bhavani Thuraisingham	National Science Foundation, USA
Hannu T.T. Toivonen	Nokia Research, Finland
Jau-Hwang Wang	Central Police University, Taiwan
Shan Wang	Renmin University of China, China
Richard Weber	University of Chile, Chile
Graham Williams	CSIRO, Australia
You-Jip Won	Hanyang University, Korea
Xindong Wu	University of Vermont, USA
Yong-Ik Yoon	Sookmyung Women's University, Korea
Philip S. Yu	IBM T.J. Watson Research Center, USA
Mohammed Zaki	Rensselaer Polytechnic Institute, USA
Ning Zhong	Maebashi Institute of Technology, Japan
Aoying Zhou	Fudan University, China
Lizhu Zhou	Tsinghua University, China

PAKDD 2003 External Reviewers

Zaher Aghbari
Lili Aunimo
Ilkka Autio
James Bailey
Vasudha Bhatnagar
Cristian Bucila
Jaehyuk Cha
Jaeyoung Chang
Li-Ging Chao
Hung-Chen Chen
Shian-Ching Chen
Yi-Wen Chen
Ru-Lin Cheng
Ding-Ying Chiu
Chung-Wen Cho
Kun-Ta Chuang
Zeng Chun
Yon-Dohn Chung
Manoranjan Dash
Prasanna Desikan
Koji Eguchi
Eric Eilertson
Levent Ertoz
Yi-Sheng Fan
Takeshi Fukuda
Deepak Gupta
Ming Hsu
Keyun Hu
Jiun-Long Huang
Noriko Imafuji
Nuanjiang Jin
Xiaoming Jin
Kunihiko Kaneko
Hung-Yu Kao
Harish Karnick
Eamonn Keogh
Ravi Khemani
Alok Khemka
Daniel Kifer
Han-Joon Kim
Yongdai Kim

Jukka Kohonen
Eunbae Kong
David W.-Y. Kwong
Kari Laasonen
Ighoon Lee
Carson K.-S. Leung
Chai Han Lin
Xueyuan Lin
Ning-Han Liu
Yi-Ching Liu
Juha Makkonen
Naoyuki Nomura
Aysel Ozgur
B. Uygar Oztekin
U.K. Padmaja
Dimitris Papadopoulos
Neoklis Polyzotis
Iko Pramudiono
Vani Prasad
V. Pratibha
Vikram Pudi
Weining Qian
Mirek Riedewald
Junho Shim
Takahiko Shintani
Sridhar
K. Srikumar
Suresha
Eng Koon Sze
Taehee
Katsumi Takahashi
Fengzhan Tian
Anthony K.H. Tung
Foula Vagena
Yitong Wang
Yew-Kwong Woon
Du Xiaoping
Hui Xiong
Lei Yu
Zhiqiang Zhang
Shuigeng Zhou

Sponsorship

We wish to thank the following organizations for their contributions to the success of this conference. We note that AFOSR/AOARD support is not intended to imply endorsement by the US federal government.

Major Sponsors

Advanced Information Technology
Research Center, KAIST

Statistical Research Center for
Complex Systems, SNU

Academic Sponsors

Korea Information Science Society

KDMS
The Korean Datamining Society

The Korean Datamining Society

Sponsors

Asian Office of Aerospace
Research & Development

Air Force Office of
Scientific Research

Korea Advanced
Institute of Science and
Technology

In Cooperation with

ACM/SIGKDD

Table of Contents

Session 2B: Bio Mining

Session 2C: Web Mining

Session 3A: Stream Mining II

Session 3B: Bayesian Networks

Session 3C: Clustering II

Session 4A: Association Rules I

Session 4B: Semi-structured Data Mining

Session 4C: Classification I

Session 5A: Data Analysis

Session 5B: Association Rules II

Session 5C: Feature Selection

Session 6A: Stream Mining III

Session 6B: Clustering III

Session 6C: Classification II

Data Mining as an Automated Service

P.S. Bradley

Bradley Data Consulting
bradleyp@acm.org

Abstract. An automated data mining service offers an out-sourced, cost-effective analysis option for clients desiring to leverage their data resources for decision support and operational improvement. In the context of the service model, typically the client provides the service with data and other information likely to aid in the analysis process (e.g. domain knowledge, etc.). In return, the service provides analysis results to the client. We describe the required processes, issues, and challenges in automating the data mining and analysis process when the high-level goals are: (1) to provide the client with a high quality, pertinent analysis result; and (2) to automate the data mining service, minimizing the amount of human analyst effort required and the cost of delivering the service. We argue that by focusing on client problems within market sectors, both of these goals may be realized.

1 Introduction

The amount spent by organizations in implementing and supporting database technology is considerable. Global 3500 enterprises spend, typically, $664,000 on databases annually, primarily focusing on transactional systems and web-based access [26]. Unfortunately, the ability of organizations to effectively utilize this information for decision support typically lags behind their ability to collect and store it. But, organizations that *can* leverage their data for decision support are more likely to have a competitive edge in their sector of the market [19].

An organization may choose from a number of options when implementing the data analysis and mining technology needed to support decision-making processes or to optimize operations. One option is to perform the analysis within the organization (i.e. "in-house"). It is costly to obtain the ingredients required for a successful project: analytical and data mining experience on the part of the team performing the work, software tools, and hardware. Hence the "in-house" data mining investment is substantial.

Another option that an organization may pursue is to out-source the data analysis and mining project to a third party. Data mining consultants have gone into business to address this market need. Consultants typically have data analysis and mining experience within one or more market sectors (e.g. banking, health-care, etc.), have analysis and mining software at their disposal, have the required hardware to process large datasets, and are able to produce customized analysis solutions. The cost of an outsourced data mining project is dependent upon a number of factors including: project complexity, data availability, and

K.-Y. Whang, J. Jeon, K. Shim, J. Srivatava (Eds.): PAKDD 2003, LNAI 2637, pp. 1–13, 2003.

dataset size. But, on average, outsourcing such projects to a third party consulting firm tends to cost on the order of tens to hundreds of thousands of dollars [21].

Unfortunately, a number of organizations simply do not have the resources available to perform data mining and analysis "in-house" or to out-source these tasks to consulting firms. But these organizations often realize that they may be able to gain a competitive advantage by utilizing their data for decision support and/or operational improvement. Hence, there is an opportunity to deliver data mining and analysis services at a reduced cost. We argue that an *automated data mining service* can provide such analysis at a reduced cost by targeting organizations (or *data mining clients*) within a given market sector (or market *vertical*) and automating the knowledge discovery and data mining process [12, 9] to the extent possible.

We note that there are similarities between the design of an automated data mining service and data mining software packages focusing on specific business needs within given vertical markets. By focusing on specific problems in a vertical market, both the software and service designer are able to address data format and preparation issues and choose appropriate modeling, evaluation and deployment strategies. An organization may find a software solution appealing, if it addresses their specific analysis needs, has the proper interfaces to the data and end-user (who may not be an analyst), and the analysis results are easily deployable. Organizations that have problems not specifically addressed by software solutions, or are attracted to a low-cost, outsourced alternative, may find an automated data mining service to be their best option.

We next present the steps of the knowledge discovery and data mining process to provide context for the discussion on automating various steps. Data mining project engagements typically follow the following process (note that the sequence is not strict and often moving back and forth between steps is required) [9]:

1. *Problem understanding*: This initial step primarily focuses on data mining problem definition and specification of the project objectives.
2. *Data understanding*: This step includes data extraction, data quality measurement, and detection of "interesting" data subsets.
3. *Data preparation*: This step includes all activities required to construct the final dataset for modeling from the "raw" data including: data transformation, merging of multiple data sources, data cleaning, record and attribute selection.
4. *Modeling*: In this step, different data mining algorithms and/or tools are selected and applied to the table constructed in the data preparation step. Algorithm parameters are tuned to optimal values.
5. *Evaluation*: The goal of this step is to verify that the data mining or analysis models generated in the modeling stage are robust and achieve the objectives specified in the problem understanding step.
6. *Deployment*: In this step, the results are put in a form to be utilized for decision support (e.g. a report) or the data mining models may be integrated with one or more of the organizations IT systems (e.g. scoring prospective customers in real-time).

Fig. 1 shows these analysis steps and the party that is primarily responsible for each individual step.

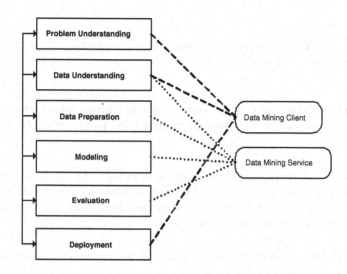

Fig. 1. Data Mining Project Steps: Problem understanding, data understanding, data preparation, modeling, evaluation and deployment. Dashed lines indicate that primary responsibility is on the part of the data mining client. Dotted lines indicate that primary responsibility is on the part of the service. Notice that both the client and service play central roles for the data understanding step.

We next specifically define the data resources and parties that form the basis of the relationship between the automated data mining service and the client.

1.1 Definitions

Definition 1 (Raw Data). *The base electronic data sources that contain or are believed to contain information relevant to the data mining problem of interest.*

Definition 2 (Metadata). *Additional information that is necessary to properly clean, join and aggregate the raw data into the form of a dataset that is suitable for analysis.*

Metadata may consist of data schemas, data dictionaries (including the number of records in each file, and the names, types and range of values for each field), and aggregation rules.

Definition 3 (Domain-specific Information). *Additional information on specific rules, conventions, practices, etc. that are limited to the problem domain.*

For example, a retail chain may indicate product returns in their transactional record with negative sales and quantity values. This information may be extremely helpful in the data preparation, evaluation and possibly the modeling steps.

Definition 4 (Third-party data sources). *Additional raw data that will likely improve the analysis result, but not collected or owned by the data mining client.*

Examples of third-party data sources include address information from the postal service, demographic information from third party data collection companies, etc.

Definition 5 (Automated Data Mining Service). *An organization that has implemented or obtained processes, tools, algorithms and infrastructure to perform the work of data understanding, data preparation, modeling, and evaluation with minimal input from a human analyst.*

The high-level goals of the automated data mining service are the following.

1. Provide the data mining client with a high quality, pertinent result. Achieving this goal ensures that the organization receiving the data mining/analysis results is satisfied (and hopefully a return customer!).
2. Remove or minimize the amount of human intervention required to produce a high-quality, pertinent result. Achieving the second goal allows the data mining/analysis service to "scale" to a large number of concurrent customers, allowing it to amortize the cost of offering the service across a large number of customers. In addition to automating as much of the operational aspects of the analysis process as possible, this goal is achievable by focusing on problems common in a few market verticals or problem domains.

For this discussion, we will assume that the set of data mining analysis problems that the automated service addresses is fixed. For example, the service may be able to offer data clustering analysis services, but is not able to offer association-rule services. We will also drop the word "automated" for the remaining discussion and it is assumed that when referring to the "data mining service", we are referring to an "automated data mining service".

Definition 6 (Data Mining Client). *An organization that possesses two items: (1) a data mining analysis problem; (2) the raw data, metadata, domain-specific information that, in combination with possibly third party data sources, are needed to solve the data mining analysis problem.*

We next specify the context of the relationship between the data mining client and the data mining service.

1.2   Relationship between Consumer and Service

The data mining service receives raw data, metadata, and possibly domain-specific information from the client. The service then performs data understanding, data preparation, modeling, and evaluation for the analyses specified by the consumer, for an agreed-upon fee. When these steps are completed, the results and possibly the provided data and other information (reports on analyses tried, intermediate aggregations, etc.) are returned to the client.

The combination of offering a fixed set of data mining and analysis solutions and focusing on clients from similar domains or market verticals enables the data mining service to perform the data understanding and deployment tasks with minimal intervention from the human data analyst working for the service. Offering a fixed set of data mining and analysis solutions allows the service to templatize problem definition documents and related information. Solution deployment is also similarly constrained by the problem domain and the focused vertical market to again allow the service to templatize deliverables. Additionally, by focusing on a particular problem domain, the data mining service analyst gains invaluable domain knowledge increasing the likelihood of successful solution delivery in future engagements.

Example: Consider the e-commerce domain. A data mining service may offer the following solutions: determining the most common paths followed by website visitors, determining the most common products purchased together and ranking products that are most likely to be purchased together. For the e-commerce domain, problem definition and specification document templates can help make the problem understanding phase clear and efficient. Additionally, the deployment steps may be nearly automated. Depending on the analysis performed, the results may take the form of an automatically generated report, a file, an executable, etc. Analysis of common paths and common products purchased together are typically best delivered in a standard report form. Product rankings may best be delivered to the client in the form of a file or an executable that takes a given product ID and returns the ranked list of product IDs.

There are often legal, security, and privacy concerns regarding data extraction on the part of the data mining client that should be addressed by the service. For more detail into these issues, please see [23].

The remainder of the paper is organized as follows. Sections 2, 3 and 4 focus on issues involved in automating and scaling the data preparation, modeling, and evaluation steps in the general KDD process. Section 5 concludes the paper.

2   Data Understanding and Preparation

Tasks involved in the data understanding step are data extraction, data quality measurement and the detection of "interesting" data subsets. The data preparation step consists of all activities required to construct the final dataset for modeling.

Responsibility for tasks in the data understanding step is typically split between the client and the service. The data mining client is responsible for extracting and providing the required raw data sources to the service for analysis. The data mining service may also augment the raw data provided by the client with third-party data sources. Data quality measurement and detection of "interesting" data subsets are performed by the service.

To efficiently address data understanding and preparation tasks, the data mining service needs to rely upon the fact that its clients come from certain specific domain areas. Ideally, these domains are chosen so that the organizations within a particular domain have similar data schemas and data dictionaries characterizing the data sources that are to be analyzed. Given a-priori knowledge about the data schemas and data dictionaries for a given domain, the data mining service can automate the operational steps of joining the appropriate raw data sources and possibly integrating third-party data sources.

Similar data formats within a market vertical or domain also justifies the building of automated domain-specific data cleaning and data quality measurement tools. In the perfect setting, domain-specific rules for data cleaning can be completely automated. But, a useful solution automatically captures and fixes a majority of data cleaning issues and only flags a small fraction of violations for human intervention and triage. The goal of data quality measurement is to justify the potential that the data may provide the required solution to the given problem [23]. Since the solutions offered by the service and the data sources themselves tend to be domain-specific, automating the data quality measurement may be done with minimal effort and requiring only configuration information.

Example: For example, suppose a data mining service is offering market-basket analysis for e-commerce companies. The typical data source of interest includes the order header, order line item information, and product catalog data. Since these sources tend to have similar schemas, it is possible to automate some data cleaning processes (ensuring that product IDs in the order information correlate with product IDs in the catalog, that rules governing the line item price information with respect to the catalog prices are respected, etc.). Additionally, initial data quality measurements may include the number of 1-itemsets that have sufficient support.

Similarly, when restricted to a small number of domains and a small set of data mining problems to address, with knowledge and expertise, it is possible to automatically apply a number data transformations and feature selection techniques that are shown to be useful for the given domain (possibly consisting of a combination of domain knowledge and automated feature selection methods). These can be automatically executed and the resulting models may be automatically scored, yielding a system that constructs a number of models with little intervention from the human analyst. For a more detailed discussion on automating the feature selection/variable transformation task, see [2].

3   Modeling

Modeling is the step in the data mining process that includes the application of one or more data mining algorithms to the prepared dataset, including selection of algorithm parameter values. The result of the modeling step is a series of either predictive or descriptive models or both. Descriptive models are useful when the data mining client is attempting to get a better understanding of the dataset at hand. Predictive models summarize trends in the data so that, if future data has the same or similar distribution to past data, the model will predict trends with some degree of accuracy.

In this section we assume that the data understanding and data preparation phases have produced a dataset from which the desired data mining solution can be derived. Although, in practice, often results of the modeling step motivate revisiting the data understanding and data preparation steps. For example, after building a series of models it may become apparent that a different data transformation would greatly aid in the modeling step.

When evaluating the utility of a given data mining algorithm for possible use in the service, the following considerations should be taken into account:

1. Assuming the prepared dataset is informative, is it possible to obtain high-quality models consistently using the given algorithm?
2. Is the algorithm capable of optimizing objectives specific to the client's organization (e.g. total monetary cost/return)?
3. Is the algorithm efficient?

There are two factors influencing the likelihood of obtaining a high quality, useful model using a given algorithm. The first factor relates to the robustness of the computed solution with respect to small changes in the input parameters or slight changes in the data. Ideally, for a majority of datasets that are encountered in a given domain, the data mining service prefers robust algorithms since model quality with respect to small parameter and data change is then predictable (and hence, the algorithm is amenable to automation).

The second factor is the ease at which the insight gained by analyzing the model can be communicated to the data mining client. Typically, prior to deployment of a model or utilizing the model in organizational decision processes, the data mining client desires to understand the insights gleaned from the model. This process is typically difficult to automate, but developing intuitive, easy to understand user interfaces aid greatly. Additionally, a process that identifies a fraction of interesting rules or correlations and reports these to a data mining service analyst is very useful quality assurance tool. These are primarily concerns during the deployment step, but the choice of modeling technique does effect this later phase.

We note that there are some data mining applications in which the client often does not analyze the model, but analyzes the computed results (e.g. results produced by product recommender systems are often analyzed, rather than attempting to understand the underlying model).

Industry standards for data mining model storage such as PMML [13] and OLE DB for Data Mining [10] enable consultants and third party vendors to build effective model browsers for specific industry problems and domains. These standards provide a basis for data mining platforms that enable data mining clients to more easily deploy and understand the models built by the service.

From the viewpoint of the data mining client, model maintenance tends to be an important issue. The data mining client may not have or may not want to invest resources to ensure that the data mining models they've received from the service are maintained and accurately model their underlying organizational processes, which may be changing over time. Techniques for incrementally maintaining data mining models are discussed in [4]. Additionally, work on identifying the fit of data mining models to data changing over time includes [7]. It may be possible for the service to incorporate these techniques into the client deliverable so that that model may maintain itself or notify the client that it is not sufficiently modeling recently collected data.

We briefly discuss some popular algorithms used in developing data mining solutions. Note that this list is not exhaustive.

3.1   Decision Trees

Decision tree construction algorithms are appealing from the perspective of the data mining service for a number of reasons. The tree's hierarchical structure enables non-technical clients to understand and effectively explore the model after a short learning period. Decision tree construction typically requires the service to provide few, if any, parameters, hence tuning the algorithm is relatively easy. Typically, small data changes result in small, if any, changes to the resulting model (although this is not true for all datasets and possible changes in data) [8]. For excellent discussions on decision tree algorithms, please see [5,18]. For a discussion on techniques used to scale decision tree algorithms to large datasets, see [4].

3.2   Association Rules

Association rule algorithms identify implications of the form $X \Rightarrow Y$ where X and Y are sets of items. The association rule model consists of a listing of all such implications existing in the given dataset. These implications are useful for data exploration and may be very useful in predictive applications (e.g. see [17]). Association rules are often presented to the user in priority order with the most "interesting" rules occurring first. Alternatively or in addition to the list of interesting rules, a browser allowing the data mining client to filter rules with specified item occurring in the set X or Y is typically useful. For an overview of association rule algorithms, see [1,16]. Approaches used to scale association rule algorithms to large databases are discussed in [4].

3.3   Clustering

Clustering algorithms aim to partition the given dataset into several groups such that records in the same group are "similar" to each other, identifying subpopulations in the data. Typically, the data mining client is not interested in the particular clustering strategy employed by the service, but is interested in a concise, effective summary of the groups that occur in their data. Although there are numerous clustering approaches available, we will focus the discussion on two methods: iterative and hierarchical methods.

Iterative clustering methods are well-known and typically straightforward to implement. But from the perspective of the data mining service, there are two challenges to automating them: obtaining a robust model that accurately characterizes the underlying data, and determining the "correct" number of clusters existing in the underlying data. Iterative clustering methods require the specification of initial clusters and the computed solution is dependent upon the quality of this initial partition. Hence to ensure a quality solution, the data mining service must implement a search for a good initial clusters [3]. This is typically done by re-running the iterative clustering algorithm from multiple random initial clusters and taking the "best" model or utilizing sampling strategies. Additionally, determining the "correct" number of clusters is challenging, but strategies such as those discussed in [24] are useful. For a general overview of iterative clustering methods, see [15,11].

Hierarchical clustering methods build a tree-based hierarchical taxonomy (*dendogram*) summarizing similarity relationships between subsets of the data at different levels of granularity. The hierarchical nature of the resulting model is a benefit for the data mining service since this structure is typically easily browsed and understood by the client. Additionally, these clustering methods are very flexible when it comes to the distance metric employed to group together similar items, making hierarchical methods applicable to a number of problems that require the use of non-standard distance metrics. The main caveat to standard hierarchical clustering implementations is in their computational complexity, requiring either $O(m^2)$ memory or $O(m^2)$ time for m data points, but automating these standard implementations is straightforward. Work on scaling these methods to large datasets includes [14,20]. For a detailed discussion of hierarchical clustering methods, see [15].

3.4   Support Vector Machines

Support Vector Machines (SVMs) are powerful and popular solutions to predictive modeling problems. SVM algorithms are typically stable and robust with respect to small changes in the underlying data. The algorithms require the specification of a parameter that effectively balances the predictive performance on the available training data with the complexity of the predictive model computed. Tuning set strategies are typically used to automate the selection of optimal values of this parameter. The SVM predictive model is a function in the space of the input or predictor attributes of the underlying data. Since this space tends to be high-dimensional, presenting the SVM model to the data mining client for the purpose of gaining insight is often a difficult proposition. The SVM is a very good predictive model, but is somewhat of a "black-box" with respect to understanding and extracting information from its functional form.

For an overview of SVMs, see [6]. For strategies on scaling SVM computation to large datasets, see [4].

4   Evaluation

Prior to delivering a data mining model or solution to the client, the service will evaluate the model or solution to ensure that it has sufficient predictive power or provides adequate insight into trends existing in the underlying data. The primary focus in the evaluation phase is ensuring that the client is being handed a high-quality result from the service.

Depending upon the project, model evaluation may involve one or two components. The first is an objective, empirical measurement of the performance of the model or of the ability of the model to accurately characterize the underlying data. The second, which may not be needed for some projects, involves a discussion or presentation of the model with the client to collect feedback and ensure that the delivered model satisfies the client's needs. This second component is typical for projects or models in which the goal is data understanding or data exploration.

We discuss the empirical measurement component of the evaluation phase in more detail for two high-level data mining tasks: predictive applications and data exploration. By exposing intuitive, well-designed model browsers to the client, the service may automate the process of presenting the model and collecting client feedback to the extent possible. As the service focuses on clients in particular domains or verticals, model browsers or report templates may be created that raise attention to "important" or "interesting" results for the specific domain or market.

4.1   Evaluating Predictive Models

The primary focus of predictive model evaluation is to estimate the predictive performance of a given model when it is applied to future or unseen data instances. Algorithms discussed that produce predictive models include decision trees (Section 3.1), association rules (Section 3.2) and support vector machines (Section 3.4).

The basic assumption underlying different predictive performance estimation strategies is that the distribution of future or unseen data is the same (or similar to) the distribution of the training data used to construct the model.

Popular methods for estimating the performance of predictive models include *cross-validation* [25] and *ROC curves* [22]. Cross-validation provides an overall average predictive performance value, given that the data distribution assumption above is satisfied. ROC curves provide a more detailed analysis of the frequency of false positives and false negatives for predictors constructed from a given classification algorithm.

From the viewpoint of the data mining service, automating cross-validation and ROC computations is straightforward. Running these evaluation techniques requires little (if any) input from a human analyst on the part of the service.

But computation of these values may be time consuming, especially when the predictive modeling algorithm used has a lengthy running time on the client's data.

In addition to evaluating a given model (or set of models) with respect to predictive performance, the data mining service may implement evaluation metrics that are more informative for clients within a given domain or vertical.

Example: Consider again the data mining service that caters to e-commerce clients. The service may provide product recommendations for e-commerce companies (i.e. when a customer is viewing sheets at the e-commerce site, recommend pillows to them also). In this case, although the recommender system is a predictive modeling system, the data mining service may evaluate different predictive models with respect to the amount of revenue expected to be generated when the recommender is placed in production.

4.2   Evaluating Data Exploration Models

The primary goal in evaluation of models that support data exploration tasks is ensuring that the model is accurately summarizing and characterizing patterns and trends in the underlying dataset. Algorithms discussed that address data exploration tasks are the clustering methods mentioned in Section 3.3. To some extent association rules (Section 3.2) are also used as data exploration tools.

Objective measures for evaluating clustering models to ensure that the model accurately captures data characteristics include Monte Carlo cross-validation [24]. This method is straightforward to automate on the part of the data mining service.

Given the nature of association rule discovery algorithms, the set of association rules found are, by definition, accurately derived from the data. So there is no need to empirically measure the "fit" of the set of association rules to the underlying dataset.

The quality of data exploration models is related to the utility of the extracted patterns and trends with respect to the client's organization. When the data mining service focuses on a particular client domain or vertical market, effective model browsers and templates can be constructed that focus the client's attention to information that is frequently deemed useful in the domain or vertical. Hence the quality of the model with respect to the particular domain is then easily evaluated by the client. Additionally, the service can use these browsers and templates to evaluate model quality prior to exposing the model to the client.

5   Conclusion

The goal of the data mining service is to effectively and efficiently produce high-quality data mining results for the data mining client for a reasonable (low) cost. We argued that a quality, low-cost data mining result may be delivered by automating the operational aspects of the data mining process and focusing on specific client domains. Upon successful execution of these tasks, the service is then an attractive option for small, medium and large organizations to capitalize on and leverage their data investment to improve the delivery of their products and services to *their* customers.

Acknowledgments. I would like to thank Usama Fayyad for numerous insightful and fruitful discussions on this topic. Thanks to Gregory Piatetsky-Shapiro for discussions regarding data mining consulting. Thanks to Max Chickering for insight into automating decision tree algorithms. I would also like to thank Mark Smucker who provided comments on an earlier draft of this work.

References

1. Rakesh Agrawal, Tomasz Imielinski, and Arun Swami. Mining association rules between sets of items in large databases. In *Proc. of the ACM SIGMOD Conference on Management of Data*, pages 207–216, Washington, D.C., May 1993.
2. J. D. Becher, P. Berkhin, and E. Freeman. Automating exploratory data analysis for efficient mining. In *Proc. of the Sixth ACM SIGKDD Intl. Conf. on Knowledge Discovery and Data Mining (KDD-2000)*, pages 424 – 429, Boston, MA, 2000.
3. P. S. Bradley and U. M. Fayyad. Refining initial points for K-Means clustering. In *Proc. 15th International Conf. on Machine Learning*, pages 91–99. Morgan Kaufmann, San Francisco, CA, 1998.
4. P. S. Bradley, J. Gehrke, R. Ramakrishnan, and R. Srikant. Scaling mining algorithms to large databases. *Comm. of the ACM*, 45(8):38–43, 2002.
5. L. Breiman, J. H. Friedman, R. A. Olshen, and C. J. Stone. *Classification and Regression Trees*. Wadsworth, Belmont, 1984.
6. C. J. C. Burges. A tutorial on support vector machines for pattern recognition. *Data Mining and Knowledge Discovery*, 2(2):121–167, 1998.
7. I. V. Cadez and P. S. Bradley. Model based population tracking and automatic detection of distribution changes. In *Proc. Neural Information Processing Systems 2001*, 2001.
8. D. M. Chickering. Personal communication, January 2003.
9. CRISP-DM Consortium. Cross industry standard process for data mining (crisp-dm). http://www.crisp-dm.org/.
10. Microsoft Corp. Introduction to ole db for data mining. http://www.microsoft.com/data/oledb/dm.htm.
11. R. Duda, P. Hart, and D. Stork. *Pattern classification*. John Wiley & Sons, New York, 2000.
12. U. M. Fayyad, G. Piatetsky-Shapiro, P. Smyth, and R. Uthurasamy. *Advances in Knowledge Discovery and Data Mining*. MIT Press, Cambridge, MA, 1996.
13. Data Mining Group. Pmml version 2.0. http://www.dmg.org/index.htm.
14. S. Guha, R. Rastogi, and K. Shim. Cure: An efficient clustering algorithm for large databases. In *Proc. ACM SIGMOD Intl. Conf. on Management of Data*, pages 73–84, New York, 1998. ACM Press.
15. A. K. Jain and R. C. Dubes. *Algorithms for Clustering Data*. Prentice Hall, 1988.
16. Heikki Mannila, Hannu Toivonen, and A. Inkeri Verkamo. Efficient algorithms for discovering association rules. In Usama M. Fayyad and Ramasamy Uthurusamy, editors, *AAAI Workshop on Knowledge Discovery in Databases (KDD-94)*, pages 181–192, Seattle, Washington, 1994. AAAI Press.

17. Nimrod Megiddo and Ramakrishnan Srikant. Discovering predictive association rules. In *Knowledge Discovery and Data Mining*, pages 274–278, 1998.
18. Sreerama K. Murthy. Automatic construction of decision trees from data: A multi-disciplinary survey. *Data Mining and Knowledge Discovery*, 2(4):345–389, 1998.
19. M. T. Oguz. Strategic intelligence: Business intelligence in competitive strategy. *DM Review*, August 2002.
20. Clark F. Olson. Parallel algorithms for hierarchical clustering. *Parallel Computing*, 21(8):1313–1325, 1995.
21. G. Piatetsky-Shapiro. Personal communication, January 2003.
22. Foster J. Provost and Tom Fawcett. Analysis and visualization of classifier performance: Comparison under imprecise class and cost distributions. In *Knowledge Discovery and Data Mining*, pages 43–48, 1997.
23. D. Pyle. *Data Preparation for Data Mining*. Morgan Kaufmann, San Francisco, CA, 1999.
24. Padhraic Smyth. Clustering using monte carlo cross-validation. In *Knowledge Discovery and Data Mining*, pages 126–133, 1996.
25. M. Stone. Cross-validatory choice and assessment of statistical predictions. *Journal of the Royal Statistical Society*, 36:111–147, 1974.
26. D. E. Weisman and C. Buss. Database functionality high, analytics lags, September 28, 2001. Forrester Brief: Business Technographics North America.

Trends and Challenges in the
Industrial Applications of KDD

Ramasamy Uthurusamy

General Motors Corporation, USA
Samy@gm.com

As an applications driven field, Knowledge Discovery in Databases and Data Mining (KDD) techniques have made considerable progress towards meeting the needs of these industrial and business specific applications. However, there are still considerable challenges facing this multidisciplinary field. Drawing from some industry specific applications this talk will cover the trends and challenges facing the researchers and practitioners of this rapidly evolving area. In particular, this talk will outline a set of issues that inhibit or delay the successful completion of an industrial application of KDD. This talk will also point out emerging and significant application areas that demand development of new KDD techniques by the researchers and practitioners. One such area is discovering patterns in temporal data. Another is the evolution of discovery algorithms that respond to changing data forms and streams. Finally, this talk will outline the emerging vertical solutions arena that is driven by business value, which is measured as a progress towards minimizing the gap between the needs of the business user and the accessibility and usability of analytic tools.

K.-Y. Whang, J. Jeon, K. Shim, J. Srivatava (Eds.): PAKDD 2003, LNAI 2637, p. 14, 2003.
© Springer-Verlag Berlin Heidelberg 2003

Finding Event-Oriented Patterns in Long Temporal Sequences

Xingzhi Sun, Maria E. Orlowska, and Xiaofang Zhou

The University of Queensland, Australia
{sun, maria, zxf}@itee.uq.edu.au

Abstract. A major task of traditional temporal event sequence mining is to find all frequent event patterns from a long temporal sequence. In many real applications, however, events are often grouped into different types, and not all types are of equal importance. In this paper, we consider the problem of efficient mining of temporal event sequences which lead to an instance of a specific type of event. Temporal constraints are used to ensure sensibility of the mining results. We will first generalise and formalise the problem of event-oriented temporal sequence data mining. After discussing some unique issues in this new problem, we give a set of criteria, which are adapted from traditional data mining techniques, to measure the quality of patterns to be discovered. Finally we present an algorithm to discover potentially interesting patterns.

1 Introduction

Finding frequent patterns in a long temporal event sequence is a major task of temporal data mining with many applications. Substantial work [1,2,3,4,5] has been done in this area, however, it often focuses on specific application requirements.

In some domains, there is a need to investigate the dependencies among selected types of events. A temporal event sequence can then be formed by integrating data sets from different domains according to common features such as the time order of each event. Naturally, in such a sequence, not every type of event may have equal impact on the overall data analysis goals. We may be particularly interested in a specific event type and any event patterns related to the occurrences of this event type. The temporal relationship between event patterns and specific type of events can be specified by temporal constraints. The temporal constraints, such as the size of time window considerations, are important in the knowledge discovery from the temporal sequence, because they closely relate to the sensibility of mining results.

An application of such kind of problem could be earthquake prediction. Many types of events, which may relate to earthquakes, are recorded by independent devices but they have one common reference point - the recorded time of event occurrence. Consequently, a temporal sequence is formed by ordering timestamps of these events. Assuming that our goal relates to ability to set up some proactive measures in an earthquake situation, the evidence of the actual earthquake is the

K.-Y. Whang, J. Jeon, K. Shim, J. Srivatava (Eds.): PAKDD 2003, LNAI 2637, pp. 15–26, 2003.

most important event type in the sequence. We would like to identify all patterns of different types of recorded events (excluding or including the earthquake type itself) which potentially indicate the possibility of an earthquake situation. The temporal constraint would be then the length of the period of interest preceding the actual earthquake occurrence.

Another application is telecommunication network fault analysis. A telecommunication network can be viewed as a hierarchical network, with switches, exchanges and transmission equipments. Various devices and software in the network monitor the network continuously, generating a large number of different types of events, which can be related to a particular physical devise at a certain level in the network, to some software modules, or to individual phone calls. The fact that an event is recorded does not necessarily mean that there is something wrong. However, technical experts strongly believe that some events might be closely related to each other, though the relationship among events is unknown to them. One type of events, trouble reports (TR), is of particular importance. A TR is put in by a consultant after a customer calling the company to complain some service difficulties, such as noisy lines, no dialling tones or abnormal busy tones. All TR must be investigated by the company, often involving sending engineers to check various parts of the network including the customer's place. The total cost for handling TR is very high for the telecommunication company due to a very large number of TR. With a database of extensive historical monitoring event data and a comprehensive monitoring network, it is highly desirable to find patterns leading to any types of TR within one or a few sensible time intervals. Such patterns, when combined with real-time event data, can help to reduce the cost of processing TR (for example, to avoid sending engineers to a customer's place if there is an event pattern suggesting the problem might be with an exchange rather than with customer's equipment or connection). Another major incentive to invest in finding such patterns is that, with such knowledge, the company can improve customer services (for example, to contact customers in an possibly affected area before they lodge complaints).

Our problem of long temporal event sequence data mining can be generalized as follows. Let k types of time-related events be given. A temporal event sequence is a list of events indexed by timestamp reflecting the time when the event was recorded (note that the intervals between adjacent events are arbitrary). Let e be a selected type of event called *target event*. Additionally, let time interval T be given. The problem is to find all frequent patterns of events (from the event sequence) which potentially lead to the occurrences of target event e within the given time interval T. We call such kind of pattern *event-oriented pattern* because it has temporal relationship with the target event. Note that the temporal relationship discussed in this paper can be extended to other types (such as after target event) or involve more temporal constraints.

The rest of the paper is organised as follows. Related work is discussed in section 2. We formalize the event-oriented mining problem and give a new pair of definitions for confidence and support in section 3. Algorithm for mining event-oriented patterns is presented in section 4. Experiment results are shown in section 5. We summarize our work in section 6.

2 Related Work

Data mining has been a topic of substantial research in the last decade. Various sequence mining problems have been investigated, some in the domain similar to that studied in this paper (for example, sequence mining in telecommunication alarm data sets [6,2,3]). A number of studies, such as [5], regard a sequence simply as an order among events, and do not consider the timestamp of each event. The problem of placing time windows to make meaningful sequence segmentation is a key problem in temporal sequence data mining. In [3], while event timestamps are considered, the sliding window is moved at a fixed time interval. This time interval, however, is difficult for the user to specify. We shall allow window placement either starting from or ending with any event in this paper. Another different feature of our work is that we introduce a specific target event type (i.e., target event e). Therefore, our goal of mining is to find those patterns related to this special event type rather than finding all frequent patterns. Our unique requirements of differentiating event types and placing windows continuously make it necessary to use different support and confidence definitions than those used in [3]. [7] mines failure plans by dealing with bad plans and good plans separately and define the support in a similar way. But it is applied to a set of sequences rather than a long temporal event sequence. In our approach, after creating the dataset, we can apply [8,9] to find occurrence patterns and [10, 11,12,13] to find sequential patterns. [4] discusses some failure prediction issues and applies genetic algorithm to compute patterns. However, our algorithm can provide more accurate and complete results.

3 Problem Statement

3.1 Event Sequences and Patterns

Let us consider k *event types* $E = \{e_i | i = 1..k\}$. For a given event type $e_i \in E$, $D(e_i)$ is the domain of the event type which is a finite set of discrete values. An *event* of type e_i can be represented as a triple $(e_i \in E, a \in D(e_i), t)$ such that a and t are *value* and *timestamp* of the event respectively. Assume for any e_i and e_j in E, $D(e_i) \cap D(e_j) = \emptyset$ if $i \neq j$. It means the value of an event can imply its type. Therefore, we simplify an event as (a, t) where $a \in Dom = \bigcup D(e_i)$ and maintain a two level hierarchical structure Γ on $Dom \cup E$. A *temporal event sequence*, or *sequence* in short, is defined as $s = \langle (a_1, t_1), (a_2, t_2), \ldots, (a_n, t_n) \rangle$ where $a_i \in Dom$ for $1 \leq i \leq n$ and $t_i < t_{i+1}$ for $1 \leq i \leq n - 1$.

As one can see from previous discussions, the definition of "patterns of interest" is very broad [10,3,5]. Unlike others, we are only interested in patterns that lead to events with a specific value of an event type. Therefore, a *target event value* e is given in our pattern discovery problem. Note that, in this paper, we discuss patterns only at the event value level, i.e., every element in a pattern corresponds to an event value. If a pattern includes both event value and event type, the method introduced in [14] can be applied to extend patterns across hierarchical levels.

Considering the pattern itself, in this paper, we only discuss the following two types of patterns:

1. *Existence pattern α with temporal constraint T* : α is a set event values, i.e., $\alpha \subseteq Dom$. Events following this pattern must happen within a time interval T.
2. *Sequential pattern β with temporal constraint T* : β is in the format of $\langle a_1, a_2, \ldots, a_l \rangle$ where $a_i \in Dom$ for $1 \leq i \leq l$. Events following this pattern must happen within a time interval T.

A rule in our problem is in the form of $r = \left\{ LHS \xrightarrow{T} e \right\}$. where e is a target event value and LHS is a pattern either of type α or β. T is the time interval that specifies not only the temporal relationship between LHS and e but also the temporal constraint of pattern LHS. The rule can be interpreted as: a pattern LHS, which occurs in interval T, will lead to occurrence of e within the same interval.

3.2   A Model for Support and Confidence

To evaluate the significance of rules, we use support and confidence. While these two measures have been used in the past by most data mining research, the way in which to give a sensible definition to these two important concepts is non-trivial for the problem investigated in this paper, as we shall explain below.

Traditional concepts for support and confidence are defined for transaction databases [8,15]. The counterpart of "transaction" in temporal event sequence mining can be defined by using a sliding window [3]. That is, a window of size T is placed along a sequence, such that each time, the sequence is segmented into a sequence fragment which comprises only the events within the window.

Formally, we define *window* as $w = [t_s, t_e)$ where t_s and t_e are start time and end time of the window respectively. We say the *window size* of w is T if $t_e - t_s = T$. A *sequence fragment* $f(s, w)$ is the part of sequence s which is in window w, i.e., $f(s, w) = \langle (a_{m1}, t_{m1}), (a_{m1+1}, t_{m1+1}), \ldots, (a_{m2}, t_{m2}) \rangle$ where $t_i < t_{i+1}$ for $m1 \leq i \leq m2 - 1$ and $f(s, w)$ includes all (a_i, t_i) in s, if $t_s \leq t_i < t_e$.

While the size of the sliding window is typically given by some domain experts (and often multiple sizes can be given according to users' interests, generating several sets of data mining results), the way to move the window at a fixed time interval, such as in [3], imposes an artificial subsequence boundary, which may lead to exclusion of some useful patterns that are across such boundaries.

To avoid imposing such artificial sequence segmentation boundaries, we move the window to the next event of interest, i.e., a window always either starts from or ends with an event. Further, windows are placed in different ways for calculating support and confidence.

Let us consider support firstly. In transaction databases, support is used to define the frequency of patterns. In our problem a frequent pattern means the pattern frequently happens in T-sized intervals before target events. Therefore, the frequent pattern should be generated from the sequence fragments before target events rather than the whole temporal event sequence. To find frequent patterns, windows are placed using the following strategy: first, we locate all the events whose value is e in the event sequence s by creating *timestamp set* $T^e = \{t \mid (e, t) \in s\}$. For any $t_i \in T^e$, we issue a T-sized window $w_i = [t_i - T, t_i)$ and get sequence fragment $f_i = f(s, w_i)$, where T is the predefined window size. The set of these sequence fragments $D = \{f_1, f_2, \ldots, f_m\}$ is called *dataset* of target event value e.

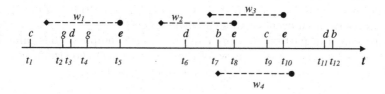

Fig. 1. Example

Example: Fig. 1 presents an event sequence. Suppose e is target event value, the timestamp set of e is $\{t_5, t_8, t_{10}\}$, w_1, w_2 and w_3 are three T-sized windows ending at timestamp t_5, t_8 and t_{10} respectively. The sequence fragments of these three windows are $\langle (g, t_2), (d, t_3), (g, t_4) \rangle$, $\langle (d, t_6), (b, t_7) \rangle$ and $\langle (b, t_7), (e, t_8), (c, t_9) \rangle$.

The support of rule r is defined as:

$$supp\,(r) = \frac{|\{f_i | LHS \sqsubseteq f_i\}|}{|D|}$$

In the formula, $\sqsubseteq$, called *include*, is a relationship between pattern LHS and sequence fragment $f(s, w)$:

1. For existence pattern α: Let D_f be a set of event values that appear in sequence fragment $f = f(s, w)$. We denote $\alpha \sqsubseteq f$ if $\alpha \subseteq D_f$.
2. For sequential pattern β: If $f(s, w) = \langle (a_{m1}, t_{m1}), \ldots, (a_{m2}, t_{m2}) \rangle$, we create a sequence $s_f = \langle a_{m1}, \ldots, a_{m2} \rangle$ by projecting the value of each event in $f(s, w)$. We denote $\beta \sqsubseteq f$ if β is the subsequence of s_f.

Example: In Fig. 1, considering rule $r = \left\{ \{d\} \xrightarrow{T} e \right\}$, there are three sequence fragments in dataset D and two of them ($f(s, w_1)$ and $f(s, w_2)$) include existence pattern $\{d\}$, so $supp\,(r) = 66.7\%$.

With support, the rule can be interpreted as pattern LHS occurs in a percentage (support) of intervals before target events.

According to the definition of support, given a threshold, we can find all frequent patterns in dataset D. However, not all the frequent patterns are valuable. For example, a pattern may happen frequently before target events, but this pattern happens everywhere on the time axis, so we still cannot conclude that this pattern relates to target events. Therefore, confidence is introduced to prune such uninteresting frequent patterns. Intuitively, for each pattern, we need find how many times this pattern happens in sequence s and out of this number, how many times the occurrence of the pattern leads to the target event.

To count the frequency that a pattern LHS occurs in sequence s with respect to window size T, we introduce the concept *occurrence window*. We think that each occurrence of an interesting pattern corresponds to a T-sized window. The formal definition of occurrence window is given as follows:

1. For existence pattern α with temporal constraint T: We say w is an occurrence window of the pattern *iif* 1) the size of w is T, 2) $\alpha \sqsubseteq f(s, w)$, and 3) w starts from an event of value a where $a \in \alpha$.
2. For sequential pattern $\beta = \langle a_1, a_2, \ldots, a_l \rangle$ with temporal constraint T: We say w is an occurrence window of the pattern *iif* 1) the size of w is T, 2) $\beta \sqsubseteq f(s, w)$, and 3) w starts from an event of value a_1.

Example: Considering event sequence s in Fig. 1, window w_4 is an occurrence window of sequential pattern $\langle b, c, e \rangle$.

Given sequence s, pattern LHS, and time interval T, we define *occurrence window set* as $W(LHS, s, T)$, whose elements are occurrence windows of pattern LHS with temporal constraint T.

Confidence of rule r is defined as:

$$conf(r) = \frac{|W(LHS \rightarrow e, s, T)|}{|W(LHS, s, T)|}$$

The numerator is how many times pattern LHS plus e occurs together within T-sized windows in the sequence s. The denominator is how many times pattern LHS itself occurs within T-sized windows in the sequence s. Confidence of rule r indicates that a percentage (confidence) of occurrences of pattern LHS will lead to events of value e within a T-sized interval.

1. For existence pattern α: $\alpha \rightarrow e$ means a new pattern in which orders of elements in α are ignored, but all elements in α must appear before e. In this case, $|W(\alpha \rightarrow e, s, T)|$ is the total number of T-sized windows that start from any element in α and include all elements of α and e (orders of elements in α are ignored, but all elements in α must appear before e).
2. For sequential pattern $\beta = \langle a_1, a_2, \ldots, a_l \rangle$: $\beta \rightarrow e$ is a new pattern which can be created by simply adding e as the last element of sequential pattern β, i.e., $\beta \rightarrow e \equiv \langle a_1, a_2, \ldots, a_l, e \rangle$.

Now, rule r with significance measures is in the format of:

$$LHS \xrightarrow{T} e\,(supp(r), conf(r))$$

3.3   Problem Definition

The formal definition of our problem is: Given sequence s, target event value e, window size T, two thresholds s_0 and c_0, find the complete set of rule $r = \left\{ LHS \xrightarrow{T} e \right\}$ such that $supp(r) \geq s_0$ and $conf(r) \geq c_0$.

4   Algorithm

In this section we propose an algorithm to solve the problem we defined previously. The basic idea is to generate candidate rules from the part of sequence and prune rules by scanning the whole sequence. High-level structure of the algorithm is shown in Fig. 2. R is a rule set and initialised as empty. We first create dataset D which is a set of sequence fragments before target events and then compute a set of frequent patterns whose support is no less than s_0. After a set of rules is generated according to these frequent patterns, we calculate confidence for each rule in R and prune those whose confidence is less than c_0. The last step of the algorithm is to output rules in R. The main subtasks in this algorithm include 1) creating the dataset, 2) computing frequent patterns and generating candidate rules, and 3) pruning rules. We will explain each subtask in detail. Considering different kinds of pattern formats, the algorithm will be different. In the rest of this section, we use the existence pattern format to demonstrate our algorithm.

Input: sequence s, target event value e, window size T, minimum
support s_0 and minimum confidence c_0
Output: the complete set of rules whose support is no less than s_0
and confidence no less than c_0
Method:
    $R = \phi$;
    $D = d(s, T, e)$; /* Create the dataset */
    $R = r(D, s_0, e)$; /* Find frequent patterns and generate rules */
    $p(R, s, T, c_0)$; /* Prune rules */
    output R;

Fig. 2. High-level structure

4.1   Create the Dataset

Fig. 3 shows a query-based approach to creating the dataset. First, a query is issued on sequence s to locate the e-valued events and retrieve their timestamps into set T^e. For each timestamp t_i in T^e, we query events whose timestamps are in window $[t_i - T, t_i)$ to get sequence fragment f_i. Here T is the pre-defined window size. Because orders of the events are ignored in existence pattern and also there are no other temporal constraints within a sequence fragment, we use a set f_i' to only store the values of events in the window. A distinct number is put into initial set to ensure each f_i' is unique. Finally, we create the dataset D whose element is f_i'.

```
Method: d(s, T, e)
    D = φ; Tᵉ = φ; id = 0;
    compute Tᵉ;
    for all tᵢ ∈ Tᵉ do
        id + +;
        f'ᵢ = {id};
        for all event (aₖ, tₖ) such that tₖ ∈ [tᵢ − T, tᵢ) do
            f'ᵢ = f'ᵢ ∪ {aₖ};
        D = D ∪ {f'ᵢ};
    return D;
```

Fig. 3. Create the dataset

4.2 Compute Frequent Patterns and Generate Candidate Rules

According to the rule format, we know that each rule corresponds to a frequent pattern. So, the key step in this subtask is to find frequent patterns from dataset D. We can apply algorithm [8,9] for mining existence patterns and [10,11,12,13] for sequential patterns.

```
Pattern
    {α; /* a set of event values */
     count; /* the number of sequence fragments that include α */ };
Rule
    {LHS; /* frequent pattern found in D */
     RHS; /* target event value */
     |RHS|; /* the number of target events */
     |fᵢ|LHS ⊑ fᵢ|; /* the number of sequence fragments that include LHS */
     supp; /* support */
     |W(LHS, s, T)|; /* the number of occurrences of pattern LHS w.r.t. T */
     |W(LHS → RHS, s, T)|; /*the number of occurrences of LHS → e w.r.t. T*/
     conf; /* confidence */ };
```

Fig. 4. Data structures

Data structures of pattern and rule are given in Fig. 4. Fig. 5 shows there are two main steps in this subtask. At the first step, function $f_r(D, s_0)$ is called to compute frequent patterns and return L, which is a set of frequent patterns with their support numbers. At the second step, for each frequent pattern, a rule is initialized.

4.3 Prune Candidate Rules

The last subtask is to prune the rules whose confidence is less than c_0. Fig. 6 shows the procedure of this subtask. The basic idea is to scan the sequence

Method: $r(D, s_0, e)$
   $L = \phi;\ R = \phi;$
   /* Compute frequent patterns */
   $L = f_r(D, s_0);$
   /* Generate candidate rule set */
   for all $l_i \in L$ **do**
      Rule r_i =**new** $Rule();$
      $r_i.LHS = l_i.\alpha;$
      $r_i.RHS = e;$
      $r_i.|RHS| = |D|;$
      $r_i.|f_i|LHS \sqsubseteq f_i| = l_i.count;$
      $r_i.supp = \frac{l_i.count}{|D|};$
      $r_i.|W(LHS, s, T)| = 0;$
      $r_i.|W(LHS \rightarrow RHS, s, T)| = 0;$
      $r_i.conf = 0;$
      $R = R \cup \{r_i\};$
   return $R;$

Fig. 5. Compute frequent patterns and generate candidate rules

Method: $p(R, s, T, c_0)$
   /* Create a set that include all elements in $LHSs$ of rules */
   $C = \phi;$
   for all $r_i \in R$ **do** $C = C \cup r_i.LHS;$
   /* Count two types of occurrences */
   for all $(a_i, t_i) \in s$ **do**
      if $(a_i \in C)$ **then**
         $w_i = [t_i, t_i + T);$
         $f_i = f(s, w_i);$
         for all $r_j \in R$ **do**
            /* Test patterns */
            if (f_i starts with the element in $r_j.LHS$ and $r_j.LHS \sqsubseteq f_i$) **then**
               $r_j.|W(LHS, s, T)| + +;$
               if $(r_j.LHS \rightarrow r_j.RHS \sqsubseteq f_i)$ **then**
                  $r_j.|W(LHS \rightarrow RHS, s, T)| + +;$
   /* Prune rules */
   for all $r_i \in R$ **do**
      $r_i.conf = \frac{r_i.|W(LHS \rightarrow RHS, s, T)|}{r_i.|W(LHS, s, T)|};$
      if $(r_i.conf < c_0)$ **then** $R = R - \{r_i\};$

Fig. 6. Prune candidate rules

once by using T-sized slide windows. According to the definition, an occurrence window of any existence pattern must start from the event whose value is an element of the existence pattern. We union all $LHSs$ of rules to get an event value set C. Only windows starting from the event, whose value is included in C, can be occurrence windows of some patterns. When such windows are set, we test two patterns (LHS and $LHS \rightarrow RHS$) for each rule. If the sequence fragment in the window includes a pattern, the number of occurrence of this pattern is increased by one. After scanning the sequence once, we can compute the confidence of each rule and prune those whose confidence is less than c_0.

5   Experiments and Discussions

We did experiments on a telecommunication event sequence which consists of 3 event types. Each event type has a number of event values. As we mentioned before, one type of events, trouble reports (TR), is of particular importance because it may indicate certain failure on telecommunication networks. Our goal is to find patterns leading to any type of TR.

Table 1. Mining results

minimum support 5% minimum confidence 20%				
Time window size(hours)	6	12	24	48
Number of rules	1	8	52	328

The telecommunication event database contains the data of a state over 2 months. There are 2,003,440 events recorded in the database. Out of this number, 412,571 events are of type TR. We specify four time intervals and set the thresholds of support and confident as 5% and 20% respectively. The mining results are given in Table 1. An example of rule is shown in Table 2. This rule means signal f_1 and e_1 together will lead to a trouble report NTN within 24 hours with some probabilities.

Table 2. Rule example

T (hours)	LHS		RHS	Support	Confidence
24	f_1	e_1	NTN	6.32%	38.72%

In the experiment, we need to find interesting patterns for each value of TR. It means a target event value is extended to a set of target event values. In this case, for each target event value, candidate rules are generated. We then combine these rules and scan the sequence once to determine their confidence.

In the second subtask, note that minimum support s_0 is a percentage. For a given target event value, the minimum support number should be $m * s_0$, where m is the population of the events with this target event value. To ensure the patterns have statistic significance, we define a threshold m_0 which is minimum population of target event value. We create the dataset for a target event value only if its population exceeds this threshold. In this case, the minimum support number of any target event value is set as $m_0 * s_0$.

Considering a set of target event values, some rules may have the same LHS. We can group rules according to their LHS and create a list for each group. The head of the list is the LHS of this group and linked by the RHS of each rule in this group. For example, if a group includes rules $\alpha \to e_1$, $\alpha \to e_2$, and $\alpha \to e_3$, we create a list which is shown in Fig. 7.

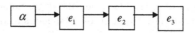

Fig. 7. Example of the list

Given a window, for each list, we first test whether a window includes the head of this list (which is the pattern LHS), if not, the test of this list terminates; otherwise, we record the last element which meets the requirement of the LHS and test whether the rest elements in the window include each RHS. We can also remove some lists if the heads of these lists are super patterns of a LHS which has been proved as not included in the window.

In the implementation, we store dataset D in the main memory. This becomes hard if the sequence is very long. To solve this problem, the sequence is divided into some segments such that the dataset of each segment can be fitted in the memory. We can first find frequent patterns from these segments separately, then combine these local frequent patterns together and scan the entire sequence once to determine their support. The candidate rules are generated only for the patterns whose global support is no less than the threshold. The principle of this approach is: if a pattern is frequent to the whole sequence, it must be frequent to at least one segment. [16] discusses this idea in detail. Because the fragmentation will break down the patterns near the borders of segments, this approach can improve efficiency but sacrifice some accuracy.

6 Summary

This paper generalizes the problem of discovering event-oriented patterns from a long temporal event sequence. We introduce speciality of events and give our definitions of support and confidence. An algorithm based on a window placement strategy is proposed to solve this problem.

Future research includes data management issues and performance issues. In this paper, we only discuss existence pattern and sequential pattern. Other pattern formats can also be computed based on our basic idea. More temporal constraints can be added into mining process. The problems of how to select the appropriate window size and evaluate rules should also be investigated. For performance issues, as a query-based approach to creating the dataset is not efficient enough, the improvement of the efficiency of the algorithm is currently under the consideration.

Acknowledgement. X. Sun would like to thank Dr. Xue Li for helpful discussions.

References

1. Wang, J.T.L., Chirn, G.W., Marr, T.G., Shapiro, B.A., Shasha, D., Zhang, K.: Combinatorial pattern discovery for scientific data: Some preliminary results. In: Proc. 1994 ACM SIGMOD Intl. Conf. on Management of Data. (1994) 115–125
2. Mannila, H., Toivonen, H.: Discovering generalized episodes using minimal occurrences. In: Knowledge Discovery and Data Mining. (1996) 146–151
3. Mannila, H., Toivonen, H., Verkamo, A.I.: Discovery of frequent episodes in event sequences. Data Mining and Knowledge Discovery **1** (1997) 259–289
4. Weiss, G.M., Hirsh, H.: Learning to predict rare events in event sequences. In: Proc. 4th Int. Conf. on Knowledge Discovery and Data Mining (KDD'98), New York, NY, AAAI Press, Menlo Park, CA (1998) 359–363
5. Yang, J., Wang, W., Yu, P.S.: Infominer: mining surprising periodic patterns. In: Proc. 7th ACM SIGKDD Conference. (2001) 395–400
6. Hatonen, K., Klemettinen, M., Mannila, H., Ronkainen, P., Toivonen, H.: Knowledge discovery from telecommunication network alarm databases. In: Proc. 12th International Conference on Data Engineering. (1996) 115–122
7. Zaki, M.J., Lesh, N., Ogihara, M.: Planmine: Sequence mining for plan failures. In: Proc. 4th Int. Conf. on Knowledge Discovery and Data Mining (KDD'98), New York, NY, ACM Press (1998) 369–373
8. Agrawal, R., Srikant, R.: Fast algorithms for mining association rules. In: Proc. 20th Int. Conf. Very Large Data Bases, Morgan Kaufmann (1994) 487–499
9. Han, J., Pei, J., Yin, Y.: Mining frequent patterns without candidate generation. In: Proc. 2000 ACM SIGMOD Intl. Conf. on Management of Data, ACM Press (2000) 1–12
10. Srikant, R., Agrawal, R.: Mining sequential patterns: Generalizations and performance improvements. In: Proc. 5th Int. Conf. Extending Database Technology. Volume 1057., Springer-Verlag (1996) 3–17
11. Zaki, M.J.: SPADE: An efficient algorithm for mining frequent sequences. Machine Learning **42** (2001) 31–60
12. Han, J., Pei, J., Mortazavi-Asl, B., Chen, Q., Dayal, U., Hsu, M.: Freespan: Frequent pattern-projected sequential pattern mining. In: Proc. 6th ACM SIGKDD Intl. Conf. on Knowledge Discovery and Data Mining, ACM Press (2000) 355–359
13. Pei, J., Han, J., Mortazavi-Asl, B., Pinto, H., Chen, Q., Dayal, U., Hsu, M.C.: PrefixSpan mining sequential patterns efficiently by prefix projected pattern growth. In: Proc. 2001 Int. Conf. Data Engineering, Heidelberg, Germany (2001) 215–226
14. Srikant, R., Agrawal, R.: Mining generalized association rules. Future Generation Computer Systems **13** (1997) 161–180
15. Agrawal, R., Srikant, R.: Mining sequential patterns. In: Proc. 11th Int. Conf. on Data Engineering, Taipei, Taiwan, IEEE Computer Society Press (1995) 3–14
16. Savasere, A., Omiecinski, E., Navathe, S.B.: An efficient algorithm for mining association rules in large databases. In: The VLDB Journal. (1995) 432–444

Mining Frequent Episodes for Relating Financial Events and Stock Trends

Anny Ng and Ada Wai-chee Fu

Department of Computer Science and Engineering
The Chinese University of Hong Kong, Shatin, Hong Kong
{ang, adafu}@cse.cuhk.edu.hk

Abstract. It is expected that stock prices can be affected by the local and overseas political and economic events. We extract events from the financial news of Chinese local newspapers which are available on the web, the news are matched against stock prices databases and a new method is proposed for the mining of frequent temporal patterns.

1 Introduction

In stock market, the share prices can be influenced by many factors, ranging from news releases of companies and local politics to news of superpower economy. We call these incidences **events**. We assume that each event is of a certain **event type** and each event has a time of occurrence, typically given by the date that the event occurs or it is reported. Each "event" therefore corresponds to a time point. We expect that events like "the Hong Kong government announcing deficit" and "Washington deciding to increase the interest rate", may lead to fluctuations in the Hong Kong stock prices within a short period of time. When a number of events occur within a short period of time, we assume that they possibly have some relationship. Such a period of time can be determined by the application experts and it is called a **window**, usually limited to a few days. Roughly speaking, a set of events that occur within a window is called an **episode instance**. The set of event types in the instance is called an **episode**.

For example, we may have the following statement in a financial report: "Telecommunications stocks pushed the Hang Seng Index 2% higher following the Star TV-HK Telecom and Orange-Mannesmann deals". This can be an example for an episode, in which all the four events, "telecommunication stocks rise", "Hang Seng Index surges" and the two deals of "Star TV-HK Telecom" and "Orange-Mannesmann", all happened within a period of 3 days. If there are many instances of the same episode it is called a *frequent episode*. We are interested to find frequent episodes related to stock movements. The stock movement need not be the last event occurring in the episode instance, because the movement of stocks may be caused by the investors' expectation that something would happen on the following days. For example, we can have a news report saying "Hong Kong shares slid yesterday in a market burdened by the fear of possible United States interest rates rises tomorrow". Therefore we do not assume an ordering of the events in an episode.

K.-Y. Whang, J. Jeon, K. Shim, J. Srivatava (Eds.): PAKDD 2003, LNAI 2637, pp. 27–39, 2003.
© Springer-Verlag Berlin Heidelberg 2003

From the frequent episode, we may discover the factors for the fluctuation of stock prices. We are interested in a special type of episodes that we call **stock-episodes**, it can be written as "$\langle e_1, e_2, ... \ e_n \ (t \ \text{days})\rangle$", where the e_1, e_2, ... e_n are event types and at least one of the events should be the event of stock fluctuation. An instance for this stock-episode is an instance where the events of the event types e_1, ... e_n appear in a window of t days. Since we are only concerned with stock-episodes, we shall simply refer to stock-episodes as episodes.

1.1   Definitions

Let $E = \{E_1, E_2, ..., E_m\}$ be a set of **event types**. Assume that we have a database that records events for days 1 to n. We call this a **event database**, we can represent this as $DB =< D_1, D_2, ..., D_n >$, where D_i is for day i, and $D_i = \{e_{i1}, e_{i2}, ..., e_{ik}\}$, where $e_{ij} \in E$ ($j \in [1, k]$), This means that the events that happen on day i have event types $e_{i1}, e_{i2}, ..., e_{ik}$. Each D_i is called a **day-record**. The day records D_i in the database are consecutive and arranged in chronological order, where D_i is one day before D_{i+1} for all $n - 1 \geq i \geq 1$. $P = \{e_{p1}, e_{p2}, ..., e_{pb}\}$, where $e_{pi} \in E$ ($i \in [1, b]$), is an **episode** if P has at least two elements and at least one e_{pj} is a stock event type. We assume that a **window size** is given which is x days, this is used to indicate a consecutive sequence of x days. We are interested in events that occur within a short period as defined by a window. If the database consists of m days and the window size is x days, there are (m) windows in the database: The first window contains exactly days $D_1, D_2, ..., D_x$ The i-th window contains $D_i, D_{i+1}, ...$, with up to x days. The second last window contains D_{m-1}, D_m, and the last window contains only D_m.

In some previous work such as [6], the frequency of an episode is defined as the number of windows which contain events in the episode. For our application, we notice some problem with this definition: suppose we have a window size of x, if an episode occurs in a single day i, then for windows that start from day $i - x + 1$ to windows starting from i, they all contain the episode, so the frequency of the episode will be x. However, the episode actually has occurred only once. Therefore we propose a different definition for the frequency of an episode.

Definition 1. *Given a window size of x days for DB, and an episode P, an* **episode instance** *of P is an occurrence of all the event types in P within a window W and where the record of the first day of the window W contains at least one of the event types in P. Each window can be counted at most once as an episode instance for a given episode.*

The **frequency of an event** *is the number of occurrences of the event in the database. The* **support** *or the* **frequency** *of an episode is the number of instances for the episode. Therefore, the frequency of an episode P is the number of windows W, such that W contains all the event types in P and the first day of W contains at least one of the event types in P. An episode is a* **frequent episode** *if its frequency is >= a given* **minimum support threshold.** ☐

Problem definition: Our problem is to find all the frequent episodes given a event database and the parameters of window size and minimum support threshold.

Let us call the number of occurrences of an event type a in DB the **database frequency** of a. Let us call the number of windows that contain an event type a the **window frequency** of a. The window frequency of a is typically greater than the database frequency of a since the same occurrence of a can be contained in multiple windows. We have the following property.

Property 1. For any episode that contains an event a, its frequency must be equal to or less than the window frequency of a. That is, the upper limit of the frequency of an episode containing a is the window frequency of a.

Lemma 1. *For a frequent episode, a subset of that episode may not be frequent.*

Proof: We prove by giving a counter example to the hypothesis that all subsets of a frequent episode are frequent. Suppose we have a database with 7 days, and the window size is 3, the records D_1 to D_7 are: $\{b\}, \{a, c\}, \{b\}, \{d\}, \{b\}, \{c, a\}, \{d\}$, respectively. If the threshold is 3, then $< abc >$, has 3 occurrences and is a frequent episode, while $< ac >$, which is a subset of $< abc >$, has only 2 occurrences and is not a frequent episode. $\square$

1.2   Related Work

The mining of frequent temporal patterns has been considered for sales records, financial data, weather forecast, and other applications. The definitions of the patterns vary in different applications. In general an episode is a number of events occurring within a specific short period of time. The restriction of the ordering of events in an episode depends on the applications. Previous related research includes discovering sequential patterns [1], frequent "episodes" [6], temporal patterns [4] and frequent patterns ([3,2]). In [6] an episode, defined as the "partially ordered" events appearing close together, is different from our definition of stock-episode. Some related work focus on stock movement [5], but we would like to relate financial events with stock movement.

When we deal with the events which last for a period of time, we may consider the starting time and ending time of the events as well as their temporal relations, such as overlap and during. [4] discovers more different kinds of temporal pattern. [9] discovers frequent sequential patterns by using a tree structure. [6] finds the frequent series and parallel episodes in a sequence of point-based events. An episode is a partially ordered of events occurring close together. An episode X is a subepisode of another episode Y if all events in X are also contained in Y and the order of events in X is the same with that in Y. The frequency of an episode is the number of windows containing the episode. Note that this definition is different from ours, since it allows the same episode instance to be counted multiple times when multiple windows happen to contain the instance.

Most of the algorithms introduced in the above are based on Apriori Algorithm [1]. Our definition of frequent episodes does not give rise to the subset closure property utilized in these methods. [3] provides a fast alternative to find the frequent pattern with a frequent pattern tree (FP-tree), which is a kind of prefix tree. However, the FP-tree is not designed for temporal pattern mining. There is some related work in applying the technique to mine frequent subsequences in given sequences [8], but the problem is quite different from ours.

2   An Event Tree for the Database

The method we propose has some similarity to that in [3]. We use a tree structure to represent the sets of event types with paths and nodes. The process is comprised of two phrases: (1) Tree construction and (2) Mining frequent episodes.

The tree structure for storing the event database is called the **event tree**. It has some similarity to the FP-tree. The root of the event tree is a null node. Each node is labeled by an event type. Each node also contains a **count**, and a **binary bit**, which indicates the *node type*.

Before the event tree is built, we first gather the frequencies of each event type in the database DB. We sort the event types by descending frequencies. Next we consider the windows in the database. For each window,

1. Find the set F of event types in the first day, and the set R of event types in the remaining days. F and R are each sorted in descending database frequency order.
2. Then the sorted list from F and that from R are concatenated into one list and inserted into the event tree. One tree node corresponds to each event type in each of F and R. If an event type is from F, the binary bit in the tree node is 0, if the event type is from R, the binary bit in the tree node is 1. Windows with similar event types may share the same prefix path in the tree, with accumulated count. Hence a path may correspond to multiple windows. If a new tree node is entered into the tree, the count is initialized to 1. When an event type is inserted into an existing node, the count in the node is incremented by 1.

In the event tree, each path from the root node to a leaf node is called a **window path**, or simply a path, when no ambiguity can arise. The event tree differs from an FP-tree in that each window path of the tree is divided into two parts. There is a cut point in the path so that the nodes above the cut point has binary bit set to 0. This is called the **firstdays part** of the path. The second part of the path, with binary bits of 1, is called the **remainingdays part** of the path.

There is a header table that contains the event types sorted in descending order of their frequencies. Each entry in the header table is the header of a linked list of all the nodes in the event tree labeled with the same event type as the header entry. Each time a tree node x is created with a label of event type e, the node x is added to the linked list from the header table at entry e. The linked list therefore has a mixture of nodes with binary bits of 0 or 1.

The advantage of the event tree structure is that windows with common frequent event types can likely share the same prefix nodes in the event tree. In each of the firstdays part and the remainingdays part, the more frequent the event is, the higher level the event node is in so as to increase the chance of reusing the existing nodes.

Before building the tree, we can do some pruning based on event type frequencies. Those event types with window frequencies less than the minimum support threshold are excluded from the tree because the event types will not appear in the frequent episodes with the reason stated in Property 1. Once an event type is excluded, it will be ignored whenever it appears in a window. This helps us to reduce the size of tree and reduce the chance of including non-frequent episodes.

Strategy 1. We remove those events with the window frequencies less than the minimum support threshold before constructing the tree,

Strategy 2. In counting the windows for the frequency of an episode, each window can be counted at most once. If an event type appears in the first day and also in the remaining days of a window simultaneously, the effect on the counting is the same as if the event type appears only in the first day. Therefore, for such a window, only the occurrence of the event type in the first day will be kept and that in the remaining part of the window is/are removed.

Example 1: Given an event database as shown in Figure 1(a), suppose the window size is set to 2 days and the minimum support is set to 5, the event database is first scanned to sum up the frequencies of each event type in the database and also the window frequencies, which are $< a : 4, b : 5, c : 3, d : 3, m : 2, x : 3, y : 2, z : 2 >$ and $< a : 6, b : 7, c : 5, d : 6, m : 4, x : 5, y : 4, z : 3 >$, respectively. Thus the frequent event types are a, b, c, d, x since their window frequencies are at least the minimum support. The frequent event types are sorted in the descending order of their database frequencies and the ordered frequent event types are $< b, a, c, d, x >$.

Day	Events
1	a, b, c
2	y, m, b, d, x
3	a, c
4	a, b
5	z, m, y, d
6	b, x, c
7	d, a, b
8	x, z

(a)

Window No.	Days included	Event-set pairs
1	1,2	$< (b, a, c), (d, x) >$
2	2,3	$< (b, d, x), (a, c) >$
3	3,4	$< (a, c), (b) >$
4	4,5	$< (b, a), (d) >$
5	5,6	$< (d), (b, c, x) >$
6	6,7	$< (b, c, x), (a, d) >$
7	7,8	$< (b, a, d), (x) >$
8	8	$< (x), () >$

(b)

Fig. 1. An event database and the corresponding windows

Next a null root node is created. The event database is then scanned for the second time to read the event types in every 2 days for inserting the windows' event types into the tree. Keeping only the frequent event types and excluding the duplicate event types, the first window can be represented by $< (b, a, c), (d, x) >$, in which the first round brackets consists of the event types in the first day of window while the second round brackets consists of the event types in the remaining days of the window (the second day of the window in this example). Both event lists are sorted in decreasing window frequency order. We call $<$ $(b, a, c), (d, x) >$ the **event-set pair** representation for the window. A first new path is built for the first window with all counts initialized to one. The nodes are created in the sorted order and the types of the nodes b, c and a are set to 0 while that of nodes d and x are set to 1.

The window is shifted one day lower and one more day of event types in the database are read to get the second window. The event types are sorted and the second window is $< (b, d, x), (a, c) >$. The tree after inserting the path for the second window is shown in Figure 2. Each tree node has a label of the form $< E : C : B >$ where E is an event type, C is the count, and B is the binary bit. In this figure, the dotted lines indicates the linked list originating from items in the header table to all nodes in the tree with the same event type.

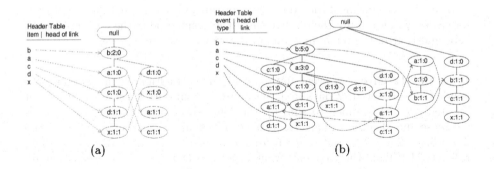

Fig. 2. (a) The tree after inserting the first two windows. (b) A rough structure of the final tree constructed in Example 1.

The remaining windows are inserted to the tree in the similar way. The rough structure of the final tree constructed is shown in Figure 2(b). (Note that some dotted lines are missing in the figure for clarity.)

3   Mining Frequent Episodes with the Event Tree

Our mining process is a recursive procedure applied to each of the linked list kept at the header table. Let the event types at the header table be $h_0, h_1, ...h_H$, in the top-down ordering of the table. We start from the event type h_H at the bottom of the header table and traverse up the header table. We have the following objective in this recursive process:

Objective A: *Our aim is that when we have finished the processing of the linked list for h_i, we should have mined all the frequent episodes that contain event types $h_i, h_{i+1}, ..., h_H$.*

Suppose we are processing the linked list for event type h_i. $\{h_i\}$ is called a **base event set** in this step. We can examine all the paths including the event type h_i from the event tree by following the linked list. Let us call the set of these paths P_i. These paths will help us to find the frequencies of episodes containing event type h_i. We have the following objective:

Objective B: From the paths in P_i, we should find all frequent episodes that contain h_i but not any of $h_{i+1}, ..., h_H$.

The reason why we do not want to include $h_{i+1}, ... h_H$ is that frequent episodes containing any of $h_{i+1}, ..., h_H$ have been processed in earlier iterations. Let us call the set of all frequent episodes in DB that contain h_i but not any of $h_{i+1}, ... h_H$, the set X_i.

We break up the Objective B into two smaller objectives:

Objective B1: *From the paths in P_i, we would like to find all frequent episodes in X_i of the form $\{a\} \cup \{h_i\}$, where a is a single event type.*

Objective B2: *From the paths in P_i, we would like to form a database of paths DB' which can help us to find the set S_i of all frequent episodes in X_i that contains h_i and at least two other event types. DB' is a* **conditional database** *which does not contain $\{h_i\}$ such that if we concatenate each conditional frequent episode in DB' with h_i, the resulting episodes will be the set we want.*

With DB' we shall build a conditional event tree T' with its header table in a similar way as the first event tree. Therefore, we can repeat the mining process recursively to get all the conditional frequent episodes in T'.

Now we consider how we can get the set of paths P_i, and from there obtain a set of **conditional paths** C_i in order to achieve Objectives B1 and B2. Naturally we examine the linked list for h_i, and locate all paths that contain h_i. Let us call the event types $h_{i+1}, ... h_H$ **invalid** and the other event types in the header table **valid**. A node labeled with an invalid(valid) event type is invalid (valid). Suppose we arrive at a node x in the linked list, there are two possibilities

1. If the node x (with event type h_i) is at the firstdays part (the binary bit is 0), we first visit all the ancestor nodes of x and include all the nodes in our **conditional path prefix**. We perform a *depth-first search* to visit all the sub-paths in the sub-tree rooted under event node x, each such path has a potential to form a conditional path. Only valid nodes are used to form paths in P_i. Note that the nodes in the firstdays part of the path below event x will be invalid and hence ignored.
2. If the node x (with event type h_i) is in the remainingdays part, we simply traverse up the path and ignore the subtree under this node. This is because all the nodes below x will be invalid. Invalid nodes may appear above x and they are also ignored.

For example, when we process the left most $< d : 1 : 1 >$ node in the tree in Figure 2(b), we traverse up the tree, include all nodes except for $< x : 1 : 0 >$, since it is invalid. When we process the left most $< c : 1 : 0 >$ node in the tree in Figure 2(b), we traverse up to node $< b : 5 : 0 >$, and then we do a depth first search. We ignore the nodes $< x : 1 : 0 >$ below it, and include $< a : 1 : 1 >$ but not $< d : 1 : 1 >$. Note that the downward traversal can be stopped when the current node has no child node or we have reached an invalid node.

Let us represent a path in the tree by $< (e_1 : c_1, e_2 : c_2, ..., e_p : c_p), (e_1' : c_1', ..., e_q' : e_q') >$, where e_i, e_i' are event types, c_i, c_i' are their respective counts, e_i are event types in the firstdays part, and e_j' are from the remainingdays part. Consider a path p that we have traversed in the above. We effectively do a few things for p:

- **Step (1)**: Remove invalid event types, namely, $h_{i+1}, ...h_H$.
- **Step (2)**: Adjust counts of nodes above h_i in the path to be equal to that of h_i
- **Step (3)**: If h_i is in the firstdays part, then move all event types in the remainingdays part to the firstdays part
- **Step (4)**: Remove h_i from the path.

The resulting path is a conditional path for h_i. After we have finished with all nodes in the linked list for h_i, we have the complete set of conditional paths C_i for h_i. For example, for Figure 2, the conditional paths for x are $< (d : 1), (b : 1, c : 1) >, < (b : 1, c : 1, a : 1, d : 1), (\phi) >, < (b : 1, a : 1, c : 1), (d : 1) >, < (b : 1, a : 1, d : 1), (\phi) >, < (b : 1, d : 1, a : 1, c : 1), (\phi) >$.

The set C_i forms our conditional database DB'. It helps us to achieve both Objectives B1 and B2. For Objective B2, we first determine those event types in DB' with a **window frequencies** which satisfies the minimum threshold requirement. This can help us to prune some event types when constructing the conditional event tree T'. The window frequency of e is the sum of the counts of nodes in C_i with a label of e.

For Objective B1, we need to find the single event types which when combined with h_i will form a frequent episode. For locating these event types we use the **first-part frequency** for event types. The first-part frequency of an event type e in the set of conditional paths C_i is the sum of the counts in the nodes with label e in C_i that are in the firstdays part.

In the above we describe how we can form a conditional database DB' with a base event set $\alpha = \{h_i\}$, and a conditional event tree T' can be built from DB'. In the header table for T', the event types are sorted in descending order of the database frequencies in DB'. Event types in the conditional paths in DB' are then sorted in the same order at both the firstdays part and the remainingdays part before they are inserted into T'.

We apply the mining process recursively on this event tree T'. T' has its own header table and we can repeat the linked list processing with each entry in the header table. When we build another conditional event tree T'' for a certain header h_j' for T', the *event base set* is updated as $h_j' \cup \alpha$. This means that frequent episodes uncovered from T'' are to be concatenated with $h_j' \cup \alpha$ as the resulting frequent episodes.

Strategy 3. When a conditional event tree contains only a single path, the frequent episodes can be generated directly by forming the set S_1 of all possible subsets of the event types in the firstdays part of path, and then the set S_2 of all possible subsets of the event types in the remainingdays part. Any element of S_1 with the event base set is a possible frequent episode. The union of any element of S_1 and any element of S_2 and the event base set is also a possible frequent episode. And the frequency of such a episode is the minimum among the event types in episode.

By the way we construct a conditional event tree, if a path contains event types in the remainingdays part, those event types corresponds to windows which contains some episode e with h_i in the remainingdays part. For such windows to be counted for the episode e, there must be some event type in e that occurs in the firstdays part. Therefore when we form episode with an element in S_2 we must combine with some element in S_1.

4   Performance Evaluation

To evaluate the performance of the proposed method, we conduct experiments on an Sun Ultra 5_10 machine running SunOS 5.8 with 512 MB Main Memory. The programs are written in C. Both synthetic and real data sets are used.

Synthetic data: The synthetic data sets are generated from a modified version of the synthetic data generator in [1]. The data is generated with the consideration of overlapping windows. That is, with the window size of x days, the program will consider what data it has generated in the previous $x-1$ days, in order to choose the suitable event types for the x-th day to maintain the target frequencies of the frequent episodes. The data generator takes six main parameters as listed in the following table:

Parameter	Description	Values
$|D|$	Number of days	1K, 2K, 3K
$|T|$	Average number of events per day	10, 20
$|I|$	Average size of frequent episodes	3, 5
$|L|$	Number of frequent episodes	1000
M	Number of event types	100 - 1000
W	Window size	2 - 10

Four datasets with different parameter settings as shown in the following table are produced. With D1 and D2, we vary the thresholds and window sizes. With D2 to D4 we vary the number of days and event types.

Dataset Name	Dataset	$|T|$	$|I|$	$|D|$	$|M|$
D1	T10.I3.M500.D1K	10	3	1K	500
D2	T20.I5.M1000.D3K	20	5	3K	1000
D3	T20.I5.M100.D2K	20	5	2K	200
D4	T20.I5.M500.D2K	20	5	2K	500

In our implementation, we used linked lists to keep the frequent episodes, one list for each episode size. Each of these lists is kept in an order of decreasing frequencies for a ranked display to the user at the end of the mining. We measure the run time as the total execution time of both CPU time and I/O time. The run time in the experiment include both tree construction and mining. Each data points in graphs are the mean time of several runs of the experiment.

The run time decreases with the support threshold as shown in Figure 3 (a). As the support threshold increases, less frequent events are found and included in the subsequent conditional trees and much less time are required to find the frequent event types in the smaller conditional trees.

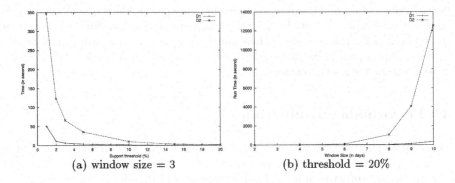

(a) window size = 3   (b) threshold = 20%

Fig. 3. Performance of synthetic datasets D1 and D2

Figure 3(b) shows the effect of different window sizes on the run time. The datasets D1 and D2 are used and the experiment run under threshold fixed to 20%. When the window size increases, the execution time increases because more items are included in window and paths of trees. The parameters of D2 are greater than D1 and therefore the sizes of the initial tree and the conditional trees are larger. So the run time for D1 is much larger than that for D2 when the window size is greater than 8 days.

To study the effect of the number of days in datasets on the execution time, the experiment on dataset D2 is conducted. The support threshold and the window size are set to 10% and 3 days respectively. The result in Figure 4(a) shows that the execution time increases linearly with the number of days.

The effect of the number of event types on the execution time is also investigated. The dataset D2 is used and the support threshold is set to 10% with 3 days of window size. The result is shown in Figure 4(b).

The curve falls exponentially as the number of event types increases. When the number of items is decreased, the distribution of event types are more concentrated and the frequencies of the event types are higher. Therefore less events are pruned when constructing the conditional trees and the run time is longer.

Real data: The real data set is the news event extraction from a internet repository of a number of local newspapers, more details of which is reported in [7]. It contains 121 event types and 757 days. For example, Cheung Kong

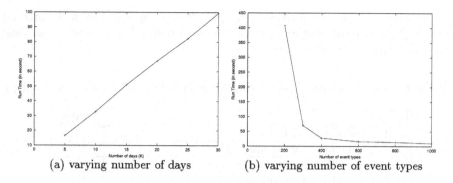

(a) varying number of days   (b) varying number of event types

Fig. 4. Synthetic dataset D2 with window size = 3 and threshold = 10%.

stock goes up is an event type. An event for an event type occurs when it is reported in the collected news. In addition we have collected stock data from the Datastream International Electronic Databse, we have retrieved Dow Jones industrial average, Nasdaq Composite Index, Hang Seng Index Future, Hang Seng Index, and prices of 12 top local companies for the same period of time.

The performance using the real dataset with different support thresholds and window sizes are shown in Figure 5 (a) and (b). In Figure 5 (a), the window size is set to 3 days. The execution time is rapidly decreased with the threshold above 15%. It is because the supports of half of the most frequent events are close together, when the threshold is below 17%, the pruning power in forming conditional trees is weak.

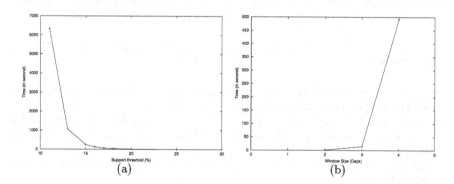

(a)   (b)

Fig. 5. Real dataset with (a) window size = 3 days. (b) support threshold = 20%.

The performance on varying the window sizes are shown in Figure 5(b) with the threshold equal to 20%. The execution time increases with the window size steeply. The run time with a window size of 5 days is too long (> 30000 seconds) and is not shown in graph. When the window size is large, the tree paths are longer and include more items. As the supports of the items are close together, the subsequent conditional trees nearly include all event types from the original trees and the sizes of conditional trees cannot be reduced. Thus the mining time increases with the window size.

Table 1. Some of the results mined with threshold = 15% and window size = 3 days.

Episode	Support
Nasdaq downs, PCCW downs	151
Cheung Kong ups, Nasdaq ups	129
Cheung Kong Holdings ups, China Mobile Ups	128
Nasdaq ups, SHK Properties flats, HSBC flats	178
Cheung Kong ups, SHK Properties flats, HK Electric flats	178
China Mobile downs, Nasdaq downs, HK Electric flats	178
China Mobile downs, Heng Sang Index downs, HSBC flats	135
US increases interest rate, HSBC flats, Dow Jones flats	100

Real Dataset Results Interpretation: Since the frequencies of the events obtained from newspapers are much less than the events of stock price movement, we have set the threshold to 15% to allow more episodes including the newspaper events to be mined. We have selected some interesting episodes mined with threshold set to 15% and window size set to 3 days in Table 1. We notice some relationship between Nasdaq and PCCW, a telecom stock. We see that Nasdaq may have little impact on SHK Properties (real estate), or HSBC (banking).

Acknowledgements. This research was supported by the RGC (the Hong Kong Research Grants Council) grant UGC REF.CUHK 4179/01E.

References

1. R. Agrawal and R. Srikant. *Mining sequential patterns.* 11th International Conf. On Data Engineering, March 1995.
2. M.S. Chen, J.S. Park, and P.S. Yu. *Efficient Data Mining for Path Traversal Patterns.* IEEE Transactions on Knowledge and Data Engineering, March/April 10:2, 1998.
3. J. Han, J. Pei, and Y. Yin. *Mining Frequent Patterns without Candidate Generation.* SIGMOD, 2000.
4. P. S. Kam and A. W. C. Fu. *Discovering Temporal Patterns for Interval-Based Events.* Proc. Second International Conference on Data Warehousing and Knowledge Discovery, 2000.
5. H. Lu, J. Han, and L. Feng. *Stock Movement Predication and N-Dimensional Inter-Transaction Association Rules.* Proc. of SIGMOD workshop on Research Issues on Data Mining and Knowledge Discovery (DMKD'98), 1998.
6. H. Mannila and H. Toivonen. *Discovering generalized episodes using minimal occurrences.* 2nd International Conf. On Knowledge Discovery and Data Mining, August 1996.

7. Anny Ng and K.H. Lee. *Event Extraction from Chinese Financial News*. International Conference on Chinese Language Computing (ICCLC), 2002.
8. J. Pei, J. Han, B. Mortazavi-Asl, H. Pinto, Q. Chen, U. Dayal, and M.C. Hsu. *PrefixSpan: Mining Sequential Patterns Efficiently by Prefix-Projected Pattern Growth*. Proceedings of the 12th IEEE International Conference on Data Engineering, 2001.
9. P. C. Wong, W. Cowley, H. Foote, and E. Jurrus. *Visualizing Sequential Patterns for Text Mining*. Proceedings IEEE Information Visualization, October 2000.

An Efficient Algorithm of Frequent Connected Subgraph Extraction*

Mingsheng Hong, Haofeng Zhou, Wei Wang, and Baile Shi

Department of Computing and Information Technology Science
Fudan University, Shanghai, 200433, P.R.China
haofzhou@fudan.edu.cn

Abstract. Mining frequent patterns from datasets is one of the key success stories of data mining research. Currently, most of the works focus on independent data, such as the items in the marketing basket. However, the objects in the real world often have close relationship with each other. How to extract frequent patterns from these relations is the objective in this paper. We use graphs to model the relations, and select a simple type for analysis. Combining the graph theory and algorithms to generate frequent patterns, a new algorithm Topology, which can mine these graphs efficiently, has been proposed. We evaluate the performance of the algorithm by doing experiments with synthetic datasets and real data. The experimental results show that Topology can do the job well. At the end of this paper, the potential improvement is mentioned.

1 Introduction

Mining association rules [1] is an important task in data mining. After nearly 10 years development, it has made great progress and expanded into many other fields, such as Web mining, bioinformatics, marketing basket analysis and so on. It is playing a more and more important role.

Frequent pattern extraction is an essential phase in association rules mining. Most works have been done on the phase of the pattern generation. From Apriori [2], DHP[3], DIC[4] to FP-Growth[5], the efficiency of the algorithm has been improved step by step. However, most of the works only process the single item, while there are many relations among objects in the real world, which form a net. It means that to deal with the items isolatedly is not suitable. Meanwhile, the fact that more and more complicated applications have been put up urge us to put more attention to the frequent pattern mining in graphs. An case in point is to analyze the structural information of chemical molecule, where there are keys between atoms. It is an interesting case and will definitely help us to reveal the potential relations from the structural information.

Because these data of structural information can be expressed as graphs, we try to begin our work with the general labeled graph, which leads to the method

* This paper was supported by the Key Program of National Natural Science Foundation of China (No. 69933010) and China National 863 High-Tech Projects (No. 2002AA4Z3430 and 2002AA231041)

K.-Y. Whang, J. Jeon, K. Shim, J. Srivatava (Eds.): PAKDD 2003, LNAI 2637, pp. 40–51, 2003.

of the structural information analysis. This work expands the analysis of the single-item data in the previous algorithms to that of the item-pair scope.

We introduce some terms and concepts in Section 2. In Section 3, some improvements are discussed, and they are implemented in the algorithm Topology described in Section 4. In section 5, the performance of Topology is shown via some experiments, and in Section 6, new algorithm is applied in real world. Finally, we review some related work in Section 7, and draw conclusions in Section 8.

2   Terms and Concepts

In this paper, we focus on general labeled graph (GLG in short).

Definition 1. *Given a simple graph $G(V, E)$, $|V| = m$, and the label set $L = \{l_i | 1 \leq i \leq n, \forall i, j\ l_i \neq l_j\}$, where l_i is the label. G is a GLG iff there exists a map $f : V \to L$, such that $\forall v_i \in V, f(v_i) = l_k, l_k \in L$ and $l_i \neq l_j \implies f'(l_i) \neq f'(l_j)$, where f' is the inverse function of f. We denote $f(v_i) = l_k$ as $l(v_i)$ and GLG as $G(V, E, L(V))$.*

All vertexes of GLG are labeled in the label set space L, and each vertex can only have one label. However each label in L can be assigned to multiple vertexes, i.e., there may be such cases that two or more vertexes share the same label in one graph. Also, because of the uncertainty, all of the other elements in the graph such as edges and subgraphs can not be uniquely represented by an unordered labeling pair $\langle l_i, l_j \rangle$ or other labeling methods.

In the background of GLG, the concepts of subgraph, support, and frequent subgraph can still be applied.

Definition 2. *Given GLGs $G(V, E)$ and $H(V', E')$, H is a subgraph of G iff i) $V' \subseteq V$, ii) $\forall v_i, v_j \in V', \langle v_i, v_j \rangle \in E \Leftrightarrow \langle v_i, v_j \rangle \in E'$, and iii) $l(v_i) = l'(v_i)$, $l(v_j) = l'(v_j)$.*

This definition requires that G must contain subgraph H in both the structure and the labels.

Definition 3. *Given GLG set $GD = \{G_1, G_2, \cdots, G_n\}$, where G_i is a GLG, the support of H is*

$$s(H) = \frac{|p(H) = \{G' | G' \in GD, H \subseteq G'\}|}{|GD|}, \tag{1}$$

where $|p(H)|$ is the frequency of H. If $s(H) \geq minsup$, where $minsup$ is a pre-defined threshold called minimum support, H is a frequent GLG.

In General, we only concern the connected graph, where we can find a path between any of the two vertices in the graph. So our goal in the paper is, given a GLS set GD and minimum support $minsup$, to find all the connected frequent general labeled (sub)graph in the original GLGs.

In the mining process of GLG, we have to face a very hard problem, isomorphism.

Definition 4. *Given GLGs $G_1(V_1, E_1.L_1(V_1))$ and $G_2(V_2, E_2, L_2(V_2))$, G_1 and G_2 are isomorphic iff there exists a one-to-one map $f : V_1 \longleftrightarrow V_2$, such that $\langle a, b \rangle \in E_1$ iff $\langle f(a), f(b) \rangle \in E_2$ and $L_1(a) = L_2(f(a)), L_1(b) = L_2(f(b))$. We denote it as $G_1 \cong G_2$.*

Till now, We don't know whether the graph isomorphism problem is $\mathcal{P}$ or $\mathcal{NP} - complete$ [6]. We can use brute force methods to check whether 2 graphs are isomorphic or not, but it is not feasible in practice.

Another concept associated with isomorphism is the canonical form [7]. As we all know, graph can be represented by adjacency matrix. For example, suppose a $n \times n$ matrix for graph G is $M = a_{ij}, (1 \leq i, j \leq n)$, then the code of G can be expressed as

$$code(G) = a_{12}a_{13}a_{23}a_{14}a_{24}a_{34} \cdots a_{n-1,n} \tag{2}$$

The canonical form is a special matrix, which is unique under certain rules, e.g., the $code(G)$ is the greatest one among all the possible codes of G in this paper. So any two graphs are isomorphic if and only if they have the same canonical form. This technique is equivalent to the isomorphism problem.

There have been some works on graph mining, and most of them are based on the Apriori framework. Currently, there are 2 types of which, one process the vertices, such as AGM [8,9] and AcGM [10], while the other treats the edges as the objects, e.g. FSG [11,12].

AcGM is based on AGM. The former focuses on the frequent connected subgraph and applies some techniques to improve the performance. It uses the concept of semi-connected graph, which is defined as a connected graph or a graph consisting of a connected part and a single vertex. Its canonical form is defined as $CODE(X) = l(v_1)l(v_2) \cdots l(v_n)code(X)$.

The join step in AcGM is the most important. AcGM also applies other techniques, such as canonical form finding and K^k tree, to improve the performance. Please refer to [10] for more details.

FSG is another Apriori-based algorithm. It uses the concept of "core" and TID list to solve the problem.

In [10], the comparison among AGM, AcGM and FSG shows that AcGM is the best one regarding performance.

3   Improvements of Performance

3.1   Principle of First-Sequence-Later-Structure(FSLS)

In GLG, each vertex in a graph has a label, and each edge can be represented by the labels of the two vertexes it connects. Thus, each frequent subgraph corresponds to a labeled vertex set and a labeled edge set. However, this mapping is not one to one. A vertex or edge label set can map to multiple frequent subgraphs. We call the graphs with the same labeled vertex set or labeled edge set as isomer-graphs(IG). The labeled set is called isomer-expression(IE). The labeled vertex/edge set is so called VIE and EIE in short respectively.

Same as all the Apriori-based methods, in graph mining, the temporary results generated in the algorithm can form lattices. By utilizing IE, we can merge elements in the lattice to form a new lattice. This kind of abstract IE lattice, without storing the detailed structure of graph, only keeps some tagged information, and can reduce the search space efficiently.

If we treat the original graph as a labeled vertex/edge set, the concepts of support and frequency threshold can also be applied to those IEs as they are in subgraph. They are defined as the number or percentage of IE in graphs. Similarly, the support of IE's any subset is equal to or greater than that of the IEs. Considering the relationship between an IE and a real subgraph, we can draw the following conclusions:

Theorem 1. *If an IE is not frequent, then any graph it corresponds with can not be frequent.*

Meanwhile, we examine the EIE in the view of connectivity. In GLG, because the label of vertex is not unique, if two edges are connected, they have at least one vertex labeled the same. We can further extend this conclusion to the field of graph, i.e., if a GLG is connected, there exists some vertex labels in its corresponding EIE which can make all labeled edges connected. The reversed is not true. So, we can find EIE facilitates the check of connectivity. If an EIE is not form-connected, then all the corresponding isomers can not be connected.

Both VIE and EIE can be used to reduce the search space, but the former emphasizes particularly on frequency, and the latter, connectivity. After that, we can identify the different structures expressed in IEs, which is more efficient than the direct mining. We refer to this principle is as First-Sequence-Later-Structure(FSLS). The term "sequence" used here will be explained in the later section, where the "sequence" and "expression" will be the same.

3.2   Improving Canonical form Generation

To improve the performance of canonical form generation, it should analyze the process of canonical form generation in AcGM.

In AcGM, the canonical form X_{ck} of a $k-$graph $G(X_k)$ is derived by the row (column) transformation of G's adjacency matrix X_k. It assumes that all the $(k-1)-$subgraphs have already been obtained. So do the canonical forms of them.

From the analysis, it is clear to see that we don't have to reply on the multiplication of matrixes in order to calculate the canonical form of a graph. If a candidate graph can survive the set generating phase of candidate graphs, all its frequent subgraphs and their corresponding canonical forms are already known in theory, and thus we can utilize these canonical forms directly to generate the objective canonical form. We can find these canonical forms efficiently if we use vertex lattice (instead of vertex sequence lattice). Moreover, to facilitate searching in the lattice, we correlate each vertex to the elements in the lattice. The process above is a process of sequence-matching in essence, and its efficiency is much higher than the multiplication of five matrixes.

It should be noted that, as is mentioned in the previous context, this method may not be sufficient to generate the canonical form of X_k. In that case, we have to perform an exhaustive search of the canonical form and its transformation matrix.

3.3 Using TID List

Solving subgraph isomorphism is an $\mathcal{NP}$−complete problem. Therefore, AcGM and FSG both take measures to minimize the cost. While AcGM uses K^k tree to reducd the cost of subtree matching, FSG uses TID list instead. In fact, these two methods can be used together, and also help the pruning of vertex sequence lattice and edge sequence lattice.

First, when the sequence lattice is generated, for each element in it we write down the TIDs that contain the element. Then, when the isomers of the lowest level in the lattice are generated, we can use these TIDs to limit the range of checking subgraph isomorphism. For each isomer, we record the TIDs corresponding to it. For those in the upper levels of the lattice, we can use join result of the TID sets of all its $(k-1)-$ subgraphs to limit the range.

If all isomers of a sequence are not frequent, the sequence itself is not frequent. To cut down the size of the sequence lattice, we can then delete all the elements containing this sequence in the lattice to further reduce the cost of subsequent candidate graph generation. On the other hand, when we calculate the intersection of TID sets, we can use the TID sets of vertex sequence lattice and edge sequence lattice. Thus, in order to facilitate the searching, we require that all elements maintain their TID sets respectively in the vertex sequence lattice and edge sequence lattice.

As for the K^k tree used in AcGM, it is a good idea to use it also. In particular, when we need to reduce the range of checking subgraph isomorphism, K^k tree's advantage is fully exploited.

So, the revision of AcGM algorithm is based on FSLS technique. In the next section, we will include these improvements in the new algorithm Topology.

4 Efficient Subgraph Extraction: Topology

We call the new algorithm Topology. The core procedures described above are the generation of sequence and the process of sequence lattice. We divide Topology into 2 phases, calculating the sequences and distinguishing the structures.

To gain the sequence, in particular, the vertex sequence, we still use PrefixSpan [13]. The sequence of graph is just used to mark the times of occurrence of the elements (vertexes or edges) in graph. In this way, we obtain a multiset of vertexes (edges). If we sort the elements in the multiset in alphabetical order, we can get a sequence. If we transform all graphs into this sequence format, and regard them as the input of PrefixSpan, we can obtain those sequences $E = (l_{start} \cdots l_{end})$, where l_i are the labels. In order to generate the final canonical form conveniently, it should better satisfy $\forall start \leq i < j \leq end, l_i \geq l_j$.

During the generation of edge sequence, we need to check whether this sequence is form-connected by applying the Warshall algorithm and so on. During the vertex/edge sequences generation, we use them to construct vertex and edge sequence lattice.

In order to use TID in the later part of the algorithm, when a sequence is generated, all its corresponding TIDs should be recorded. That is to say, every time when the projected database is constructed, the IDs of a projected sequence are unique in the global range.

The first phase of Topology can be described as in figure 1.

Algorithm 1 (the First Phase of Topology).

Input: GD, $minsup$
Output: Vertex sequence lattice L_V, edge sequence lattice L_E
Method:
Transform all the graphs in GD into the form ($list_{lv}$, $list_{le}$);
//$list_{lv}$ and $list_{le}$ are lists of labeled vertexes and edges respectively;
Call GPrefixSpan($\langle\rangle$, 0, GD, 0);
Call GPrefixSpan($\langle\rangle$, 0, GD, 1);
GPrefixSpan(α, l, $S|_\alpha$, m)
{   Scan $S|_\alpha$ once to find all frequent labeled elements b such that
        (a) b can be assembled to the last element of α
            to form a sequential pattern; or
        (b) $\alpha\langle b\rangle$ is a sequential pattern;
        record the TID of each transaction containing b to $b.tid\_list$;
        for each frequent labeled element b do
        { $\alpha' = \alpha b$;
            $\alpha'.tid\_list = b.tid\_list$;
            if ($m == 1$) then $\alpha'.connected$=ISConnected(α');
            if ($m == 0$) then add α' to L_V;
            else add α' to L_E;
            construct α'-projected database $S|'_\alpha$
                from $S|_\alpha$ by keeping the TID in $S|_\alpha$;
            call GPrefixSpan(α', $l + 1.S|'_\alpha$, m);
        }
}

Fig. 1. the First Phase of Topology

After the sequence lattices are generated, we can do the real mining job of frequent connected subgraphs.

In the step of joining, we use sequence lattice as the guide. The basic conditions of join are the same as those in AcGM. However, before 2 $k-$graphs join, we first check whether there exists the vertex sequence of this new coming $(k + 1)-$graph in the lattice. The determination of $z_{k,k+1}$'s value can use edge

Algorithm 2 (the Second Phase of Topology).

Input: GD, $minsup$, vertex sequence lattice L_V, edge sequence lattice L_E
Output:Frequent connected graph set F
Method:
F_k is a set of adjacency matrix of $k-$frequent graphs
C_k is a set of adjacency matrix of $k-$candidate graphs
GAcGM(GD, $minsup$, L_V, L_E)
{   scan L_V to generate F_1 and F_2^v with all adjacency
                matrices of one and two vertices respectively;
       scan L_E to generate F_2 with all adjacency matrices with one edge;
       $F_2' = F_2 \cup F_2^v$;
       $k = 2$; C_{k+1}=Generate-Candidate(F_k',L_E,L_V);
       while ($C_{k+1} \neq \emptyset$)
       {   for all $c_k \in C_k$ do counting c_k by scanning $\{G|G.tid \in c_k.tid\_list\}$;
            $F_k' = \{c_k|c_k \in C_k$ and $s(G(c_k)) \geq minsup\}$;
            $F_k = \{c_k|c_k \in F_k'$ and $G(c_k)$ is connected$\}$;
            Revise L_V and L_E according to F_k and F_k';
            C_{k+1}=Generate-Candidate(F_k', L_E, L_V);
            $k + +$;

       }
       return $\bigcup_k\{f_k|f_k \in F_k$ and f_k is canonical$\}$;
}
Generate-Candidate(F_k', L_E, L_V)
{   for all pair $f_i, f_j \in F_k$ that satisfied Conditions do
       {   refer to L_V and L_E do $c_{k+1} = f_i \oplus f_j$;
            flag=1; $CF = 0$;
            for all $k-$subgraph G_c of $G(c_{k+1})$ do
            {   if $(s(G_c) < minsup)$ then $flag = 0$ and break;
                    $CF = max(CF$, combining CF and $CF(G_c))$;
                    $c_{k+1}.tid\_list = c_{k+1}.tid\_list \cap G_c.tid\_list$;
            }
            if $(flag \neq 0)$ then $C_{k+1} = C_{k+1} \cup \{c_{k+1}$with CF and $tid\_list\}$;
       }
       return C_{k+1};
}

Fig. 2. the Second Phase of Topology

sequence lattice to see whether the edge sequence of newly generated graph is in it or not. Following steps are the same as those in AcGM.

The calculation of canonical forms has been combined with the filtering conditions of frequent subgraph in the process of candidate join. When the frequent subgraphs of a graph G are being searched, the canonical form of G can be found in the same way as the winner-select in an elimination game. In some cases, we will have to use exhaustive searching. Meanwhile, the searching process needs to work on the actual (structure) lattice based on vertex sequence lattice.

As for the frequency counting, it also utilizes the vertex structure lattice. First, it executes intersection of all subgraphs' *tid_lists*. Next, it performs subgraph matching by using K^k tree. Once a failure occurs, we change the searching path in the K^k tree by selecting different subtrees. If none of the matching is successful, we use exhaust searching to match the subgraph with this K^k tree.

In general, lattice plays an important role in the implementation. With the progress of the algorithm, the original vertex sequence lattice and edge sequence lattice are replaced by vertex/edge structure lattice.

5   Performance Study of Topology

In this section, we evaluate the perform of Topology by doing experiments in comparison with AcGM. All experiments are run in a Pentium III 733MHz PC, with 384MB RAM. The platform is Windows XP and the algorithm is coded in C++.

5.1   Experimental Data

We wrote a data generator to produce all the data required in the experiments. The generator can be adjusted by 6 parameters: the size of label set S, the possibility of an edge existing between two vertexes p, the number of basic patterns (graphs) L, the average scale of basic patterns(defined as the number of vertexes) I, the size of graph database N, and the average scale of original graphs T.

Given the size of a graph, the label of a vertex can be selected from the label set with an equal possibility. Whether there exists one edge between two vertexes depends on a random variable. If this variable is larger than p, we add an edge there. Generally, we require that all graphs are connected.

The basic patterns (graphs) are generated first. Then, the generation of original graphs is divided in two phases. The first phase is the same as that of basic patterns. In the second phase, we select a pattern P from basic patterns with an equal possibility, and overlap it in the original data graph randomly. When overlapping, if the vertex number of pattern P is N, we randomly select N vertexes in the original graph, and map them with those in P randomly. Then we replace those original vertexes with those in P, and rebuild the relationship between them according to their relationship in P.

5.2   Performance Evaluation

We evaluated the performance of Topology by comparing it with AcGM. We tested the execution time of the algorithm under the circumstances of different sizes and scales of graph sets, and different possibilities of edge existing respectively. The default minimum support is $minsup = 0.1$.

First, we analyze the relationship between the number of label graphs and the execution time. We set $S = 5$, $p = 0.7$, $L = 10$, $I = 5$ and $T = 10$. N ranges from 50 to 1000. The final experimental results are shown in figure 3. In

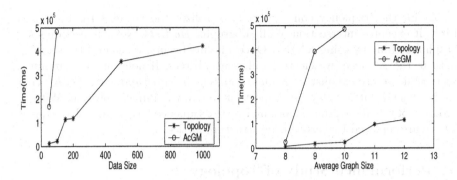

Fig. 3. The N vs. execution time

Fig. 4. The T vs. execution time

this figure, the execution time of both Topology and AcGM increases with the increase in the number of labeled graphs. However, it is quite evident that the slope of Topology is smaller than that of AcGM.

The relationship between the scale of labeled graph and the execution time is similar, as shown in Figure 4, where we set $S = 5$, $p = 0.7$, $L = 10$, $I = 5$, $N = 100$, and T varies from 8 to 12.

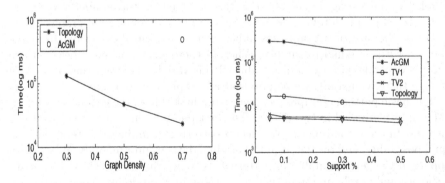

Fig. 5. The p vs. execution time

Fig. 6. The factors improvement

The trend is the same when we analyze the relationship between the possibility of edge existing in labeled graph and the execution time. We set $S = 5$, $L = 10$, $I = 5$, $N = 100$ and $T = 10$. The p ranges from 0.3 to 0.7. The results are shown in figure 5. It is the same case that Topology can be applied to a wider scale, while AcGM can run normally only when $p = 0.7$.

We also did test on checking the contribution of the improvements. Based on the original version of AcGM, we wrote a new version TV_1 applying vertex lattice to help the isomorphism checking. Next, we used the principle of same-sequence-different-structures to form another version TV_2 based on TV_1. Finally, we incorporated the techniques such as TID list and formed the final version, Topology.

We set $S = 5$, $p = 0.7$, $L = 10$, $I = 5$, $N = 100$ and $T = 10$. We examine the degree of contribution of each part in the algorithm by varying $minsup$ from 0.05 to 0.5. We also choose AcGM as the opponent. Figure 6 shows the results. Clearly, among the three improvements, using vertex lattice to improve the efficiency of checking isomorphism contributes most. Using the principle of same-sequence-different-structures to further limit the searching space ranks secondly. Other techniques such as TID list contribute relatively small.

6   Analyzing Molecules

We use the dataset PTE1 of the PTE project (Predictive Toxicology Evaluation) from British Cambridge University to do the analysis. The dataset PTE1 collects 340 compounds in total, among which 24 are atoms. Totally 66 atom types are discovered. There exist many chemical bonds among these atoms. We simplify it and suppose that these bonds are just the connections between atoms without any other meaning. These molecules have been marked by some basic properties such as whether they cause disease, etc.

First we convert PTE1 into the format of graph, where an atom is mapped to a vertex, and a chemical bond to an edge. The 66 atom types correspond to the labels in the label set. We set the $minsup = 0.1$, and start the analysis. Part of the result is shown in figure 7. The molecule structures displayed in the left of figure 7 occur frequently in non-disease-causing compounds, while that displayed in the right of figure 7 occur frequently in disease-causing compounds. So, next time when we try to analyze a certain compound whose properties are unknown, and if we find that it contains the molecule structure shown in the figure 7, we may pay much attention to testing whether it is disease-causing or not.

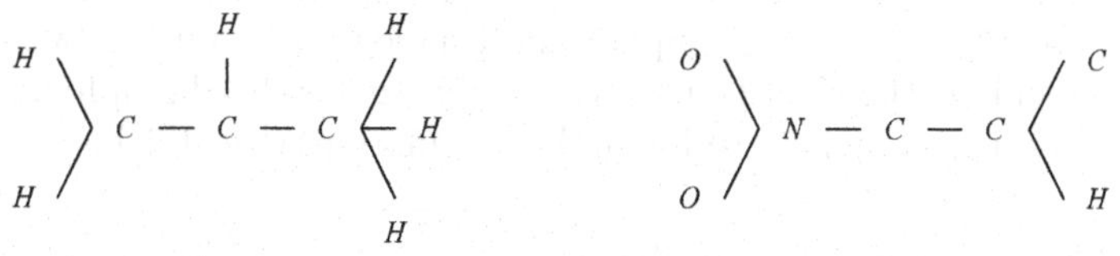

Fig. 7. Analysis of chemical molecule structures

7   Related Work

Extracting frequent subgraphs from a graph dataset is a rather challenging job. Some heuristic based approaches, such as SUBDUE [14], proposed by Cook in

1994, and GBI [15,16] Yoshida and Motoda proposed and revised respectively in 1995 and 1997, are introduced in recent years. However, both methods cannot avoid losing some isomorphic patterns by adopting exhaustive searching. Hence, in year 2002, this algorithm is revised again [18]. Another method proposed at the same time as heuristic method is De Raedt's method [19]. Yet another method that adopted the same exhaustive searching method as AGM, AcGM and FSG is WARMR [20].

From the description above, we can see that Apriori-based AcGM, FSG and AGM may be inferior to SUBDUE in efficiency, but considering all other factors such as the results, this method based on analyzing frequent patterns has obvious advantages.

Concerning the works done in analyzing molecule structures, besides the techniques mentioned above, there is the way proposed in [21]. It uses SMILES [22] in computational chemistry and the coding technique of SMARTS [23] to analyze the HIV. Nevertheless, the method can only be applied to some special fields, which is not suitable for GLG. Recently, X. Wang proposed a method to use 3D graph mining techniques to analyze the patterns in proteins [24].

8   Conclusion

In this paper, we paid much attention to the problem of how to efficiently extract frequent patterns from a graph dataset. We proposed a down-to-earth scheme to revise AcGM, and also implemented the Topology algorithm. The experimental results show that our efforts paid off. Among the several improvements, the technique of using vertex lattice to reduce the cost of checking isomorphism improves the performance most. The method of incorporating the principle of same-sequence-different-structures to further limit the searching space ranks secondly. Other techniques such as TID list contribute relatively small.

We also noticed that gSpan [25] was proposed. It encodes the edge set of graph and obtains an edge sequence similar to that in PrefixSpan, and addresses isomorphism directly in sequence analysis. One of our future work is to analyze and compare the efficiency of gSpan and Topology. In addition, we have been working on making the algorithm process a more generalized graph model, which will probably help Topology to be applied to more practical fields.

References

1. Agrawal, R., et al: Mining association rules between sets of items in large databases. In: Proc. of ACM SIGMOD (1993) 207–216
2. Agrawal, R., et al: Fast algorithms for mining association rules in large databases. In: Proc. of VLDB (1994) 487–499.
3. Park, J.S., et al: An Effective Hash Based Algorithm for Mining Association Rules. In: Proc. of ACM SIGMOD (1995) 175-186
4. Brin, S., et al:Dynamic itemset counting and implication rules for market basket data. In: Proc. of ACM SIGMOD (1997) 255–264

5. Han, J., et al: Mining frequent patterns without candidate generation. In: Proc. of ACM SIGMOD (2000) 1–12
6. Read, R. C., et al: The graph isomorphism disease. J. of Graph Theory, 4 (1977) 339–363
7. Babai, L., et al: Canonical labeling of graphs. In: Proc. of ACM STOC (1983) 171–183.
8. Inokuchi, A., et al: An apriori-based algorithm for mining frequent substructures from graph data. In: Proc. of PKDD, LNCS, Vol. 1910, Springer (2000) 13–23
9. Inokuchi, A., et al: Applying algebraic mining method of graph substructures to mutageniesis data analysis. In: KDD Challenge,PAKDD (2000) 41–46.
10. Inokuchi, A., et al: A fast algorithms for mining frequent connected subgraphs. Research Report RT0448, IBM Research, Tokyo Research Laboratory (2002)
11. Kuramochi, M., et al: Frequent subgraph discovery. In: Proc. of IEEE ICDM (2001) 313–320.
12. Kuramochi, M., et al: An efficient algorithm for discovering frequent subgraph. Technical Report 02-026, Dept. of Computer Science, University of Minnesota (2002)
13. Pei, J., et al: PrefixSpan: Mining sequential patterns by prefix-projected growth. In: Proc. of ICDE (2001) 215–224.
14. Cook, D. J., et al: Substructure discovery using minimum description length and background knowledge. J. of Artificial Intelligence Research 1 (1994) 231–255
15. Yoshida, K., et al: CLIP: Concept learning from inference patterns. Artificial Intelligence 1 (1995) 63–92
16. Motoda, H., et al: Machine learning techniques to make computers easier to use. In: Proc. of IJCAI, Vol. 2 (1997) 1622–1631
17. Matsuda, T., et al: Extension of graph-based induction for general graph structured data. In: Proc. of PAKDD, LNCS, Vol.1805, Springer (2000) 420–431
18. Matsuda, T., et al: Knowledge discovery from structured data by beam-wise graph-based induction. In: Proc. of PRICAI, LNCS, Vol.2417, Springer (2002) 255–264
19. Raedt, L. De, et al: The levelwise version space algorithm and its application to molecular fragment finding. In: Proc. of IJCAI Vol. 2 (2001) 853–862.
20. Dehaspe, L., et al: Finding frequent substructures in chemical compounds. In: Proc. of KDD (1998) 30–36
21. Kramer, S., et al: Molecular feature mining in HIV data. In: Proc. of ACM SIGKDD (2001) 136–143
22. Weininger D.: SMILES, a chemical language and information system. J. of Chemical Information and Computer Sciences 1 (1988) 31–36
23. James,C. A., et al: Daylight Theory Manual — Daylight 4.71
24. Wang, X., et al: Finding patterns in three dimensional graphs: Algorithms and applications to scientific data mining. IEEE TKDE 4 (2002) 731–749
25. Yan, X., et al: gSpan: Graph-based substructure pattern mining. In: Proc. of IEEE ICDM (2002)

Classifier Construction by Graph-Based Induction for Graph-Structured Data

Warodom Geamsakul, Takashi Matsuda, Tetsuya Yoshida, Hiroshi Motoda,
and Takashi Washio

Institute of Scientific and Industrial Research
Osaka University
8-1, Mihogaoka, Ibaraki, Osaka 567-0047, Japan

Abstract. A machine learning technique called Graph-Based Induction (GBI) efficiently extracts typical patterns from graph-structured data by stepwise pair expansion (pairwise chunking). It is very efficient because of its greedy search. Meanwhile, a decision tree is an effective means of data classification from which rules that are easy to understand can be obtained. However, a decision tree could not be produced for the data which is not explicitly expressed with attribute-value pairs. In this paper, we proposes a method of constructing a classifier (decision tree) for graph-structured data by GBI. In our approach attributes, namely substructures useful for classification task, are constructed by GBI on the fly while constructing a decision tree. We call this technique Decision Tree - Graph-Based Induction (DT-GBI). DT-GBI was tested against a DNA dataset from UCI repository. Since DNA data is a sequence of symbols, representing each sequence by attribute-value pairs by simply assigning these symbols to the values of ordered attributes does not make sense. The sequences were transformed into graph-structured data and the attributes (substructures) were extracted by GBI to construct a decision tree. Effect of adjusting the number of times to run GBI at each node of a decision tree is evaluated with respect to the predictive accuracy. The results indicate the effectiveness of DT-GBI for constructing a classifier for graph-structured data.

1 Introduction

In recent years a lot of chemical compounds have been newly synthesized and some compounds can be harmful to human bodies. However, the evaluation of compounds by experiments requires a large amount of expenditure and time. Since the characteristics of compounds are highly correlated with their structure, we believe that predicting the characteristics of chemical compounds from their structures is worth attempting and technically feasible. Since structure is represented by proper relations and a graph can easily represent relations, knowledge discovery from graph structured data poses a general problem for mining from structured data. Some other examples amenable to graph mining are finding typical web browsing patterns, identifying typical substructures of

K.-Y. Whang, J. Jeon, K. Shim, J. Srivatava (Eds.): PAKDD 2003, LNAI 2637, pp. 52–62, 2003.
© Springer-Verlag Berlin Heidelberg 2003

chemical compounds, finding typical subsequences of DNA and discovering diagnostic rules from patient history records.

Graph-Based Induction (GBI) [9,3] is a technique which was devised for the purpose of discovering typical patterns in a general graph data by recursively chunking two adjoining nodes. It can handle a graph data having loops (including self-loops) with colored/uncolored nodes and links. There can be more than one link between any two nodes. GBI is very efficient because of its greedy search. GBI does not lose any information of graph structure after chunking, and it can use various evaluation functions in so far as they are based on frequency. It is not, however, suitable for graph structured data where many nodes share the same label because of its greedy recursive chunking without backtracking, but it is still effective in extracting patterns from such graph structured data where each node has a distinct label (*e.g.*, World Wide Web browsing data) or where some typical structures exist even if some nodes share the same labels (*e.g.*, chemical structure data containing benzene rings etc).

On the other hand, decision tree construction method [6,7] is a widely used technique for data classification and prediction. One of its advantages is that rules, which are easy to understand, can be induced. Nevertheless, as is the case in the majority of data mining methods, to construct a decision tree it is usually required that data is represented by attribute-value pairs beforehand. Since it is not trivial to define proper attributes for graph-structured data beforehand, it is difficult to construct a classifier represented by decision tree for the classification of graph-structured data.

This paper proposes a method for constructing a decision tree for graph-structured data while constructing the attributes themselves for classification task using GBI simultaneously. A pair extracted by GBI, which consists of nodes and the links between them, is treated as an attribute and the existence/non-existence of the pair in a graph is treated as its value for the graph. Thus, attributes (pairs) that divide data effectively are extracted by GBI while a decision tree is being constructed. To classify graph-structured data after the construction of a decision tree, attributes are produced from data before the classification. We call this technique as Decision Tree - Graph-Based Induction, DT-GBI for short.

Section 2 briefly describes the framework of GBI. Section 3 describes the proposed method for constructing a classifier with GBI for graph-structured data. Evaluation of the proposed method is explained in Section 4. Section 5 concludes the paper with summary of the results and the planned future work.

2   Graph-Based Induction Revisited

GBI employs the idea of extracting typical patterns by stepwise pair expansion as shown in Fig. 1. ' In the original GBI an assumption is made that typical patterns represent some concepts/substructure and "typicality" is characterized by the pattern's frequency or the value of some evaluation function of its frequency. We can use statistical indices as an evaluation function, such as frequency itself, Information Gain [6], Gain Ratio [7] and Gini Index [2], all of which are based on frequency. In Fig. 1 the shaded pattern consisting of nodes 1, 2, and 3 is thought typical because it occurs three times in the graph. GBI first finds the 1→3 pairs based on its frequency, chunks them into a new node 10, then in the next iteration finds the 2→10 pairs, chunks them into a new node 11. The resulting node represents the shaded pattern.

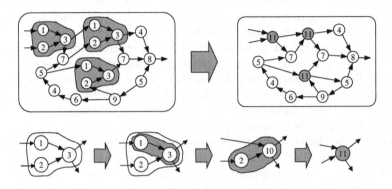

Fig. 1. The basic idea of the GBI method

It is possible to extract typical patterns of various sizes by repeating the above three steps. Note that the search is greedy. No backtracking is made. This means that in enumerating pairs no pattern which has been chunked into one node is restored to the original pattern. Because of this, all the "typical patterns" that exist in the input graph are not necessarily extracted. The problem of extracting all the isomorphic subgraphs is known to be NP-complete. Thus, GBI aims at extracting only meaningful typical patterns of a certain size. Its objective is not finding all the typical patterns nor finding all the frequent patterns.

As described earlier, GBI can use any criterion that is based on the frequency of paired nodes. However, for finding a pattern that is of interest any of its subpatterns must be of interest because of the nature of repeated chunking. In Fig. 1 the pattern 1→3 must be typical for the pattern 2→10 to be typical. Said differently, unless pattern 1→3 is chunked, there is no way of finding the pattern 2→10. Frequency measure satisfies this monotonicity. However, if the criterion chosen does not satisfy this monotonicity, repeated chunking may not find good patterns even though the best pair based on the criterion is selected at each iteration. To resolve this issue GBI was improved to use two criteria, one for frequency measure for chunking and the other for finding discriminative patterns after chunking. The latter criterion does not necessarily hold monotonicity property. Any function that is discriminative can be used, such as Information Gain [6], Gain Ratio [7] and Gini Index [2], and some others.

The improved stepwise pair expansion algorithm is summarized in Fig. 2. It repeats the following four steps until chunking threshold is reached (normally minimum support value is used as the stopping criterion).

GBI(G)
    Enumerate all the pairs P_{all} in G
    Select a subset P of pairs from P_{all} (all the pairs in G)
        based on typicality criterion
    Select a pair from P_{all} based on chunking criterion
    Chunk the selected pair into one node c
    G_c := contracted graph of G
    while termination condition not reached
        $P := P \cup$ GBI(G_c)
    return P

Fig. 2. Algorithm of GBI

Step 1 Extract all the pairs consisting of connected two nodes in the graph.

Step 2a Select all the typical pairs based on the typicality criterion from among the pairs extracted in Step 1, rank them according to the criterion and register them as typical patterns. If either or both nodes of the selected pairs have already been rewritten (chunked), they are restored to the original patterns before registration.

Step 2b Select the most frequent pair from among the pairs extracted in Step 1 and register it as the pattern to chunk. If either or both nodes of the selected pair have already been rewritten (chunked), they are restored to the original patterns before registration. Stop when there is no more pattern to chunk.

Step 3 Replace the selected pair in Step 2b with one node and assign a new label to it. Rewrite the graph by replacing all the occurrence of the selected pair with a node with the newly assigned label. Go back to Step 1.

The output of the improved GBI is a set of ranked typical patterns extracted at Step 2a. These patterns are typical in the sense that they are more discriminative than non-selected patterns in terms of the criterion used.

3   Decision Tree Graph-Based Induction

3.1   Feature Construction by GBI

Since the representation of decision tree is easy to understand, it is often used as the representation of classifier for data which are expressed as attribute-value pairs. On the other hand, graph-structure data are usually expressed as nodes and links, and there is no obvious components which corresponds to attributes and their values. Thus, it is difficult to construct a decision tree for graph-structured data in a straight forward manner. To cope with this issue we regard the existence of a subgraph in a graph as an attribute so that graph-structured data can be represented with attribute-value pairs according to the existence of particular subgraphs.

However, it is difficult to extract subgraphs which are effective for classification task beforehand. If pairs are extended in a step-wise fashion by GBI and discriminative ones are selected and extended while constructing a decision tree, discriminative patterns (subgraphs) can be constructed simultaneously during the construction of a decision tree. In our approach attributes and their values are defined as follows:

- attribute: a pair in graph-structured data.
- value for an attribute: existence/non-existence of the pair in a graph.

When constructing a decision tree, all the pairs in data are enumerated and one pair is selected. The data (graphs) are divided into two groups, namely, the one with the pair and the other without the pair. The selected pair is then chunked in the former graphs. and these graphs are rewritten by replacing all the occurrence of the selected pair with a new node. This process is recursively applied at each node of a decision tree and a decision tree is constructed while attributes (pairs) for classification task is created on the fly. The algorithm of DT-GBI is summarized in Fig. 3.

DT-GBI(D)
   Create a node DT for D
   if termination condition reached
     return DT
   else
     $P := $ GBI(D) (with the number of chunking specified)
     Select a pair p from P
     Divide D into D_y (with p) and D_n (without p)
     Chunk the pair p into one node c
     $D_{yc} :=$ contracted data of D_y
     for $D_i := D_{yc}, D_n$
       $DT_i :=$ DT-GBI(D_i)
     Augment DT by attaching DT_i as its child along yes(no) branch
   return DT

Fig. 3. Algorithm of DT-GBI

Since the value for an attribute is yes (with pair) and no (without pair), the constructed decision tree is represented as a binary tree.

The proposed method has the characteristic of constructing the attributes (pairs) for classification task on-line while constructing a decision tree. Each time when an attribute (pair) is selected to divide the data, the pair is chunked into a larger node in size. Thus, although initial pairs consist of two nodes and the link between them, attributes useful for classification task are gradually grown up into larger pair (subgraphs) by applying chunking recursively. In this sense the proposed DT-GBI method can be conceived as a method for feature construction, since features, namely attributes (pairs) useful for classification task, are constructed during the application of DT-GBI.

3.2   Illustrating Example of DT-GBI

For instance, suppose DT-GBI receives a set of graphs in the upper left-hand side of Fig. 4. There are 13 kinds of pair in data, i.e., a→a, a→b, a→c, a→d, b→a, b→b, b→c, b→d, c→b, c→c, d→a, d→b, d→c. As for the pair a→a, the graphs with class A and that with class B have the pair; the one with class C does not have it. The existence/non-existence of the pairs in each graph is converted into the ordinal table representation of attribute-value pairs, as shown in Table 1. After that, the pair with the highest evaluation is selected and used to divide the data into two groups, namely, the one with the pair and the other without the pair at the root node. In the former group the selected pair is then chunked in each graph and the graphs are rewritten (the table representation of attribute-value pairs for this group is shown in Table 2). DT-GBI applies the above process recursively at each node to grow up the decision tree while constructing the attributes (pairs) useful for classification task at the same time. The attribute-value pairs at the last step in Fig. 4 is shown in Table 3.

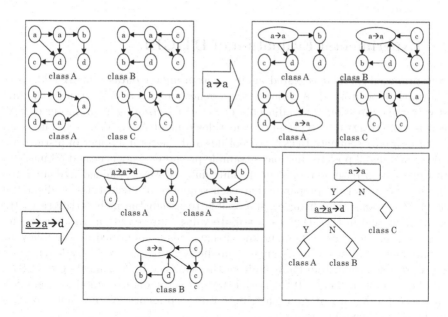

Fig. 4. The basic idea of DT-GBI

Table 1. Attribute-value pairs at the first step

Graph	a→a	a→b	a→c	a→d	b→a	b→b	b→c	b→d	c→b	c→c	d→a	d→b	d→c
1 (class A)	1	1	0	1	0	0	0	1	0	0	0	0	1
2 (class B)	1	1	1	0	0	0	0	0	0	1	1	1	0
3 (class A)	1	0	0	1	1	1	0	0	0	0	0	1	0
4 (class C)	0	1	0	0	0	1	1	0	1	0	0	0	0

Table 2. Attribute-value pairs at the second step (the graph in the lower right-hand side is eliminated)

Graph	a→a→b	a→a→c	a→a→d	b→a→a	b→b	...	d→c
1 (class A)	1	0	1	0	0	...	1
2 (class B)	1	1	0	0	0	...	0
3 (class A)	0	0	1	1	1	...	0

Table 3. Attribute-value pairs at the last step

Graph	a→a→d→a→a→d	a→a→d→b	a→a→d→c	b→a→a→d	b→d
1 (class A)	1	1	1	0	0
2 (class A)	0	1	0	1	1

4   Experimental Evaluation of DT-GBI

The proposed method is tested against the promoter dataset in UCI Machine Learning Repository[1]. A promoter is a genetic region which initiates the first step in the expression of an adjacent gene (*transcription*). The promoter dataset consists of strings that represent nucleotides (one of A, G, T, or C). The input features are 57 sequential DNA nucleotides and the total number of instances is 106 including 53 positive instances (sample promoter sequence) and 53 negative instances (non-promoter sequence). This dataset was explained and analyzed in [8]. The data is so prepared that each sequence of nucleotides is aligned at a reference point, which makes it possible to assign the n-th attribute to the n-th nucleotide in the attribute-attribute value representation. In a sense, this dataset is encoded using domain knowledge. This is confirmed by the following experiment. Running C4.5[7] gives a predictive error of 16.0% by leaving one out cross validation. Randomly shifting the sequence by 3 elements gives 21.7% and by 5 elements 44.3%. If the data is not properly aligned, standard classifiers such as C4.5 that use attribute-attribute value representation does not solve this problem, as shown in Fig. 5.

One of the advantage of graph representation is that it does not require the data to be aligned at a reference point. In this paper, each sequence is converted to a graph representation assuming that an element interacts up to 10 elements on both sides (See Fig. 6). Each sequence, thus, results in a graph with 57 nodes and 515 lines.

In the experiment, frequency was used to select a pair to chunk in GBI and information gain [6] was used in DT-GBI to select a pair from the pairs returned by GBI. A decision tree was constructed in either of the following two ways: 1) apply chunking for n_r times only at the root node and for only once at other nodes of a decision tree, 2) apply chunking for n_e times at every node of a decision

	Prediction Error (C4.5, LVO)
aacgtcgattagccgat gtccatggtcaagtccg tccaggtgcagtcagtc	
Original data	16.0%
Shift randomly by	
≦ 1 element	16.0%
≦ 2 element	21.7%
aacgtcgattagccgat gtccatggtcaagtccg ccaggtgcagtcagtc	
≦ 3 element	26.4%
≦ 5 element	44.3%

Fig. 5. Error

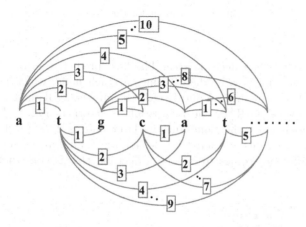

Fig. 6. Conversion of DNA Sequence Data to a graph

tree. In 1) n_r was set to 1, 2, 3, 4, 5, 6, 7, 8, 9, and 10. In 2), n_e was set to 1, 2, 3, 4, 5, 6, and 7. The termination condition in DT-GBI in Fig. 3 was set to whether the number of graphs in D is equal to or less than 10. The prediction error rate of a decision tree constructed by DT-GBI was evaluated by 5 fold cross-validation in both experiments. The results are summarized in Fig. 7. In this figure the solid lines indicate the error rate of decision trees for 1) and 2). On the other hand, the dotted lines indicate the error rate of decision trees when a pair is selected only from the final contracted graphs after the repeated application of chunking. The increase of error rate in dotted lines indicates that the repeated application of chunking can worsen the predictive accuracy of the constructed decision tree if only the remaining pairs in the rewritten graphs are available. Our current conjecture is that the discriminative power (i.e., information gain) of a pair can decrease when it is chunked to constitute a large pair.

[4] reports another approach to construct a decision tree for the promoter dataset. The extracted patterns (subgraphs) by B-GBI, which incorporates beam search into GBI to enhance search capability, were treated as attributes and C4.5

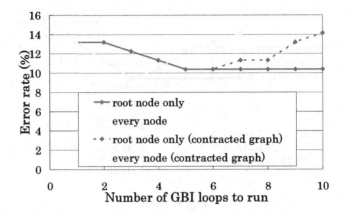

Fig. 7. Result of experiment for the promoter dataset.

was used to construct a decision tree. The best predictive error of DT-GBI is 9.43% when chunking was applied 3 times at each node, which is superior to 16.04% and 11.32% by in C4.5 and B-GBI (beam-wise GBI) in[4]. The induced decision tree by DT-GBI for the best result is shown in Fig. 8. However, the best result reported by the same authors later in[5] is 2.8% (with Leave-One-Out) using the patterns extracted by B-GBI and C4.5. Our current conjecture is that the difference comes from how a decision tree is constructed, namely,

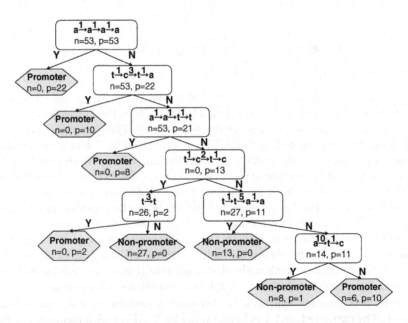

Fig. 8. Decision Tree (chunking applied 3 times at every node)

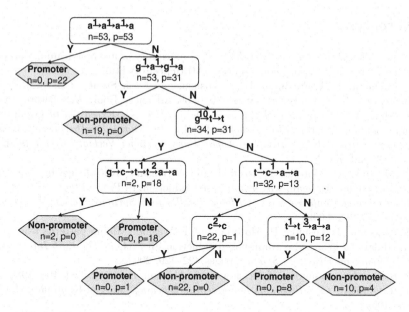

Fig. 9. Decision Tree (chunking applied 5 times only at the root node)

information gain ratio for selecting attributes (patterns) and more sophisticated pruning used in C4.5. On the other hand, the best prediction error for 1), namely, when chunking was applied only at the root node, was 10.4% for $n_r = 5$. The induced decision tree in this case is shown in Fig. 9.

5 Conclusion

This paper proposes a method called DT-GBI, which constructs a classifier (decision tree) for graph-structured data by GBI. Especially, DT-GBI constructs attributes, namely substructures useful for classification task, by applying chunking in GBI on the fly while constructing a decision tree. DT-GBI was evaluated on a classification problem of DNA promoter sequences and the results indicate that it is possible to extract discriminative patterns which otherwise are hard to extract.

Immediate future work includes to incorporate more sophisticated method for determining the number of cycles to call GBI at each node to improve prediction accuracy. Utilizing the rate of change of information gain by successive chunking is a possible way to automatically determine the number.

Acknowledgement. This work was partially supported by the grant-in-aid for scientific research on priority area "Active Mining" (No. 13131101, No. 13131206) funded by the Japanese Ministry of Education, Culture, Sport, Science and Technology.

References

1. C. L. Blake, E. Keogh, and C.J. Merz. Uci repository of machine leaning database, 1998. http://www.ics.uci.edu/~mlearn/MLRepository.html.
2. L. Breiman, J. H. Friedman, R. A. Olshen, and C. J. Stone. *Classification and Regression Trees.* Wadsworth & Brooks/Cole Advanced Books & Software, 1984.
3. T. Matsuda, T. Horiuchi, H. Motoda, and T. Washio. Extension of graph-based induction for general graph structured data. In *Knowledge Discovery and Data Mining: Current Issues and New Applications, Springer Verlag, LNAI 1805*, pages 420–431, 2000.
4. T. Matsuda, H. Motoda, T. Yoshida, and T. Washio. Knowledge discovery from structured data by beam-wise graph-based induction. In *Proc. of the 7th Pacific Rim International Conference on Artificial Intelligence, Springer Verlag, LNAI 2417*, pages 255–264, 2002.
5. T. Matsuda, T. Yoshida, H. Motoda, and T. Washio. Mining patterns from structured data by beam-wise graph-based induction. In *Proc. of The Fifth International Conference on Discovery Science*, pages 422–429, 2002.
6. J. R. Quinlan. Induction of decision trees. *Machine Learning*, 1:81–106, 1986.
7. J. R. Quinlan. *C4.5:Programs For Machine Learning.* Morgan Kaufmann Publishers, 1993.
8. G. G. Towell and J. W. Shavlik. Extracting refined rules from knowledge-based neural networks. *Machine Learning*, 13:71–101, 1993.
9. K. Yoshida and H. Motoda. Clip : Concept learning from inference pattern. *Journal of Artificial Intelligence*, 75(1):63–92, 1995.

Comparison of the Performance of Center-Based Clustering Algorithms

Bin Zhang

Hewlett-Packard Research Laboratories, Palo Alto, 94304, USA
bzhang@hpl.hp.com

Abstract. Center-based clustering algorithms like K-means, and EM are one of the most popular classes of clustering algorithms in use today. The author developed another variation in this family – K-Harmonic Means (KHM). It has been demonstrated using a small number of "benchmark" datasets that KHM is more robust than K-means and EM. In this paper, we compare their performance statistically. We run K-means, K-Harmonic Means and EM on each of 3600 pairs of (dataset, initialization) to compare the statistical average and variation of the performance of these algorithms. The results are that, for low dimensional datasets, KHM performs consistently better than KM, and KM performs consistently better than EM over a large variation of clustered-ness of the datasets and a large variation of initializations. Some of the reasons that contributed to this difference are explained.

Keywords: Clustering, K-means, K-Harmonic Means, Expectation-Maximization, Data Mining

1 Introduction

Clustering is a family of techniques in the field of data mining [FPU96], statistical data analysis [KR90], data compression and vector quantization [GG92], and many others. K-means (*KM*) (MacQueen 1967), and the Expectation Maximization (*EM*) with linear mixing of Gaussian density functions (Dempster at. el. 1977) are two of the most popular clustering algorithms. See also Gersho and Gray (1992) for more complete references on K-means and McLachlan and Krishnan (1997), and Rendner and Walker (1984) on EM. K-Harmonic Means (KHM) is another center-based clustering algorithm, developed by this author (Zhang at. el. 2000, Zhang 2001).

Among many clustering algorithms, the center-based clustering algorithms stand out on two important aspects – a clearly defined objective function that the algorithm minimizes (comparing with agglomerative algorithms for example) and a low runtime cost, comparing with many most other types of clustering algorithms. The time complexity per iteration for all three algorithms is *linear* in the size of the dataset N, the number of clusters K, and the dimensionality of data D. The number of iterations it takes for the algorithms to converge is very insensitive to N. With guarantee of convergence to only a local optimum,

K.-Y. Whang, J. Jeon, K. Shim, J. Srivatava (Eds.): PAKDD 2003, LNAI 2637, pp. 63–74, 2003.
© Springer-Verlag Berlin Heidelberg 2003

the quality of the converged results, measured by the performance function of the algorithm, could be far from its global optimum. Other researchers explored better initializations to achieve the convergence to a better local optimum (Pena 1999, Meila & Heckerman 2001, Bradley & Fayyad 1998). KHM takes a very different approach and tries to address directly the source of the problem – a single cluster is capable of trapping many centers far larger than its fair share. With the introduction of a *dynamic weighting* function of the data, KHM is much less sensitive to the initialization, demonstrated through the experiments in this paper. The dynamic weighting function reduces the ability of a single data cluster trapping many centers.

Understanding the conditions under which these algorithms do or do not converge to a good local optimum is very important to both the researchers and the practitioners. Such understanding provides hints on selecting or even improving the algorithms. This paper provides the comparisons of the three algorithms under different dimensionalities of datasets, very different quality of initializations, and different degree of clustering in the datasets.

With the following notations, we will briefly introduce the details of each algorithm.

Let $X = \{x_i | i = 1, ..., N\}$ be a dataset (iid sampled from a hidden distribution for example) with K clusters, and $M = \{m_k | k = 1, ..., K\}$ be a set of K centers to mark the location of the clusters. All three algorithms minimize a function of the following form over M (details later),

$$Perf_{KM}(X, M) = \sum_{x \in X} d(x, M),\qquad(1)$$

where $d(x, M)$ measures the "distance" from a data point to the set of centers.

1.1   K-Means (KM)

K-means is the simplest of all three. K-means' performance function is the within-partition variance,

$$Perf_{KM}(X, M) = \sum_{k=1}^{K} \sum_{x \in S_k} \|x - m_k\|^2,\qquad(2)$$

where $S_k \subset X$ is the subset of $x's$ that are closer to m_k than to all other centers (the Voronoi partition). Another form of (2) is

$$Perf_{KM}(X, M) = \sum_{i=1}^{N} MIN\{\|x_i - m_l\|^2 | l = 1, ..., K\},\qquad(3)$$

which provides easier comparison with KHM and EM's performance functions.

K-means algorithm calculates its centers' locations in the following iterative steps:

Step 0: Initialize the centers in M (randomly match them with K data points in the dataset is commonly used)

Step 1: Partition the dataset X into K partitions $S_k = \{x \mid argmin_k \| x\text{-}m_k \| = k\}$ (Voronoi partition)

Step 2: Calculate new m_k as the centroid of S_k

$$m_k = \sum_{x \in S_k} x \Big/ |S_k|. \tag{4}$$

Repeat Step 1 and Step 2 until there is no more changes to the partitions. The local optimum K-means converges to depends (heavily) on the initialization of the centers.

1.2   K-Harmonic Means (KHM)

The performance function of KHM_p is defined by:

$$Perf_{KGM}(X, M) = \sum_{i=1}^{N} \frac{K}{\sum_{l=1}^{K} \frac{1}{\|x_i - m_l\|^p}} = \sum_{i=1}^{N} HA\{\|x_i - m_l\|^p | l = 1, \dots, K\}, \tag{5}$$

where $p > 2$. Different values of p gives different KHM performance functions and different KHM algorithms. It is observed that *best p* increases with the dimensionality of the dataset. The quantity under the outer summation in (5) is the harmonic average of the distances from data point x to all the centers. Harmonic average of K numbers is sandwiched by the *MIN()* function:

$$MIN(\|x_i - m_l\|^p | l = 1, \dots, K\} = \leq \frac{K}{\sum_{l=1}^{K} \frac{1}{\|x_i - m_l\|^p}}$$

$$\leq K * MIN(\|x_i - m_l\|^p | l = 1, \dots, K\}, \tag{6}$$

which makes it serving a similar purpose as the K-means' performance function (3) does, but leading to an algorithm much less sensitive to initialization (see the experimental results in this paper).

Taking partial derivatives of the KHM_p's performance function (5) with respect to the center positions m_k, $k=1,\dots,K$, and set them to zero, we have

$$\frac{\partial Perf_{KGM}(X, M)}{\partial m_k} = -K \sum_{i=1}^{N} \frac{p(x_i - m_k)}{d_{i,k}^{p+2} \left(\sum_{l=1}^{K} \frac{1}{d_{i,l}^{p}} \right)^2} = \vec{0} \tag{7}$$

where $d_{i,l} = ||x_i - -m_l||$. "Solving" for the centers from (7), we get the KHM$_p$ recursive algorithm:

$$m_k = \sum_{i=1}^{N} \frac{1}{d_{i,k}^{p+2} \left(\sum_{l=1}^{K} \frac{1}{d_{i,l}^p} \right)^2} x_i \Bigg/ \sum_{i=1}^{N} \frac{1}{d_{i,k}^{p+2} \left(\sum_{l=1}^{K} \frac{1}{d_{i,l}^p} \right)^2}, \tag{8}$$

the center locations in the next iteration are calculated from the distances $d_{i,l}$, which are calculated from the center locations in the current iteration. For more details, see (Zhang 2001).

KHM differs most from KM and EM on having a weighting function $w(x)$ that changes in each iteration (i.e. dynamic). This weighting function determines the percentage of x that participate in the next iteration. The data points in the clusters that have already trapped a center will have a lower weight. Therefore, they will not be able to trap more centers in the next iteration or will lose some of the extract centers happened to be in there. This weighting function reduces the ability of a single data cluster trapping many centers, which is a common cause to the problem of both KM and EM's convergence to bad local optima.

Here is how the weighting function is extracted:

$$m_k = \sum_{i=1}^{N} \frac{1/d_{i,k}^{p+2}}{\sum_{l=1}^{K} 1/d_{i,l}^{p+2}} w(x_i) x_i \Bigg/ \sum_{i=1}^{N} \frac{1/d_{i,k}^{p+2}}{\sum_{l=1}^{K} 1/d_{i,l}^{p+2}} w(x_i),$$

where

$$w(x_i) = \left(\sum_{i=1}^{N} 1/d_{i,l}^{p+2} \right) \Bigg/ \left(\sum_{l=1}^{K} 1/d_{i,l}^{p+2} \right)^2.$$

Both KM and EM have all data points participate in all iterations 100%. They do not have such a dynamic weighting function.

Since the dynamic weighting function lowers the weight on the data points that are close to a center, will it magnify the effects of noise (i.e. the data points not in any cluster)? We have not seen an observable effect in the experiments. It is very important to see that this weighting function is not a fixed function; it changes in each iteration. The weight on a data cluster is lowered only if it is identified (i.e. it trapped a center). As soon as the center is influenced by any other forces to leave the cluster, the weight on the data points in that cluster will go back up.

1.3   Expectation-Maximization (EM)

The third algorithm we investigate is the linear mixture of Gaussian based Expectation-Maximization. We limit ourselves to the linear mixture of K identical spherical (Gaussian density) functions so that it matches with the spherical shaped functions used by KM and KHM for a fair comparison. The performance function of EM is the log-likelihood,

$$Perf_{EM}(X,M) = -\log\left\{\prod_{i=1}^{N}\left[\sum_{l=1}^{K}p_l * \frac{1}{(\sqrt{\pi})^D}EXP(-\|x_i - m_l\|^2)\right]\right\} \quad \text{(see (1))}$$

$$= \sum_{i=1}^{N} -\log\left[\sum_{l=1}^{K}p_l * \frac{1}{(\sqrt{\pi})^D}EXP(-\|x_i - m_l\|^2)\right] \quad (9)$$

EM algorithm is a recursive algorithm with the following two steps:

E-Step:

$$p(m_l|x_i) = \frac{p(x_i|m_l) * p(m_l)}{\sum\limits_{i=1}^{N} p(x_i|m_l) * p(m_l)}, \quad (10)$$

where $p(x|m)$ is the prior probability with Gaussian distribution, $p(m_l)$ is the mixing probability.

M-Step:

$$m_l = \sum_{i=1}^{N}p(m_l|x_i) * x_i \Big/ \sum_{i=1}^{N}p(m_l|x_i) \quad (11)$$

and

$$p(m_l) = \frac{1}{N}\sum_{i=1}^{N}p(m_l|x_i), \quad (12)$$

where N is the size of the whole dataset. For more details, see McLachlan and Krishnan (1997) and the references there.

2   Setup of the Experiments

We compare the three algorithms on 2, 5, and 8 dimensional datasets. For each dimensionality, we run all three algorithms on 400 randomly generated datasets with different *degrees of clustering*. For each dataset we use three different initializations – bad, good and better, which are also randomly generated. A total of 10800 individual experiments were conducted. The details of the setup are given in the following sections.

2.1   Clustered-Ness of a Dataset

To compare the algorithms over a variety datasets with different degrees of clustered-ness, we used the ratio of the inter-cluster variance over the within cluster variance to control the clustered-ness. Intuitively, this ratio can be compared with the ratio of the average distance between the centers of the clusters over the average diameter of the clusters.

For a partition of the dataset, $S_k \subset X$, $k=1,\ldots,K$, the (estimated) variance of the hidden distribution from the dataset X decomposes into the sum of the within partition-variance and the inter-partition variance (see Duda & Hart 1972 for some similarity),

$$\sigma^2(X) = \sum_{i=1}^{N}(x_i - m)^2 = \sum_{k=1}^{K} p_k \sigma^2(S_k) + \sum_{k=1}^{K} p_k(m_k - m)^2 \qquad (13)$$

where

$$p_k = |S_k| \Big/ |X|$$

$$m_k = \sum_{x \in S_k} x \Big/ |S_k|, \text{ the centroids of } S_k,$$

$$m = \sum_{x \in X} x \Big/ |X|, \quad \text{ the centroids of } X,$$

and

$$\sigma^2(S_k) = \frac{\sum_{i=1}^{N}(x_i - m_k)^2}{|S_k|}.$$

The K-clustered-ness or simply clustered-ness of the dataset is defined to be

$$C(X) = \max_{M} \frac{\sum_{k=1}^{K} p_k(m_k - m)^2}{\sum_{k=1}^{K} p_k \sigma^2(S_k)}. \qquad (14)$$

The larger the *clustered-ness* the better separated are the clusters. We compare all three algorithms under a large variation of the clustered-ness of the datasets.

2.2   Datasets

We set the size of the datasets $N = 2500$, which is large enough to create non-trivial clusters and small enough to be affordable to run 10800 experiments. We set the number of clusters $K = 50$, also large enough to be interesting and small enough to control the total cost. The number of data points in each cluster varies. The dimensionality of the datasets are $D = 2, 5$, and 8. For each dimensionality, we randomly generated 400 datasets, $Dataset(d,i), d=2,5,8, i=1, \ldots, 400$, with their clustered-ness uniformly distributed in $[0, 60]$. This range of clustered-ness covers the datasets from having all clusters overlapping deeply to having very wide empty space between the clusters. The following Matlab function is used to generate all the datasets.

function [dataset,centers] = ClusGen(K, N, D, r)
% K = # clusters, N = # data points, D = dimensionality,
% r = a parameter to control the within cluster variance/ inter-cluster variance.

% Step 1: Generate cluster centers.

centers = r * **rand**(K,D); % K center locations are scaled up by r.

% Step 2: Generate the random sizes of the K clusters.

s = 2***rand**(K,1)+1;

s = **round**(N*s/**sum**(s));

N1 = **sum**(s);

diff = **abs**(N-N1);

s(1:diff) = s(1:diff) + **sign**(N-N1); % adjust the size so that they add up to N.

% Step 3: Generate clusters one-by-one.

for k=1:K,

    cluster = **randn**(s(k),D); % normal distribution.

    mean = **sum**(cluster)/s(k);

    Sk = **repmat**(centers(k,:)-Mean,s(k),1)+cluster;

    dataset = [dataset' Sk']'; % merge the cluster into the dataset.

end; % of for loop.   % End of the cluster dataset generator.

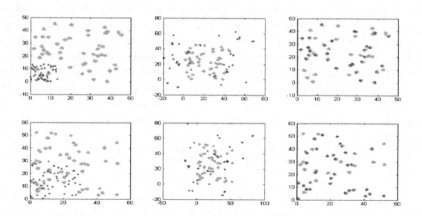

Fig. 1. Two samples of each type of initializations are shown here. Ordered from left to right: Type-1: bad, Type-2 : good, and Type-3: better. The light green backgrounds are the data points and the red dots are the initial center locations.

2.3   Initialization

We use three types of random initializations with each dataset, *Init(d,i,j), j=1, 2, 3, for d=2,5,8 and i=1,. . . ,400*. Separating initializations into three bins helps us to see the comparison of the three algorithm's performance on the quality of initialization.

    Type-1: Bad – all 50 centers are initialized to be within a small region relative to the data.

    Type-2: Good – all centers are randomly initialized, covering the region of data.

    The centers have a bigger spread than the data itself.

Type-3: Better – the centers are initialized to 50 randomly chosen data points from the dataset.

2.4   A Common Performance Measure

To compare different algorithms under the same measure, we use the *square-root* of K-means' performance function

$$Perf(X, M) = \sqrt{Perf_{KM}(X, M)}$$

to measure the quality of the clusters (centers) derived by all three algorithms. We choose K-means' performance function because it is more popular, more intuitive and simpler than others. Taking square root removes the quadratic behavior of the function and restores the linear behavior. This choice is in favor of the KM and put the other two at a lightly disadvantage because their algorithms are designed to optimize their performance functions not KM's. But this choice does not change the overall picture of the comparison results. The performance values on the datasets with different clustered-ness are expected to be in different ranges. We cannot aggregate by calculating averages of them. Instead, we calculate the quality ratio (15), the performance value at the local optimum divided by the global optimal performance, which is also measured by the same function (the global optimum is derived by running K-means starting from the location of the "true" centers of the clusters returned by the ClusGen() function, except for the data sets with very small clustered-ness. When the clustered-ness is very small, the can only promise a very good approximation to the global optimal.)

$$QR(M_{local\_opt}) = \frac{Perf(X, M_{local\_opt})}{\min_{M} Perf(X, M)} \qquad (15)$$

We partitioned the 400 datasets, with their clustered-ness uniformly distributed in $[0, 60]$, into 12 equal sized bins, B_l, $l = 1, ..., 12$. Each bin has about 33 (=400/12) datasets. Aggregating the performance ratios over each small bin allows us to keep the comparison's dependency on the clustered-ness of the dataset. With about 33 datasets in each bin, we can calculate both the average and the coefficient of standard deviation of the quality ratios,

$$avg_{i,j} = \frac{\sum\limits_{i \in B_l} QR_{i,j}}{|B_l|}, \quad \sigma_{l,j}^2 = \frac{\sum\limits_{i \in B_l} (QR_{i,j} - avg_{l,j})^2}{|B_l|}, \quad \theta_{l,j} = \frac{\sigma_{l,j}}{avg_{l,j}}. \qquad (16)$$

3   Results and Observations

Each algorithm, KM, KHM $(p = 3, 4)$ and EM, was run up to 45 iterations on each pair *(Dataset(d, i), Init(d, i, j))*, which is sufficient for the centers to stabilize. The results is presented in a "four" dimensional plot in 2D in Figure 2. The average quality ratio is plotted against the average clustered-ness of the dataset in each small figure in Figure 2, and 9 small figures are arranged in 2D indexed by dimensionality and initialization type. There is no significant reason for this choice other than its compactness. We could have used either the dimensionality or the initialization type as the horizontal axis in each small figures and the remaining two as the indices of the 2D arrangements.

Observation 1:
For low dimensional datasets, the performance ranking of three algorithms is KHM > KM > EM (">" means better) . As the dimensionality becomes higher the diferences becomes less significant. We know that for high dimensional data sets, between KM and KHM, either one could happen to be better depending on the data distribution and the initialization (often all of them are bad for very high dimensional datasets).

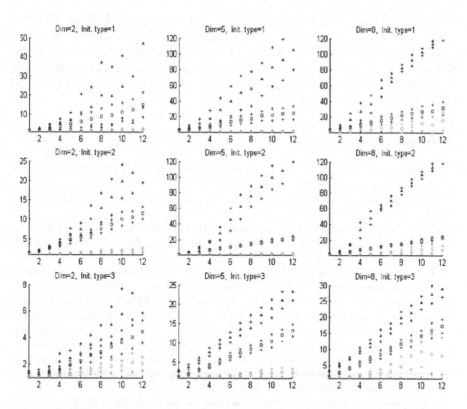

Fig. 2. Comparison of the quality ratios of the performance of KM, KHM, and EM. KM's performance measure is used as the common measure for all three algorithms. The color and icon coding are: KM – red square, KHM – green circle, and EM – blue triangles. One-sigma (on each side) confidence intervals are printed in the same color. For KHM (*p=3 for Dim=2,5; p=4 for Dim=8*). The horizontal axis in each small plot is the bin number. The vertical axis is the quality ratio.

Observation 2:
KHM's performance has the smallest variation in each bin of experiments, which implies that KHM's clustering results are more re-producible by re-runs of KHM on similar datasets with different initializations. EM's performance has the biggest variation of all. Its results are most sensitive to initializations.

Observation 3:
Reproducible results become even more important when we use these algorithms to different datasets that are sampled from the same hidden distribution. The results from KHM better represents the properties of the distribution and less dependent on a particular sample set. EM's results are more dependent on the sample set.

4   Conclusion

On low dimensional datasets, KHM consistently outperforms KM and KM consistently out performs EM over a very wide range of clustered-ness of datasets. Under different initializations, KHM's performance has the lowest variation and EM performance has the highest variation among the three. It is also interesting to see that KHM optimizes KM's performance function often better that KM algorithm itself.

References

1. Bradley, P., Fayyad, U. M., C.A., "Refining Initial Points for KM Clustering", MS Technical Report MSR-TR-98-36, May (1998)
2. Duda, R., Hart, P., "Pattern Classification and Scene Analysis", John Wiley & Sons, (1972)
3. Dempster, A. P., Laird, N.M., and Rubin, D.B., "Maximum Likelyhood from Incomplete Data via the EM Algorithm", Journal of the Royal Statistical Society, Series B, 39(1):1-38, (1977)
4. Fayyad, U. M., Piatetsky-Shapiro, G. Smyth, P. and Uthurusamy, R., "Advances in Knowledge Discovery and Data Mining", AAAI Press (1996)
5. Fink, L.J., Fung, M., McGuire, K.L., Gribskov, M, "Elucidation of Genes Involved in HTLV-I-induced Transformation Using the K-Harmonic Means Algorithm to Cluster Microarray Data", follow the link http://www.ismb02.org/posterlist.htm to find an extended abstract. Software tools at http://array.sdsc.edu/. (2002)
6. Gersho & Gray, "Vector Quantization and Signal Compression", KAP, (1992)
7. Kaufman, L. and Rousseeuw, P. J., "Finding Groups in Data : An Introduction to Cluster Analysis", John Wiley & Sons, (1990)
8. MacQueen, J., "Some Methods for Classification and Analysis of Multivariate Observations". Pp. 281-297 *in:* L. M. Le Cam & J. Neyman [eds.] Proceedings of the fifth Berkeley symposium on mathematical statistics and probability, Vol. 1. University of California Press, Berkeley. xvii + 666 p, (1967)
9. McKenzie, P. and Alder, M., "Initializing the EM Algorithm for Use in Gaussian Mixture Modeling", The Univ. of Western Australia, Center for Information Processing Systems, Manuscript

10. McLachlan, G. J. and Krishnan, T., "The EM Algorithm and Extensions.", John Wiley & Sons, (1997)
11. Meila, M., Heckerman, D., "An Experimental Comparison of Model-based Clustering Methods", Machine Learning, 42, 9-29, (2001)
12. Pena, J., Lozano, J., Larranaga, P., "An Empirical Comparison of Four Initialization Methods for the K-means Algorithm", Pattern Recognition Letters, 20, 1027-1040, (1999)
13. Rendner, R.A. and Walker, H.F., "Mixture Densities, Maximum Likelihood and The EM Algorithm", SIAM Review, vol. 26 # 2, (1984)
14. Tibshirani, R., Walther, G., and Hastie, T., "Estimating the Number of Clusters in a Dataset via the Gap Statistic", Available at
    http://www-stat.stanford.edu/~tibs /research.html. March, (2000)
15. Zhang, B., Hsu, M., Dayal, U., "K-Harmonic Means", Intl. Workshop on Temporal, Spatial and Spatio-Temporal Data Mining, Lyon, France Sept. 12, (2000)
16. Zhang, B., "Generalized K-Harmonic Means – Dynamic Weighting of Data in Unsupervised Learning,", the First SIAM International Conference on Data Mining (SDM'2001), Chicago, USA, April 5-7, (2001)

Appendix A. Implementation of K-Harmonic Means

```
function [gkhm_centers,gkhm_perf] = ...
Generalized_K_HarmonicMeans_p(p, data, gkhm_centers, X2)
EPS = 10^(-16);
[N D] = size(data);
K = size(gkhm_centers, 1);
gkhm_c = gkhm_centers .* gkhm_centers; % components squared
gkhm_m = sum(gkhm_c')'; % vector length of the centers
gkhm_d_2 = X2 - 2*gkhm_centers*data' + repmat(gkhm_m,1,N);
gkhm_d_p = gkhm_d_2 .^ (p/2); % Lp distance matrix
alpha1 = 1 ./ max(EPS,gkhm_d_p);
alpha = 1 ./ sum(alpha1);
gkhm_perf = sum(alpha);
alpha_2 = alpha .* alpha;
temp = 1 ./ max(EPS, gkhm_d_p .* gkhm_d_2);
temp = repmat(alpha_2,K,1) .* temp;
gkhm_centers_n = temp * data;
gkhm_de = max(EPS, repmat(sum(temp')',1,D));
gkhm_centers = gkhm_centers_n ./ gkhm_de;


function [gkhm_centers,gkhm_perf] = Driver(data, K, p, MaxIter)
[N D] = size(data); % User: make sure the matrix is NxD.
init_type = 1; % or 2,3.
r = 60*rand(1,1);
gkhm_centers = initialize(data,K,r,init_type);
data_squared = data .* data;
vector_length_squared = sum(data_squared')';
```

```
X2 = repmat(vector_length_squared',K,1);
for iter = 1:MaxIter,
    % Fixed number of iterations is used
    % but it could be replaced by other stopping rules.
        [gkhm_centers,gkhm_perf] = ...
      Generalized_K_HarmonicMeans_p(p,data,gkhm_centers,
X2);
end;
% Add your code to print or plot the results here.

function centers = initialize (data,K,r,init_type)
[N,D] = size(data); mean = sum(data)/N;
if init_type == 1, % bad initialization
    rr = r*rand(1,1);
    centers =repmat(rr,[K D]) .* rand(K,D); end;
if init_type == 2, id = floor((N-1)*rand(K,1)+1);
    centers=2*(data(id,:)-repmat(mean,K,1))+
    repmat(mean,K,1); end;
if init_type == 3 % better initialization
    id = floor((N-1)*rand(K,1)+1);
    centers = data(id,:); end;
```

Automatic Extraction of Clusters from Hierarchical Clustering Representations[1]

Jörg Sander, Xuejie Qin, Zhiyong Lu, Nan Niu, and Alex Kovarsky

Department of Computing Science, University of Alberta
Edmonton, AB, Canada T6G 2E8
{joerg, xuq, zhiyong, nan, kovarsky}@cs.ualberta.ca

Abstract. Hierarchical clustering algorithms are typically more effective in detecting the true clustering structure of a data set than partitioning algorithms. However, hierarchical clustering algorithms do not actually create clusters, but compute only a hierarchical representation of the data set. This makes them unsuitable as an automatic pre-processing step for other algorithms that operate on detected clusters. This is true for both dendrograms and reachability plots, which have been proposed as hierarchical clustering representations, and which have different advantages and disadvantages. In this paper we first investigate the relation between dendrograms and reachability plots and introduce methods to convert them into each other showing that they essentially contain the same information. Based on reachability plots, we then introduce a technique that automatically determines the significant clusters in a hierarchical cluster representation. This makes it for the first time possible to use hierarchical clustering as an automatic pre-processing step that requires no user interaction to select clusters from a hierarchical cluster representation.

Keywords: Hierarchical clustering, OPTICS, Single-Link method, dendrogram, reachability-plot

1 Introduction and Related Work

One of the primary data mining tasks in a Knowledge discovery (KDD) process is cluster analysis. It is often a first and important step in analyzing a data set, understanding its properties, and preparing it for further analysis.

There are many different types of clustering algorithms, and the most common distinction is between partitioning and hierarchical algorithms (see e.g. [6]).

Partitioning algorithms create a "flat" decomposition of a data set. Examples of partitioning algorithms are the k-means [7] and the k-medoids algorithms PAM, CLARA [6], and CLARANS [8], or density-based approaches such as [4], [2], [9], [3]. These algorithms need in general some input parameters that specify either the number of clusters that a user wants to find or a threshold for point density in clusters. Correct parameters, which allow the algorithm to reveal the true clustering structure of a data set, are in general difficult to determine, and they may not even exist.

Hierarchical clustering algorithms, on the other hand, do not actually partition a data set into clusters, but compute a hierarchical representation, which reflects its

---

[1] Research partially supported by NSERC Canada.

K.-Y. Whang, J. Jeon, K. Shim, J. Srivatava (Eds.): PAKDD 2003, LNAI 2637, pp. 75–87, 2003.

possibly hierarchical clustering structure. The Single-Link method [10] is a well-known example. Other algorithms such as Average-Link or Complete-Link, which produce similar hierarchical structures have also been suggested (see, e.g., [4]) and are widely used. The result of these algorithms is a dendrogram, i.e., a tree that iteratively splits a data set into smaller subsets until each subset consists of only one object. A different hierarchical clustering algorithm, which generalizes density-based clustering, is OPTICS [1]. This algorithm produces another type of output, a so-called reachability plot, which is a bar plot showing clusters as "dents" in the plot.

Hierarchical clustering algorithms have several advantages: they are robust with respect to input parameters, they are less influenced by cluster shapes, they are less sensitive to largely differing point densities of clusters, and they can represent nested clusters. It is an important property of many real-data sets, that clusters of very different point densities may exist in different regions of the data space and that clusters may be nested, which makes it very hard to detect all these clusters using a partitioning algorithm. The reason is that in these cases global parameters, which would characterize all clusters in the data space, do not exist. The data sets depicted in Fig. 1 illustrate the problems (nested clusters and largely differing densities). In cases like these, hierarchical algorithms are a better choice for detecting the clustering structure.

There are, however, also problems with hierarchical clustering algorithms, and alleviating those problems is the focus of this paper. The first problem is that hierarchical clustering representations are sometimes not easy to understand, and depending on the application and the user's preferences, one of the representations, dendrogram or reachability-plot, may be preferable, independent of the algorithm used. For very small data sets, a dendrogram may give a clearer view of the cluster membership of individual points, and in some application areas such as biology, domain experts may prefer tree representations because they are more used to it. However, dendrograms are much harder to read than reachability plots for large data sets. Fig. 1 shows reachability plots and Single-Link dendrograms for the depicted data sets. Roughly speaking, in a reachability plot, clusters are indicated by a "dent" in the plot; in a dendrogram, the nodes of the tree represent potential clusters. The second problem with hierarchical clustering algorithms is that clusters are not explicit and have to be determined somehow from the representation. There are two common approaches. 1.) A user selects manually each single cluster from a dendrogram or reachability plot, guided by visual inspection. 2.) A user first defines a horizontal cut through the clustering representation, and the resulting connected components are automatically extracted as clusters. The latter approach has two major drawbacks: when clusters have largely differing densities, a single cut cannot determine all of the clusters, and secondly, it is often difficult to determine where to cut through the representation so that the extracted clusters are significant. The resulting clusters for some cut-lines through the given reachability plots are illustrated in Fig. 1. The circles and ellipse describe the extracted clusters that correspond to the cut-line having the same line style.

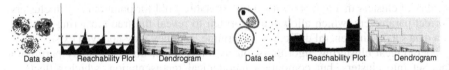

Fig. 1. Illustration of nested clusters, largely differing densities and sizes of clusters

Both approaches have the disadvantage that they are unsuitable for an automated KDD process in which the output of a hierarchical clustering algorithm is the input of a subsequent data mining algorithm. A typical application of such a data mining process is, e.g., the description and characterization of detected clusters in a geographic database by higher-level features such as the diameter, the number neighbors of a certain type etc (see e.g., [5]). The ability to automatically extract clusters from hierarchical clustering representations will also make it possible to track and monitor changes of a hierarchical clustering structure over time in a dynamic data set.

To our knowledge, the only proposal for a method to extract automatically clusters from a hierarchical clustering representation can be found in [1]. The authors propose a method for reachability plots that is based on the steepness of the "dents" in a reachability plot. Unfortunately, this method requires an input parameter, which is difficult to understand and hard to determine. Small variations in the parameter value can result in drastic changes to the extracted hierarchical clustering structure. Therefore, it is unsuitable as an automatic preprocessing step in a larger KDD process.

Our paper contributes to a solution of these problems in the following ways. In section 2, we analyze the relation between hierarchical clustering algorithms that have different outputs, i.e. between the Single-Link method, which produces a dendrogram, and OPTICS, which produces a reachability plot. We will show that the clustering representations have essentially the same properties, and that they can be even considered equivalent in a special case. Based on this analysis, we develop methods to convert dendrograms and reachability plots into each other, making it possible to choose the most advantageous representation, independent of the algorithm used. In section 3, we introduce a new technique to create a tree that contains only the significant clusters from a hierarchical representation as nodes. Our new tree has the additional advantage that it can be directly used as input to other algorithms that operate on detected clusters. Selecting only the leaves from our cluster tree corresponds to selecting the most significant clusters simultaneously from different levels of a dendrogram or reachability plot (rather than applying a simple cut through a hierarchical clustering representation, which may miss important clusters). This method is based on reachability plots. To extract clusters from a dendrogram, it has to be converted first, using the method proposed in section 2. Section 4 concludes the paper.

2   Converting between Hierarchical Clustering Representations

2.1   The Relation between the Single-Link Method and OPTICS

The Single-Link method and its variants create a hierarchical decomposition of a data set. Starting by placing every object in a unique cluster; in every step the two closest clusters in the current clustering are merged until all points are in one cluster. The methods Single-Link, Average-Link and Complete-Link differ only in the distance function that is used to compute the distance between two clusters.

The result of such a hierarchical decomposition is then represented by a dendrogram, i.e., a tree that represents the merges of the data set according to the algorithm. Fig. 2 (*left*) shows a very simple data set and a corresponding Single-Link dendrogram. The root represents the whole data set, a leaf represents a single object, an internal node represents the union of all the objects in its sub-tree, and the height of an internal node represents the distance between its two child nodes.

OPTICS is a hierarchical clustering algorithm that generalizes the density-based notion of clusters introduced in [2]. It is based on the notions of core-distance and reachability-distance for objects with respect to parameters *Eps* and *MinPts*. The parameter *MinPts* allows the core-distance and reachability-distance of a point p to capture the point density around that point. Using the core- and reachability-distances, OPTICS computes a "walk" through the data set, and assigns to each object p its core-distance and the smallest reachability-distance *reachDist* with respect to an object considered before p in the walk (see [1] for details).

For the special parameter values *MinPts* = 2 and *Eps* = ∞, the core-distance of an object p is always the distance of p to its nearest neighbor, and the reachability-distance of an object p relative to an object q is always the distance between p and q. The algorithm starts with an arbitrary object assigning it a reachability-distance equal to ∞. The next object in the output is then always the object that has the shortest distance d to *any* of the objects that were "visited" previously by the algorithm. The reachability-value assigned to this object is d. We will see in the next two subsections that a similar property holds for the sequence of objects and the height of certain ancestor nodes between objects in a dendrogram.

Setting the parameter *MinPts* in OPTICS to larger values will weaken the so-called single-link effect, and also smoothen the reachability plot. If the dimensionality of the data is not too high, OPTICS is more efficient than traditional hierarchical clustering algorithms since it can be supported by spatial index structures. The output is a *reachability plot*, which is a bar plot of the reachability values assigned to the points in the order they were visited. Fig. 2 (*right*) illustrates the effect of the *MinPts* parameter on a reachability plot. Such a plot is interpreted as following: "Valleys" in the plot represent clusters, and the deeper the "valley", the denser the cluster. The tallest bar between two "valleys" is a lower bound on the distance between the two clusters. Large bars in the plot, not at the border of a cluster represent noise, and "nested valleys" represent hierarchically nested clusters.

Although there are differences between dendrograms and reachability plots, both convey essentially the same information. In fact, as we will see in the next subsections, there is a close relationship between a dendrogram produced by the Single-Link method and a reachability plot of OPTICS for *Eps* = ∞ and *MinPts* = 2, which allows us to convert between these results without loss of information.

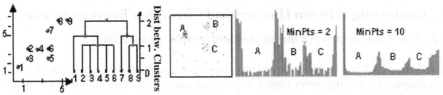

Fig. 2. *Left*: Illustration of a Dendrogram – *Right*: Illustration of Reachability Plots

For other algorithms such as Average-Link, Complete-Link, and OPTICS using higher values for *MinPts*, there is no strict equivalence of the results (although in practice the result will be similar since the detected clustering structures are a property of the data set and not a property of the algorithm used). Nevertheless, the proposed methods for converting between dendrograms and reachability plots are also applicable in those cases. Converting, e.g., a dendrogram produced by the Average-Link method into a reachability plot will generate a reachability plot that contains the same cluster information as the dendrogram (even though there may be no *MinPts* parameter so that OPTICS would generate a strictly equivalent result). Vice versa: converting a reachability plot of OPTICS for a higher *MinPts* value into a dendrogram will result in a tree representation of the same clustering information as the reachability plot (even though no other hierarchical method may produce a strictly equivalent result). Thus, these conversion methods allow a user to choose the best representation, given the properties of the data set, the user's preferences, or some application needs, e.g., algorithms that process a clustering result further and expect a certain input format – one example of such a method is our technique for automatic cluster extraction from a hierarchical clustering representations, proposed in section 3.

Neither a dendrogram nor a reachability plot for a given data set is unique. Intuitively: by rotating the two children of any node in a particular dendrogram, we get another dendrogram, which is also a valid representation of the same clustering structure. For instance, in the dendrogram from Fig. 2 (left), the right sub-tree of the root can as well be drawn first to the left, followed by the left sub-tree, drawn to the right. Using different starting points in the OPTICS algorithm will also produce different reachability plots, which however represent the same clustering structure. For instance, in Fig. 2(right), if we would start with a point in cluster C, the "dent" for that region would then come before the "dents" for clusters A and B.

2.2 Converting Dendrograms into Reachability Plots

It is easy to see a dendrogram satisfies the following condition: for any two points in leaf nodes it holds that the height of their smallest common ancestor node in the tree is a lower bound for the distance between the two points. For instance, using the dendrogram depicted in Fig 3, the minimum distance between T (or any of the four points T, S, V, U) to I (or any other point among A, through R), is at least 3 since the smallest common ancestor node, which separates T and I, is the root at height 3.

We get an even stronger condition for a dendrogram that is organized in the following way: for any internal node, the right child is drawn in a way that the point in this sub-tree which is closest to the set of points in the left child (the point that determined the distance between the two children in the Single-Link method) is always the leftmost leaf in the sub-tree of the right child. This property can always be achieved via suitable rotations of child nodes in a given dendrogram. In such a dendrogram the height of the smallest common ancestor between a point p and p's left neighbor q is the minimum distance between p and the whole set of points to the left of p.

Recall that in a reachability plot (using $MinPts = 2$ and $Eps = \infty$) the height of a bar (the reachability value) for a point p is also the smallest distance between p and the set of points to the left of p in the reachability plot. This relationship allows us to easily convert a dendrogram into a reachability plot. We iterate through the leaves of a given dendrogram from left to right and assign each point its reachability value: the reachability value of the first point is (always) infinity in any reachability plot. For each other point p in the dendrogram the reachability value is simply the height of the smallest common ancestor of p and the left neighbor of p. In pseudo code:

```
leafs[0].r_dist = ∞;
for (i = 1; i < n; i++)
    leafs[i].r_dist = height_of_smallest_common_ancestor(leafs[i], leafs[i-1]);
```

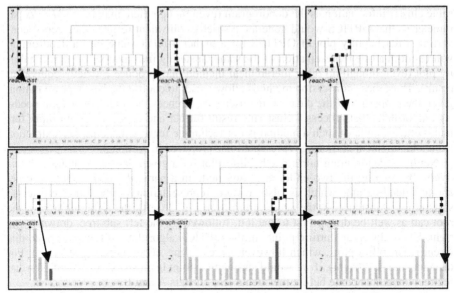

Fig. 3. Illustration of the conversion of a dendrogram into a reachability plot

The transformation of our example dendrogram is illustrated in Fig. 3: For the first point A: r_dist = ∞. For the next point B, the height of the smallest common ancestor of A and B is 2. The height of the smallest common ancestor of the next point I, and the previous point B is also 2. This is the smallest distance between I and the group of points {A, B}. Point I is the first point in a cluster, but its reachability value is still high. This is correct since in reachability plots the value for "border" objects of a cluster indicates the distance of that cluster to the previous points. For the next point J, the reachability value will be set equal to 1, since this is the height of the smallest common ancestor with previous point I. This is again the smallest distance between the current point J and the set of point to the left of J. In the same way we can derive the reachability values for all points in the dendrogram. When considering point T in the dendrogram, we see that it is the first point of a cluster (containing T, S, V, and U) with an internal "merging distance" of 1. Its reachability distance is correctly set to 3, since the smallest common ancestor of T and H is the root at height 3.

A reachability plot for *MinPts* = 2 and *Eps* = ∞ contains more information than an arbitrary Single-Link dendrogram. Our transformation will generate a reachability plot in a strict sense only when the dendrogram satisfies the additional condition above. However, converting an arbitrary dendrogram we will produce a bar plot that reflects the hierarchical clustering structure. Only the order of the points and the border points may be different, since this information is not represented in a dendrogram.

2.3 Converting Reachability Plots into Dendrograms

Consider a reachability plot generated by OPTICS (using *MinPts* = 2 and *Eps* = ∞). Assume that the order generated by OPTICS is $(p_1, p_2, ..., p_{i-1}, p_i, p_{i+1}, ..., p_n)$. The reachability value d_i of point p_i is the shortest distance between p_i and the set of points $S1 = \{p_1, p_2, ..., p_{i-1}\}$ to the left of p. Furthermore, p_i is the point that is closest to the set

S1 among all points in the set of points $S2=\{p_i, p_{i+1}, ..., p_n\}$ to the right of p (this is guaranteed by the way OPTICS algorithm select the next point after processing p_{i-1}).

Now consider applying the Single-Link method to this data set. Obviously, at some point (possibly after some iterations) a node *N2* containing p_i (and possibly other points from *S2*) will be merged with a node *N1* containing only points from *S1*. The distance used for the merging (the height of the resulting node) will be the reachability value d_i of p_i. The node *N1* (containing points to the left of p_i) will always include p_{i-1} (the left neighbor of p_i). The reason is that in fact a stronger condition on the distance between p_i and the points to its left holds in a dendrogram: there is a point p_{i-m} to the left of p_i with the largest index such that its reachability value is larger than d_i, and d_i is the smallest distance to the set of points $S1^*=\{p_{i-m}, ..., p_{i-1}\}$. When the node *N2* containing p_i will be merged in the Single-Link method at the distance d_i, it will be merged with the node *N1* containing its nearest neighbor in $S1^*$. But, since the reachability values of the points in $S1^*$ are smaller than d_i, – except for d_{i-m}, the points in $S1^*$' will have been merged already in previous steps of the Single-Link method. Hence they will be contained in *N1*, including p_{i-1}. Based on this observation, we can transform a reachability plot into a dendrogram using the following method:

```
// RPlot=[p1, ..., pn] is the input reachability plot; for each element RPlot[i]=pi:
//          RPlot[i].r_dist = reachability distance of pi
//          RPlot[i].current_node = the current node in the dendrogram that pi belongs to
// PointList = [q1, ..., qn] is a sequence of references to the points in ascending order of
// the corresponding reachability values (excluding p1 since always: p1.r_dist = ∞).

for (i = 1; i < n; i++)
          create a leaf node N_i for RPlot[i];
          RPlot[i].current_node = N_i;  // initially each point has its own current_node

for (i = 1; i < n; i++)
          p = PointList[i]; // pick the point with the next smallest r_dist
          q = left neighbor of p in Rplot;
          create a new node N in the dendrogram;
          N.height            = p.r_dist;
          N.left_child        = q.current_node;
          N.right_child       = p.current_node;
          p.current_node      = q.current_node = N;
```

Fig. 4 illustrates the algorithm. In the first step, the algorithm will connect a point with the smallest reachability value, here J, with its left neighbor, here I. It will create a new node N1 with its height equal to the reachability value of J. The current_node for both J and I will become the new node N1. In the second step, it selects the next point (with the next smallest reachability value) from the sorted list of points, say L. Again, a new node N2 is created that connects the current_node of L (which is the leaf containing L) with the current_node of L's left neighbor. The point is J, and J's current_node is the node N created in the previous step. After processing the points in our example with a reachability value of 1, the intermediate dendrogram looks like the one depicted. The points I, J, L, M, K, and N at this stage will have the same current node, e.g. Nr, and similar for the other points. The next smallest reachability value in the next step will be 2. Assume the next point found in PointList (with a reachability value of 2) is point B. B is merged with B's left neighbor A. The current nodes for A and B are still the leaves that contain them. A new node N_u is created with the reachability value of B as its height. The current node for both A and B is set to N_u. In the next step, if I is selected, the current_node of I (=N_r) will be merged with current_node of I's left neighbor B (=N_u). After that, node N_s will be merged with this new node in the next step and so on until all points in our example with a reachability value of 2 are processed; the intermediate result looks like depicted. The only point remaining in PointList at this point is T with a reachability value of 3. For point T the current_node is N_t, the current node of T's left neighbor H is N_v. Hence N_t and N_v are merged into a new node N_w at height 3 (the reachability value of T).

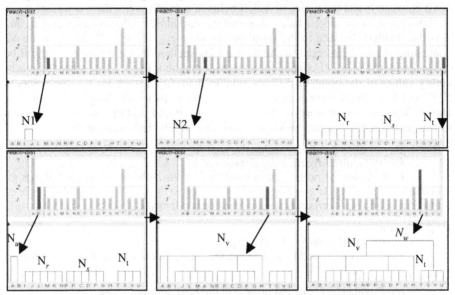

Fig. 4. Illustration of the conversion of a reachability plot into a dendrograms

Fig. 5 demonstrates our method using the data set in Fig. 2(right), where we have a larger variation of distances than in the above illustration example. Obviously, if we have a reachability plot for parameters other than MinPts = 2, our method generates a dendrogram where the merging distances are based on the minimum reachability distances between groups of points instead of point-wise distances.

Fig. 5. Transformation between Dendrogram and Reachability plot for the data set in Fig. 2

3 Extracting Clusters from Dendrograms and Reachability Plots

As argued in section 1, it is desirable to have a simplified representation of a hierarchical clustering that contains only the significant clusters, so that hierarchical clustering can be used as an automatic pre-processing step in a larger KDD process.

We base our method on reachability plots. To apply it to dendrograms, we have to transform a dendrogram into a reachability plot first, using the method introduced in the previous section. Fig. 6(a) illustrates our goal of simplifying a hierarchical clustering structure. Given a hierarchical representation such as the depicted reachability plot, the goal of our algorithm is to produce a simple tree structure, whereby the root represents all of the points in the dataset, and its children represent the biggest clusters in the dataset, possibly containing sub-clusters. As we proceed deeper into the tree the nodes represent smaller and denser clusters. The tree should only contain nodes for significant clusters as seen by a user who would identify clusters as "dents" or "valleys" in a reachability plot. The dents are roughly speaking regions in the reachability plot, having relatively low reachability values, and being separated by higher reachability values. Our automatic cluster extraction method first tries to find the reachability values that separate clusters. Those high reachability values are local maxima in the reachability plot. The reachability values of the immediate neighbors to the left and right side of such a point p are smaller than the reachability value of p, or do not exist (if p is the leftmost or rightmost point in the reachability plot). However, not all local maxima points are separating clusters, and not all regions enclosed by local maxima are prominent clusters. A local maximum may not differ significantly in height from the surrounding values, and local maxima that are located very close to each other create regions that are not large enough to be considered important clusters. This is especially true when using a small value for MinPts in OPTICS or a reachability plot converted from a Single-Link dendrogram (see Fig. 6(b)).

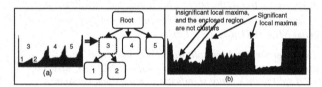

Fig. 6. (a) Desired transformation; (b) illustration of insignificant local maxima and clusters

To ignore regions that are too small in the reachability plot we assume (as is typically done in other methods as well) a minimum cluster size. In our examples, we set the minimum cluster size to 0.5% of the whole data set, which still allows us to find even very small clusters. The main purpose is the elimination of "noisy" regions that consists of many insignificant local maxima, which are close to each other.

The significance of a separation between regions is determined by the ratio between the height of a local maximum p and the height of the region to the left and to the right of p. The ratio for a significant separation is set to 0.75, i.e. for a local maximum p to be considered a cluster separation, the average reachability value to the left and to the right has to be at least 0.75 times lower than the reachability value of p. This value represents approximately the ratio that a user typically perceives as the minimum required difference in height to indicate a relevant cluster separation.

In the first step of our algorithm, we collect all points p whose reachability value is a local maximum in a neighborhood that encloses minimum-cluster-size many point to the left of p and to the right of p (if this is possible – points close to the border of the reachability plot may have less points on one side). The local maxima points are sorted in descending order of their reachability value, and then processed one by one to construct the final cluster tree. The pseudo-code for a recursive construction of a cluster tree is shown in Fig. 7. Intuitively, the procedure always removes the next largest local maximum s from the maxima list (until the list is empty), determines whether this split point justifies new nodes in the cluster tree – and if yes, where those nodes have to be added. This is done in several steps: First, two new nodes are created by distributing the points in the current node into two new nodes according to the selected split point s: All points that are in the reachability plot to the left of s are inserted into the first node; the second node contains all the points that are right of s (for the recursion, also the list of all local maxima points for the current node are distributed accordingly into two new lists). The two new nodes are possible children of the current node or the parent node, depending on the subsequent tests.

```
cluster_tree(N, parent_of_N, L)
    // N is a node; the root of the tree in the first call of the procedure
    // parent_of_N is the parent node of N; nil if N is the root of the tree
    // L is the list of local maxima points, sorted in descending order of reachability
    if (L is empty) return; // parent_of_N is a leaf

    // take the next largest local maximum as possible separation between clusters
    N.split_point = s = L.first_element(); L = L - {s};

    // create two new nodes and add them to a list of nodes
    N1.points = {p ∈ N.points | p is left of s in the reachability plot};
    N2.points = {p ∈ N.points | p is right of s in the reachability plot};
    L1 = {p ∈ L | p lies to the left of s in the reachability plot);
    L2 = {p ∈ L | p lies to the right of s in the reachability plot);
    Nodelist NL = [(N1, L1), (N2, L2)];

    // test whether the resulting split would be significant:
    if ((average reachability value in any node in NL) / s.r_dist > 0.75)
        // if split point s is not significant, ignore s and continue
            cluster_tree(N, parent_of_N, L); //
    else //remove clusters that are too small
            if (N1.NoOfpoints) < min_cluster_size) NL.remove(N1, L1);
            if (N2.NoOfpoints) < min_cluster_size) NL.remove(N2, L2);
            if (NL is empty) return; // we are done and parent_of_N will be a leaf
            // check if the nodes can be moved up one level
            if (s.r_dist and parent_of_N.split_point.r_dist are approximately the same)
                let parent_of_N point to all nodes in NL instead of N;
            else
                let N point to all nodes in NL;
            // call recursively for each new node
            for each element (N_i, L_i) in NL
                cluster_tree(N_i, parent_of_N, L_i)
```

Fig. 7. Algorithm for constructing a cluster tree from a reachability plot

We first check that the difference in reachability between the newly created nodes and the current local maximum indicates a relevant separation (as discussed above). If the test is not successful, the current local maximum is ignored and the algorithm continues with the next largest local maximum. If the difference is large enough, we have to check whether the newly created nodes contain enough points to be considered a cluster. If one of these nodes does not have the minimum cluster size it is deleted. If the nodes are large enough, they will be added to the tree. There are two possibilities to add the new nodes to the tree: as children of the current node, or as children of the parent node, replacing the current node. If we would always add the nodes as children of the current node, we would create a binary tree. This may not result in the desired representation as illustrated in Fig. 8. To determine whether two reachability values are approximately the same, we use a simple but effective test: we assume that the reachability value of the parent node (after scaling) is the true mean of a normed normal distribution ($\sigma=1$). If the (scaled) reachability value of the current node is within one standard deviation of the parent's value, we consider the two values as similar enough to attach the newly created nodes to the parent node of the current node instead of the current node itself. This heuristic gives consistently good results for all classes of reachability plots that we tested.

We performed a large number of experiments to evaluate our method, but due to space limitations, we can only show a limited number of representative results in Fig. 9. For comparison, we also show the results of the ξ-cluster method that was proposed in the original OPTICS publication [1]. We also show the data sets clustered with OPTICS. This should only help to understand the extracted clusters; the compared methods are, strictly speaking, not dependent the actual data set, but only on the properties of the reachability plot from which clusters are extracted.

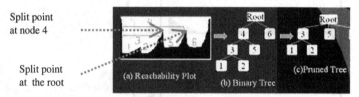

Fig. 8. Illustration of the "tree pruning" step

The results for our new method are shown in the middle column, and the results for ξ-clusters are given in the right column. For both methods, the extracted cluster trees are depicted underneath the reachability plots. The nodes are represented as horizontal bars that extend along the points that are included in this node, and which would be used to generate cluster descriptions such center, diameter, etc for each node. This tree itself is drawn in reverse order, i.e., the leaves are closest to the reachability plot. The root of the tree is always omitted since it always contains all points.

The experimental results we show are for three different reachability plots, each representing an important class of reachability plots that have the following characteristis: (a) relatively smooth plot with nested clusters, (b) noisy plot and fuzzy cluster separation, (c) smooth plot with clusters having largely differing densities. Examples (a) and (c) are also examples where simply cutting through the dendrogram will not reveal all existing clusters at the same time, since such a cut does not exist.

The results clearly show the superiority of our method: in all three different cases, our cluster tree that is very similar to what a user would extract as clusters. The ξ-cluster method, on the other hand, always constructs rather convoluted and much less useful cluster trees – using its default parameter value $\xi=0.03$ (for each of the examples there are some ξ-values, which give slightly better results than those shown above, but those best ξ-values are very different for each example, and even then, none of them matched the quality of the results for our proposed new method). Note that we did not have to change any parameter values to extract the clusters from largely differing reachability plot (whereas finding the best ξ- values for the each example would involve an extensive trial and error effort). The internal parameter value of a 75% ratio for a significant cluster separation works well in all our experiments. Our method is in fact very robust with respect to this ratio, as shown in Fig. 10 where we present the results for the same data as above, but for different ratios (70%, and 80%). In fact, we only suggest in extreme cases that a user might consider changing this value to a higher or lower value. The rule to set the value for this ratio is: the lower the value, the less deep a "valley" can be (relative to its surrounding areas) to be considered as a significant cluster region in a reachability plot.

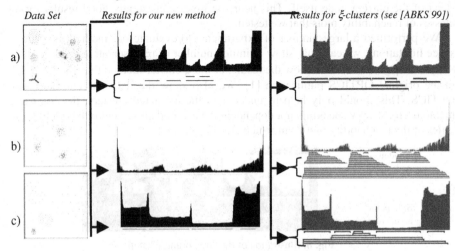

Fig. 9. Comparison of our cluster extraction method with ξ-clusters [1]

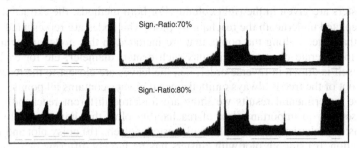

Fig. 10. Results for different reachability ratios

4 Conclusions

In this paper we first showed that dendrograms and reachability plots contain essentially the same information and introduced methods to convert them into each other. We then proposed and evaluated a new technique that automatically determines the significant clusters found by hierarchical clustering methods. This method is based on a reachability plot representation, but also applicable to dendrograms via the introduced transformation. The major purpose of these methods is to allow hierarchical clustering as an automatic pre-processing step in a data mining process that operates on detected clusters in later steps. This was not possible before, since hierarchical clustering algorithms do not determine clusters, but compute only a hierarchical representation of a data set. Without our methods, a user typically has to manually select clusters from the representation based on visual inspection. Our algorithm generates a cluster tree that contains only the significant clusters from a given representation.

There are some directions for future work: currently, a cluster from a hierarchical clustering result is represented simply as the set of points it contains. For an automatic comparison of different clustering structures, we will investigate the generation of more compact cluster descriptions. Using those descriptions, we will also investigate methods to compare different cluster trees generated by our method. This is an important prerequisite for data mining applications like monitoring the clustering structure and its changes over time in a dynamic data set.

References

1. Ankerst M., Breunig M. M., Kriegel H.-P., Sander J.: "OPTICS: Ordering Points To Identify the Clustering Structure", Proc. ACM SIGMOD, Philadelphia, PA, 1999, pp 49–60.
2. Ester M., Kriegel H.-P., Sander J., Xu X.: "A Density-Based Algorithm for Discovering Clusters in Large Spatial Databases with Noise", Proc. KDD'96, Portland, OR, 1996, pp. 226–231.
3. Hinneburg A., Keim D.: "An Efficient Approach to Clustering in Large Multimedia Databases with Noise", KDD'98, New York City, NY, 1998.
4. Jain A. K., Dubes R. C.: "Algorithms for Clustering Data," Prentice-Hall, Inc., 1988.
5. Knorr E. M., Ng R.T.: "Finding Aggregate Proximity Relationships and Commonalities in Spatial Data Mining," IEEE Trans. on Knowledge and Data Engineering, Vol. 8, No. 6, December 1996, pp. 884–897.
6. Kaufman L., Rousseeuw P. J.: "Finding Groups in Data: An Introduction to Cluster Analysis", John Wiley & Sons, 1990.
7. MacQueen J.: "Some Methods for Classification and Analysis of Multivariate Observations", Proc. 5th Berkeley Symp. Math. Statist. Prob., 1967, Vol. 1, pp. 281–297.
8. Ng R. T., Han J.: "Efficient and Effective Clustering Methods for Spatial Data Mining", Proc. VLDB'94, Santiago, Chile, Morgan Kaufmann Publishers, San Francisco, CA, 1994, pp. 144v155.
9. Sheikholeslami G., Chatterjee S., Zhang A.: "WaveCluster: A Multi-Resolution Clustering Approach for Very Large Spatial Databases", Proc. VLDB'98, New York, NY, 1998, pp. 428–439.
10. Sibson R.: "SLINK: an optimally efficient algorithm for the single-link cluster method", The Computer Journal Vol. 16, No. 1, 1973, pp. 30–34.

Large Scale Unstructured Document Classification Using Unlabeled Data and Syntactic Information

Seong-Bae Park and Byoung-Tak Zhang

School of Computer Science and Engineering
Seoul National University
Seoul 151-742, Korea
{sbpark,btzhang}@bi.snu.ac.kr

Abstract. Most document classification systems consider only the distribution of content words of the documents, ignoring the syntactic information underlying the documents though it is also an important factor. In this paper, we present an approach for classifying large scale unstructured documents by incorporating both lexical and syntactic information of documents. For this purpose, we use the co-training algorithm, a partially supervised learning algorithm, in which two separated views for the training data are employed and the small number of labeled data are augmented by a large number of unlabeled data. Since both lexical and syntactic information can play roles of separated views for the unstructured documents, the co-training algorithm enhances the performance of document classification using both of them and a large number of unlabeled documents. The experimental results on Reuters-21578 corpus and TREC-7 filtering documents show the effectiveness of unlabeled documents and the use of both lexical and syntactic information.

1 Introduction

Automatic document classification is an important research area in information retrieval and has a great potential for many applications handling text such as routing and filtering. Its aim is to assign a given document to the predefined category to which it belongs. Up to now, various kinds of algorithms based on machine learning or statistics have been applied to this problem and showed relatively high performance. However, most of them have been using a simple *bag of words* representation of documents [7]. That is, they use only the distribution of content words, assuming that the words are independent one another. But, this representation ignores *linguistic information* underlying the documents, which is another important factor in filtering texts.

Each document has its own traits in the style, and the syntactic information is one of the best measures to capture the stylistic divergence among the different kinds of documents [1]. Although the syntactic features can give much information in classifying the documents, they are not widely used due to their lack

K.-Y. Whang, J. Jeon, K. Shim, J. Srivatava (Eds.): PAKDD 2003, LNAI 2637, pp. 88–99, 2003.

of formal definition and complicated representation. In addition, unfortunately, the current NLP (natural language processing) techniques are not able to provide accurate results in syntax analyzing. However, some studies show that text chunks can give enough information on syntax analysis instead of full parsing [14].

Another problem in document classification is that there are a great number of inexpensive unlabeled documents while there are a few labeled documents. It is very expensive to obtain the labeled documents, since labeling of documents must be done by human experts. The co-training algorithm [2] is one of the successful algorithms handling unlabeled data. It is in general applied to the problems where there are two distinct views of each example in the dataset. For example, web pages can be considered to have two views: one is for contents of them and the other is for link information. It learns separate classifiers over each of the views, and augments a small number of labeled examples incorporating unlabeled examples.

In order to incorporate unlabeled examples with labeled examples, a metric to measure the confidence of labeling an unlabeled example is required in the co-training algorithm. And, recently the concept of margin is introduced with the emergence of Support Vector Machine [13], where the margin is defined as the product of true label (assumed to be -1 or +1) and the classifier's output. If the prediction of the classifier is correct, the margin is positive, otherwise negative. When the output of the classifier is real value, it can be considered as a confidence on prediction.

In this paper, we propose a co-trained Support Vector Machines (SVM) for document classification. It is based on the co-training algorithm, so that it effectively uses not only given small number of labeled documents but a great number of unlabeled documents. For the two views of the co-training algorithm, we use both lexical information and syntactic information. Thus, the proposed method can be applied to classify the unstructured normal documents.

This paper is organized as follows. Section 2 explains several approaches to incorporating unlabeled documents and linguistic information in document classification. Section 3 describes the co-training algorithm using both the lexical and the syntactic information. Section 4 reports the experimental results on Reuters-21578 corpus and TREC-7 filtering documents. Finally, Section 5 draws conclusions.

2   Related Work

Incorporating unlabeled examples is not a new scheme in information retrieval. It has been a major research topic under the name of *pseudo relevance feedback*. The idea of relevance feedback is to examine a portion of the retrieved documents and assign relevance values to each of them. The contents of the documents are analyzed to make the query closer towards the examined relevant documents and away from the examined non-relevant documents. The most common model for relevance feedback is proposed by Rocchio [11]. The terms in the new query are ranked by the weighted sum of the term weights in the current query, the known relevant documents, and the known non-relevant documents. That is, the new query is constructed with several top terms ranked by:

$$w_i(Q') = \alpha \cdot w_i(Q) + \beta \cdot w_i(R) + \gamma \cdot w_i(N),$$

where $w_i(Q')$ and $w_i(Q)$ are the weights of a term w_i in the new and current queries, $w_i(R)$ and $w_i(N)$ are the weights of w_i in the relevant and non-relevant documents, and α, β, and γ are the parameters to be determined. Many experiments on relevance feedback empirically show that it improves retrieval performance dramatically [12]. Pseudo relevance feedback typically adds new terms to the initial query by assuming that the several top documents in the initial ranked output are relevant.

In most discussions of relevance feedback, it is usually assumed that the initial query is of high quality, so that a major portion of the returned documents by the initial query are highly relevant. However, in many situations such an assumption is impractical. In document classification, we always can not trust all the labels of unlabeled documents estimated by the current classifier, since usually there are a small number of labeled documents while there are a great number of unlabeled documents. If the trained classifier does not coincide to the intrinsic model which generates the documents, the performance will be hurt by the unlabeled documents. Thus, more sophisticated methods are required for document classification using unlabeled ones.

One way to overcome the limit of traditional term weighting method is to use syntactic and semantic information. The merits of syntactic information is investigated by Hull [6]. Other studies also show that it is quite useful as content-identifier [15]. They state that using non-NLP generated phrases as terms in vector space retrieval is helpful at low recall level, while it is not helpful at high recall level. However, it is required to obtain more accurate syntactic information than just predefined window-size word sequences for its general use.

Because the current NLP techniques do not provide accurate information enough to be used in information retrieval, text chunking is considered to be an alternative to full parsing [14]. Text chunking is to divide text into syntactically related non-overlapping segments of words. Since this task is formulated as estimating an identifying function from the information (features) available in the surrounding context, various techniques have been applied to it [4].

3   Co-training Algorithm for Classifying Unstructured Documents

3.1   Co-training Algorithm

The co-training algorithm is one of the successful algorithms handling unlabeled examples [2]. It is in general applied to the problems where there are two distinct views of each example in the dataset. It learns separate classifiers over each of the views, and augments a small set of labeled examples by incorporating

Given L: a set of labeled examples, and U: a set of unlabeled examples.

Do Until there is no unlabeled example.
1. Train classifier h_1 on view V_1 of L.
2. Train classifier h_2 on view V_2 of L.
3. Allow h_1 to determine labels of examples in U.
4. Allow h_2 to determine labels of examples in U.
5. Determine $U' \subset U$, whose elements are most confidently labeled by h_1 and h_2.
6. $U = U \setminus U'$
7. $L = L + U'$

Fig. 1. An abstract of the co-training algorithm.

unlabeled examples. Its final prediction is made by combining their predictions to decrease classification errors. The larger is the variance of the classifiers when both classifiers are unbiased, the better is the performance of the algorithm. Since the co-training uses two classifiers with distinct views, its performance is better than any single classifier.

Figure 1 outlines the co-training algorithm. It uses two distinct views V_1 and V_2 when learning from labeled and unlabeled data, and incrementally upgrades classifiers (h_1 and h_2) over each view. Each classifier is initialized with a few labeled examples. At every iteration, each classifier chooses unlabeled examples and adds them to the set of labeled examples, L. The selected unlabeled examples are those which each classifier can determine their labels with the highest confidence. After that, the classifiers are trained again using the augmented labeled set. The final output of the algorithm is given as a combination of the two classifiers. Given an example $\mathbf{x}$ to be classified, the probability of the possible class c_j is determined by multiplying two posterior probabilities. The class c^* of $\mathbf{x}$ is set to the one with the highest probability:

$$c^* = \arg \max_{c_j \in C} \left(P(c_j|\mathbf{x}) = P_{h_1}(c_j|\mathbf{x}) P_{h_2}(c_j|\mathbf{x}) \right),$$

where C is the set of all possible classes.

Blum and Mitchell formalized the co-training settings and provided theoretical guarantee under certain assumptions [2]. Their first assumption is that the data distribution is compatible with the target function. That is, the target functions over each view predict the same class for most examples. The second assumption is that the views are conditionally independent with each other. If this assumption holds, the added examples at each iteration will be at least as informative as random examples. Thus, the iteration can progress though there are some mislabeled examples during the learning. However, these assumptions are somewhat unrealistic in practice, since the views from the same data are inclined to be related with each other in some way. Nigam and Ghani performed thorough empirical investigation on the conditional independence of the views [10]. Their experiments showed that the view independence affects the performance of the co-training algorithm, but the algorithm is still more effective than other algorithms incorporating unlabeled examples even when some dependence exist between the views.

For the classifiers in the co-training algorithm, we adopt Support Vector Machines that show significant improvement over other machine learning algorithms when applied to document classification [7]. At each iteration of the co-training, the most confident $|U'|$ examples should be selected from U. The SVMs provide a natural way to calculate the confidence of labeling unlabeled examples by a margin. The margin m for an unlabeled example $\mathbf{x}_i$ is defined as

$$m = y_i(\mathbf{w} \cdot \mathbf{x}_i + b),$$

where $y_i \in \{-1, +1\}$ is the label predicted by the hyperplane with the trained parameters $\mathbf{w}$ and b of SVMs. That implies, the margin can be considered to be a distance from $\mathbf{x}_i$ to the hyperplane, assuming that the predicted label is correct. The more distant $\mathbf{x}_i$ lies from the hyperplane, the more confident it is to predict the label of $\mathbf{x}_i$. Since SVMs are not probabilistic models, the final prediction of the co-training can not be made by multiplying the probabilities of each classifier. Instead, the final prediction is made only by the classifier whose margin is larger than the other.

3.2   Two Views

One possible view for document classification is to treat each document as a vector whose elements are the weight to the vocabulary. Most machine learning algorithms applied to document classification adopt this representation. The main drawbacks of this representation are that (i) it assumes that each word in the documents is independent one another, and (ii) it ignores much linguistic information underlying in the documents.

Stamatatos et al. showed experimentally that the syntactic information among various kinds of linguistic information is a reliable clue for document classification [14]. One additional benefit in using syntactic information for document classification by the co-training algorithm is that it is somewhat independent from term weights. Unfortunately, the current natural language processing techniques are not able to provide accurate syntactic analysis results. However, the *text chunks* instead of full parsing are good features enough to provide syntactic information for document classification. The chunks can be obtained in high accuracy with superficial investigation.

Therefore, we can define two distinct views for unstructured documents, so that the co-training algorithm can be naturally applied to classifying them. The two views are:

- **Lexical Information**
  Most machine learning algorithm applied to automatic document classification are based on $tf \cdot idf$, a commonly used term weighting scheme in information retrieval. The tf factor is an estimation of the occurrence probability of a term if it is normalized, and the idf is the amount of information related to the occurrence of the term.

– **Syntactic Information**
  Each document is represented in a vector in which the elements are syntactic features, and the features are derived from text chunking. This information can support finding particular or specific style of the documents.

3.3   Syntactic Features

To represent the documents in vectors whose elements are syntactic information, all documents need to be represented with chunk information. Since the unstructured documents are normally raw, all sentences in the documents must be chunked in the preprocessing step of classification. The purpose of chunking is to divide a given sentence into nonoverlapping segments. Let us consider the following sentence.

Example 1. He reckons the current deficit will narrow to only # 1.8 billion in September .

This sentence, composed of 15 words including a period, is grouped into nine segments after chunking as follows:

Example 2. [**NP** He] [**VP** reckons] [**NP** the current deficit] [**VP** will narrow] [**PP** to] [**NP** only # 1.8 billion] [**PP** in] [**NP** September] [**O** .]

In order to chunk the sentences in the documents, the lexical information and the POS (part-of-speech) information on the contextual words are required. Brill's tagger [3] is used to obtain POS tags for each word in the documents. The chunk type of each word is determined by Support Vector Machines trained with the dataset of CoNLL-2000 shared task[1].

Although there are 12 types of phrases in CoNLL-2000 dataset, we consider, in this paper, only five types of phrases: NP[2], VP, ADVP, PP, and O, where O implies none of NP, VP, ADVP, and PP. This is reasonable because those five types take 97.31% of the dataset and are major phrases for constituting a sentence. Each phrase except O has two kinds of chunk types: B-XP and I-XP. For instance, B-NP represents the first word of the noun phrase, while I-NP is given to other words in noun phrase. Thus, we consider 11 chunk types.

The use of Support Vector Machines showed the best performance in the shared task [8]. The contexts used to identify the chunk type of the i-th word w_i in a sentence are:

$$w_j, POS_j \ (j = i - 2, \ i - 1, \ i, \ i + 1, \ i + 2)$$
$$c_j \ (j = 1 - 2, \ i - 1)$$

where POS_j and c_j are respectively the POS tag and the chunk type of w_i. Since SVMs are basically binary classifiers and there are 11 types of chunks, SVMs are extended to multi-class classifiers by *pairwise classification* [13].

---

[1] http://lcg-www.uia.ac.be/conll2000/chunking

[2] NP represents a noun phrase, VP a verb phrase, ADVP an adverb phrase, and PP a prepositional phrase.

Table 1. Syntactic features for document classification.

Feature	Description
SF1	detected NPs / total detected chunks
SF2	detected VPs / total detected chunks
SF3	detected PPs / total detected chunks
SF4	detected ADVPs / total detected chunks
SF5	detected Os / total detected chunks
SF6	words included in NPs / detected NPs
SF7	words included in VPs / detected VPs
SF8	words included in PPs / detected PPs
SF9	words included in ADVPs / detected ADVPs
SF10	words included in Os / detected Os
SF11	sentences / words

Table 1 shows the features used to represent documents using text chunks. Top five features represent how often the phrases are used in a document, the following five features imply how long they are, and the final feature means how long a sentence is on the average. That is, every document is represented in a 11-dimensional vector.

4   Experiments

4.1   Datasets

Reuters-21578. The Reuters-21578 corpus is the most commonly used benchmark corpus in text classification. It consists of over 20,000 the Reuters newswire articles in the period between 1987 to 1991, and has 135 kinds of topics while only 10 of them are used for experiments. There are three versions to divide this corpus into a training set and a test set: "ModLewis", "ModApte", and "Mod-Hayes". Among them, "ModApte" which is most widely used is employed in this paper. In this version, there are 9603 training documents, 3299 test documents, and 27863 unique words after stemming and stop word removal.

TREC-7. The dataset for the TREC-7 filtering track consists of the articles from AP News for the years between 1988 to 1990, Topics 1-50. The 1988 AP documents are used for a training set, and the 1989-1990 AP documents for a test set. We do not use any information from the topic, such as 'description'. For documents representation using lexical information, we stem them and remove words from the stop-list. No thesaurus or other datasets are used. Though there are more than 160,000 words in this dataset, we choose only 7289 most frequently appearing words. There are 79,898 articles in the training set, but only 9,572 articles, just about 12% of them are labeled. Thus, we regard those 9,572 articles as a labeled document set, and the remaining 88% of the articles as an unlabeled document set.

Table 2. The accuracy improvement by using syntactic information on Reuters-21578 corpus. *Lexical* implies when we use only $tf \cdot idf$, *Syntactic* when we use only text chunks, and *Both* when we use both of them. The performance measure is LF1.

Class	Relevant Documents	Lexical	Syntactic	Both
earn	2877	2876	2504	**2895**
acq	1650	1587	1028	**1642**
money-fx	538	188	147	**193**
grain	433	**26**	14	**26**
crude	389	292	150	**312**
trade	369	153	114	**155**
interest	347	130	112	**141**
wheat	212	113	88	**116**
ship	197	98	70	**108**
corn	181	86	68	**87**

4.2 Performance Measure

To evaluate the classification performance of the proposed method, we use *utility* measure, for example LF1. Let R_+ be the number of relevant and retrieved documents, N_+ the number of non-relevant but retrieved documents, R_- the number of relevant but non-retrieved documents, and N_- the number of non-relevant and non-retrieved documents. Then, *linear utility* is defined as

$$\text{Linear Utility} = aR_+ + bN_+ + cR_- + dN_-,$$

where a, b, c and d are constant coefficients. The F1 and F3 measures are defined:

$$\text{LF1} = 3R_+ - 2N_+,$$
$$\text{LF2} = 3R_+ - N_+.$$

The drawback of linear utility is that it is meaningless to average through all topics because a few topics will have dominant effects on the results. Thus, to show the whole performance effectively we use the *scaled utility* [5] defined as

$$\text{Scaled Utility} = \frac{\max\{u(S,T), U(s)\} - U(s)}{MaxU(T) - U(s)},$$

where $u(S,T)$ is the utility of system S for topic T, $MaxU(T)$ is the maximum possible utility score for topic T, and $U(s)$ is the utility of retrieving s non-relevant documents. Since we consider only 10 topics in Reuters-21578 corpus, we use LF1 measure. However, there are 50 topics in TREC-7, so that we use the scaled linear utility.

4.3 Effect of Syntactic Information

Reuters-21578. When we do not consider unlabeled examples on Reuters-21578 dataset, the effect of using additional syntactic information is given in

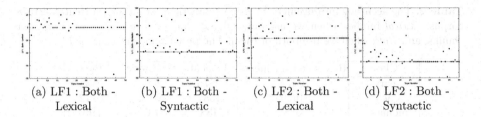

|(a) LF1 : Both -|(b) LF1 : Both -|(c) LF2 : Both -|(d) LF2 : Both -|
|Lexical|Syntactic|Lexical|Syntactic|

Fig. 2. The performance improvement by using both lexical and syntactic information. Two different measures are used: LF1 and LF2.

Table 3. The result on TREC-7 dataset. The performance measure is the averaged scaled LF1 and LF2 when $U(s)$ is set to 0.

Measure	Lexical	Syntactic	Both
LF1	0.2005	0.1680	**0.2192**
LF2	0.2010	0.2010	**0.2155**

Table 2. The classification performance using both lexical and syntactic information outperforms the one using any single information for most classes. We find that the larger is the number of relevant documents in the training set, the higher is the LF1 for using lexical information. This is because the SVMs tend to achieve lower error rate and higher utility value when a class has a large number of relevant documents. This trend is kept except *grain* class even when we use both information. For *grain* class, the extremely low LF1 for the syntactic information causes the performance using both information to be dominated by the lexical information.

TREC-7. Figure 2 shows the advantage of using both lexical and syntactic information. These results are obtained in considering both labeled and unlabeled documents. The X-axis is the topic numbers, while the Y-axis is the difference of performance that we obtain by using both kinds of information rather than single information. Thus, positive values in these graphs mean that using both information outperforms using single one. We find by these figures that using both kinds of information improves the performance for both LF1 and LF2. In addition, the difference between *Both–Syntactic* is larger than *Both–Lexical* for LF1. This result coincides with Table 2 that using only lexical information is better than using only syntactic information.

The averaged performance of Figure 2 is given in Table 3. Since the results in the table did not enter TREC pools, all the unjudged documents in the test set are ignored. The use of both information gives the best result for both measures. Using only lexical information is better than using only syntactic information in LF1, but achieves the same performance in LF2. This coincides with the work of Mitra et al. [9] that the performance of a phrase-based retrieval system is not superior than that of a term-based retrieval system. Even though we do not use the phrases for the index of documents, a similar result is obtained that

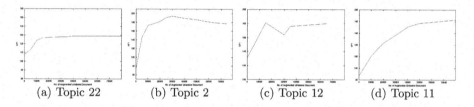

(a) Topic 22   (b) Topic 2   (c) Topic 12   (d) Topic 11

Fig. 3. The analysis of the effects of unlabeled documents for the topics with a great number of relevant documents in TREC-7. The performance measure is LF1.

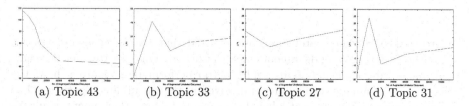

(a) Topic 43   (b) Topic 33   (c) Topic 27   (d) Topic 31

Fig. 4. The analysis of the effects of unlabeled documents for the topics with a small number of relevant documents in TREC-7. The performance measure is LF1.

the information derived from phrases are not better than terms. We have not studied this phenomenon in greater details, but we presume three reasons for it. First, we do not consider any content words in the document for syntactic information. Thus, it overgeneralizes the documents with styles. Second, there are only 1,468 reporters in the training set of TREC-7, though there are more than 79,000 articles. One reporter should write about 50 articles on the average, but more than a half of them write less than 10 articles. That is, a few reporters wrote a large portion of the articles. This forces the syntactic character of each reporter not to reveal and the performance of using only syntactic information to be low. The other reason is that there are a great number of non-sentence forms such as tables and listings in the training set since they are news articles. They make the syntactic analysis failed.

4.4   Effect of Unlabeled Documents

TREC-7. Figure 3 and Figure 4 show that the performance of the proposed method is enhanced by unlabeled documents even though there are a small number of relevant documents. The figures in Figure 3 are on the topics with a great number of relevant documents, while the figures in Figure 4 are on the topics with a small number of relevant documents. For example, *Topic 22* has 358 relevant documents, but *Topic 31* has only 1 relevant ones. Most topics except *Topic 43* show the similar trends that the performance gets higher by using unlabeled documents and gets higher and higher as more unlabeled documents are used in the learning. This is a good point of Support Vector Machines since they tend to correctly classify the relevancy with the larger number of documents.

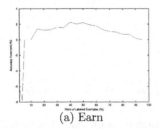
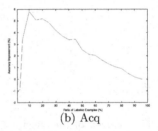

(a) Earn   (b) Acq

Fig. 5. The improvement in accuracy by using additional unlabeled documents.

Reuters-21578. Many previous work asserted that unlabeled documents improve the performance of document classification [16]. Our experiments also show that unlabeled documents are useful for better performance. Figure 5 shows the effectiveness of unlabeled documents involved in the co-training algorithm. The X-axis represents the ratio of labeled documents to total documents, while Y-axis is the accuracy improvement by additional unlabeled documents. For *earn*, the unlabeled documents play a positive role when more than 10% of training documents are labeled. So is for *acq*, when more than 7% of training documents are labeled.

However, even when we obtain the highest improvement by unlabeled documents, it does not reach to the best performance when we know the labels of all the training examples beforehand. For example, the improvement of accuracy is 5.81% when 10% of documents are labeled in *acq*. In this case, the accuracy is just 89.93% while the accuracy with 100% labeled documents is 95.21%. This implies that some of the unlabeled documents are mislabeled during the co-training process and have a negative effect on the classifiers. The effectiveness of unlabeled documents can be maximized when the number of labeled ones is small. To fill a gap of this difference, human intervention is needed. But, it is still an open problem when to intervene in the process.

5   Conclusion

We proposed the co-trained Support Vector Machines for document filtering. This method uses both the traditional lexical information and the syntactic information for the co-training algorithm and makes the algorithm applied to unstructured document classification, not only web page classification. In addition, with the algorithm we can incorporate ubiquitous unlabeled documents naturally to augment a small number of given labeled documents. The experiments on Reuters-21578 corpus and TREC-7 filtering documents show that using the syntactic information with the traditional lexical information improves the classification performance and the unlabeled documents are good resources to overcome the limited number of labeled documents. This is important because we can construct the classifier for large scale documents in high performance with just a small number of labeled documents. While the effectiveness of un-labeled documents is empirically proved, another problem is revealed that we need a method to overcome the mislabeling during the co-training algorithm.

Acknowledgement. This research was supported by the Korean Ministry of Education under the BK21-IT program and by the Korean Ministry of Science and Technology under BrainTech program.

References

1. D. Biber. *Dimensions of Register Variation: A Cross-Linguistic Comparison.* Cambridge University Press, 1995.
2. A. Blum and T. Mitchell. Combining labeled and unlabeled data with co-training. In *Proceedings of CONLT-98*, pages 92–100, 1998.
3. E. Brill. A simple rule-based part-of-speech tagger. In *Proceedings of ANLP-92*, pages 152–155, 1992.
4. CoNLL. *Shared Task for Computational Natural Language Learning (CoNLL).* http://lcg-www.uia.ac.be/conll2000/chunking, 2000.
5. D. Hull. The TREC-7 filtering track: Description and analysis. In *Proceedings of TREC-7*, pages 33–56, 1998.
6. D. Hull, G. Grefenstette, B. Schulze, E. Gaussier, H. Schutze, and J. Pedersen. Xerox TREC-5 site report: Routing, filtering, nlp, and spanish tracks. In *Proceedings of TREC-7*, 1997.
7. T. Joachims. Text categorization with support vector machines: Learning with many relevant features. In *Proceedings of ECML-98*, pages 137–142, 1998.
8. T. Kudo and Y. Matsumoto. Use of support vector learning for chunk identification. In *Proceedings of CoNLL-2000 and LLL-2000*, pages 142–144, 2000.
9. M. Mitra, C. Buckley, A. Singhal, and C. Cardie. An analysis of statistical and syntactic phrases. In *Proceedings of RIAO-97*, pages 200–214, 1997.
10. K. Nigam and R. Ghani. Analyzing the effectiveness and applicability of Co-training. In *Proceedings of CIKM-2000*, pages 86–93, 2000.
11. J. Rocchio. Relevance feedback in information retrieval. *The SMART Retrieval System: Experiments in Automatic Document Processing*, pages 313–323, 1971.
12. G. Salton and C. Buckley. Improving retrieval performance by relevance feedback. *Journal of the American Society for Information Science*, pages 288–297, 1990.
13. B. Scholkopf, C. Burges, and A. Smola. *Advances in Kernel Methods: Support Vector Machines.* MIT Press, 1999.
14. E. Stamatatos, N. Fakotakis, and G. Kokkinakis. Automatic text categorization in terms of genre and author. *Computational Linguistics*, 26(4):471–496, 2000.
15. A. Turpin and A. Moffat. Statistical phrases for vector-space information retrieval. In *Proceedings of SIGIR-1999*, pages 309–310, 1999.
16. T. Zhang and F. Oles. A probability analysis on the value of unlabeled data for classification problems. In *Proceedings of ICML-2000*, pages 1191–1198, 2000.

Extracting Shared Topics of Multiple Documents

Xiang Ji and Hongyuan Zha

Department of Computer Science and Engineering
The Pennsylvania State University, University Park
PA, 16802, USA
{xji, zha}@cse.psu.edu

Abstract. In this paper, we present a weighted graph based method to simultaneously compare the textual content of two or more documents and extract the shared (sub)topics of them, if available. A set of documents are modelled with a set of pairwise weighted bipartite graphs. A generalized mutual reinforcement principle is applied to the pairwise bipartite graphs to calculate the *saliency* scores of sentences in each documents based on pairwise weighted bipartite graphs. Sentences with advantaged saliency are selected, and they together convey the dominant shared topic. If there are more than one shared subtopics among the documents, a spectral min-max cut algorithm can be used to partition a derived sentence similarity graph into several subgraphs. For a subgraph, if all documents contribute some sentences(nodes) to it, then these sentences(nodes) in the subgraph may convey a shared subtopic. The generalized mutual reinforcement principle is applied to them to verify and extract the shared subtopic.

1 Introduction

The amount of textual information is increasing explosively, especially triggered by the growth of World Wide Web. The size of online texts or documents almost doubles every year. Inevitably, there are large number of texts and documents are correlated in content, such as various newswire presenting news materials on the same events. The topical correlation among documents exhibits a variety of patterns: they can present the same or related topics and/or events, or describe the same or related topics and/or events from different points of view. Discovering and utilizing the correlation among them is a very interesting research issue. It provides the first step towards advanced multi-documents summarization [6,8] as well as assist the concept retrieval, which is to find content-related web pages from a text database based on a given query web page or link [1,4]. Last but not least, correlating the textual information of hit list is a promising approach for search engine users navigating through search results. Since returned webpages for a query are related in content to some extent, comparing their textual information explicitly discloses their similarities and differences in content and thus help users select their intended pages.

Recently, many papers on multiple document clustering and summarization have emerged [3,5,8]. Their approaches exploited meaningful relations between

K.-Y. Whang, J. Jeon, K. Shim, J. Srivatava (Eds.): PAKDD 2003, LNAI 2637, pp. 100–110, 2003.

units based on the analysis of text cohesion and the context in which the comparison is desired. Another document topic related research direction, which is known as Topic Detection and Tracking (TDT), is to discover and thread together topically related material in streams of data [9,10]. However, few of them addressed the correlation of textual content of multiple documents and extraction of shared topics. Their algorithms do not fit the above application demand. Recently, Zha et.al. discussed the approach of correlating the content a pair of multilingual documents [13].

In this paper, we present an approach to automatically correlate the textual contents of two or more documents simultaneously through comparison of textual content of documents and highlight their similarities and shared topical features. A document consists of a consecutive sequence of text units, which can be paragraph or sentence. For our purpose, we consider sentence as the text unit, which can be substituted by paragraph. A document usually addresses a coherent topic. However, a long document could address several subtopics, and the sentences on a common subtopic can be at several location inside the document. Thus, it is necessary to cluster sentences into topical groups before studying the topical correlation of multiple documents. The essential idea of our approach is to apply the generalized mutual reinforcement principle on k documents in question to compute the *saliency* of each sentence and then extract k sets of sentences with high saliency. Sentences in each set are from only one document, and the k sets of sentences together convey the dominant shared topics. These extracted sentences in each set are ranked based on their importance in embodying the shared dominant topics. Thus, we can produce variable size of multiple documents summaries by extracting a certain number of sentences from high rank to low rank. For k documents with several shared subtopics, a spectral min-max cut algorithm is first employed to partition sentences of the k documents into several clusters each containing sentences from the k documents. Each of these sentence clusters corresponds to a shared subtopic, and within each cluster, the mutual reinforcement algorithm can be used to extract topic sentences.

The major contributions of this paper are the following: We extend previous research on correlating the content of a pair of documents into simultaneously correlating the content of two or more documents. We propose the generalized mutual reinforcement principle to find the correlated topics of more than two documents. Our method extracts sentences that convey the shared (sub)topics of them. Thus, it discloses more explicitly the similarities and differences in contents of two or more documents than approaches in previous research. In detail, our method is able to extract a single-document's topics as well as shared topics of two or more documents. Our method can also discover and compare both the dominant topic of multiple documents and a series of subtopics inside them.

We now give an outline of the paper. Section 2 presents our method, which contains the weighted bipartite graph model, the generalized mutual reinforcement principle, and spectral min-max cut algorithm. In Section 3, we present detailed experimental results on different sets of data, and demonstrate that our algorithm is effective in correlating summarization. We conclude the paper in Section 4 and discuss possible future work.

2   Methodology

For our purpose, a document consists of a consecutive sequence of sentences. However, other text units such as paragraphs or text lines can also be used to substitute for sentences. As a preprocessing step on the multiple documents in question, we first remove generic stop words, which are usually uninformative.

2.1   Weighted Bipartite Graph Model and Generalized Mutual Reinforcement Principle

We first model a document with a set of vectors each corresponding to a sentence. The attributes of vectors correspond to words appearing in the sentence and are valued with the word frequencies. For simplicity of mathematic notation, we disclose the construction of weighted bipartite graph model beginning with a pair of documents. For a pair of documents A and B, we construct a sentence cross-similarity bipartite graph $G(A, B, W)$. The texts in the two documents are split into two sets of sentences of $A = \{a_1, \ldots, a_m\}$ and $B = \{b_1, \ldots, b_n\}$, respectively. Sentences in A and B correspond to the vertices in G, and W is the adjacency matrix representing the edge weights. We denote $W = [w_{i,j}]$ of the m-by-n weight matrix with $w_{i,j}$ measuring the edge weight between a_i in A and b_j in B. We compare sentences in A and B and use the sentence cross-similarities, which are nonnegative, as the edge weights. For our experiments, we use the dot-product of a pair of sentence vectors as the measure of sentence similarity

$$w_{i,j} = \frac{a_i^T b_j}{\|a_i\|\|b_j\|}, \tag{1}$$

where $1 \leq i \leq m$ and $1 \leq j \leq n$. Figure 1 illustrates a weighted bipartite graph of a pair of documents. In the context of analyzing k documents, $k(k-1)/2$ cross-similarity bipartite graphs are constructed for all pairs of documents.

As a supplement, we denote the k documents in question by $\{D_1, D_2, \ldots, D_k\}$, and there are n_i sentences in document D_i. Thus, $D_i = \{s_{i1}, s_{i2}, \ldots, s_{in_i}\}$, where s_{ij} is a sentence in document D_i and $1 \leq j \leq n_i$. The dominant shared topic of the k documents, if available, is embodied by k subsets of sentences, denoted as $\{S_1, \ldots, S_k\}$, where $S_i \subseteq D_i$ and $i = 1, \ldots, k$. Sentences in these subsets are correlated well in content, and they together embody the dominant shared topic of the set of documents. For the jth sentence s_{ij} in D_i, there is a *saliency score* $u(s_{ij})$. The saliency represents the capability that a sentence conveys or contributes the dominant shared topic of the k documents. Those sentences that embody the dominant shared topic well have high saliency scores, while sentences do not convey the shared topic tend to have very low saliency score. We compute the saliency scores based on the sentence cross-similarity bipartite graphs $G(D_i, D_j, W_{ij})$, $1 \leq i, j \leq k$ and select the sentences with high saliency scores. We have the observation of

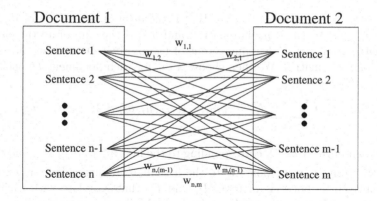

Fig. 1. A weighted bipartite graph of a pair of documents

A sentence in document D_i should have a high saliency score if it is correlated well to many sentences in D_j, $1 \leq j \leq k$ and $j \neq i$, with high saliency scores.

In another words,

the saliency score of a sentence in document D_i is determined by the saliency scores of the sentences in D_j, $1 \leq j \leq k$ and $j \neq i$, that are correlated to the sentence in D_i.

Mathematically, the above statement is rendered as

$$u(s_{mn}) \propto \sum_{i=1}^{k} \sum_{j=1}^{n_i} w_{n,j} u(s_{ij}). \tag{2}$$

where $i \neq j$, the symbol $\propto$ stands for "proportional to", and $w_{n,j}$ is the sentence similarity between the nth sentence in document D_m and the jth sentence in document D_i. The saliency scores for each sentence in all documents can be calculated by repeating the above procedure a certain number of times, which is usually less than 20.

Sentences with advantaged saliency scores are selected from the each sentence set D_i. These sentences and cross edges among them construct a dense sub-graph of the original weighted bipartite graph, which are well correlated in textual content.

Based on the above results, we propose the following criterion to determine the number of sentences included in the dense sub-graph. We first reorder the sentences in each document according to their corresponding saliency scores, respectively. Then for a pair of documents D_i and D_j, we obtain a permuted weight matrix $\hat{W}_{ij}$ from W_{ij}. For $\hat{W}_{ij} = [\hat{W}_{ij}^{st}]_{s,t=1}^{2}$, we compute the quantity of

$$J_{ij}^{mn} = h(\hat{W}_{ij}^{11})/n(\hat{W}_{ij}^{11}) + h(\hat{W}_{ij}^{22})/n(\hat{W}_{ij}^{22})$$

for $m = 1, \ldots, n_i$ and $n = 1, \ldots, n_j$. $\hat{W}_{ij}^{11}$ is the m-by-n sub-matrix of $\hat{W}_{ij}$, $\hat{W}_{ij}^{22}$ is the $(n_i - m)$-by-$(n_j - n)$ sub-matrix, and $h(\cdot)$ is the sum of all the elements of a matrix and $n(\cdot)$ is the square root of the product of the row and column dimensions of a matrix. We then choose a m_i^* for each document D_i such that

$$J = \max\{ \sum_{i=1}^{k} \sum_{j=1}^{k} J_{ij}^{m_i m_j} \mid m_i = 1, \ldots, n_i, m_j = 1, \ldots, n_j, i \neq j \}. \quad (3)$$

The m_i^* sentences from D_i corresponding to the first m_i^* rows of $\hat{W}_{ij}$ and the m_j^* sentences from D_j corresponding to the first m_j^* columns of $\hat{W}_{ij}$ are extracted and displayed as portions of texts D_i and D_j that correlate with each other the most. This sentence selection criteria avoids local maximum solution and extremely unbalanced bipartition of the graph.

We take a post-evaluation step to verify the existence of shared topics in the pair of texts in question. The first m_i^* rows of $\hat{W}_{ij}$ and the first m_j^* columns of $\hat{W}_{ij}$ compose a sub-matrix of $\hat{W}_{ij}$. Only when the $\sum_{i=1}^{k} \sum_{j=1}^{k} \hat{W}_{ij}^{11}/(m_i^* m_j *), i \neq j$, is greater than an empirical threshold, there is shared topic across the k documents and the extracted $m_i^*, i = 1, 2, \ldots, k$, sentences sentences efficiently convey the dominant shared topic. If it is below the threshold, there is no shared non-trivial topic among the set of documents. Ideally the threshold should be chosen using methods such as cross-validation to optimize certain performance criterion. In this paper they are chosen empirically by experimentation as $0.7(k-1)^2$. When there are two documents in question, the mutual reinforcement procedure is demonstrated to equal to the problem of singular value decomposition [12].

The major computational cost in the implementation of the generalized mutual reinforcement principle comes from computation of the pairwise sentence similarities, saliency scores, and searching the maximum J. Thus, the computation complexity is about $O(n^2)$, where n is the total number of sentence in k documents.

2.2   Spectral Min-Max Cut Algorithm

The above approach usually extracts a dominant topic that is shared by multiple documents. However, the set of documents may be very long and contain several shared subtopics besides the dominant shared topic. In order to extract these less dominant shared topics, we can apply a spectral min-max cut algorithm [2,11] to a sentence similarity graph $\bar{W}$ before applying the generalized mutual reinforcement principle for shared topic extraction.

The sentence similarity graph $\bar{W}$ is constructed as following: When there are k documents in question and $k \geq 2$, these documents are split into k sets of sentences, each documents containing n_i sentences, $D_i = \{s_{i1}, \ldots, s_{in_i}\}, i = 1, \ldots, k$. We generate D with $D = D_1 \cup D_2 \cup \ldots D_k$, and then $D = \{s_1, \ldots, s_n\}$, where $n = n_1 + n_2 + \ldots + n_k$. The $\bar{W}$ is constructed as $G(D, \bar{W})$, where $\bar{W} = [w_{i,j}]_{n \times n}$ and $w_{i,j}$ measures the edge weight between sentence s_i and s_j in D. The $w_{i,j}$ is computed with equation 1.

The spectral min-max cut algorithm will cut the sentence similarity graph into two subgraphs, and each subgraph contains a subset of sentences, which corresponds to one subtopic of all sentences. We may recursively partition each subset for different levels of topical granularity. Inside each subset, there are sentences from one or many documents. We exam the source document of each sentence. If a subset contains a certain number of sentences from each of the k documents, then the subset of sentences may embody a shared subtopic of the k documents. We name this kind of subset as *candidate subset*. For each subgraph corresponding to a candidate subset, we apply the generalized mutual reinforcement principle to verify the existing of shared subtopic and extract topical sentences. The traditional K-means algorithms can be an alternative approach for the problem. However, one of its major drawbacks is that it is prone to local minima giving rise to some clusters with very few data points (sentences).

Given the sentence similarity graph $G(D, \bar{W})$, where D is the vertex set and $\bar{W}$ is the edge set of the graph. We denote a vertex partition with $P(X)$ and $D = X \cup X^c$. Borrowed from [2,11], we utilize the minimum cut criteria for finding cut points in the graph and a spectral min-max cut algorithm based on the cut criteria.

As a supplement, we define the cross subgraph weight W: for any two subsets of vertices $A \subset D$ and $B \subset D$, $W(A, B) = \sum w_{ij}$, where $i \in A, j \in B$. We seek a partition of $G(D, \bar{W})$ such that the cross similarity between unmatched vertices is as small as possible. The following variant of edge cut is used:

$$Ncut(X) = \frac{W(X, X^c)}{W(X, D)} + \frac{W(X, X^c)}{W(X^c, D)}.$$

Therefore the bi-clustering problem is equivalent to the problem of

$$min_{P(X)} Ncut(X).$$

This is to partition the vertex sets D to minimize the normalized cut of the bipartite graph $G(D, \bar{W})$. Let x and y be vectors partitioned with X and X^c, i.e., $x = (1 \ldots 1, 0 \ldots 0)^T$ and $y = (0 \ldots 0, 1 \ldots 1)^T$. With $\bar{W}e = Se$ and $\hat{\bar{W}} = S^{-1/2} \bar{W} S^{-1/2}$, the optimization problem can be converted into problem of

$$min \frac{x^T(I - \hat{\bar{W}})x}{x^T \hat{\bar{W}} x} + \frac{y^T(I - \hat{\bar{W}})y}{y^T \hat{\bar{W}} y}.$$

We denote λ_1 and λ_2 as the largest two eigenvalues of $\hat{\bar{W}}$, and $\lambda_1 = 1$ by construction. The optimal solution is achieved when

$$\frac{x^T(I - \hat{\bar{W}})x}{x^T \hat{\bar{W}} x} = \frac{y^T(I - \hat{\bar{W}})y}{y^T \hat{\bar{W}} y} = \frac{2}{\lambda_1 + \lambda_2} - 1.$$

We can calculate the second largest eigenvector of $\hat{\bar{W}}$ to find the optimal solution. The solution indicates that X and X^c tend to have similar weights and are well balanced. Based on the criteria, we get clusters of X, X^c with small edge cut, we also get the two subgraphs formed between the matched vertices to be as dense as possible and balanced.

In the context of sentences clustering, D is clustered into sentence clusters X and X^c, and sentences in X and X^c have relatively low cross-similarities. However, the cross-similarities between X and X, and between X^c and X^c are relatively high.

Based on the above discussion, the spectral min-max cut algorithm presented as:

1. Compute S with $\bar{W}e = Se$. Calculate the scaled weight matrix $\hat{\bar{W}} = S^{-1/2}\bar{W}S^{-1/2}$.
2. Compute the second largest eigenvectors of $\hat{\bar{W}}$, $\hat{\lambda}$.
3. Calculate $\lambda = S^{-1/2}\hat{\lambda}$, and then permute the matrix $\bar{W}$ based on λ.
4. Search for the optimal cut point according to $minNcut$.
5. Recursively partition the sub-graphs $G(X)$ and $G(X^c)$ if necessary.

To seek all shared subtopics, we recursively partition the subgraphs. The following observation provides a good idea to stop the recursive process: If the average edge weight of a subgraph is greater than a certain threshold, then the sentences in the subgraph correspond to a subtopic and we stop clustering the subgraph; if the average edge weight of the subgraph is below the threshold, then we continue the recursive partition procedure until the size of the subgraph shrinks to a certain degree. Consequently, we get a serial of subgraphs which correspond to subtopics of the k documents. By applying the generalized mutual reinforcement principle to those subgraphs containing sentences from all documents, sentences conveying each shared subtopic are extracted.

The implementation of the spectral min-max cut algorithm contains eigenvectors computing. The eigenvectors can be obtained by Lanczos process, which is an iterative process each involving the matrix-vector multiplications. The computation cost is proportional to the number of nonzero elements of the object matrix $\bar{W}$, $nnz(\bar{W})$, and the total number of sentence in all documents in question. During searching the minimum cut point in $\bar{W}$, $\bar{W}(X + u, X^c - u)$ can be calculated based on $\bar{W}(X + u, X^c - u) = \bar{W}(X, X^c) + \bar{W}(u, X^c - u) - \bar{W}(u, A)$. Thus, We can implement it in order of $nnz(\bar{W})$.

3   Experimental Results

In this section, we present our experimental results. Some researchers use users' feedback to evaluate search results summarization, others use human-generated summarization or extraction to evaluate their text mining algorithms. However, most manual sentence extraction and summarization results tend to differ significantly, especially for long texts with several subtopics. Hence, the performance evaluation for our shared topic extraction method becomes a very challenging task. We try to provide some quantitative evaluation on our algorithms with experimental results on the two data sets.

3.1 Experiment Procedure

We utilize two data sets to evaluation our methods. The first data set is a synthetic data set, which contains 60 samples. Each sample is generated by concatenating 4 text segments. Each text segments is any n sentences of one of the randomly selected 20 documents from the Brown corpus. The n ranges from six to eleven. We deem that sentences of one document are cohesive in textual content. There is a great possibility that several samples have sentences from the same document and thus have shared topics. The second data set is a collection of ten pair of news articles that are randomly selected from some news websites. Each news article contains one or more coherent topics, and any two news articles in a pair have at least one shared topics. In average, there are 25 sentences and 1.4 topics in each news article. We remove generic stop words and applied Porter's stemming [7] to all texts.

For easiness in understanding our method, we report the experimental results in two sections. The first section demonstrates the spectral min-max cut algorithm for sentences clustering based on subtopics sharing. The second section demonstrates the generalized mutual reinforcement principle for extracting the dominant shared topic of multiple documents.

3.2 The Spectral Min-Max Cut Algorithm to Find Shared Subtopics

The first section of our experiments is on spectral min-max cut algorithm clustering sentences based on the first data set. Each time, k samples are selected as a set of samples, and k varies from 2 to 6. The spectral min-max cut algorithm is applied to the set of documents to cluster sentences corresponding to shared topics into groups.

The sentences selected as conveying shared topics by our clustering algorithm are named as retrieved sentences, and the sentences actually contribute the shared topics are named as relevant sentences. We define the precision measure as the fraction of the retrieved sentences that are relevant, and the recall measure as the fraction of the relevant sentences that are retrieved. High precision and recall means that the retrieved sentence set matches the relevant sentence set well. The average precision and recalls are listed in Table 1. It indicates that when the size of sample set varying, the algorithm's performance is fairly stable. We vary the ratio of the number of shared subtopics to the total number of subtopics in each article for a pair of samples. The results in Table 2 show that the spectral min-max cut algorithm is effective in clustering sentences for different number of shared subtopics.

Table 1. Precision and recall of clustering a set of samples with different sample size

Number of samples	2	3	4	5	6
Precision	0.72	0.76	0.70	0.75	0.74
Recall	0.79	0.74	0.76	0.75	0.69

Table 2. Precision and recall of clustering a set of samples with different number of shared subtopics

Number of shared subtopic(s)	1	2	3	4
Precision	0.72	0.74	0.75	0.76
Recall	0.69	0.76	0.75	0.74

3.3   The Generalized Mutual Reinforcement Principle to Extract Topical Sentences

We performed the automatic shared topic extraction on each pair of the news articles, and we also manually extract sentences that embody the dominant shared topic. The automatically extracted sentences for each news article is named as retrieved sentences, and the manually extracted sentences for each news article is named as relevant sentences. Then we define the precision measure as the fraction of the retrieved sentences that are relevant, and the recall measure as the fraction of the relevant sentences that are retrieved. The precisions and recalls are listed in Table 3. We also tried to find the dominant shared topic between two un-correlated news articles. The average cross-similarities between extracted sentences are less than our empirical threshold, 0.7. Therefore, there is no shared topic between each pair of news articles, as we have known in advanced. The results are listed in the last three row of Table 3.

We also apply the generalized mutual reinforcement method to the first data set after the spectral min-max cut process. The size of sample set varies and the average precisions and recalls are presented in Table 4.

Table 3. Dominant shared topic extraction from pair of news article

Article Pair ID	Average Similarity of Extracted Sentences	Precision	Recall
1	3.40	0.81	0.93
2	1.07	0.76	0.82
3	2.57	0.79	0.89
4	1.50	0.59	0.81
5	1.25	0.61	0.87
6	0.74	0.57	0.82
7	0.73	0.71	0.73
8	1.39	0.61	0.83
9	1.28	0.62	0.89
10	1.12	0.74	0.91
11	0.54	N/A	N/A
12	0.29	N/A	N/A
13	0.31	N/A	N/A

Table 4. Precision and recall of topical extraction on a set of samples with different size

Number of samples	2	3	4
Precision	0.62	0.65	0.57
Recall	0.59	0.56	0.50

4 Conclusions and Future Work

In this paper, we present the generalized mutual reinforcement principle for simultaneously extracting sentences, which embodies the dominant shared topic, from multiple documents. A simplified spectral relation partition algorithm to cluster the sentences in two or more documents is also discussed to find the shared subtopics of them. We test our algorithm with news articles from different topics and some artificial data sets. The experimental results suggest that our algorithms are effective in extracting dominant shared topic and other shared subtopics of a pair of multilingual documents.

In the future, we would like to investigate the criteria for threshold selection for stopping the recursive partition. We will also try to improve the accuracy of topical sentences selection by improving text topical information capturing. Other co-clustering algorithms can be applied to study their performance in text document topic matching.

Acknowledgment. The research was supported in part by NSF grant CCR-9901986. The authors also want to thank the referees for many constructive comments and suggestions.

References

1. J. Dean and M. Henzinger. Finding Related Pages in the World Wide Web. *Proceedings of WWW8*, pp. 1467–1479. 1999.
2. C. Ding, X. He, H. Zha, M. Gu, and H. Simon. Spectral Min-max Cut for Graph Partitioning and Data Clustering. *Proceedings of the First IEEE International Conference on Data Mining*, pp. 107–114, 2001.
3. L. Ertoz, and M. Steinbach and V. Kumar. Finding Topics in Collections of Documents: A Shared Nearest Neighbor Approach. *Proceedings of 1st SIAM International Conference on Data Mining*, 2001.
4. T. H. Haveliwala, A. Gionis, D. Klein, and P. Indyk. Evaluating Strategies for Similarity Search on the Web. *Proceedings of WWW11*, 2002.
5. J. Mani and E. Bloedorn. Multi-document Summarization by Graph Search and Matching. *Proceedings of the Fourteenth National Conference on Artificial Intelligence*, pp. 622–628, 1997.
6. I. Mani and M. Maybury. Advances in automatic text summarization. *MIT Press*, 1999.
7. M. Porter. The Porter Stemming Algorithm.
   www.tartarus.org/ martin/PorterStemmer

8. D. R. Radev and K. R. McKeown. Generating Natural Language Summaries from Multiple On-Line Sources. *Computational Linguistics*, Vol. 24, pp. 469–500, 1999.
9. Topic Detection and Tracking. http://www.nist.gov/speech/tests/tdt/
10. C. Wayne. Multilingual Topic Detection and Tracking: Successful Research Enabled by Corpora and Evaluation. *Language Resources and Evaluation Conference*, pp. 1487–1494, 2000.
11. H. Zha, X. He, C. Ding, M. Gu, and H. Simon. Bipartite Graph Partitioning and Data Clustering. *Proceedings of ACM CIKM*, pp. 25–32, 2001.
12. H. Zha. Generic Summarization Keyphrase Extraction Using Mutual Reinforcement Principle and Sentence Clustering. *Proceedings of the 25th Annual International ACM SIGIR Conference on Research and Development in Information Retrieval*, pp. 113–120, 2002.
13. H. Zha and X. Ji. Correlating Multilingual Documents via Bipartite Graph Modeling. *Proceedings of the 25th Annual International ACM SIGIR Conference on Research and Development in Information Retrieval*, pp. 443–444, 2002.

An Empirical Study on Dimensionality Optimization in Text Mining for Linguistic Knowledge Acquisition

Yu-Seop Kim[1], Jeong-Ho Chang[2], and Byoung-Tak Zhang[2]

[1] Division of Information and Telecommunication Engineering, Hallym University
Kang-Won, Korea 200-702
yskim01@hallym.ac.kr
[2] School of Computer Science and Engineering, Seoul National University
Seoul, Korea 151-744
{jhchang, btzhang}@bi.snu.ac.kr***

Abstract. In this paper, we try to find empirically the optimal dimensionality in data-driven models, Latent Semantic Analysis (LSA) model and Probabilistic Latent Semantic Analysis (PLSA) model. These models are used for building linguistic semantic knowledge which could be used in estimating contextual semantic similarity for the target word selection in English-Korean machine translation. We also facilitate k-Nearest Neighbor learning algorithm. We diversify our experiments by analyzing the covariance between the value of k in k-NN learning and accuracy of selection, in addition to that between the dimensionality and the accuracy. While we could not find regular tendency of relationship between the dimensionality and the accuracy, however, we could find the optimal dimensionality having the most sound distribution of data during experiments.

Keywords: Knowledge Acquisition, Text Mining, Latent Semantic Analysis, Probabilistic Latent Semantic Analysis, Target Word Selection

1 Introduction

Data-driven models in this paper are much beneficial in natural language processing application because the cost for building new linguistic knowledge is very expensive. But only raw text data, called untagged corpora, are needed in data-driven models of this paper. LSA is construed as a practical expedient for obtaining approximate estimates of meaning similarity among words and text segments and is applied to various application. LSA also assumes that the choice of dimensionality can be of great importance[Landauer 98]. The PLSA model is based on a statistical model which has been called aspect model[Hofmann 99c]. In this paper, we ultimately have tried to find out regular tendency of covariance

*** This work was supported by the Korea Ministry of Science and Technology under the BrainTech Project

K.-Y. Whang, J. Jeon, K. Shim, J. Srivatava (Eds.): PAKDD 2003, LNAI 2637, pp. 111–116, 2003.
© Springer-Verlag Berlin Heidelberg 2003

between dimensionality and soundness of acquired knowledge in natural language processing application, especially in English-Korean machine translation. We utilized k-Nearest neighbor learning algorithm to resolve the meaning of unseen word or data sparseness problem. Semantic similarity among words should be estimated and the semantic knowledge for similarity could be built from the data-driven models to be shown in this paper[Kim 02b]. We also extended our experiment by computing covariance values between k in k-NN learning and selectional accuracy.

2   Knowledge Acquisition Models

Next two subsections will explain what LSA and PLSA are and how these two models can build the linguistic knowledge for measuring semantic similarity among words.

2.1   Latent Semantic Analysis Model

LSA can extract and infer relations of expected contextual usage of words in passages of discourse[Landauer 98]. LSA applies singular value decomposition (SVD) to the matrix explained in [Landauer 98]. SVD is defined as

$$A = U \Sigma V^T \tag{1}$$

,where Σ is a diagonal matrix composed of nonzero eigen values of AA^T or $A^T A$, and U and V are the orthogonal eigenvectors associated with the r nonzero eigenvalues of AA^T and $A^T A$, respectively. The details of above formula are described in [Landauer 98]. There is a mathematical proof that any matrix can be so decomposed perfectly, using no more factors than the smallest dimension of original matrix. The singular vectors corresponding to the $k(k \leq r)$ largest singular values are then used to define k-dimensional document space. Using these vectors, $m \times k$ and $n \times k$ matrices where m and n is the number of rows, U_k and V_k may be redefined along with $k \times k$ singular value matrix $\sum_k$. It is known that $A_k = U_k \Sigma_k V_k^T$ is the closest matrix of rank k to the original matrix. The term-to-term similarity is based on the inner products between two row vectors of A, $AA^T = U\Sigma^2 U^T$. One might think of the rows of $U\Sigma$ as defining coordinates for terms in the latent space. To calculate the similarity of two words, $\mathbf{V_1}$ and $\mathbf{V_2}$, represented in the reduced space, cosine computation is used:

$$\cos \phi = \frac{\mathbf{V_1} \cdot \mathbf{V_2}}{\| \mathbf{V_1} \| \cdot \| \mathbf{V_2} \|} \tag{2}$$

Knowledge acquired from above procedure has a characteristics like that of people before and after reading a particular text and that conveyed by that text.

2.2   Probabilistic Latent Semantic Analysis

PLSA is based on *aspect model* where each observation of the co-occurrence data is associated with a latent class variable $z \in Z = \{z_1, z_2, \ldots, z_K\}$ [Hofmann 99c]. A word-document co-occurrence event, (d, w), d for documents and w for words, is modelled in a probabilistic way where it is parameterized as in

$$P(d, w) = \sum_z P(z)P(d, w|z)$$
$$= \sum_z P(z)P(w|z)P(d|z), \tag{3}$$

$P(w|z)$ and $P(d|z)$ are topic-specific word distribution and document distribution, respectively. The objective function of PLSA, unlike that of LSA, is the likelihood function of multinomial sampling. The parameters $P(z), P(w|z)$, and $P(d|z)$ are estimated by maximizing the log-likelihood function

$$L = \sum_{d \in D} \sum_{w \in W} n(d, w) \log P(d, w), \tag{4}$$

and this maximization is performed using the EM algorithm. Details on the parameter estimation are referred to [Hofmann 99c]. To estimate the similarity of two words, w_1 and w_2, we compute the similarity of w_1 and w_2 in the constructed latent space. This is done by

$$sim = \frac{\sum_k P(z_k|w_1)P(z_k|w_2)}{\sum_k P(z_k|w_1)P(z_k|w_1) \sum_k P(z_k|w_2)P(z_k|w_2)} \tag{5}$$

$$P(z_k|w) = \frac{P(z_k)P(w|z_k)}{\sum_l P(z_l)P(w|z_l)} \tag{6}$$

3   Target Word Selection Process

We used grammatical relations stored in the form of a dictionary for target word selection. The structure of the dictionary is as follows [Kim 01]:

$$T(S_i) = \begin{cases} T_1 \text{ if } Cooc(S_i, S_1) \\ T_2 \text{ if } Cooc(S_i, S_2) \\ \quad \cdots \\ T_n \text{ otherwise,} \end{cases} \tag{7}$$

where $Cooc(S_i, S_j)$ denotes grammatical co-occurrence of source words S_i and S_j, which one means an input word to be translated and the other means an argument word to be used in translation, and T_j is the translation result of the source word. $T(\cdot)$ denotes the translation process. One of the fundamental difficulties is the problem of data sparseness or unseen words. We used k-nearest neighbor method that resolves this problem. The linguistic knowledge acquired from latent semantic spaces is required when performing k-NN search to select the target word. The nearest instance of a given word is decided by selecting a word having the highest semantic similarity, which is extracted from the knowledge. The k-nearest neighbor algorithm for approximating a discrete-valued target function is given in [Cover 67].

4   Experimental Result

For constructing latent space, we indexed 79,919 documents in 1988 AP news text from TREC-7 data to 21,945,292 words. From this, 19,286 unique words with above 20 occurrence was collected and the resulting corpus size is 17,071,211. Kim *et. al*[Kim 02b] built a 200 dimension space in SVD of LSA and 128 latent dimension of PLSA. We, however, made the dimensionality of LSA and PLSA ranging from 50 to 300 individually because we tried to discover the relationship between the dimensionality and the adequateness of semantic knowledge acquired. We utilized a single vector Lanczos method of SVDPACK[Berry 93] when constructing LSA space. The similarity of any two words could be estimated by performing cosine computation between two vectors representing coordinates of the words in the spaces. We extracted 3,443 example sentences containing predefined grammatical relations, like *verb-object*, *subject-verb* and *adjective-noun*, from Wall Street Journal corpus and other newspapers text of totally 261,797 sentences. 2,437,188, and 818 examples were utilized for *verb-object*, *subject-verb*, and *adjective-noun*, respectively.

In the first place, we selected an appropriate target word using a default meaning, which is selected when there is no target word found in a built-in dictionary. About 76.81% of accuracy was acquired from this method. Secondly, we also evaluated the selection from non-reductive dimensionality which is same as the number of documents used in knowledge acquisition phase, 79,919. Up to 87.09% of selection was successful. Finally, the reduction was performed on the original vector using LSA and PLSA. Up to 87.18% of successful selection was acquired by using LSA and PLSA. However, time consumed in selection by non-reductive dimensionality was much longer about from 6 to 9 times than those of LSA and PLSA reduction methods.

And we have also tried to find out regular tendency of covariance between dimensionality and distributional soundness of acquired knowledge. For this, we have computed the co-relationship between the dimensionality and the selection accuracy and between k in the k-NN learning method and the accuracy. Figure 1 shows the relationship between the dimensionality and selectional accuracy on each k value. In figure 2, the relationship between k in k-NN learning and accuracy of target word selection over the various dimensionality was shown.

From this experiment, we could calculate the covariance values among several factors [Bain 87]. Table 1 shows the covariance between dimensionality and accuracy and between k and accuracy. As shown in the table, the covariance value between k and accuracy is much higher than that between the dimension-

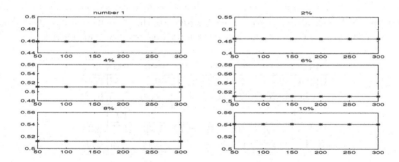

Fig. 1. This figure shows the tendency of selectional accuracy with various dimensionality over each k value of k-NN learning. 6 figures are shown for each k value. The line with 'X' (-X-) represents the result from SVD latent space, the line with 'O' (-O-) represents the result from PLSA space, and the other line (-*-) represents the result from non-reductive space. X-axis represents the dimensionality from 50 to 300 and y-axis represents selectional accuracy.

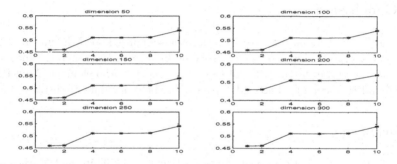

Fig. 2. This figure shows the tendency of selectional accuracy with various k over each dimensionality. Six figures are shown for each dimensionality value. The line with 'X' (-X-) represents the result from SVD latent space, the line with 'O' (-O-) represents the result from PLSA space, and the other line (-*-) represents the result from non-reductive space. X-dimension represents the k from 1 and from 2% to 10% of samples and y-dimension represents selectional accuracy.

ality and accuracy. It can be said that the value of k can affect the selectional accuracy much more than the dimensionality of a reduced vector. In PLSA, the soundness of space distribution is in the highest position when the dimensionality is 150, which could be inferred from the fact that the covariance between k and the accuracy is located at the top in the case of 150 dimensionality. It is known that the larger the k value is, the more robust to noisy data the sample data space is. From this, we can have an analogy that higher covariance value with k could make more sound distribution of latent space. In contrast, LSA has the most sound space in its 200 dimensionality. On average, PLSA has a little higher covariance then LSA. As a matter of fact, PLSA with 150 dimensionality selected accurate target words the most and LSA with 200 dimensionality se-

116   Y.-S. Kim, J.-H. Chang, and B.-T. Zhang

Table 1. The 3 upper rows of the table shows covariance values among k of k-NN learning and selectional accuracy in accordance to the dimensionality. And 3 lower rows show the covariance between dimensionality and accuracy for each k.

dim	50	100	150	200	250	300	avg
PLSA	52.89	58.35	59.83	32.45	49.02	34.84	47.90
SVD	43.54	29.20	44.52	60.09	47.90	19.75	40.83
k	1	2	4	6	8	10	
PLSA	7.89	23.91	6.46	0.12	1.79	-2.15	6.34
SVD	0.60	2.15	-5.50	-2.75	-3.11	-5.26	-2.31

lected the most accurately. Consequently, the distributional soundness and the selection accuracy could be said to have a strongly shared characteristics to each other.

5 Conclusion

LSA and PLSA were used for constructing the semantic knowledge in this paper. The dimensionality does not have a specific linkage to the semantic knowledge construction. However, the value of k could be the essential element. And, PLSA could build the semantic knowledge robust to the noisy data than LSA.

References

[Bain 87] Bain, L., M. Engelhardt, "Introduction to Probability and Mathematical Statistics," *PWS publishers*, pp. 179,190, 1987.
[Berry 93] Berry, M., T. Do, G. O'Brien, V. Krishna, and S. Varadhan, "SVDPACKC: Version 1.0 User's Guide," *University of Tennessee Technical Report*, **CS–93–194**, 1993.
[Cover 67] Cover, T., and P. Hart, "Nearest Neighbor Pattern Classification," *IEEE Transactions on Information Theory*, **13**, pp. 21–27, 1967.
[Hofmann 99c] Hofmann, T., "Probabilistic latent semantic indexing," *Proceedings of the 22th Annual International ACM SIGIR conference on Research and Developement in Information Retrieval (SIGIR99)*, pp. 50–57, 1999.
[Kim 01] Kim, Y., B. Zhang and Y. Kim, "Collocation Dictionary Optimization using WordNet and k-Nearest Neighbor Learning," *Machine Translation* **16**(2), pp. 89–108, 2001.
[Kim 02b] Kim, Y., J. Chang, and B. Zhang, "A comparative evaluation of data-driven models in translation selection of machine translation," *Proceedings of the 19th Internation Conference on Computational Linguistics (COLING-2002)*, Taipei, Taiwan, pp. 453–459, 2002.
[Landauer 98] Landauer, T. K., P. W. Foltz, and D. Laham, "An Introduction to Latent Semantic Analysis," *Discourse Processes*, **25**, pp. 259–284, 1998.

A Semi-supervised Algorithm for Pattern Discovery in Information Extraction from Textual Data

Tianhao Wu and William M. Pottenger

Computer Science and Engineering, Lehigh University
{tiw2, billp}@lehigh.edu

Abstract. In this article we present a semi-supervised algorithm for pattern discovery in information extraction from textual data. The patterns that are discovered take the form of regular expressions that generate regular languages. We term our approach 'semi-supervised' because it requires significantly less effort to develop a training set than other approaches. From the training data our algorithm automatically generates regular expressions that can be used on previously unseen data for information extraction. Our experiments show that the algorithm has good testing performance on many features that are important in the fight against terrorism.

1 Introduction

Criminal Justice is an important application domain of data mining. There are thousands of incident reports generated daily around the world, many of them in narrative (unstructured) textual form. Information extraction techniques can be used to identify relational data in such unstructured text, which in turn is used in a variety of computational knowledge management applications such as link analysis.

After studying hundreds of incident reports, we find that regular expressions can be readily employed to express patterns of features. For example, a suspect's height might be recorded as "{CD} feet {CD} inches tall", where {CD} is the part of speech tag for a numeric value. We have developed a semi-supervised algorithm for automatic discovery of regular expressions of this nature.

We term our approach 'semi-supervised' because it requires significantly less effort to develop a training set than other approaches. Instead of labeling the exact location of features in a training set, the training-set developer need only record whether a specific feature of interest occurs in a sentence segment. For instance, if a segment is "A 28 year old woman was walking from the store to her car.", and the feature of interest is Age, then the training-set developer need only assign this segment the label Age. Using this training data, our algorithm discovers a regular expression for a person's age. For example, "{CD} year old" might be found as a regular expression[1] for a person's age in this particular example. The automatically generated regular expressions can be used to extract various features from previously unseen data. Our experi-

---

[1] Currently, only words and English part of speech tags are used in the discovery process.

K.-Y. Whang, J. Jeon, K. Shim, J. Srivatava (Eds.): PAKDD 2003, LNAI 2637, pp. 117–123, 2003.

ments show that the algorithm has good testing performance on many features that are important in homeland defense.

Although much work has been done in the field of information extraction, relatively little has focused on the automatic discovery of regular expressions. Stephen Soderland developed a supervised learning algorithm, WHISK [1], which uses regular expressions as patterns to extract features from semi-structured and narrative text. Eric Brill [2] applied his transformation-based learning framework to learn reduced regular expressions for correction of grammatical errors in text. A crucial difference between these two approaches and ours is that WHISK and Brill's approach require the user to identify the precise location of features for labeling while our approach requires only that instances (segments) be labeled. Moreover, our regular expressions are more general since our approach supports the inclusion of the logical "OR" operator in regular expressions, while Brill's approach does not.

Michael Chau, Jennifer J. Xu, and Hsinchun Chen have published results of research on extracting entities from narrative police reports [5]. They employed a neural network to extract names, addresses, narcotic drugs, and items of personal property. Their cross-validation accuracy results vary from a low of 46.8% to a high of 85.4%. In our approach, however, we achieve significantly better results without limiting ourselves to noun phrases. In addition, we are able to extract a greater variety of textual features that can be used, for example, in link analysis of modus operandi.

The article is organized as follows. In section 2, we provide a framework for understanding our algorithm. Our approach is described in section 3, and in section 4 we detail preliminary experimental results. Finally, we present our conclusions, discuss future work, and give our acknowledgements in section 5.

2   Definitions

In this section, we start with the standard definition of a regular expression, and then define a reduced regular expression as used in our algorithm.

Regular expression: "Given a finite alphabet Σ, the set of regular expressions over that alphabet is defined as:

$\forall \alpha \in \Sigma$, α is a regular expression and denotes the set $\{\alpha\}$.

if r and s are regular expressions denoting the languages R and S, respectively, then (r+s), (rs), and (r*) are regular expressions that denote the sets $R \cup S$, RS and R* respectively." [2] [7]

Reduced regular expression (RRE): Our reduced regular expression is at first glance similar to that defined in [2]. However, there are some significant differences. Given a finite alphabet Σ, our reduced regular expression is defined as a set:

$\forall \alpha \in \Sigma$, α is a RRE and denotes the set $\{\alpha\}$.

$\sim\alpha^*$ is a RRE and denotes the Kleene closure of the set $\Sigma - \alpha$.

(^, $, \s \S \w \W) $\subset \Sigma$, where ^ is the start of a line, $ is the end of a line, \s is any white space, \S is any character except white space, \w is any alphanumeric character, and \W is any non-alphanumeric character.

(\w)* is a RRE denoting the Kleene closure of the set {\w}.

(\w){i,j} is a RRE denoting that \w is repeated between i and j times, where $0 \le i \le j$.

α? is a RRE and denotes that α is an optional part of the RRE.

if r and s are RREs denoting the languages R and S, respectively, then (r+s) and (rs) are RREs that denote the sets R $\cup$ S and RS, respectively.

Some examples of regular expressions that are not RREs are: "a*", "(ab)*", and "a+". We have not found it necessary to support these regular expressions to achieve high accuracies.

3 Approach

In this section we present our approach to the discovery of RREs from a small set of labeled training segments. The process begins with the processing of datasets. Next, a greedy algorithm is applied. Finally, RREs for the same feature are combined to form a single RRE.

Pre-Processing. Pre-processing includes segmentation, feature identification, segment labeling, and part of speech tagging. Each incident report is split into segments at this stage and becomes an instance in our system. We assume that no features cross segments. This assumption is practical for a number of important features, including those listed in Table 1. A domain expert must identify features that will be extracted such as 'Height', 'Weight', 'Eye Color', etc. Each segment is then assigned labels that correspond to the set of features present. After labeling, each feature has a *true set* corresponding to true positive segments and *false set* corresponding to true negative segments. Finally, each word is (automatically) assigned its part of speech [3].

Learning Reduced Regular Expressions. The goal of our algorithm is to discover sequences of words and/or part of speech tags that, for a given feature, have high frequency in the true set of segments and low frequency in the false set. The algorithm first discovers the most common element of an RRE, termed the *root* of the RRE. The algorithm then extends the 'length' of the RRE in an "AND" learning process. During the "OR" learning process, the 'width' of the RRE is extended. Next, optional elements are discovered during the "Optional" learning process. The algorithm then proceeds with the "NOT" learning process, and finally discovers the start and the end of the RRE. Figure 1 depicts the entire learning process.

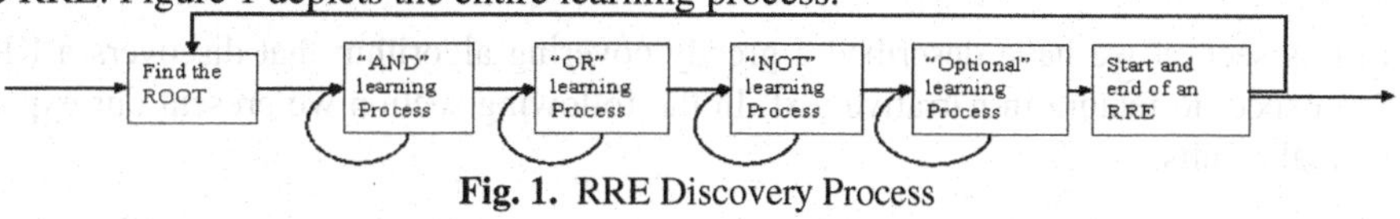

Fig. 1. RRE Discovery Process

Our approach employs a covering algorithm. After one RRE is generated, the algorithm removes all segments covered by the RRE from the true set. The remaining segments become a new true set and the steps in Figure 1 repeat. The learning process stops when the number of segments left in the true set is less than or equal to a threshold δ. We describe the details of the first three steps of the algorithm in what follows. [6] describes the details of the other steps.

To discover the root of a RRE, the algorithm matches each word and/or part of speech tag (specified as a simple RRE) in each true set segment against all segments. The performance of each such RRE in terms of the metric $F_\beta = ((\beta^2 + 1)PR)/(\beta^2 P + R)$ (from [4]) is considered. In this formula, $P = precision = TP/(TP+FP)$ and $R = recall = TP/(TP+FN)$, where TP are true positives, FP are false positives, and FN are false negatives. The parameter β enables us to place greater or lesser emphasis on precision, depending on our needs. The word or part of speech tag with the highest score is chosen as the root of the RRE. In other words, the algorithm discovers the word or part of speech tag that has a high frequency of occurrence in segments with the desired feature (the true set). Meanwhile, it must also have low frequency in segments that do not contain the desired feature (the false set).

After the root is discovered, the algorithm places new candidate elements immediately before and after the root, thereby forming two new RREs. Any word or part of speech tag (other than the root itself) can be used in this step. The RRE with the highest score will replace the previous RRE. The new RRE is then extended in the same way. Adjacency implies the use of an "AND" operator. As before, candidate words and parts of speech are inserted into all possible positions in the RRE. The algorithm measures the performance of each new RRE and the one with the highest score is selected if its score is greater than or equal to the previous best score. In this sense our algorithm is greedy. The RRE learned after this step is termed R_{AND}. The overall complexity of this step is $O(N^2)$, where N is the number of elements in R_{AND} [6].

After the "AND" learning process is complete, the algorithm extends the RRE with words and part of speech tags using the "OR" operator. For each element in R_{AND}, the algorithm uses the "OR" operator to combine it with other words and/or part of speech tags. If the newly discovered RRE has a better F-measure than the previous RRE, the new RRE will replace the old one. The complexity of "OR" learning process is $O(N)$.

The "Optional" learning process and the "NOT" learning process are then applied. Each of them has complexity $O(N)$. Finally the algorithm discovers the start and the end of the current RRE. For details on these three steps, please refer to [6].

Post Processing During each iteration in Figure 1, one RRE is generated. This RRE is considered a sub-pattern of the current feature. After all RREs have been discovered for the current feature (i.e., all segments labeled by the feature are covered), the system uses the "OR" operator to combine the RREs.

In this section we have described a greedy covering algorithm that discovers a RRE for a specific feature in narrative text. In the following section we present our experimental results.

4 Experimental Results

In this section, we describe the datasets for training and testing. We use domain expert labeled segments for training. We employ two different methods to evaluate the training results. The first method tests whether segment labels are correctly predicted. We term this segment evaluation. The second method evaluates the performance of the model with respect to an exact match of the feature of interest. The metric F_{β} (with $\beta=1$ to balance precision and recall) is used to evaluate the test performance with both methods. We use the widely employed technique of 10-fold cross-validation to evaluate our models.

Our training set consists of 100 incident reports obtained from Fairfax County, USA. These reports were automatically segmented into 1404 segments. The first column of Table 1 depicts 10 features supported by our system. *Eye Color* and *Hair Color* are not well represented in our dataset due to their infrequent appearance in the Fairfax County data.

The result of the training process is one RRE for each feature. For example, "(CD (NN)? old)|(in IN (MALE)? CDS)" is a high-level abstraction of the RRE for the feature *Age*. In this example, "NN" is the part of speech tag for noun, "MALE" is the tag for words of male gender such as "his", and "CDS" is the tag for the plural form of a number (e.g., "twenties" or "teens").

After completing 10-fold cross-validation, there are 10 test results for each feature. The average precision, recall and F_{β} ($\beta=1$) for these ten results is depicted in Tables 1 and 2. Table 1 contains the results based on segment evaluation, and Table 2 depicts the result of testing for an exact match. We also include a column for the average absolute number of true positives covered by each RRE.

Table 1. 10-fold cross-validation test performance based on segment evaluation

Feature	Precision	Recall	F_{β}	Avg. TP
Age	97.27%	92.38%	94.34%	13
Date	100%	94.69%	97.27%	8.8
Time	100%	96.9%	98.32%	8.9
Eye Color	100%	100%	100%	1
Gender	100%	100%	100%	33.6
Hair Color	60%	60%	60%	0.8
Height	100%	98%	98.89%	2.4
Race	95%	96.67%	94.67%	3.3
Weekday	100%	100%	100%	9.8
Weight	90%	90%	90%	1.9

In Table 1, *Eye Color, Gender* and *Weekday* have perfect test performance (100%) in part because we have modified the lexicon used in part of speech tagging to label these features during pre-processing. The performance of *Age, Date, Time, Height, Race,* and *Weight* are also excellent (F_{β} •90%). Although we also modified the lexicon to include a special "month" tag, which is a part of *Date*, the performance of *Date* is not perfect. This is a result of the fact that "2001" and "2002" cover over 95% of the years in our dataset, so the algorithm discovers "2001" or "2002" as a sub-pattern in *Date*. As a result, years other than "2001" and "2002" were not recognized as such during testing. This caused a slight drop in performance for the *Date* feature.

In order to address this issue we developed a more complete training set and re-evaluated our algorithm on the *Date* feature. In the new training set, there were ten "2002", ten "2001", ten "2000", ten "1999" elements, as well as a few "none" year elements. In this case our algorithm discovered the RRE "^ ([0-9a-zA-Z]){4} CD $". This is a more general pattern for detection of the year in the *Date* feature.

The performance of *Hair Color* is, however, not as good. As noted, this is due to the lack of *Hair Color* segments in test sets (2, 4, 5, 8). However, the test performance on sets (1, 3, 6, 7, 9, 10) is perfect for *Hair Color* (F_β of 100%). Therefore, we conclude that the RRE discovered for the feature *Hair Color* is optimal.

In a practical application, a user is interested in the exact feature extracted from narrative text rather than the segment. Therefore, it is necessary to evaluate our RREs based on their ability to exactly match features of interest. An exact match is defined as follows: "If a sub-string extracted by an RRE is exactly the same as the string that would be identified by a domain expert, then the sub-string is an exact match." An example of an exact match is as follows: if a human expert labels "28 year old" as an exact match of the feature *Age* in "*A 28 year old woman was walking from the store to her car.*", and an RRE also discovers "28 year old" as an age, then "28 year old" is an exact match. The result of our experiments in exactly matching features of interest is depicted in Table 2.

Table 2. 10-fold cross-validation test performance based on exact match

Feature	Precision	Recall	F_β	Avg. TP
Age	92.61%	88%	89.83%	12.4
Date	100%	94.69%	97.27%	8.8
Time	87.87%	85.01%	86.32%	7.8
Eye Color	100%	100%	100%	1
Gender	100%	100%	100%	33.6
Hair Color	60%	60%	60%	0.8
Height	95%	93.5%	94.17%	2.2
Race	90%	91.67%	89.67%	3
Weekday	100%	100%	100%	9.8
Weight	82.5%	82.5%	82.5%	1.7

In the exact match results in Table 2, *Age, Time, Height, Race, and Weight* have slightly lower performance than in segment evaluation. Sometimes the string accepted by a RRE is a sub-string of the actual feature. For example, suppose that the correct value for a particular instance of the *Time* feature is "*from/IN 9/CD :/: 00/CD p/NN ./. m/NN ./. until/IN 3/CD :/: 00/CD a/DT ./. m/NN ./NN*", but the string accepted by the RRE is "*from/IN 9/CD :/: 00/CD p/NN ./. m/NN ./.*". In this case, the algorithm failed to match the feature exactly. Nevertheless, these features still have very good performance, with high precision and recall (both of them are greater than 80%). On the other hand, *Date, Gender, Eye Color, Hair Color, and Weekday* all have the same performance as that achieved during segment evaluation. In fact, it turns out that the RREs discovered automatically for these five features are exactly the same patterns developed manually by human experts who studied this same dataset. Based on these exact match results, we conclude that our approach to RRE discovery has good performance on all ten features supported in our current system.

5 Conclusion

We have presented a semi-supervised learning algorithm that automatically discovers Reduced Regular Expressions based on simple training sets. The RREs can be used to extract information from previously unseen narrative text with a high degree of accuracy as measured by a combination of precision and recall. Our experiments show that the algorithm works well on ten features that are often used in police incident reports.

One of the tasks ahead is to develop more sophisticated segment boundary detection techniques (similar to existing sentence boundary detection techniques). Other future work includes the application of our techniques to extract additional features of interest such as those used in describing modus operandi. We plan to focus on these two tasks in the coming months.

This work was funded by the Pennsylvania State Police (PSP) under a subcontract with the Lockheed-Martin Corporation. We gratefully acknowledge the Lockheed-Martin/PSP team, co-workers, and our families for their support. Finally, we gratefully acknowledge our Lord and Savior, Yeshua the Messiah (Jesus Christ), for His many answers to our prayers. Thank you, Jesus! ☺

References

[1] S. Soderland: Learning information extraction rules for semi-structured and free text. Machine Learning, 34(1-3):233-272. (1999)

[2] Eric Brill: Pattern-Based Disambiguation for Natural Language Processing. Proceedings of Joint SIGDAT Conference on Empirical Methods in Natural Language Processing and Very Large Corpora. (EMNLP/VLC-2000)

[3] Christopher D. Manning and Hinrich Schütze: Foundations of Statistical Natural Language Processing, MIT Press. (2000)

[4] Van Rijsbergen: Information Retrieval. Butterworths, London. (1979)

[5] Michael Chau, Jennifer J. Xu, Hsinchun Chen: Extracting Meaningful Entities from Police Narrative Reports. Proceedings of the National Conference for Digital Government Research. Los Angeles, California. (2002)

[6] Tianhao Wu, and William M. Pottenger: An extended version of "A Semi-supervised Algorithm for Pattern Discovery in Information Extraction from Textual Data". Lehigh University Computer Science and Engineering Technical Report LU-CSE-03-001. (2003)

[7] Hopcroft, J. and J. Ullman: Introduction to Automata Theory, Languages and Computation. Addison-Wesley. (1979)

Mining Patterns of Dyspepsia Symptoms Across Time Points Using Constraint Association Rules

Annie Lau[1], Siew Siew Ong[2], Ashesh Mahidadia[2], Achim Hoffmann[2], Johanna Westbrook[1], and Tatjana Zrimec[1]

[1] Centre for Health Informatics, The University of New South Wales,
Sydney NSW 2052, Australia
anniel@student.unsw.edu.au, j.westbrook@unsw.edu.au, tatjana@cse.unsw.edu.au
[2] School of Computer Science and Engineering, The University of New South Wales,
Sydney NSW 2052, Australia
{ssong, ashesh, achim}@cse.unsw.edu.au

Abstract. In this paper, we develop and implement a framework for constraint-based association rule mining across subgroups in order to help a domain expert find useful patterns in a medical data set that includes temporal data. This work is motivated by the difficulties experienced in the medical domain to identify and track dyspepsia symptom clusters within and across time. Our framework, Apriori with Subgroup and Constraint (ASC), is built on top of the existing Apriori framework. We have identified four different types of phase-wise constraints for subgroups: *constraint across subgroups*, *constraint on subgroup*, *constraint on pattern content* and *constraint on rule*. ASC has been evaluated in a real-world medical scenario; analysis was conducted with the interaction of a domain expert. Although the framework is evaluated using a data set from the medical domain, it should be general enough to be applicable in other domains.

Keywords: association rule with constraints, domain knowledge, human interaction, medical knowledge discovery

1 Introduction

The motivation behind this work is the difficulty encountered in analysing medical data using traditional statistical analysis. We have a data set that looks at patients' dyspepsia symptoms at three points in time. Dyspepsia is a term used to describe a group of upper gastrointestinal symptoms that may be caused by underlying organic disease such as peptic ulcer. Most commonly, dyspepsia is defined as persistent or recurrent abdominal pain, or abdominal discomfort centred in the upper abdomen. The domain expert[1], who is an epidemiologist, is interested in finding relationships between symptoms within and across time points. She has used statistical approaches such as principal component analysis to identify clusters of symptoms that occur together at a point in time [1]. However, it is difficult to track how these symptom clusters change over time because there are many different combinations and

---

[1] One of the authors, Johanna Westbrook

K.-Y. Whang, J. Jeon, K. Shim, J. Srivatava (Eds.): PAKDD 2003, LNAI 2637, pp. 124–135, 2003.
© Springer-Verlag Berlin Heidelberg 2003

possibilities. In this paper, we focus on applying association rule mining [2, 3, 4] to assist the domain expert in this analysis. Initially, we used the Apriori algorithm discussed in [4] to mine patterns. Examples of these rules are shown below:

```
Rule 1. SAT2=0 ==> ANOREX2=0
Rule 2. ANOREX2=0 VOMIT2=0 NAUSEA3=0 ANOREX3=0 VOMIT3=0==> DYSPH2=0
```

The term <symptom>N=V represents a symptom and its value V at time point N. The attribute value 1 or 0, represents the presence or absence of the symptom. For example, SAT2=0 refers to the absence of the early satiety symptom at the second time point; ANOREX refers to anorexia, DYSPH refers to dysphagia (i.e. difficulty in swallowing), VOMIT refers to vomiting and NAUSEA refers to nausea.

Many of the rules returned from Apriori are of limited clinical value to the domain expert for three reasons:
1. Relationships between symptoms from different time points in many rules have little immediate usefulness in the absence of other clinical information.
2. The data set is sparse; many of the interesting patterns that involve the presence of a symptom at a single time point or at multiple time points are not likely to be discovered because they are not relatively frequent enough. For example, Rules 1 and 2 describe combinations of symptom absences, which carry little clinical value to the domain expert.
3. As a consequence, the domain expert is interested in following the behaviour of a symptom or a symptom cluster over time; however, an unguided mining approach is not likely to find these rules.

To address these issues, we need a mechanism to express the temporal nature of the data. We are interested in the behaviour of a collection of symptoms at a given point in time. Hence, we use the term "subgroup" to denote a set of symptoms at a single time point. Since the data can be divided into subgroups, with all data within a subgroup corresponding to the same time point, additional methods of associating data with a time component, such as those discussed in [5], are not required.

We also allow users to specify constraints to direct the mining. Work on constraints in association rule mining has been done in a number of directions. One direction concerns constraints on support and confidence of the mined rules, example [6]. Another direction allows users to perform exploratory mining by specifying SQL-style aggregate constraints, example [7]. Other work deals with constraints (meta-rules) on association rules, such as [8] or [9]. In more recent work, hierarchical structures among items can be defined to allow generalisation; for example, allowing one to refer to beer (a generalisation) instead of enumerating all kinds of beer available in the store [10, 11]. In contrast to item-sets, which are largely motivated by market basket analyses, in our work we are dealing with possibly multi-valued or numerical attributes and groups of attributes related to different points in time. The idea of constraints is also known in the field of Inductive Logic Programming, where meta-rules have been proposed to define the syntactical form of rules that can be learned, examples [12] and [13]. Inductive Logic Programming falls into the category of directed knowledge discovery and operates in a supervised learning framework where a specific target concept is to be learned. This is different to our scenario where we are interested in undirected knowledge discovery in which we want to discover interesting associations among the various attributes.

Furthermore, we are interested in mining rules that reveal interesting sequential patterns. Work in that direction can be found in [5, 14, 15, 16]; it is essentially concerned with mining sequential patterns on a single signal, such as a sequence of web page visits [15], i.e. at every point in time only a single value is to be considered.

In this paper, we draw ideas from the three areas mentioned above, and focus on finding interesting patterns across subgroups using combinations of constraints and pattern templates. Based on these considerations, we extended the Apriori algorithm and implemented a framework for mining association rules across subgroups using constraints, called Apriori with Subgroups and Constraints (ASC). The implementation of ASC is built on top of the Apriori algorithm from WEKA [17]. ASC mining is performed using phase-wise constraints. We denote a phase-wise constraint in the following manner – a phase is a stage in association-rule mining across subgroups and a constraint is a set of criteria that must be satisfied at that mining phase. The constraints outline the intra- and inter- subgroup criteria in an association rule. There are four types of phase-wise constraints in ASC, they are: *constraint across subgroups*, *constraint on subgroup*, *constraint on pattern content* and *constraint on rule*.

We describe our ASC framework in Section 2, followed by the use of ASC in the dyspepsia symptom mining in Section 3. We discuss the limitations of our current work and future work in Section 4, and conclude in Section 5.

2   Apriori with Subgroup and Constraint (ASC)

In this section, we describe the architecture of Apriori with Subgroup and Constraint (ASC). Figure 1 shows an overview of the ASC mining – a domain expert has a question, this question is translated into a user constraint by knowledge engineers and inputted into the ASC mining. Results are then returned to the expert for analysis. We describe the composition of a user constraint in Section 2.1 and demonstrate an example of ASC mining, looking at how a user constraint is integrated into the phase-wise mining process, in Section 2.2.

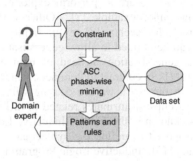

Fig. 1. Overview of ASC mining

2.1 Constraints for Subgroups

There are four types of phase-wise constraints in ASC, which are constructed from various basic elements. Table 1 lists a description and examples for each basic element. The user specifies these phase-wise constraints using a set of basic elements available in that phase (see Table 2 for details). These phase-wise constraints are responsible for different stages of the mining process. The following is a description of the mining process in each phase:

- Phase 1: *constraint across subgroups* specifies the order of subgroups in which they are to be mined.
- Phase 2: *constraint on subgroup* describes the intra-subgroup criteria of the association rules. It describes a minimum support for subgroups and a set of constraints for each subgroup. The mining process builds item-sets in each subgroup, prunes away those that do not meet the minimum support and the criteria in the subgroup, and merge together item-sets from different subgroups.
- Phase 3: *constraint on pattern content* outlines the inter-subgroup criteria on association rules. It describes the criteria on the relationships between subgroups. The constraint checks the merged item-sets and prunes away those that do not meet the criteria. For example, the constraint (a-1=1, a-2=0, a-3=1) describes the behaviour of symptom (a) at time points 1, 2 and 3 (a-1, a-2, a-3). The symptom (a) must be present at the initial time point (a-1=1), absent at time point 2 (a-2=0) and present again at time point 3 (a-3=1). Item-sets that do not contain at least one symptom with this behaviour are pruned away.
- Phase 4: *constraint on rule* outlines the composition of an association rule; it describes the attributes that form the antecedents and the consequents, and calculates the confidence of an association rule. It also specifies the minimum support for a rule and prunes away item-sets that do not meet this support at the end of each subgroup-merging step.

2.2 ASC Mining Process

In this section, we demonstrate an example of the ASC mining process with the following user constraint:

```
[1,2,3][1, a=A1&n<=2][2, a=B1&n<=2][3, v=1][rule, (s1 s2) => s3]
```

The user constraint is interpreted as follows: looking at subgroups 1, 2 and 3 ([1,2,3]) – from subgroup 1, extract patterns that contain the attribute A1 (a=A1) and contain no more than 2 attributes (n<=2); from subgroup 2, extract patterns that contain the attribute B1 (a=B1) and contain no more than 2 attributes (n<=2); then from subgroup 3, extract patterns with at least one attribute that has a value of 1 (v=1). Then, attributes from subgroups 1 and 2 form the antecedents in a rule, and attributes from subgroup 3 form the consequents ([rule, (s1 s2) => s3]).

We now proceed with the phase-wise ASC mining process that is illustrated in Figure 2. Initially, the system checks for *constraint across subgroups* ([1,2,3]), and selects subgroup 1, 2 and 3 in that order for the mining. In Step 1, the Apriori algorithm is used to find the frequent item-sets in subgroup 1; in Step 2, item-sets that do not satisfy the *constraint on subgroup* for subgroup 1, i.e. `[1, a=A1&n<=2]` are pruned away. This process is repeated for subgroup 2 (see Steps 3 and 4 in Figure 2).

Table 1. Basic elements of constraints in ASC

Elements of constraints	Data types	Description	Examples
Attribute name (a)	String	Attribute name	`a = VOMIT1`
Attribute value (v)	String, Numeric	Value of the attribute	`v = 1`
Attribute with value (av)	String, Numeric	Attribute name with value	`av:VOMIT1=1`
Number of attribute (n)	Numeric	Number of attributes in a subgroup	`n >= 3`
Weight (w)	Numeric	Selection of attributes with respect to degree of importance specified in this constraint	`w = 0.3`
Support (s)	Numeric	The number of occurrences of a pattern (this is applicable on two levels: subgroup and rule)	`s > 0.9`
Confidence (c)	Numeric	The number of occurrences of a pattern given a conditional pattern	`c > 0.7`
Ordering of subgroups	Numeric	The order of subgroups in building association rules	`[1,2,3]` `[3,1,2]`
Subgroups as Antecedents and consequences in a rule	String	The subgroups (i.e. the attributes in those subgroups) as antecedents and consequences in an association rule	`s1 =>(s2 s3)`
Pattern (p)	String, Numeric	The attribute behaviour across subgroups	`a-1=1` `a-2=0` `a-3=1`

Table 2. Basic elements available in each phase-wise constraint

Phase-wise constraint	Available basic elements
Constraint across subgroups	Ordering of subgroups
Constraint on subgroup	Attribute name, Attribute value, Number of attribute Attribute with value, Minimum support on subgroup, Weight
Constraint on pattern content	Pattern
Constraint on rule	Subgroups as antecedents and consequences in a rule, Minimum support on rule, Confidence

In Step 5, the pruned item-sets from subgroups 1 and 2 are merged together, which will be pruned again based on the minimum support for a rule in Step 6. The processes in Steps 1 and 2 are repeated on subgroup 3 (Steps 7 and 8). The resulting pruned subgroup 3 item-sets are merged with the combined item-sets from subgroups 1 and 2 (Step 9). Once again, these item-sets are pruned based on the minimum support for a rule (Step 10). The last stage of the mining process prunes away item-sets that do not meet *constraint on pattern content* (if applicable) and orders attributes in rules according to *constraint on rule* (Step 11).

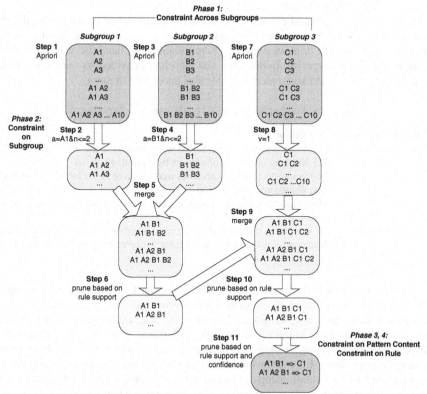

Fig. 2. ASC phase-wise mining

3 Mining Dyspepsia Symptom Patterns with Domain Expert

We applied the ASC framework in mining dyspepsia symptoms across time points with the domain expert. The medical domain data set contains records of 303 patients treated for dyspepsia. Each record represented a patient, the absence or presence of 10 dyspepsia symptoms at three time points (initial presentation to a general practitioner, 18 months after endoscopy screening, and 8–9 years after endoscopy) and the endoscopic diagnosis for the patient. The 10 dyspepsia symptoms are: anorexia (ANOREX), stomach bloating (BLOAT), moderate to severe burping (BURP), dysphagia (DYSPH), epigastric pain (EPI), heartburn (HEART), nausea (NAUSEA), acid regurgitation (REFLUX), early satiety (SAT) and vomiting (VOMIT). Each of these symptoms can have one of the following three values: *symptom present*, *symptom absent*, *missing (unknown)*. At each of the three time points, a symptom can take any of these three possible values.

The iterative data mining session with the domain expert proceeded in the following manner: the domain expert asked a question, the knowledge engineers translated it into a constraint for ASC mining, and then results were presented to the domain expert for further directions in the mining. The expert's first question is described in Task 1.

Task 1: What is the behaviour of a symptom over time?
Initially, we used a general constraint to track a symptom's behaviour across the three time points. However, the rules returned were too scattered and the number of rules was too large for the domain expert to comprehend. In fact, she was interested in finding rules with a specific hypothesis; she therefore provided us with some pattern templates (Table 3) and suggested to mine rules according to them.

Table 3. Pattern templates supplied by the expert to direct the mining

Pattern template	Description
Quick recovery	All symptoms under observation were positive at the initial time point, and disappeared at the subsequent time points (1 0 0)
Long recovery	All symptoms under observation were positive at the initial and second time points, and disappeared at the last time point (1 1 0)
No recovery	All symptoms under observation were positive at the three time points (1 1 1)
Unstable recovery	All symptoms under observation were positive at the initial time point, disappeared at the second time point but came back at the last time point (1 0 1)

Mining was then performed using the above pattern templates for each symptom. The domain expert found the results very useful and they were used to trigger discussion with medical colleagues. For example, the results for vomiting – where most patients belonged to the *quick recovery* group and there were no patients in the *unstable recovery* group – suggested that patients without vomiting at 18 months post-endoscopy have fully recovered from that symptom. This demonstrates that experts tend to find relationships between different rules when interpreting results. Hence, rules with zero or low support may be important when they are interpreted in context with other rules.

Task 2: Do certain symptoms tend to appear together?
After the analysis from Task 1, the domain expert wanted to know whether patients suffered from particular combinations of symptoms. She suggested focusing on one symptom at a time to avoid the return of the large number of rules as had been experienced previously. The symptom chosen was acid regurgitation (REFLUX). We looked at groups of three symptoms to compare and verify our results with the expert's statistical analysis. We used a constraint to extract patterns of three positive symptoms at the initial time point ([1, n=3&v3-eq=1]). The component v3-eq=1 extracts patterns with exactly 3 symptoms in which their values equal to 1. One of these symptoms was required to be REFLUX (av:REFLUX1=1). Overall, the constraint is:

        [1][1, n=3&v3-eq=1&av:REFLUX1=1]

Table 4. Correlated symptoms with acid regurgitation at the initial time point (REFLUX1):

ASC Results:
BURP1=1 HEART1=1 REFLUX1=1 (support=68)
EPI1=1 HEART1=1 REFLUX1=1 (support=48)
BURP1=1 BLOAT1=1 REFLUX1=1 (support=47)

The results (Table 4) show that the most frequent symptoms associated with acid regurgitation at the initial time point are severe burping (BURP1) and heartburn (HEART1). These results supported findings from previous studies that have used techniques such as principal components analysis to identify symptom clusters [15].

Task 3: How does the group of symptoms BURP1=1 HEART1=1 REFLUX1=1 *behave over time?*
After confirming that patients often suffered severe burping (BURP), heartburn (HEART) and acid regurgitation (REFLUX) together, the expert wanted to know whether these patients recovered from these symptoms over time. We used a constraint to track the behaviour of these 3 positive symptoms across time points 2 and 3 (results in Table 5). The constraint extracts three symptoms at each time point ([1,2,3] n=3), where the first time point must display a presence of severe burping (av:BURP1=1), heartburn (av:HEART1=1) and acid regurgitation (av:REFLUX1=1). The next two time points must also display the three symptoms, but there is no restriction on their presence or absence (time point 2: [2, a=BURP2&a=HEART2&a=REFLUX2&n=3]; time point 3: [3, a=BURP3&a=HEART3&a=REFLUX3&n=3]). The rules must only have symptoms at the first two time points as antecedents, and symptoms at the last time point as consequents ([rule, (s1 s2) => s3]):

```
[1,2,3]
[1,av:BURP1=1&av:HEART1=1&av:REFLUX1=1&n=3]
[2,a=BURP2&a=HEART2&a=REFLUX2&n=3]
[3,a=BURP3&a=HEART3&a=REFLUX3&n=3][rule, (s1 s2)=>s3]
```

Table 5. The presence of severe burping, heartburn and acid regurgitation across time
(BURP1=1 HEART1=1 REFLUX1=1) (support = 68)

ASC results
Quick recovery
BURP1=1 HEART1=1 REFLUX1=1
BURP2=0 HEART2=0 REFLUX2=0 ==> BURP3=0 HEART3=0 REFLUX3=0
support=16; conf=0.73; population coverage=24%
No recovery
BURP1=1 HEART1=1 REFLUX1=1
BURP2=1 HEART2=1 REFLUX2=1 ==> BURP3=1 HEART3=1 REFLUX3=1
support=7; conf=0.73; population coverage=10%
Long recovery
BURP1=1 HEART1=1 REFLUX1=1
BURP2=1 HEART2=1 REFLUX2=1 ==> BURP3=0 HEART3=0 REFLUX3=0
support=6; conf=0.3; population coverage=9%
Unstable recovery
BURP1=1 HEART1=1 REFLUX1=1
BURP2=0 HEART2=0 REFLUX2=0 ==> BURP3=1 HEART3=1 REFLUX3=1
support=3; conf=0.14; population coverage=4%

The results shown in Table 5 were useful to the expert as the technique tracked the behaviour of a symptom combination over time. These results triggered new research questions about the associations between symptom patterns and patients' diagnoses and demographics (examples include age, smoking habit and medication use).

Task 4: Do symptom groups change into other symptom groups over time?
Triggered by the results from the previous analyses, the knowledge engineers suggested following patients who experienced a *quick recovery* from heartburn and acid regurgitation to see whether they developed new symptoms. We formed a set of constraints to find whether, given a *quick recovery* pattern from heartburn (HEART) and acid regurgitation (REFLUX), there is any symptom that did not appear initially, but did appear at the second and/or last time point. A *quick recovery* from heartburn and reflux is represented by the following constraint:

```
[1,av:HEART1=1 & av:REFLUX1=1]
[2,av:HEART2=0 & av:REFLUX2=0]
[3,av:HEART3=0 & av:REFLUX3=0]
```

A *constraint on pattern content* is used in combination with the above constraint. A new symptom (a) that did not appear at the initial time point ([pattern, a-1=0]), but appeared at time points 2 and 3 (& a-2=1 & a-3=1]) is illustrated in the following constraint:

```
[pattern, a-1=0 & a-2=1 & a-3=1]
```

Table 6. Development of a new symptom when patients have a *quick recovery* from heartburn (HEART) and acid regurgitation (REFLUX) (support = 30)

ASC results
New symptom at the last two time points
[1,2,3]
[1,av:HEART1=1&av:REFLUX1=1&n=3]
[2,av:HEART2=0&av:REFLUX2=0&n=3]
[3,av:HEART3=0&av:REFLUX3=0&n=3]
[pattern,a-1=0&a-2=1&a-3=1][rule, (s1 s2)=>s3]
Number of patients=0; population coverage=0%
New symptom at the second time point only
[1,2,3]
[1,av:HEART1=1&av:REFLUX1=1&n=3]
[2,av:HEART2=0&av:REFLUX2=0&n=3]
[3,av:HEART3=0&av:REFLUX3=0&n=3]
[pattern,a-1=0&a-2=1&a-3=0][rule, (s1 s2)=>s3]
Number of patients=4; population coverage=13%
New symptom at the last time point only
[1,2,3]
[1,av:HEART1=1&av:REFLUX1=1&n=3]
[2,av:HEART2=0&av:REFLUX2=0&n=3]
[3,av:HEART3=0&av:REFLUX3=0&n=3]
[pattern,a-1=0&a-2=0&a-3=1][rule, (s1 s2)=>s3]
Number of patients=5; population coverage=17%

From this analysis (see Table 6 for results), we see that nearly one-fifth (17%) of patients who had a *quick recovery* from heartburn and acid regurgitation developed a new symptom 8–9 years later. We now want to know how many patients developed a new symptom at later stages, in general. We found that out of the 303 patients, 36% of patients developed a new symptom 18 months later, 38% of patients developed a new symptom 8–9 years later, and 23% of the patients had the same new symptom at the last two time points. This result suggests the hypothesis that patients who recover quickly from heartburn and acid regurgitation are less likely to develop a new symptom than those patients who develop a new symptom in general.

The ASC technique has overcome some of the limitations of traditional epidemiological statistical analysis by identifying complex patterns of symptoms over time. For example, to identify the outcomes of individual symptoms or symptom combinations over time is very difficult and to date such analyses have not been published.

4 Discussion

One of the major limitations of the existing framework is the expressiveness of the constraints that can be specified by the user. The current implementation only considers a disjunction of conjunct constraints for each subgroup. An extension of this would be to include a conjunction of disjunct constraints as well as nested constraints. Also, there are limitations on *constraint on pattern content* – users can only specify at most one attribute in a constraint that looks for particular attribute behaviour in a pattern.

In addition to addressing the above-mentioned limitations, there are various aspects of the system that can be improved and new functionality can be provided in the future. One of the aims is to provide an interactive framework that supports users to actively participate in the mining process. This approach is similar to [10], which suggests "an architecture that opens up the black-box, and supports constraint-based, human-centred exploratory mining of associations". Also, the provision of a GUI is necessary to provide a user-friendly interface that abstracts the current expression of constraints to allow users to interact with the mining process more intuitively. A further development of this would be a machine learning application that identifies commonly used constraints and uses them as the basis for forming new templates. We also plan to allow users to indicate the degree of importance of each attribute and use it to determine *weight*, i.e. priority in the mining process, as mentioned in Section 2.1. This feature can be easily incorporated into the framework.

5 Conclusion

We have developed a framework for constraint-based association rule mining for data that exhibits a subgroup nature and where sequential patterns are of interest. Through our empirical study with an epidemiologist in the medical domain, we have shown that this framework provides interesting and useful results. In addition, we have demonstrated the flexibility and power of using different combinations of constraints to extract complicated patterns. We are aware of a number of limitations in the existing framework. However, we envisage that if these limitations are overcome, the framework will provide a hybrid approach of knowledge acquisition and machine learning techniques to facilitate better understanding of data sets with subgroups. Although the framework is evaluated using data sets from the medical domain, it is developed with flexibility in mind to be applicable in other domains.

Acknowledgement. The authors would like to thank Enrico Coiera and the anonymous reviewers for their valuable comments.

References

1. Westbrook, J.I., Talley, N.J.: Empiric Clustering of Dyspepsia into Symptom Subgroups: a Population-Based Study. Scand. J. Gastroenterol. **37** (2002) 917–923
2. Agrawal, R., Imielinski, T., Swami, A.N.: Mining Association Rules Between Sets of Items in Large Databases. In: Buneman, P., Jajodia, S. (eds.): Proceedings of the 1993 ACM SIGMOD International Conference on Management of Data. Washington, DC (1993) 207–216
3. Agrawal, R., Mannila, H., Srikant, R., Toivonen, H., Verkamo, A.I.: Fast Discovery of Association Rules. In: Fayyad, U.M., Piatetsky-Shapiro G., Smyth P., Uthurusamy, R. (eds.): Advances in Knowledge Discovery and Data Mining. Menlo Park, CA (1996) 307–328
4. Agrawal, R., Srikant, R.: Fast Algorithms for Mining Association Rules. In: Bocca, J.B., Jarke, M., Zaniolo, C. (eds.), Proceedings of the Twentieth International Conference on Very Large Data Bases (VLDB'94). Santiago (1994) 487–499
5. Srikant, R., Agrawal, R.: Mining Sequential Patterns: Generalizations and Performance Improvements. In: Proceedings of the Fifth International Conference on Extending Database Technology (EDBT'96). Avignon (1996) 3–17
6. Bayardo, R.J., Agrawal, R., Gunopulos, D.: Constraint-Based Rule Mining in Large, Dense Databases. In: Proceedings of the Fifteenth International Conference on Data Engineering. (1997) 188–197
7. Ng, R., Lakshmanan, L.V.S., Han, J., Pang, A.: Exploratory Mining and Pruning Optimizations of Constrained Associations Rules. In: Proceedings of ACM SIGMOD Conf. on Management of Data (SIGMOD'98). Seattle (1998) 13–24
8. Fu, Y., Han, J.: Meta-Rule-Guided Mining of Association Rules in Relational Databases. In: Proceedings of the First International Workshop on Integration of Knowledge Discovery with Deductive and Object-Oriented Databases (KDOOD'95). Singapore (1995) 39–46
9. Klemettinen, M., Mannila, H., Ronkainen, P., Toivonen, H., Verkamo, A.I.: Finding Interesting Rules from Large Sets of Discovered Association Rules. In: Proceedings of the Third International Conference on Information and Knowledge Management (CIKM'94). Maryland (1994) 401–407
10. Han, J., Kamber, M.: Data Mining: Concepts and Techniques. Morgan Kaufmann (2000)
11. Srikant, R., Vu, R., Agrawal, R.: Mining Association Rules With Item Constraints. In: Proceedings of the Third International Conference on Knowledge Discovery and Data Mining (KDD'97). (1997) 67–73
12. Cohen, W.W.: Grammatically Biased Learning: Learning Logic Programs Using an Explicit Antecedent Description Language. Artif. Intell. **68** (1994) 303–366
13. Morik, K., Wrobel S., Kietz, J., Emde, W.: Knowledge Acquisition and Machine Learning. Academic Press (1994)

14. Agrawal, R., Srikant, R.: Mining Sequential Patterns. In: Yu, P.S., Chen, A.S.P. (eds.): Proceedings of the Eleventh International Conference on Data Engineering (ICDE'95). Taipei (1995) 3–14
15. Garofalakis, M., Rastogi, R., Shim, K.: Mining Sequential Patterns with Regular Expression Constraints. IEEE Trans. Knowl. Data Eng. **14** (2002) 530–552
16. Pei, J., Han, J., Wang, W.: Mining Sequential Patterns with Constraints in Large Databases. In: Proceedings of the Eleventh International Conference on Information and Knowledge Management (CIKM'02). McLean, VA (2002) 18-25
17. University of Waikato: Weka 3 – Data Mining with Open Source Machine Learning Software. http://www.cs.waikato.ac.nz/~ml/weka. (1999–2000)

Predicting Protein Structural Class from Closed Protein Sequences

N. Rattanakronkul, T. Wattarujeekrit, and K. Waiyamai

Knowledge Discovery from very Large database research group: KDL,
Computer Engineering Department, Kasetsart University, Thailand
{Rnatthawan, wtuangthong}@kdl.cpe.ku.ac.th,
fengknw@ku.ac.th

Abstract. ProsMine is a system for automatically predicting protein structural class from sequence, based on using a combination of data mining techniques. Contrary to our previous protein structural class prediction system, where only enzyme proteins can be predicted, ProsMine can predict the structural class for all of the proteins. We investigate the most effective way to represent protein sequences in our new prediction system. Based on the lattice theory, our idea is to discover the set of Closed Sequences from the protein sequence database and use those appropriate Closed Sequences as protein features. A sequence is said to be "closed" for a given protein sequence database if it is the maximal sub-sequence of all the protein sequences in that database. Efficient algorithms have been proposed for discovering closed sequences and selecting appropriate closed sequence for each protein structural class. Experimental results, using data extracted from SWISS-PROT and CATH databases, showed that ProsMine yielded better accuracy compared to our previous work even for the most specific level (Homologous-Superfamily Level) of the CATH protein structure hierarchy, which consists of 637 possible classes.

Keywords: Data Mining, Protein Structural Class, Bioinformatics, Closed Sequence, CARs

1 Introduction

Protein structure is one of essential protein information. Analysis of protein 3d-structure gives rises to understanding evolution, development and relationships of proteins. Further, knowledge of a protein's structure can also help in understanding its function, and has enormous implications in computer-aided drug design and protein engineering. However, the ever-increasing number of proteins has become too large to analyze their structure in deep detail. Therefore, protein structure classification methods are needed. Existing protein structural classification systems are manual [6], semi-automatic [15,16] and automatic [10]. Furthermore, input of those systems which are the 3d-structures, are still very complex. Consequently, those systems are not able to classify proteins with unknown 3D-structure. Automatic methods that classify proteins with unknown 3d-structure into structural classes are necessary for preliminary discovering of 3d-structure. In our previous work [18], we have developed a method of automatic protein structural class prediction combining data mining techniques. However, this method is capable of predicting the structural class

K.-Y. Whang, J. Jeon, K. Shim, J. Srivatava (Eds.): PAKDD 2003, LNAI 2637, pp. 136–147, 2003.

of only enzyme proteins. Our objective here is to propose a new method for automatic protein structural class prediction without the use of protein 3d-structures.

From empirical observation, we strongly deem that protein sequences and structures are closely related. In this paper, we develop a novel method of automatic protein structural class prediction from information available on the protein amino sequences. We call this method **Pro**tein **S**tructural Class **Min**ing (**ProsMine**). ProsMine can predict the structural class for all of the proteins, not limited to the enzyme proteins, eliminating the limitation of our previous work. The core idea of ProsMine is to investigate the most effective way to represent protein sequences. In order to represent protein sequences, it is possible to use the Hidden Markov Model (HMM), however this method generates a more complex topology than a profile [4,11,12]. In this paper, we propose a new approach for representing protein sequence using *Closed Sequences*. The concept of closure theory has been introduced in the lattice theory, which has been shown to provide a theoretical framework for a number of data mining techniques [13,19]. Our idea is to discover the set of Appropriate Closed Sequences (ACS) from the protein sequence database which share common structural class and use those ACS for representing proteins. Efficient algorithms are proposed for discovering all the closed sequences and ACS for each protein structural class. ProsMine constructs the protein structure prediction model by using a combination of association rule discovery and data classification techniques. Instead of using directly the protein sequences, we then use the set of ACS as protein features in the ProsMine prediction system. We ran our experiments, using data extracted from SWISS-PROT [2] and CATH databases, for each level of the protein structure hierarchy. Having compared to our previous prediction model and the general decision tree model, the results indicated that ProsMine has higher accuracy in every level of the protein structure hierarchy. ProsMine is much more accurate, with accuracy rate between 83-88%.

The rest of the paper is organized as follows. We give description of the CATH protein structural classes in section 2. In section 3, we explain the overview of ProsMine protein structural class prediction system. In section 4, we present some basic definitions. In section 5, we explain the algorithms for discovering all the closed sequences and appropriate closed sequences. In section 6, we explain how to combine association rule discovery and data classification for building the prediction model. The algorithm for building ProsMine prediction model is then given. Section 7 details experimental results and section 8 concludes the paper.

2   CATH Protein Structural Classes

The CATH database is a hierarchical domain classification of the PDB protein structures. There are four major levels in the CATH protein structural classes: Class-Level (C-Level), Architecture-level (A-Level), Topology-Level (T-level) and Homologous-Superfamily-Leve (H-Level). Table 1 shows the number of classes at each level in the protein structure hierarchy. Hierarchically, C-Level is the most general level and H-Level is the most specific level. A unique number called CATH number is assigned for each level. Fig. 1 gives an example of a CATH number 1.10.490.20 of a protein '1cpcA0'where the protein's structure is Alpha at the C-

Level, Orthogonal Bundle at the A-Level, Globin-Like at the T-level, and Phycocynins at the H-Level.

Table 1. The number of class labels at each level of protein structure hierarchy

Each level of protein structural class	Number of class labels
C-Level	4
A-Level	30
T-Level	379
H-Level	637

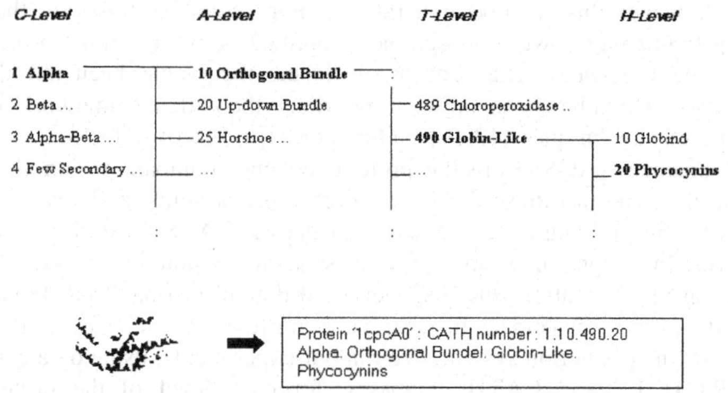

Fig. 1. An example of class labels at each level in the CATH hierarchy of the protein '1cpcA0'

3 System Overview

ProsMine protein structural class prediction is composed of six principle steps. Fig. 2 illustrates the flow chart of ProsMine. In the following, we explain each step in further details:

1. **Protein database integration**: Protein structural classes are retrieved from the CATH database and protein sequences are retrieved from the SWISS-PROT database. The two data sources are combined using the PDB index. Table 2 shows output example resulting from protein database integration. We use an estimated ¾ of the data source as training set and an estimated ¼ of data source as testing set. Equally, we select the training and testing data from each structure class.
2. **Closed sequences (CS) discovery**: All the closed sequences are discovered for each protein structural class. From this set of closed sequences, we derive the appropriate closed sequence (ACS) for each structural class. This appropriate closed sequence will represent the feature of each structural class.
3. **Table generation for model training and testing**: Two tables are generated from the previously discovered appropriate closed sequences (ACS). The first one is for model training and the second one is for model testing.
4. **Class association rule generation**: Class association rules that relate appropriate closed sequences with protein structural classes are generated from the training table.

5. **Prediction model construction**: Protein structural class prediction model is constructed from the previous generated class association rules.
6. **Model evaluation**: Protein structural class prediction model is evaluated against the previously generated testing table from step 3.

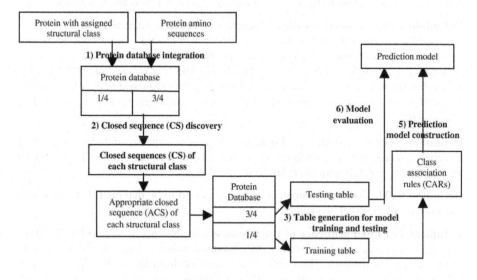

Fig. 2. ProsMine flow chart

Table 2. The database resulting from integrating CATH and SWISS-PROT

Protein-Index	Protein Sequences	CATH Number
102L	MNIFEMNIFEMLNIFENRIDEGLRLKIY...	1.10.530.40
117E	FENIFMPIEMLRIDEGLRLKIYFIDEGLR...	3.90.80.10
128L	MPIFEMNIFEMLLLIDEGLPKIYL IDLR...	1.10.530.40
...	...	...

4   Basic Definitions of Closed Sequences

In this section, we start by setting the groundwork of our approach by briefly resume some definitions, which are relevant in our context.

Definition 1 (Item). Let i be an item, which corresponds to an amino acid. There are 20 possible kinds of amino acids which are A, C, D, E, F, G, H, I, K, L, M, N, P, Q, R, S, T, V, W and Y. Hence, each item i represents one of the above amino acids.

Definition 2 (Sequence). Let $S = \{i_1 \rightarrow i_2 \rightarrow ... \rightarrow i_n\}$ be a protein sequence, where each element i_j corresponds to an item. The symbol '->' represents a happen-after relationship. If a protein sequence element i_j occurs at the position before the protein sequence element i_k, it will be denoted as $i_j \rightarrow i_k$.

Definition 3 (Sub-Sequence). A protein sequence $S_1 = \{i_1 \rightarrow i_2 \rightarrow ... \rightarrow i_n\}$ is a sub-sequence of another sequence $S_2 = \{j_1 \rightarrow j_2 \rightarrow ... \rightarrow j_n\}$, if there exist integers $k_1 < k_2 < ... < k_n$ such that $i_1 = j_{k1}, i_2 = j_{k2}, i_n = j_{kn}$. For example, the sequence $S_1 = \{C \rightarrow D \rightarrow E \rightarrow C\}$ is a sub-sequence of $S_2 = \{G \rightarrow C \rightarrow H \rightarrow D \rightarrow E \rightarrow C \rightarrow F\}$ because there exist $k_1 = 2 < k_2 = 4 < k_3 = 5 < k_4 = 6$ such that $i_1 = j_2 = C, i_2 = j_4 = D, i_3 = j_5 = E$ and $i_4 = j_6 = C$.

Definition 4 (Sub-sequence with max-gap). Let a protein sequences $S_1 = \{i_1 \rightarrow i_2 \rightarrow ... \rightarrow i_n\}$ be a sub-sequence of another sequence $S_2 = \{j_1 \rightarrow j_2 \rightarrow ... \rightarrow j_n\}$. S1 is sub-sequence **with max-gap** of S_2, if there exists also $k_2 - k_1 \leq max\text{-}gap$, ... , $k_n - k_{n-1} \leq max\text{-}gap$. For example, given $S_1 = \{C \rightarrow D \rightarrow E \rightarrow F\}$ and $S_2 = \{G \rightarrow C \rightarrow H \rightarrow D \rightarrow E \rightarrow F\}$. S_1 is sub-sequence of S_2 with $max\text{-}gap = 2$ since S_1 is sub-sequence of S_2 (cf. definition 3) and there exists $k_2 - k_1 \leq max\text{-}gap$ $(4\text{-}2 \leq 2)$, $k_3 - k_2 \leq max\text{-}gap$ $(5\text{-}4 \leq 2)$, and $k_4 - k_3 \leq max\text{-}gap$ $(6\text{-}5 \leq 2)$.

Definition 5 (Protein sequence database). A protein sequence database is a set of protein sequences of where each record is identified by a protein-index. Table 3 gives an example of a protein sequence database.

Definition 6 (Maximal sequence). A sequence is said to be maximal if it is not sub-sequence of all the other sequences.

Definition 7 (Closed sequence of the database). A sequence is said to be closed for a protein sequence database if it is the maximal sub-sequence of all the protein sequences in that database. This definition is derived from the lattice theory [13,19, 20]. The set of all closed sequences derived from the protein sequence database in table 3 is given in Table 4.

Table 3. Protein sequence database example

Table 4. Set of closed sequences derived from the database in Table 3

Protein Index	Protein-Sequence	Closed Sequence
1	C -> G -> A -> C -> D -> E -> F	G -> A -> F
2	D -> G -> C -> E -> A -> G -> F -> A	D -> E -> F
3	C -> G -> A -> D -> E -> F	C -> G
4	G -> C -> D -> A -> E -> D -> F -> G	
5	A -> D -> C -> E -> G -> A -> F	

5   Discovering the Appropriate Closed Sequences for Representing Protein Characteristics

Our method for generating all the appropriate closed sequences from a protein sequence database consists of the three following phases (see also Fig. 3):

1. In the pre-processing phase, a class at the H-level is assigned for each protein sequence. We choose the H-level because it is the most specific protein structure consisting of 637 distinct classes. Once a class at the H-level has been predicted, all the other more general classes at the C-Level, A-Level and T-Level can be automatically derived. At the end of this phase, we obtain a set of protein sequences for each class at the H-level.

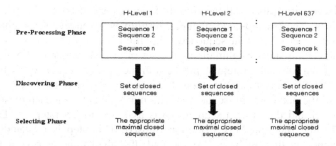

Fig. 3. Method of discovering the appropriate closed sequence (ACS)

Algorithm 1. Discovering phase: discovering all the closed sequences in each structural class.
Input: Protein sequences in each structural class in H-Level ($d_{H\text{-}Level}$).
Output: A set of closed sequences for the class (*CS*).
1) $CS \leftarrow \varnothing$
2)   For all protein sequences $p \in d_{H\text{-}Level}$ do begin
3)   if($CS = \varnothing$) then
4) $CS = p$
5)   else
6) $CS = CS \cup Max\_Common\_Subseq(CS,$ p, *max-gap*); {for finding common sub-sequence with max-gap between *CS* and p}
7)   end if
8)   end for
9)   Return *CS*;

Fig. 4. Algorithm for discovering the set of closed sequences

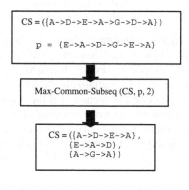

Fig. 5. Execution example of the *Max_Common_Subseq* function

Algorithm 2. Selecting phase: selecting the appropriate closed sequences from the closed sequences in each structural class.
Input: A set of closed sequences in each structural class (*CS*).
Output: The appropriate closed sequences of the class (*ACS*).
1)   for each $cs \in CS$ do begin
2)   cs.error = 0
3)   for each $o\text{-}seq \in$ set of all protein sequences in other structural classes do begin
4)   if ($Sub\text{-}Seq(cs, o\text{-}seq, max\text{-}gap)$) then
5)   cs.error $\leftarrow$ cs.error + 1 ;
6)   end if
7)   end for
8)   end for
9)   ACS $\leftarrow \varnothing$
10)   for each $cs \in CS$ do begin
11)   if (($ACS = \varnothing$) or (ACS.error < CS.error)) then begin
12) ACS.closure = CS.closure
13) ACS.error = CS.error
14)   end if
15)   end for
16)   Return ACS.closure

Fig. 6. Algorithm for selecting the ACS for each structural class

2. In the discovering phase, all the closed sequences are discovered for each protein structural class. The pseudo-code of the algorithm is given in Fig. 4. Input of this algorithm is the set of protein sequences classified into different classes at the H-level. The algorithm starts by initializing the set of closed sequences CS to an empty set (step 1). Then for each protein sequence p in the protein sequence database D, CS is updated. If CS is an empty set, then the sequence p is assigned to it (step 3-4). Otherwise, all the maximal sub-sequences with max-gap between each existing closed sequence in CS and the protein sequence p is assigned to CS (step 6). This task is performed by the function Max_Common_Subseq. This step is repeated for every protein sequence in the database. At the end, the algorithm returns all the closed sequences for each protein structural class. Fig. 5 gives the execution of function Max_Common_Subseq. Suppose that the max-gap value is set to 2. At the end, the function returns 3 closed sequences {A->D->E->A},{E->A->D} and {A->G->A} which are stored in the CS set, and will be used to search for all the other closed sequences against other protein sequences in the database.

3. In the selecting phase, we use the algorithm given in Fig. 6 to select the appropriate closed sequence (ACS) for each protein structural class. The algorithm works as follows. For each closed sequence *cs* in the set of closed sequences of a class *CS*, we start by initializing its error value to 0 (step 2). Then we increase error value of each closed sequence *cs* if the closed sequence *cs* is sub-sequence with max-gap of each protein sequence *o-seq* of all protein sequences in other structural classes (step 3-6). Final, (step 8-14) we obtain the appropriate closed sequence *ACS* by choosing the closed sequence *cs,* which holds the least error value.

6   Combining Association Rule Discovery and Data Classification for Building a Prediction Model

From our previous work, we noticed that a prediction model built from class association rules (CARs) method [14] yields better accuracy than a prediction model built from a traditional decision tree (C 4.5). Therefore, in this work, we continue using CARs method for constructing the prediction model. Two following steps are needed for the prediction model construction.

6.1   Generating All the Class Association Rules (CARs)

When we obtain the appropriate closed sequence (ACS) of proteins for each protein structural class in H-Level   (637 ACS), we use all of them as features to predict protein structural class. Our method builds the prediction model from the class association rules. In order to discover relations between ACS and protein structural classes in the H-Level, we identify each ACS by numbering it varying from ACS1 to ACS637. Class association rules generation can be performed as follows:

First, we generate the training table, which constitutes the input for the association rules discovery step. The table contains information of structural class represented by structural class number, and information of being sub-sequence with max-gap of ACS represented by value "1". Fig. 7 illustrates the training table and an example of rules output from the association rules discovery step.

Then, we discover all the Class Association Rules (CARs) to build our prediction model. CARs are association rules that satisfy minimum support and minimum confidence. Each CAR is of the form: *{ACS1,...,ACSk} -> Structural Class Number = Hid (Support, Confidence)* where *{ACS1, ..., ACSk}* is a set of ACS, *Hid* corresponds to the class label. We call each ACS as *item*. All the items that satisfy the minimum support are called *frequent items*. We call *{ACS1,...,ACSk} -> Structural Class Number = Hid a ruleitem*. All the ruleitems that satisfy the minimum support are called *frequent ruleitems*. All the frequent ruleitems are then used to generate all the class association rules (CARs).

The key operation of the class association rule generation is to find all the frequent ruleitems. In the following, we explain our method to generate all the frequent ruleitems at the *H-Level*. First, we classify the training protein data into different classes. For each structural class, we generate all the frequent ruleitems within that class. After this step, all the frequent ruleitems for each class are generated. Then, we use the Apriori algorithm (Agrawal et al. 1996) to generate all the frequent ruleitems. Fig. 8 gives different steps of generating all the frequent ruleitems with the minimum support equal to 1 (number of proteins) and the minimum confidence equal to 50% of the class 2.40.15.10 in H-Level. The used notations are similar to the ones used in Apriori algorithm where L is the set of frequent itemsets. We obtain many class association rules that satisfy minimum confidence (50%) and minimum support (1 protein) such as: *{ACS1,ACS637} -> Structural Class Number = 2.40.15.10*. The meaning of this rule is as follow: If ACS1 and ACS637 are sub-sequences with max-gap of protein sequence S and then the structural class of S is 2.40.15.10.

Protein Index	ACS1	ACS2	...	ACS637	Structural Class Number
1	"1"	"0"	...	"1"	1.10.8.10
2	"1"	"0"	...	"1"	1.10.8.40
3	"0"	"1"	...	"0"	2.40.15.10
4	"0"	"0"	...	"1"	3.40.80.10
:	:	:	:	:	:
3.141	"0"	"1"	...	"0"	2.40.15.10

Association rules discovery

{ACS1,ACS637} -> Structural Class Number is 1.10.8.10

Fig. 7. Training table and an example of class association rule

L₁	
Itemsets	Support Counts
ACS1	100
ACS2	62
:	:
ACS637	256

L₂	
Itemsets	Support Counts
ACS1, ACS2	59
ACS2, ACS3	35
:	:
ACS636, ACS637	250

L₃	
Itemsets	Support Counts
ACS1, ACS2, ACS3	30
ACS2, ACS3, ACS4	28
:	:
ACS635, ACS636, ACS637	240

Fig. 8. Different steps in generating class association rules at the H-Level for the class 2.40.15.10

6.2   Constructing a Prediction Model Based on the Generated CARs

In this sub-section, we present the step of creating our prediction model using all the class association rules (CARs) generated from previous step. To produce the best prediction model out of the whole set of rules would involve evaluating all the possible subsets of CARs on the training data and selecting the subsets with the right rule sequence that gives the least number of errors. The algorithm for model construction is given in Fig. 9. First, we arrange CARS from the previous phase in descendent order of the precedence value in sequence from the highest precedence value (step 1). The precedence value is considered from confidence value and support value by using the following logic:

- If rule1's confidence value is more than rule2's, rule1 has a higher precedence value.
- If rule1's confidence value is equal to rule2's and rules1's support of items is higher than rule2's, rule1 has a higher precedence value.
- If rules' confidence and support values are equal, the prior rule has a higher precedence value.

Algorithm 3. **Prediction model construction**: Building the prediction model from all the class association rules.
Input: A set of the generated class association rules.
Output: ProsMine prediction model.

1) $CARs$ = sort($CARs$)
2)   Forall case $d \in D$ do
3) d.update = 0
4) $n = 0$
5)   For each rule $r \in R$ in sequence do begin
6) $n \leftarrow n+1$
7) $Model_n = Model_{n-1}$
8)   For each case $d \in D$ do begin
9)   if (d satisfies the conditions of r) and (d.update = 0) then begin
10) d.update = 1
11)   insert r at the end of model $Model_n$
12)   end if
13)   end for
14) $Model_n$.default = Select a default class of $Model_n$
15) $Model_n$.errors = Compute the total errors of $Model_n$
16)   end for
Return $Model = _{j=1 ->j=n}Model_j$ {that has the lowest total number of errors}

Fig. 9. Algorithm for constructing ProsMine prediction model based on the generated CARs

After CARs have been sorted, we create some sets of CARs to represent as prediction models. The sorting process is necessary to ensure that we would choose the highest precedence rules for each prediction model. Each prediction model is a set of CARs with the following format: <$rule_1$, $rule_2$..., $rule_n$, default_class>, where a default class is selected from the remaining classes where $rule_1$, $rule_2$..., $rule_n$ cannot apply. At the end of the algorithm, we select the prediction model that has the lowest total number of errors (last step). Each prediction model is constructed as follows.

For each rule R, we went through the database D to find those cases covered by rule R (step 5). If rule R could correctly at least one case, it would be a potential rule in our prediction model (step 11). Those cases it covered were then removed from database D (step 10). A default class is also selected, which means the fact that if we stop selecting more rules for our prediction model this class would be the default class of our prediction model (step 14). When there is no rule or no training case left, the rule selection process is completed.

7 Experimental Results

In this section, we study the performance of our algorithms. We implemented the algorithms in C language. The platform we used was a PC-Pentium III with CPU clock rate of 733 MHz, 512 MB of main memory. We ran our experiments using 3,896 protein sequences extracted from two different database sources, which are integrated using PDB entry names as keys. In order to perform the prediction task, we separated the proteins into two sets: training data (3,632 sequences) and testing data (755 sequences). For training, the 3,896 protein sequences were classified into different structural classes at the Homologous Super-family level.

Table 5. Average CPU time of discovering all the closed sequences and all the appropriate closed sequences for various values of max-gap

Max–gap value	Execution time for discovering all the CS (in seconds)	Execution time for selecting all the ACS (in seconds)
1	31	697
2	79	699
3	220	695

Table 6. ProsMine prediction accuracy at the four levels of the protein structural hierarchy using different values of max-gap: 1, 2 and 3

Each level of protein structure hierarchy	% Accuracy		
	max-gap = 1	max-gap = 2	max-gap = 3
C-Level	60.9	65.6	88.21
A-Level	46.6	63.6	85.43
T-Level	42.8	53.4	84.37
H-Level	38.9	53.3	83.97

Table 5 shows the average CPU time (including disk access) for, discovering all the closed sequences (CS) and all the appropriate closed sequences (ACS) given in seconds respectively. The CPU time is given using different Max-Gap values. In order to study the behaviors of the algorithms in various max-gap values, table 6 compares the accuracy of our protein structure prediction model using different values of max-gap: 1, 2 and 3. At the Homologous Super-family level, we have found that increasing the max-gap value also increases average execution time for discovering all CS. Furthermore, this increasing time is non-linear with respect to the max-gap value. The execution time of discovering all the CS exponentially increases when the value of max-gap is larger. Large value of max-gap will generate more interesting closed sequences (i.e. longer specific closed sequences), and the prediction model constructed from those sequences will be more accurate, however this is very time-consuming. On the contrary to the discovering time of CS, the discovering time of ACS is almost constant with respect to the different max-gap values.

Table 7 compares the accuracy of our protein structure prediction model with our previous protein structure prediction model and the traditional decision tree model (C4.5). We choose the max-gap value equal to 3 since it produces the most accuracy for our model. We ran our experiments for each level of the protein structure hierarchy. Having compared to the two other models, the results indicated that our new prediction model has higher accuracy in every level of the protein structural hierarchy. Especially, if the number of classes to be predicted is large (at the H-level, where the number of classes is 637 classes), the new prediction model was much more accurate.

Table 7. Comparison of ProsMine prediction accuracy at the four levels of the protein structural hierarchy with our previous prediction model and the general decision tree model

Each level of protein structure hierarchy	% Accuracy		
	The general decision tree model (C 4.5)	Our previous model	Our model (max-gap = 3)
C-Level	87.8	87.8	88.21
A-Level	75.2	76.4	85.43
T-Level	60.8	65.8	84.37
H-Level	54.6	63.6	83.97

8 Concluding Remarks

In this paper, we propose a new system for automatic protein structural class prediction combining the lattice theory and two data mining techniques. Experimental results, showed that ProsMine yielded better accuracy than our previous prediction model. Obtaining promising results, we conclude that closed sequence is an appropriate method for representing protein sequences that have common characteristics. In this study, we combine the lattice theory and two data mining techniques, which are association rules discovery and data classification. In order to have appropriate rule length and acceptable execution time, we set up the minimum confidence value to 50% and the minimum support value to 1, during our experiments in the association discovery step. It would be interesting to investigate appropriate values of minimum confidence and minimum support that allow the most accurate protein structural class prediction model. The same observation can be also applied to the max-gap value. During our experiments, we have varied the max-gap values: 1,2, and 3. Other values of max-gap should be investigated in order to obtain a more accurate prediction model. We have shown that our method is robust and the results promising.

References

1. Agrawal R., Mannila H., Srikant R., and Verkamo A.I. (1996) Fast Discovery of Association Rules, In Fayyad U.M., Piatetsky-Shapiro G., Smyth P., and Uthurusamy R., editors, Advances in Knowledge Discovery and Data Mining, *AASAi/MIT Press*, p. 307-328.
2. Bairoch A. and R. Apweiler. (1996) The SWISS-PROT protein sequence data bank and its new supplement TREMBL, *Nucleic Acids Research, Vol. 24.*
3. Berman H.M., Westbrook J., Feng Z., Gilliland G., Bhat T.N., Weissig H., Shindyalov I.N., Bourne P.E. (1996) The Protein Data Bank, *Nucleic Acids Research*, p. 37-57.
4. Birney E. (2001) Hidden Markov models in biological sequence analysis, *IBM J. RES. & DEV.* Vol. 45, No 3/4.
5. Bucher P., Bairoch A. (1994) A generalized profile syntax for biomolecular sequences motifs and its function in automatic sequence interpretation, (In) *ISMB-94, Proceedings 2nd International Conference on Intelligent Systems for Molecular Biology.* Altman R., Brutlag D., Karp P., Lathrop R., Searls D., Eds., *AAAIPress*, p. 53-61, Menlo Park.
6. Conte L.L., Ailey B., Hubbard T.J.P., Brenner S.E., Murzin A.G. and Chothia C. (2000) SCOP: a Structural Classification of Proteins database, *Nucleic Acids Research.*, Vol. 28, No. 1, p. 257-259.
7. Falquet L., Pagni M., Bucher P., Hulo N., Sigrist C.J, Hofmann K., Bairoch A. (2002) The PROSITE database, its status in 2002, *Nucleic Acids Research*, p. 235-238.
8. Fayyad U.M., Piatetsky-Shapiro G., Smyth P., and Uthurusamy R. (1996) Advances in Knowledge Discovery and Data Mining, *AAAI/MIT Press*, p. 37-57.
9. Han J., Kamber M. (2000) Data Mining Concepts and Techniques, *Morgan Kaufmann.*
10. Holm L., and Sander C. (1996) Fold Classification based on Structure-Structure alignment of Proteins (FSSP), *Mapping the protein universe,. Science 273*, p. 595-602.
11. Joshua S. (2000) Identification of Protein Homology By Hidden Markov Models. *MB&B*
12. Kevin K., Christian B., Melissa C., Mark D., Leslie G., Richard H. (1999) Predicting Protein Structure using only Sequence Information, *ProteinsL Struncture, Function, and Genetics copyright 1999.*
13. Lakhal L. et al. (1999) Efficient Mining of Association Rules using Closed Itemset Latticed, *Information systems*, Vol. 24, No. 1, p. 25-46.
14. Liu B., Hsu W., and Ma Y. (1998) Integrating Classification and Association Rule Mining, *In Proc. Int. Conf. KDD'98.*, p. 80-86.
15. Orengo C.A., Michie A.D., Jones S., Jones D.T., Swindells M.B., and Thornton J.M. (1997) CATH- A Hierarchic Classification of Protein Domain Structures, *Structure.*, Vol. 5, No. 8, p. 1093-1108.
16. Pearl F.M.G, Lee D., Bray J.E, Sillitoe I., Todd A.E., Harrison A.P., Thornton J.M. and Orengo C.A. (2000) Assigning genomic sequences to CATH, *Nucleic Acids Research.*, Vol 28, No 1, p. 277-282.
17. Quinlan L.R. (1992) C4.5: program for machine learning, *Morgan Kaufmann.*
18. Rattanakronkul N., Yuvaniyama J., Waiyamai K. (2002) Combining Association Rule Discovery and Data Classification for Protein Structural Class Prediction, *The International Conference on Bioinformatics*, North-South Network, Bangkok, Thailand.
19. Waiyamai K., and Lakhal L. (2000) Knowledge Discovery from very Large Databases Using Frequent Concept Lattices, *Springer-Verlag Lecture Notes in Artificial Intelligence*, Number 1810, p. 437-445.
20. Wille, R. (1992) Concept lattices and conceptual knowledge systems. *In Computers Math. Applications.* Vol 23 , p. 493-515.

Learning Rules to Extract Protein Interactions from Biomedical Text

Tu Minh Phuong, Doheon Lee [*], and Kwang Hyung Lee

Department of BioSystems, KAIST
373-1 Guseong-dong Yuseong-gu Daejeon 305-701, KOREA
{phuong, dhlee, khlee}@bioif.kaist.ac.kr

Abstract. We present a method for automatic extraction of protein interactions from scientific abstracts by combing machine learning and knowledge-based strategies. This method uses sample sentences, which are parsed by a link grammar parser, to learn extraction rules automatically. By incorporating heuristic rules based on morphological clues and domain specific knowledge, this method can remove the interactions that are not between proteins and improve the performance of extraction process. We present experimental results for a test set of MEDLINE abstracts. The results are encouraging and demonstrate the feasibility of our method to perform accurate extraction without need of manual rule building.

1 Introduction

The study of protein interactions is one of the most important issues in recent molecular biology research, especially for system biology. Mining biological literature for extracting protein interactions is essential for transforming discoveries reported in the literature into a form useful for further computational analysis [5]. However, most of the interaction data are stored in free text format with irrelevant and confusing text bodies, which makes automatic querying for specific information inefficient. At the same time, manual identification and collection of protein interaction data for storing in databases is time consuming and laborious. This situation has made automatic extraction of protein interaction an attractive application for information extraction researches.

The methods proposed for protein interaction extraction differ from each other by their degrees of using natural language techniques, reliance on pre-specified protein names, statistical or knowledge-based strategies, and ability to identify interaction type.

Marcotte et. al. [5] used Bayesian approach to score the probability that an abstract discusses the topic of interest. The score is computed based on the frequencies of discriminating words found in the abstracts. However, this method allows only ranking and selecting abstracts. Actual extraction must be performed manually.

---

[*] Corresponding author

K.-Y. Whang, J. Jeon, K. Shim, J. Srivatava (Eds.): PAKDD 2003, LNAI 2637, pp. 148–158, 2003.

A number of researchers have taken simple pattern-matching approach to extract interactions. Blaschke et. al. [1] used the simple pattern "protein A...interaction verb...protein B" to detect and classify protein interactions in abstracts related to yeast. They simplified the task assuming that protein names are pre-specified by users. Ono and colleagues [7] presented a similar approach. By using part-of-speech tags and simple rules, this approach transforms complex and compound sentences into simple ones before applying pattern matching. However, this approach cannot automatically detect new protein names. Instead, it uses lexical lookup and requires protein name dictionaries.

Several publications proposed using natural language processing techniques for the extraction task. Rindflesch et. al. [10] described EDGAR, a system that used a stochastic part-of-speech tagger, a syntactic parser, and extensive knowledge source (Unified Medical Language System) to identify names and binding relationships. Other linguistic motivated approaches use different grammar formalisms with shallow or full syntactic parsing. Yakushiji et. al. [15] applied a full syntactic parser to produce so called *argument structure* of whole sentences before mapping the argument structure to extraction slots. Park and his colleagues [8] used a syntactic parser with combinatory categorical grammar to analyze sentences. However, without special protein name verification this approach gives recall and precision rates of only 48 and 80 respectively. Thomas et. al. [14] adapted general-purpose information extraction system Highlight to protein interaction extraction by tuning the vocabulary and the part-of-speech tagger. This system groups sequences of words into phrases by cascaded finite state machines. The authors added some kind of protein names verification to the system and reported the interesting fact that more restrict name verification rules lead to higher recall and precision.

All of the mentioned approaches and systems use hand-crafted extraction rules. Tuning rules manually requires significant time and domain-specific knowledge. To overcome this knowledge-engineering bottleneck, a number of automatic rule learning systems have been made. One of such systems, RAPIER [2], employs inductive logic programming techniques to learn so-called extraction fillers. Another system, WHISK [12], learns extraction rules by gradually adding lexical and/or semantic terms to an empty rule. However, these systems are designed for extracting only independent pieces of data from general text. It is difficult to adapt these systems to extracting relationships between terms – interactions between proteins in our case.

In this paper, we show how extraction rules can be automatically learned from training samples. We use the link grammar parser to parse sentences. The sentences with links are tagged by the user and serve input for the learning procedure. The learned rules are then used to detect potential protein interactions. We describe the heuristic rules that verify detected nouns and noun phrases if they are protein names or not. The results and evaluation of an experiment on a test set of Medline abstracts related to *Saccharomyces cerevisiae* are presented.

The remainder of the paper is organized as follows. A brief description of a link grammar is given in subsection 1.1. Section 2 describes extraction process and heuristic rules for protein name verification. The rule learning algorithm is described in section 3. Experimental results are presented and discussed in section 4. Section 5 draws our conclusion.

1.1 Link Grammar

Link grammar is a dependency grammatical system introduced by D. Sleator, D. Temperley, and J. Lafferty [11]. Rather than examine the constituents or categories a word belongs to, the link grammar is based on a model that words within a text form links with one another. A sample parse result by a link grammar parser for sentence "The boy ran away from school" is shown in Fig.1.

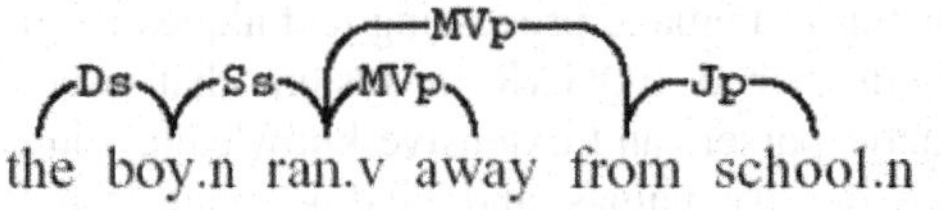

D: connects determiner to nouns
S: connects subject-nouns to verbs
MV: verbs (adjectives) to modifying phrases
J: connects prepositions to their objects

Fig. 1. A sample parse with links

The arcs between words are called "links" with the labels showing the link types. For example, the verb "ran" is connected to "from school" identifying a prepositional phrase by the link "MV". The part-of-speech tags (nouns, verbs and so on) are also added to some words as suffixes.

The uppercase letters of the link labels indicate the primary types of links (there are 107 primary link types for the link parser version 4.1), and the lowercase letters detail the relationships of words. The meanings of several link labels uppercase letters are given in the right part of Fig.1.

Each word in the link grammar must satisfy the linking requirements specifying which types of links it can attach and how they are attached. These linking requirements are stored in a dictionary. It is easy to express the grammar of new words or words with irregular usage. We can just define the grammar for these words and add them to the dictionary. It is how we deal with domain-specific terms and jargon in our work.

The reason why we have decided to adopt the link grammar in our work is that the grammar provides the simple and effective way to express relationships between terms in a sentence. This feature is very important in detecting protein interactions. We can follow the related links to find participants of an interaction without concerning the rest of the sentence.

2 Extraction of Protein Interactions

In this section, we describe how to extract protein interactions using extraction rules and heuristic rules for protein name verification. First, sentences are parsed by the link parser; words are stemmed by the Porter stemming algorithm [9] to solve the problem of inflection. Next, our method looks for terms involved in interactions by following the links described in the extraction rules. Finally, the heuristic rules are applied to verify if the terms detected from the previous step are protein names or not. We describe the details for each step below.

2.1   Rule Representation

In our method, rules are subsets of links that connect terms describing an interaction within a sentence. Our method requires three terms for an interaction: two protein names and a keyword that indicates the type of relationship betweens the proteins. We use the following keywords: "interact", "bind", "associate", and "complex" as well as their inflections.

A rule always begins with a keyword and contains all intermediate links that connect the keyword to the first and the second proteins. Consider a sample sentence, which has the parse result shown in Fig.2a. In this example, the keyword is "binds", the pairs of proteins are "Ash1p" - "HO" and "Ash1p" - "Swi5p". The links that connect "binds" with the pairs of proteins names are shown in Fig. 2b. We can think of the parsed sentence as a graph where words are vertices and links are edges. Then the connection between the keyword and a protein name is the shortest path between these two vertices. This connection is recorded as a rule.

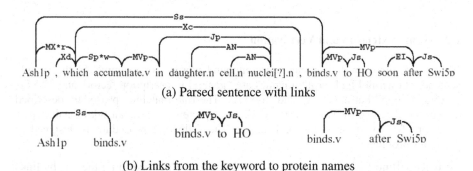

(a) Parsed sentence with links

(b) Links from the keyword to protein names

Fig. 2. A sample of connections between a keyword and protein names

In order to store and process rules, we have adopted textual presentation for rules. A rule for the above example is shown in Fig. 3.

```
bind S- @NAME1
bind MV+ to J+ @NAME2
=>bind(@NAME1,@NAME2)
```

Fig. 3. An example of rule representation

Each rule consists of three lines. The first and second lines specify the links to follow for detecting the first and second protein names respectively. The third line is a template for interaction output. For example, the second line of the sample rule in Fig.3 looks for keyword "bind". If the keyword is found, the rule looks further for a word "to" which is on the right of "bind" and which is connected to "bind" by a link "MV". The next step is looking rightward for a word connected to "to" by "J". This word will be extended and considered as a candidate for protein name. The procedure for name extension will be described in the next subsection. Our rules contain only the part of link labels that consists of uppercase letters. The signs "+" and "-" next to the link labels represent search directions for right and left respectively. We call intermediate words in link paths between keywords and protein names (for example word "to" in the above rule) nodes. A rule line can have any number of nodes.

Given a sentence, a rule is said to be satisfied if we can find all links and words specified in the rule within this sentence. A rule can be applied many times to each sentence to find all interactions satisfying the rule. For instance, applying the above rule to the sentence depicted in Fig.2 will output two candidates of interactions =>*bind(Ash1p,HO)* and =>*bind(Ash1p,Swi5p)*.

Our rules allow a form of disjunction as well as use of wildcard "*". Rule lines can look like the following:

```
          bind MV+ to|after J+ @NAME2
or        bind MV+ * J+ @NAME2
```

the first line looks for either "to" or "after", connected by "MV" to "bind", whereas the second lines allows any word on the right of "bind" which are connected to "bind" with "MV" link.

2.2   Name Extension and Verification

In practice, many protein names are compound words. For example, in the sample sentence shown in Fig.4, protein names are "general transcription factor" and "TATA binding protein" but not "IIA" and "TATA". The rule matching procedure described above can detect only two words "IIA" and "protein". Thus, an additional step is necessary for capturing compound names. The following procedure is designed to solve this problem.

(a) If the leftmost word of a name is connected to the next word on the left by links "G" or "GN" then extend the name to the left one word. From the above example, we have: *IIA=>factor IIA*
(b) If the leftmost word of a name is connected to word or words on the left by links "A" or "AN" then extend the name to the left by adding all the connected words: *factor IIA => general transcription factor IIA*.
(c) If the rightmost word of a name is connected to a word on the right with link "G" or "GN" then extend the name by adding the connected word to the right.

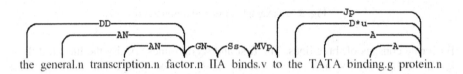
the general.n transcription.n factor.n IIA binds.v to the TATA binding.g protein.n

Fig. 4. An example of protein names which are compound words

Having an interaction with names detected as described above, we have to verify that it is a desired interaction between two proteins rather than an interaction between two non-proteins. The most obvious approach is to use a protein name dictionary for checking whether the detected names are protein names or not. Unfortunately, the number of new proteins is growing rapidly, which challenges maintaining the dictionary up to date. In this work, we verify detected names by applying several heuristic rules (part of rules are adopted from [4], [13] and [14]) that give some score. Examples of rules are:

- If a name contains word or words with uppercase letters, digits, some special symbols, the Greek letters then give the name score 1.0.
- If a name ends with one of the following words (*molecule, gene, bacteria, base*) then reject the name.
- If a name ends with a word that has suffixes "ole, ane, ate, ide, ine, ite, ol, ose" then reject the name.
- If a name is compound words containing functional descriptor (*adhesion, channel, filament, junction*), activity descriptors (*regulated, releasing, promoting, stimulating*), other keywords (*receptor, factor, protein*) then give the name score 1.0.
- If a name is compound words without special words described above then give the name score 0.5.
- If a name is single word with suffix "-in" then give score 0.5.

If a rejecting rule is triggered, the name is removed. If several scoring rules are triggered, the highest score is recorded. Names with the scores higher than a predefined threshold will be accepted as protein names. By adjusting the threshold, we can emphasize the importance of precision (with higher threshold) or the importance of recall (with lower threshold).

3   Learning Extraction Rules

In this section we describe how to learn extraction rules automatically. Our algorithm requires a set of hand-tagged sentences. During the tagging process the user must explicitly point out the first protein name, the second protein name, and the keyword of each interaction. There may be more than one protein pair linked by one keyword as well as more than one keyword within a sentence. For example, the sentence presented in Fig. 4 would be tagged as

```
The   /n1   general   transcription   factor   IIA/n0   /v1
binds/v0 to the /n2 TATA binding protein/n0
```

Our algorithm begins with creating a rule for each interaction being tagged in the training set. To create the first line of a rule, the algorithm looks in the link-parsed sentence for the shortest path from the keyword to any word of the first name. The rule line for the second name is created in the similar way. Applying this processing to all the examples of a training set, we get a set of rules, each per a tagged interaction. Some of rules can have duplicates. These duplicates are removed by a pruning procedure. The rules retained after pruning are referred to in this paper as specific rules.

There are two general design approaches for rule learning systems: compression (or bottom-up) and covering (top-down). Compression-based systems begin with a set of highly specific rules, typically one for each example, and gradually compress rule sets by constructing more general rules, which subsume more specific ones. This approach has been chosen in designing our algorithm. The reason of our choice is our preference of overly specific rules to overly general ones. In information extraction, there is a trade-off between high precision and high recall. For the potential application of our algorithm – populating protein interactions into databases – precision must be emphasized. The bottom-up approach tends to learn more specific rules, which also are more precise.

The learning algorithm is shown in Fig. 5. Starting with a set of specific rules, the algorithm generalizes rules by repeatedly calling two procedures GENERALIZE_TERM and GENERALIZE_FRAGMENT. Each of the procedures produces a set of more general rules CandidateSet. The rules used to build CandiadetSet are stored in BaseSet. From the CandidateSet returned by GENERALIZE_TERM, the algorithm looks for the best rule r_0. This rule will subsume the rules in BaseSet if its evaluation is not worse than the overall evaluation of BaseSet. In the case of calling GENERALIZE_FRAGMENT, the whole CandidateSet will subsume the whole RuleSet if the former performs not worse than the latter.

To evaluate a rule r we use the Laplace estimation, given by:

$$Score(r) = \frac{p+1}{n+1}$$

where n is the number of extractions made on the training set by r, and p is the number of correct extractions.

The GENERALIZE_TERM procedure. This procedure performs generalization over rule terms. We use word "term" to refer to any word staying at a node of the link path given by the first line or the second line of a rule. For example, the rule shown in Fig.3 has term "to" in the second line. The keyword and the variables @NAME are special terms and are not considered during term generalization.

The procedure looks for rules that are different only by terms at one node in one of first two lines. Such rules can be found easily by regular expression matching. Consider the following rule lines:

```
interaction M+ between J+ @NAME2
interaction M+ of J+ @NAME2
interaction M+ with J+ @NAME2
```

These rule lines are different only by terms at the second node (denoted by i). If such rules are found, they are added to BaseSet. The procedure then performs generalization by creating two rules, one by replacing term at node i with disjunctions of terms i from rules in BaseSet, and another by replacing term at node i by wildcard '*'. These two rules will form CandidateSet.

From the example above we have the following rules after generalization

```
        interaction M+ of|between|with J+ @NAME2
and     interaction M+ * J+ @NAME2
```

RuleSet ← the set of specific rules from examples
loop
    GENERALIZE_TERM (RuleSet, BaseSet, CandidateSet)
    if CandidateSet is not empty
        find rule $r_0 \in$ CandidateSet with Score(r_0)=max $_{r \in CandidateSet}$ Score(r)
        if Score(r_0) >= $\Sigma_{r \in BaseSet}$ Score(r)
            RuleSet = $r_0 \cup$(RuleSet−BaseSet)
until CandidateSet is empty
loop
    GENERALIZE_FRAGMENT(RuleSet, CandidateSet)
    if CandidateSet is not empty
        if $\Sigma_{r \in CandidateSet}$ Score(r) >= $\Sigma_{r \in RuleSet}$ Score(r)
            RuleSet ← CandidateSet
until CandidateSet is empty

GENERALIZE_TERM (RuleSet, BaseSet, CandidateSet)
    BaseSet ← {}
    CandidateSet ← {}
    for rule r∈RuleSet
        for i=1 to number of nodes of r
        find all rules∈RuleSet that differ from r by only the terms at node i
        if such rules found
            BaseSet ← the found rules
            CandidateSet ← disjuntions of the found rules
            CandidateSet = CandidateSet∪r with term i replaced by '*'
            Return

GENERALIZE_FRAGMENT(RuleSet, CandidateSet)
    CandidateSet ← {}
    For rule r∈RuleSet
        Find a rule r′∈RuleSet that differ from r only by the suffix of one line
        if r′ is found
            s ← suffix of r
            s′ ← suffix of r′
            for each rule p∈RuleSet with suffix s_p
                if s_p=s OR s_p=s′
                    replace s_p by s|s′
                    CandidateSet = CandidateSet ∪ p

Fig. 5. The learning algorithm

The GENERALIZE_FRAGMENT procedure. We call *rule fragment* (or just *fragment*) any part of a rule line (the first or the second line) that begins with a term, which is not a keyword, and ends with a term (another or the same). In the example above a fragment can be "of J+ @NAME2" or "@NAME2". We call *suffix* any fragment that contains the rightmost term. The procedure looks for a pair of rules that differ from each other only by the suffixes of one line. If such pair is found, the suffixes of the rules are recorded. There may be more than one pair of suffixes for each pair of rules. Then the procedure looks for all rules in RuleSet with suffixes identical to one of the recorded suffixes. Is such rules are found the procedure builds more general rules from them by replacing their suffixes with the disjunction of the recorded suffixes. The new rules are then added to CandidateSet. In the example below, the first two rule lines have two pairs of suffixes (`to J+ @NAME2`; `to J+ domain M+ of J+ @NAME2`) and (`@NAME2`; `domain M+ of J+ @NAME2`)

```
bind MV+ to J+ @NAME2
bind MV+ to J+ domain M+ of J+ @NAME2
bind O+  to J+ @NAME2
```

Generalization using these suffixes produces the following rules

```
bind MV+ (to J+ @NAME2)|(to J+ domain M+ of J+ @NAME2)
bind O+  (to J+ @NAME2)|(to J+ domain M+ of J+ @NAME2)

bind MV+ to J+ (@NAME2)|(domain M+ of J+ @NAME2)
bind O+  to J+ (@NAME2)|(domain M+ of J+ @NAME2)
```

The underlying assumption of fragment generalization is that if two different suffixes are found in the same position of two similar rules, the suffixes probably can appear in similar contexts and therefore can replace each other in other rules.

Example. As an example of the learning process, consider generalizing the rules based on the following three sentences (only fragments are shown)

"While *Scd2* interacted with the R1 N-terminal domain of *Shk1*..."
"The interaction between *Sec1p* and *Ssop* is..."
"... we observed an interaction of *Sp1* and *ZBP-89*..."

The following specific rules are created after parsing and processing links for these sentences (for the purpose of this example, we consider only the second lines of the rules produced)

```
interact M+ with J+ domain M+ of J+ @NAME2
interact M+ between J+ @NAME2
interact M+ of J+ @NAME2
```

During term generalization, the second and third lines are found to have only different terms "between" and "of". Thus, the lines are generalized to:

```
interact M+ with J+ domain M+ of J+ @NAME2
interact M+ * J+ @NAME2
```

During fragment generalization, these lines give two suffixes "`@NAME2`" and "`domain M+ of J+ @NAME2`", which are used to build the following general rule

```
interact M+ * J+ (@NAME2)|(domain M+ of J+ @NAME2)
```

This rule can be used, for example, to find the interaction between *Spc72p* and *Kar1p* from the sentence:

"Here we show that the interaction between yeast protein *Spc72p* and the N-terminal domain of *Kar1p*...."

which cannot be detected by the initial specific rules.

4   Experimental Results

We present here experimental results on a set of abstracts from Medline, a literature database available through PubMed [6]. The abstracts were obtained by querying Medline with the following keywords: "Saccharomyces cerevisiae" and "protein" and "interaction". We filtered the returned 3343 abstracts and retained 550 sentences containing at least one of four keywords "interact", "bind", "associate", "complex" or one of their inflections.

We adopt the standard *cross validation* methodology to test our algorithm. The data collection is partitioned several times into a training set and a testing set, rules are learned using the training set and then are evaluated using the test set. In our experiment, ten-fold cross validation was done on the set of sentences. An extracted interaction is considered correct if both extracted protein names are identical to those tagged by the user. The order of protein names (which is the first, which is the second) is not taken into consideration although experimental results show that the order is retained.

We evaluate performance in terms of *precision*, the number of correct extracted interactions divided by the total number of extracted interactions, and *recall*, the number of correct extracted interactions divided by the total number of interactions actually mentioned in the sentences.

In order to analyze the effect of term generalization and fragment generalization on the results, four versions of the learning algorithm were tested. The first version uses only specific rules without any generalization. The other two versions use either term or fragment generalization. The full version uses both type of generalization as shown in Fig. 5. Recalls and precisions of the four versions are given in Table. 1.

Table 1. Results of different versions of the learning algorithm

	Recall (%)	Precision (%)
Without generalization	41	93
With term generalization only	48	89
With fragment generalization only	49	91
Full algorithm	60	87

These results show that whereas the generalization slightly decreases precision, it leads to valuable improvement on recall. The results also show a little advantage of fragment generalization over term generalization.

There are a number of publications addressing the similar task of extracting protein interaction. Unfortunately, it is not simple to quantitatively compare our method with these alternatives because they use different text corpora, different assumptions about protein names, and different treatment of errors. For instance, Ono et.al. [7] describe an extraction method with high recall and precision of 86% and 94% respectively. However, they required the presence of protein name dictionaries, which are not always available. It is more reasonable to compare our approach with those that do not require pre-specified protein names or dictionaries. Thomas and his colleague [14] present one of such systems with 58% recall and 77% precision. Park et. al. [8] describe an extraction method using a combinatory categorical grammar for parsing and detecting interactions. They report recall and precision rates of 48 and 80 respectively. Both of the systems require hand-crafted patterns or rules for detecting protein interactions.

5   Conclusion

We have described an algorithm that automatically learns information extraction rules from training sentences. The rules learned by our algorithm can be used in combination with heuristic rules, which verify whether a noun phrase is protein name or not, to extract protein interactions from scientific abstracts. The learning and extraction algorithms exploit the link grammar parser to parse input sentences. The grammar has been shown appropriated for expressing relationships between words. This makes it possible to design a relative simple learning algorithm that can achieve accurate extraction performance without the need of manual rule building.

Acknowledgement. We thank Dokyun Na and Hyejin Kam for tagging training examples. The first author was supported by The Korea Foundation for Anvanced Studies and The Natural Science Council of Vietnam.

References

1. Blaschke, A., Andrade, M.A., Ouzounis, C., Valencia, A.: Automatic extraction of biological information from scientific text: protein-protein interactions. In: Proceedings of the 5th Int. Conference on Intelligent Systems for Molecular Biology. AAAI Press (1999).
2. Califf, M.E.: Relational learning techniques for natural language information extraction. PhD thesis. University of Texas, Austin (1998).
3. Freitag, D.:Machine learning for information extraction in informal domains. In: Machine learning, 39. Kluwer Academic Publishers.(2000).
4. Fukuda, K., Tamura, A., Tsunoda, T., Takagi, T.: Toward information extraction: identifying protein names from biological papers. In: Proceedings of the Pacific Symposium on Biocomputing. (1998).
5. Marcotte, E.M., Xenarios, I., Eisenberg, D.: Mining literature for protein-protein interactions. Bioinformatics. 17(4). Oxford University Press (2001)
6. Medline Pubmed: http://www.ncbi.nlm.nih.gov/entrez/query.fcgi?db=PubMed
7. Ono, T., Hishigaki, H., Tanigami, A., Takagi, T.: Automated extraction of information on protein-protein interactions from the biological literature. In: Bioinformatics. 17(2) (2001).
8. Park, J.C., Kim, H.S., Kim, J.J.: Bidirectional Incremental Parsing for Automatic Pathway Identification with Combinatory Categorial Grammar. In: Proceedings of the Pacific Symposium on Biocomputing (2001).
9. Porter, M.F.: An algorithm for suffix stripping. In: Program 14 (1980).
10. Rindflesch, T.C.,Tanabe, L.,Weinstein, J.,Hunter, L.: EDGAR: extraction of drugs, genes and relations from the biomedical literature. In: Proceedings of the Pacific Symposium on Biocomputing (2000).
11. Sleator, D., Temperley, D.: Parsing English with a Link Grammar. In: Proceedings of 3d International Workshop on Parsing Technologies (1993).
12. Soderland, S.: Learning information extraction rules for semi-structured and free text. In: Machine learning, 34. Kluwer Academic Publishers.(1999).
13. Tanabe, L., Wilbur, W.: Tagging gene and protein names in biomedical text. In: Bioinformatics. 18(8). Oxford University Press (2002).
14. Thomas, J., Milward, D., Ouzounis, C., Pulman, S., Carroll, M.: Automatic extraction of protein interactions from scientific abstracts. In: Proceedings of the Pacific Symposium on Biocomputing (2000).
15. Yakushiji, A., Tateisi, Y., Miyao, Y., Tsujii, J.: Event Extraction from Biomedical Papers Using a Full Parser. In: Proceedings of the Pacific Symposium on Biocomputing (2001).

Predicting Protein Interactions in Human by Homologous Interactions in Yeast[§]

Hyongguen Kim[1], Jong Park[2], and Kyungsook Han[1,*]

[1] School of Computer Science and Engineering, Inha University,
Inchon 402-751, Korea
[2] MRC-DUNN, Human Nutrition Unit, Hills Road, Cambridge, CB2 2XY UK
http://wilab.inha.ac.kr/

Abstract. As the genes of the human genome become known, one of the challenges is to identify all their interactions. Protein interactions are known for several organisms due to recent improvements in the detection methods for protein interaction, but they are limited to low-order species only. Direct determination of all the interactions occurring between the complete set of human genes remains difficult, even with current large-scale detection methods for protein interaction. This paper presents protein interactions between all human genes using the concept of homologous interaction. We believe this is the first attempt to map a whole human interactome.

1 Introduction

The last three years have seen a rapid expansion of protein interaction data due to the recent development of high-throughput, interaction detection methods, whereas traditional biochemical experiments had generated a small set of data for individual protein-protein interactions. However, our understanding of protein-protein interactions has not kept pace with the rapidly expanding amount of protein interaction data, which provides a rich source of data mining. Genome-wide protein interactions in several organisms such as yeast and Helicobacter pylori have been detected [2, 7] by experimental methods such as yeast two-hybrid [4] and mass spectrometry techniques, but they are limited to low-order species only. Direct determination of the interactions between the complete set of human proteins remains difficult, even with current, large-scale detection methods for protein interaction. Therefore, little is known about protein interactions between human proteins.

The primary focus of the work described in this paper is to discover the interactions between all human proteins using sequence similarities between human proteins and yeast proteins, whose interactions have already been determined. A network of human protein interactions will define the size of the problem of mapping interactome and

---

[§] This work was supported by the Ministry of Information and Communication of Korea under grant IMT2000-C3-4.
[*] To whom correspondence should be addressed. khan@inha.ac.kr

K.-Y. Whang, J. Jeon, K. Shim, J. Srivatava (Eds.): PAKDD 2003, LNAI 2637, pp. 159–165, 2003.

will be beneficial for many biomedical studies. It is also essential to compare the interactomic networks in evolutionary terms. It had been widely conjectured, but never demonstrated, that core protein interactions are conserved among different organisms. As our first step towards comparing all protein interactions across different organisms, we have built a network of human protein interactions by analogy with yeast protein interactions and have compared this network with another of yeast protein interactions. We believe this is the first attempt to map a whole human interactome. The rest of this paper describes our approach to mapping a whole human interactome and our results.

2 Inferring Protein Interactions by Analogy

2.1 Protein Interaction Data

The data of protein interactions in yeast is available in several databases, although in different forms. We extracted the data of protein-protein interactions in yeast from the MIPS   (http://mips.gsf.de/proj/yeast/tables/interaction/),   DIP   (http://dip.doe-mbi.ucla.edu/) and BIND (http://www.binddb.org/) databases. As of June 2002, these databases had 3929, 14703, and 12492 interactions, between 2271, 3134, and 2295 yeast proteins, respectively. After eliminating redundant data, we obtained 13777 interactions between 3751 yeast proteins. Amino acid sequences of the 3877 proteins were also extracted from the databases. We obtained amino acid sequences of 27049 human proteins from the Ensembl database (http://www.ensembl.org/). The total number of amino acids in the 27049 human proteins is 12118427.

2.2 Homology Search and Parsing the Result

We constructed a composite database with the amino acid sequences of human proteins, yeast proteins, SCOP proteins [6] and a non-redundant protein sequence database, NRDB90 [3]. We executed PSI-BLAST [1] for the homology search with yeast protein sequences as the query sequences. Human proteins became the target during the homology search. We parsed the homology search results with our own program written in C#. The parsing program extracts human proteins matched to yeast proteins and the start and end positions of each matched part. The program allows the user to select the e-value and identity when parsing. Small e-values indicate high matching rates, and large identity values indicate high homologous rates.

For the data of human protein interactions, we constructed a database using SQL 2000 on a Windows 2000 server. The relation between human proteins and yeast proteins requires two tables, one for the result of PSI-BLAST, another for selected data from the parsing results. Since the same sequence fragment of a human protein can be matched to multiple yeast proteins, we selected yeast proteins with the highest score based on the following criteria:

1.   Ratio of the length of a matched sequence fragment to the length of a yeast protein sequence.

2. Ratio of overlapped sequence fragment with previously matched sequences in that position.
3. Identity value of PSI-BLAST.

The number of hits decreases as the matched part of a yeast protein becomes longer and the identity value required becomes higher.

3 Results and Discussion

To predict the interactions between human proteins, we assigned yeast proteins to matched human proteins and generated three different sets of interaction data according to the following criteria. Both the first and third criteria are stricter than the second criterion. But there is no direct relation between the first and third criteria.

1. Minimum matched part: 70%, maximum overlap: 0%, identity: 0%.
2. Minimum matched part: 70%, maximum overlap: 10%, identity: 0%.
3. Minimum matched part: 70%, maximum overlap: 10%, identity: 10%.

In Table 1, the total number of yeast proteins matched to human proteins is 2184, 2190, and 2171, in data sets 1, 2, and 3, respectively.

Table 1. Number of yeast proteins matched to human proteins.

data set 1		data set 2		data set 3	
yeast pro-teins	human proteins	yeast pro-teins	human proteins	yeast proteins	human proteins
7	2	7	2	9	1
6	4	6	4	8	1
5	6	5	9	7	3
4	31	4	32	6	3
3	185	3	205	5	10
2	1467	2	1550	4	52
1	11030	1	10923	3	175
				2	1476
				1	9876
total	12725		12725		11597

From the three data sets, the protein-protein interaction networks predicted for the human genome contained 474164, 483635, and 324305 interactions between 11795, 11794, and 10659 proteins, respectively. Fig. 1 displays the number of human proteins and their interactions we found in the three data sets. In the first data set, the protein of ENSP00000300860 had the largest number of interactions, 715. In the second and third data sets, ENSP00000233607 and ENSP00000291900 exhibited the largest number of interactions, 765 and 703, respectively. Table 2 presents the number of nodes and edges in the interaction networks of yeast proteins and human proteins. While the interaction network for the yeast genome is composed of 8 connected components, we found 19, 16, and 14 connected components in the interaction network of human proteins from data sets 1, 2, and 3, respectively.

The presence of basic protein interactions that are conserved evolutionarily has been an important research question. Comparing protein interaction networks across different organisms can solve this question. Fig. 2 presents the network of human protein interaction obtained by analogy with the largest connected component of the network of yeast protein interaction. While the largest connected component of the network of yeast protein interaction is a single connected component, the network obtained by analogy with it is a disconnected graph with 14 connected components.

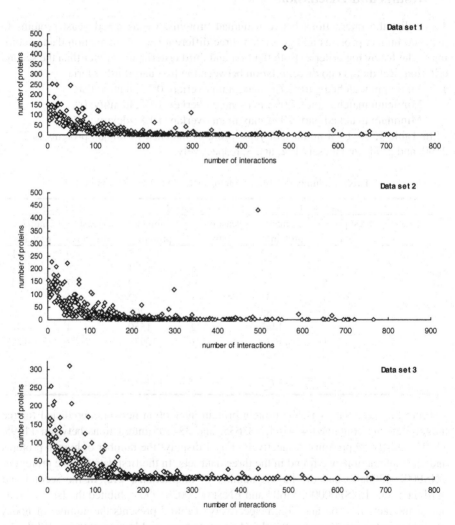

Fig. 1. Number of proteins and protein-protein interactions in the three data sets.

4 Conclusion and Future Work

A network of human protein interactions will define the size of the problem of mapping interactome and will be beneficial for many biomedical studies. It is also essential to compare the interactomic networks in evolutionary terms. We have built a network of human protein interactions by analogy with yeast protein interactions.

Table 2. Number of proteins and their interactions in the interaction network of yeast proteins and human proteins.

conn. comp.	yeast data		human data set 1		human data set 2		human data set 3	
	#nodes	#edges	#nodes	#edges	#nodes	#edges	#nodes	#edges
1	3736	13765	11699	473863	11723	483409	10601	324167
2	3	2	8	36	8	36	5	14
3	2	2	7	12	13	90	2	3
4	2	2	13	90	5	14	20	51
5	2	2	5	14	2	3	4	10
6	2	2	2	3	20	51	8	35
7	2	1	20	51	3	6	3	6
8	2	1	3	6	2	3	4	4
9			9	45	4	4	1	1
10			2	3	1	1	2	2
11			9	18	3	6	3	6
12			4	4	2	2	3	2
13			1	1	1	1	2	3
14			3	6	3	6	1	1
15			2	2	3	2		
16			1	1	1	1		
17			3	6				
18			3	2				
19			1	1				
total	3751	13777	11795	474164	11794	483635	10659	324305

When assessing the quality of interaction data, coverage and accuracy are often considered. How many protein-protein interactions were predicted in humans by our work? From the three data sets, we predicted 474164, 483635, and 324305 interactions between 11795, 11794, and 10659 human proteins, respectively. Since the total number of human proteins is 27049, we predicted interactions between 42% of human proteins on average. The remaining 58% of human proteins had no significant homology with yeast proteins, and hence interactions between them could not be predicted. Higher coverage of proteins could have been achieved by relaxing the parameters during the homology search and assignment. However, the interaction data of high coverage is not useful if its accuracy is low, i.e. if many false positives are contained in the data. A better way to improve coverage would be to identify additional interactions by homology with other organisms whose protein interactions have been elucidated. In the interactions predicted for 42% of human proteins, there may be false positives. We are currently working on reducing potentially false positives by removing interactions between proteins from distinct subcellular locations.

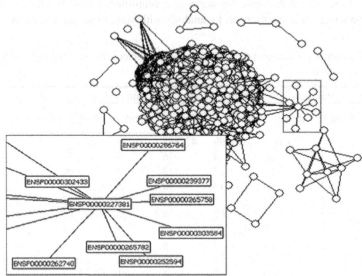

Fig. 2. The network of human protein interaction, obtained by analogy with the largest connected component of the network of yeast protein interaction. This network, visualized by InterViewer [5], consists of 14 connected components, and contains 470077 interactions between 11742 proteins. The network enclosed in a red box is a blowup of the marked subnetwork.

The results from three data sets revealed the following common property: the ratio of the number of interactions to the number of proteins was much higher than that in the network of yeast protein interaction. The high ratio of interactions in human can be explained firstly by the homologous interaction method producing artifacts in the process of homology, and secondly by the fact that the complexity of the higher organism is a result not only of the increase of proteins, but also of more complex interactions.

References

1. Altschul, S.F., Madden, T.L., Schaffer, A.A., Zhang, J., Zhang, Z., Miller, W., Lipman, D.J.: Gapped BLAST and PSI-BLAST: a new generation of protein database search programs. Nucl. Acids. Res. 25 (1997) 3389-3402
2. Gavin, A.-C., Bosche, M., Krause, R., et al.: Functional organization of the yeast proteome by systematic analysis of protein complexes. Nature 415 (2002) 141-147
3. Holm, L. Sander, C.: Removing near-neighbour redundancy from large protein sequence collections. Bioinformatics 14 (1998) 423-429

4. Ito, T., Tashiro, K., Muta, S., Ozawa, R., Chiba, T., Nishizawa, M., Yamamoto, K., Kuhara, S., Sakaki, Y.: Toward a protein-protein interaction map of the budding yeast: A comprehensive system to examine two-hybrid interactions in all possible combinations between the yeast proteins. Proc. Natl. Acad. Sci. USA 97 (2000) 1143-1147

5. Ju, B.-H., Park, B.G., Park, J.H., Han, K.: Visualization of Protein-Protein Interactions. Bioinformatics 19 (2003) 317-318

6. Lo Conte, L. Brenner, S.E. Hubbard, T.J.P., Chothia, C., Murzin, A.G.: SCOP database in 2002: refinements accommodate structural genomics. Nucl. Acids. Res. 30 (2002) 264-267

7. Rain, J.-C., Selig, L., De Reuse, H., Battaglia, V., Reverdy, C., Simon, S., Lenzen, G., Petel, F., Wojcik, J., Schachter, V., Chemama, Y., Labigne, A., Legrain, P.: The protein-protein interaction map of Helicobacter pylori. Nature 409 (2001), 211-215

Mining the Customer's Up-To-Moment Preferences for E-commerce Recommendation

Yi-Dong Shen[1], Qiang Yang[2], Zhong Zhang[3], and Hongjun Lu[2]

[1] Lab. of Comp. Sci., Institute of Software, Chinese Academy of Sciences, China
ydshen@ios.ac.cn
[2] Hong Kong University of Science and Technology, Hong Kong, China
{qyang, luhj}@cs.ust.hk
[3] School of Computing Science, Simon Fraser University, Burnaby, Canada
zzhang@cs.sfu.ca

Abstract. Most existing data mining approaches to e-commerce recommendation are past data model-based in the sense that they first build a preference model from a past dataset and then apply the model to current customer situations. Such approaches are not suitable for applications where fresh data should be collected instantly since it reflects changes to customer preferences over some products. This paper targets those e-commerce environments in which knowledge of customer preferences may change frequently. But due to the very large size of past datasets the preference models cannot be updated instantly in response to the changes. We present an approach to making real time online recommendations based on an up-to-moment dataset which includes not only a gigantic past dataset but the most recent data that may be collected just moments ago.

1 Introduction

E-commerce recommendation is aimed at suggesting products to customers and providing consumers with information to help them decide which products to purchase. Several approaches have been applied to making such recommendations, such as nearest-neighbor collaborative filtering algorithms [7], Bayesian networks [4], classifiers [3], and association rules [9]. Observe that most existing recommendation approaches are *past data model-based* in the sense that they walk through the following process:

Historical Data ⇒ Preference Model ⇒ Preferences ⇒ Recommendation.

That is, they first build a preference model from a set of historical transaction data. Then they measure customer preferences over a set of products based on the model. The preferences establish a partial order on the products. Therefore, recommendation is made by choosing products with top preferences. For example, Bayesian approach builds its model − a Bayesian network, by learning the prior and posterior probabilities of products from the given dataset. Then, given a customer's current shopping information, the customer's preferences over some

K.-Y. Whang, J. Jeon, K. Shim, J. Srivatava (Eds.): PAKDD 2003, LNAI 2637, pp. 166–177, 2003.
© Springer-Verlag Berlin Heidelberg 2003

products are derived by computing the posterior probabilities of these products. Similarly, the association-rule approach takes a set of association rules as its model. An association rule [1] is of the form $p_1, ..., p_m \rightarrow p_0$ ($s\%, c\%$), which, in the context of super-marketing, may mean that there would be $c\%$ possibility that a customer who has bought products $p_1, ..., p_m$ will buy the product p_0. Let M be a preference model and $\{p_1, ..., p_m\}$ be the set of products in a customer's current shopping cart. p_0 is a *candidate* product for recommendation to the customer if M has an association rule of the form $p_1, ..., p_m \rightarrow p_0$ ($s\%, c\%$). p_0 will be recommended to the customer if the confidence $c\%$ is among the top of all candidate products.

Apparently, the bottleneck of past data model-based approaches is in the development of their models. In many e-commerce applications, the historical dataset would be very large and it may take a long time to build a model from it. This suggests that the model needs to be built off-line (or in batch). As a result, such approaches are suitable only for applications where knowledge of customer preferences is relatively stable.

However, in many e-commerce applications the business patterns may change frequently. In such situations, new precious data should be collected instantly since it reflects changes (increase or decrease) to customer preferences over some products. Since past data model-based approaches are unable to catch up with such frequent changes, their models may not fully cover the up-to-moment status of the business, and therefore applying them alone would lower down the recommendation accuracy.

The above discussion suggests that methods of making use of up-to-moment (not just up-to-date) data need to be developed for e-commerce recommendation. Such methods should keep abreast of the up-to-moment changes to customer preferences. This motivates the research of this paper.

1.1 Problem Statement

The problem can be generally stated as follows. Let SM_{past} be a model built from the data which was collected between time points t_0 and t_1 as depicted below. Assume that no updated model is available until time t_2 and that there would be a certain amount of fresh data available during the period $t_2 - t_1$. Suppose a customer logs into the system for online shopping and asks for product recommendation at time t_i ($t_1 < t_i < t_2$). For convenience, the data collected between t_0 and t_1 is called *past data*, whereas the data collected between t_1 and t_i is called *recent data*. We are then asked to design a system that can make real time online recommendations based on the *up-to-moment data* which includes both the past and recent data. By "real time" we mean a response time with which customers can tolerate during online shopping (e.g., a few seconds).

Here are the characteristics and challenges of the problem: (1) The past data would be massive, say Megs or Gigs of records, so that it cannot be directly used in real time. A model needs to be built off-line (or in batch) from the data. Model building is quite time-consuming, though applying a model to make recommendations can be done in real time. (2) The volume of the recent data is much smaller than that of the past data, possibly in the hundreds of orders. For example, in our super-store data mining project, the past dataset always contains at least three months of past transaction data whereas the recent dataset only contains the data of the current day. The recent data was most recently generated and includes some data that was collected just moments ago. (3) Thus we are faced with the following challenge in applying the up-to-moment data. The dataset is so huge that we cannot apply it to make online real time recommendations unless a model is built from it off-line or in batch in advance. However, in reality the time between the most recent data being collected and the customer requesting a recommendation may be very short (possibly less than one second), so it is impossible for us to prepare such a model in advance or build it instantly. Thus it appears that we are in a dilemma situation. Effectively resolving such a dilemma then presents a challenging task.

1.2   Our Solution

In its most general case, this problem seems too difficult to be tractable. Our study shows, however, that it could be effectively handled when some practical constraints are applied. We choose association rules as our preference model. Since the population of the recent data within the up-to-moment data is much smaller than that of the past data, statistically the recent data can only bring minor changes (increase or decrease) to the past preferences. In other words, the recent data can only play a role of small adjustments to the customer's past preferences in response to the recent minor changes. This implies that the large majority of the supporting records of any up-to-moment frequent pattern may come from the past data. Therefore, in this paper we restrict ourselves to these situations where any up-to-moment frequent pattern occurs in at least one record of the past data.

The key idea of our approach stems from the following important observation: A few patterns that do not frequently occur in the past data may become frequent in the up-to-moment data, whereas a few other patterns that frequently occur in the past data may not be frequent in the up-to-moment data. However, due to the rather small population of the recent data within the up-to-moment data, the preferences of all such inversable patterns in the past data must be close to the minimum preference that is required for recommendation.

Based on the above observation, we introduce a concept of *expanded past preference models*, as opposed to the standard preference models [1]. An expanded past preference model is built off-line from the past data and contains not only standard association rules but also a few expanded association rules that represent those patterns that are less frequent in the past data but quite likely to be frequent in the up-to-moment data. Then, given the customer's current shopping information, we apply a criterion, called *Combination Law*, to derive the customer's up-to-moment preferences directly from the expanded past model and the recent data. Let SM_{utm} denote a standard preference model built from the up-to-moment data. We will prove that applying our approach will generate the same recommendations to the customer as SM_{utm} does, though we do not need to build SM_{utm} at all.

1.3 Related Work

For typical recommendation approaches and applications, see the 1997 special issue of the Communications of the ACM [10] and a recent special issue of Data Mining and Knowledge Discovery [8]. In particular, Shafer et al. [11] made an elegant survey on major existing systems and approaches to e-commerce recommendation. To the best of our knowledge, however, there is no report in the literature on how to use the up-to-moment data to make real time recommendations. Most existing approaches are past data model-based, which make recommendations by first building a (standard) past preference model off-line against a previously prepared training dataset and then applying the model to current customer situations. The model is updated periodically. As a typical example, the SmartPad system developed at IBM makes recommendations using a model of association rules [9]. The model is built on the past eight weeks of data from Safeway Stores and is updated weekly or quarterly to reflect seasonal differences.

Our work focuses on doing interactive and online preference mining (in real time). So it is essentially different from the incremental update of association rules, a topic that addresses how to efficiently update a (standard) past preference model that was built from an old dataset (DB) based on an incremental dataset (db) [2,5,6,12]. Existing incremental updating algorithms make use of some knowledge of an old model to build a new model from $DB \cup db$. Although applying these algorithms to update a model is faster than building a new model from scratch, it is still quite time-consuming since scanning the old dataset cannot be avoided. When DB is massive, such update cannot be completed online in real time. Moreover, if we applied an incremental updating procedure to mine the up-to-moment preferences, we would have to update the model every time a new record is added to the dataset, which is quite unrealistic in real online processing.

2   System Architecture

We assume readers are familiar with association rule mining, especially with the widely used Apriori algorithm [1]. A *dataset DS* is a set of records each of which is of the form $\{p_1, ..., p_m\}$. We use $|DS|$ to denote the total number of records in DS and $count(\{p_1, ..., p_m\}, DS)$ to denote the number of records in DS that contain all the items $p_1, ..., p_m$. A (standard) *preference model* built from DS consists of a set of association rules. A set of items, $\{p_1, ..., p_m\}$, is a *frequent pattern* in DS if its support $s\% = \frac{count(\{p_1, p_2, ..., p_m\}, DS)}{|DS|} \geq ms\%$, where $ms\%$ is a pre-specified parameter called *minimum support*. Let $\{p_1, ..., p_m\}$ be a frequent pattern. For each $i \leq m$, $p_1, ..., p_{i-1}, p_{i+1}, ..., p_m \rightarrow p_i$ ($s\%, c\%$) is an *association rule* in the preference model if $c\% \geq mc\%$, where $mc\%$ is another pre-specified parameter called *minimum confidence*, and $c\%$ is the confidence of the rule given by $c\% = \frac{count(\{p_1, ..., p_m\}, DS)}{count(\{p_1, ..., p_{i-1}, p_{i+1}, ..., p_m\}, DS)}$. We will frequently use three terms: past, recent and up-to-moment. Let x be one of them. Here is some of the notation used throughout this paper. $ms\%$: the minimum support; $mc\%$: the minimum confidence; DS_x: an x dataset; SM_x: a standard x preference model built from DS_x with $ms\%$ and $mc\%$; and EM_{past}: an expanded past preference model built from DS_{past} with $ms\%$ and $mc\%$.

The architecture of our recommendation system is shown in Figure 1. The system works on two datasets: the *past dataset* DS_{past} that stores the past data, and the *recent dataset* DS_{recent} that stores the recent data. DS_{past} is much larger than DS_{recent}, possibly in the hundreds of orders. The up-to-moment dataset $DS_{utm} = DS_{past} \cup DS_{recent}$. Since DS_{past} is gigantic, the system builds off-line or in batch an expanded past preference model from it (see Definition 1 in the following section). Then, given a customer's current shopping information,[1] the customer's up-to-moment preferences are computed online by combining the knowledge about the customer's past preferences that is derived from the expanded past model EM_{past} and the knowledge about his/her recent preferences that is derived from DS_{recent}. When a customer finishes shopping by logging out the system, a new record built from the products in the customer's shopping cart is added to DS_{recent} immediately. Since we only keep most recent data in DS_{recent}, the "old" data in DS_{recent} will be moved to DS_{past} from time to time. Moreover, when DS_{past} gets too big, some data will be removed. All such data transfers are monitored by the Database management facility.

3   Building an Expanded Past Model

Our goal is to compute the customer's up-to-moment preferences without building an up-to-moment preference model. To this end, we need to build a so-called expanded past preference model EM_{past} from DS_{past}. Two factors are considered in developing such an expanded model: *completeness* and *efficiency*. EM_{past} is complete if for any standard association rule $X \rightarrow Y(s\%, c\%)$ that is derivable from DS_{utm}, there is an expanded association rule $X \rightarrow Y(s'\%, c'\%)$ in EM_{past}. One might consider defining EM_{past} as a standard preference model which is built from DS_{past} by setting the minimums $ms\% = 0$ and $mc\% = 0$. Although doing so can guarantee the completeness, it is definitely too inefficient, if not impossible, because of the nature of the association mining problem [1]. The efficiency factor then requires that both the number of rules in EM_{past} and the time to build EM_{past} be as close as possible to those of SM_{past} − the standard

---

[1] It may contain the customer's profile as well as the products in his current shopping cart. To simplify our presentation, we omit the profile.

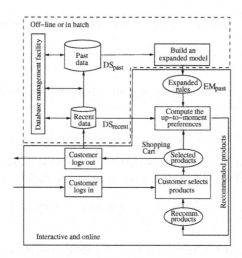

Fig. 1. A recommendation system based on the up-to-moment data.

past preference model built from DS_{past} with the pre-specified minimums. In particular, EM_{past} is said to be *scalable* if the larger the past dataset DS_{past} is, the closer EM_{past} is to SM_{past} w.r.t. the number of their rules and the running time.

In order to get high efficiency while guaranteeing the completeness, we introduce a new parameter, $\mathcal{N}_{max}$, called *one day's maximum*. Intuitively, $\mathcal{N}_{max}$ represents the one day's maximum sales of a single product, which can be obtained from past experience. More formally, let P be the set of all products, then at any time we have $\mathcal{N}_{max} \geq max\{count(\{p\}, DS_{recent})|p \in P\}$. It is based on this new parameter that our expanded model is defined. In the sequel, both $\mathcal{N}_{max}$ and the two minimums $ms\% > 0$ and $mc\% > 0$ are supposed to be provided by the e-commerce manager.

Definition 1. *An* expanded past preference model EM_{past} of DS_{past} consists of all expanded association rules defined as follows: (1) $\{p_1, ..., p_m\}$ is an expanded frequent pattern if $count(\{p_1, ..., p_m\}, DS_{past}) > 0$ and $\frac{count(\{p_1,...,p_m\},DS_{past})+\mathcal{N}_{max}}{|DS_{past}|+\mathcal{N}_{max}} \geq ms\%$. (2) Let $\{p_1, ..., p_m\}$ be an expanded frequent pattern. For each $i \leq m$, $p_1, ..., p_{i-1}, p_{i+1}, ..., p_m \to p_i$ $(s\%, c\%)$ is an expanded association rule if $\frac{count(\{p_1,...,p_m\},DS_{past})+\mathcal{N}_{max}}{count(\{p_1,...,p_{i-1},p_{i+1},...,p_m\},DS_{past})+\mathcal{N}_{max}} \geq mc\%$ where $s\% = \frac{count(\{p_1,...,p_m\},DS_{past})}{|DS_{past}|}$ and $c\% = \frac{count(\{p_1,...,p_m\},DS_{past})}{count(\{p_1,...,p_{i-1},p_{i+1},...,p_m\},DS_{past})}$.

Note that the support $s\%$ and confidence $c\%$ of an expanded association rule are the same as those defined for a standard association rule. However, due to the inclusion of $\mathcal{N}_{max}$, the thresholds for an expanded frequent pattern or an expanded association rule are a little lower than the two pre-specified minimums. This allows EM_{past} to catch those patterns/association rules that are unable to be included in SM_{past} but may possibly appear in SM_{utm}. We have the following.

Theorem 1. EM_{past} *is complete.*

The following theorem shows that any standard association rule in SM_{past} is an expanded association rule in EM_{past}.

Theorem 2. $SM_{past} \subseteq EM_{past}$.

The following result assures us that the expanded association rules can be derived using standard association mining methods such as Apriori.

Theorem 3. *Any sub-pattern of an expanded frequent pattern is an expanded frequent pattern.*

Before moving to the next section, we briefly answer the question readers may pose about how to keep $\mathcal{N}_{max}$ to be the maximum count in DS_{recent}. This can be realized simply by treating DS_{recent} in batches. If at time t_i there is a product in DS_{recent} whose sales are about to reach the maximum $\mathcal{N}_{max}$ (10% below $\mathcal{N}_{max}$ in our project. The exact measure depends on applications), we add the data of DS_{recent} to the past dataset DS_{past} and build in batch a new expanded past preference model from DS_{past} in advance. As a result, when it happens that the sales of an item in DS_{recent} exceed $\mathcal{N}_{max}$, the new expanded past model must be available, which covers the past data collected up to time t_i. Thus the new expanded past model EM_{past}, together with a new recent dataset DS_{recent} that consists of the data collected from t_i up to the current moment, is used to compute the customer's up-to-moment preferences.

4   Computing the Up-To-Moment Preferences

Let $SC = \{p_1, ..., p_m\}$ be the set of products in a customer's current shopping cart. We make online product recommendations for the customer by computing his up-to-moment preferences over those products related to SC. That is, we derive from EM_{past} and DS_{recent} all rules of DS_{utm} of the form $p_1, ..., p_m \rightarrow p$ $(s\%, c\%)$ with $s\% \geq ms\%$ and $c\% \geq mc\%$. This is done via the three steps summarized in the following table.

Given $SC = \{p_1, ..., p_m\}$.
Step 1: Derive R_{past} (rules about the customer's past preferences) from EM_{past}.
Step 2: Derive R_{recent} (rules about the customer's recent preferences) and $count_{p_1-pm}$ from DS_{recent} based on R_{past} (see Algorithm 1).
Step 3: Combine R_{past} and R_{recent} to obtain PL_{utm} (the customer's up-to-moment preferences; see Algorithm 2).

The first step is to extract from EM_{past} the knowledge about the customer's past preferences over those products related to SC. We search EM_{past} to obtain the set R_{past} of rules of the form $p_1, ..., p_m \rightarrow p$ $(s\%, c\%)$. By Theorem 1, all rules of DS_{utm} related to SC are in R_{past}. The second step is to extract from DS_{recent} the knowledge about the customer's recent preferences related to SC. By Theorem 1, it is unnecessary to derive from DS_{recent} all rules with a left-hand side $p_1, ..., p_m$; it suffices to derive only the set R_{recent} of rules each of which is of the form $p_1, ..., p_m \rightarrow p$ $(s_2\%, c_2\%)$ such that there is a rule of the form $p_1, ..., p_m \rightarrow p$ $(s_1\%, c_1\%)$ in R_{past}. Such new rules reflect changes to the customer's past preferences over those products related to SC. Here is an algorithm.

Algorithm 1 Extracting rules about the customer's recent preferences.
Input: SC, DS_{recent} and R_{past}.
Output: R_{recent} and $count_{p_1-p_m} = count(SC, DS_{recent})$.
Procedure:
1. $R_{recent} = \emptyset$;
2. $Head = \{p | p_1, ..., p_m \rightarrow p$ $(s_1\%, c_1\%)$ is in $R_{past}\}$;
3.   Select all records containing $p_1, ..., p_m$ into $TEMP$ from DS_{recent};
4.   **for each product** $p \in Head$ **do begin**
5.   Compute $count_p = count(\{p\}, TEMP)$;
6.   **if** $count_p > 0$ **then do begin**
7. $s_2\% = \frac{count_p}{|DS_{recent}|} * 100\%$; $c_2\% = \frac{count_p}{|TEMP|} * 100\%$;
8.   Add the following rule to R_{recent}: $p_1, ..., p_m \rightarrow p(s_2\%, c_2\%)$
9.   **end**
10. **end**
11. Return R_{recent} and $count_{p_1-p_m}$ with $count_{p_1-p_m} = |TEMP|$.

Although both the support and confidence of a rule are computed in Algorithm 1 (see line 7), neither the minimum support nor the minimum confidence is used because we are not ready to remove any rules at this stage. We will do it next in conjunction with those rules about the past preferences.

Theorem 4 (Combination Law). SM_{utm} has a rule $p_1, ..., p_m \rightarrow p$ $(s_3\%, c_3\%)$ if and only if one of the following holds: (1) There are two rules $p_1, ..., p_m \rightarrow p$ $(s_1\%, c_1\%)$ in R_{past} and $p_1, ..., p_m \rightarrow p$ $(s_2\%, c_2\%)$ in R_{recent}. Then $s_3\% = \frac{s_1\%*|DS_{past}|+s_2\%*|DS_{recent}|}{|DS_{utm}|} \geq ms\%$ and $c_3\% = \frac{s_1\%*|DS_{past}|+s_2\%*|DS_{recent}|}{\frac{s_1*|DS_{past}|}{c_1}+count_{p_1-p_m}} \geq mc\%$. (2) Only R_{past} has a rule $p_1, ..., p_m \rightarrow p$ $(s_1\%, c_1\%)$. Then $s_3\% = \frac{s_1\%*|DS_{past}|}{|DS_{utm}|} \geq ms\%$ and $c_3\% = \frac{s_1\%*|DS_{past}|}{\frac{s_1*|DS_{past}|}{c_1}+count_{p_1-p_m}} \geq mc\%$.

Theorem 4 is quite significant. It allows us to derive the customer's up-to-moment preferences directly from R_{past} and R_{recent}, without building any up-to-moment preference model. In the following algorithm, by CL(i) we refer to the i-th case of the Combination Law ($i = 1$ or 2).

Algorithm 2 Deriving the customer's up-to-moment preferences.
Input: R_{past}, R_{recent}, $|DS_{past}|$, $|DS_{recent}|$, $|DS_{utm}|$, $count_{p_1-p_m}$, $ms\%$ and $mc\%$.

Output: an up-to-moment preference list, PL_{utm}.
Procedure:
1. $PL_{utm} = \emptyset$;
2. **while** $(R_{past} \neq \emptyset)$ **do begin**
3.   Remove from R_{past} a rule $p_1, ..., p_m \to p$ $(s_1\%, c_1\%)$;
4.   **if** R_{recent} has a rule $p_1, ..., p_m \to p$ $(s_2\%, c_2\%)$ **then**
5.   Compute $s_3\%$ and $c_3\%$ by applying CL(1);
6.   **else** Compute $s_3\%$ and $c_3\%$ by applying CL(2);
7.   **if** $s_3 \geq ms$ and $c_3 \geq mc$ **then**
8. $PL_{utm} = PL_{utm} \cup \{p\ (s_3\%, c_3\%)\}$
9. **end**
10. Return PL_{utm} which is sorted by confidence.

Algorithm 2 computes the customer's up-to-moment preferences based on the Combination Law. In particular, it computes the up-to-moment preferences by applying the recent preferences to adjusting the past preferences. Let $PP = p\ (s_1\%, c_1\%)$ be a past preference item; i.e. R_{past} has a rule $p_1, ..., p_m \to p$ $(s_1\%, c_1\%)$. There may be three cases. First, both s_1 and c_1 are sufficiently high to survive any recent changes, so that the product p is guaranteed to be included in the up-to-moment preference list PL_{utm}. Second, PP is *positively marginal* in the sense that s_1 or c_1 is just a little higher than ms or mc. In this situation, the recent changes may invalidate the past preference by lowering it down below either of the two minimums so that the product p is excluded from PL_{utm}. The third case is that PP is *negatively marginal*; i.e. either s_1 or c_1 is a little lower than ms or mc. In this case, the past preference may be lifted above the minimums by the recent changes so that the product p is included in PL_{utm}.

Theorem 5. *Algorithm 2 is correct in the sense that for any customer's current shopping cart $SC = \{p_1, ..., p_m\}$, SM_{utm} has a rule $p_1, ..., p_m \to p$ $(s_3\%, c_3\%)$ if and only if p $(s_3\%, c_3\%) \in PL_{utm}$.*

5   Experimental Evaluation

The correctness of our approach is guaranteed by Theorem 5. We now show its efficiency by empirical experiments. Our experiments were conducted on a Pentium III 600 MHz PC with 512M memory. We applied the Apriori algorithm to mine association rules with the minimums $ms\% = mc\% = 1\%$.

Our dataset for the experiments came from a retail store. The dataset consists of eight weeks of daily transaction data. To facilitate our experiments, we fixed a past dataset DS_{past} that consists of 687571 records. We then collected five recent datasets DS_{recent}^1 (4415 records), DS_{recent}^2 (5830 records), DS_{recent}^3 (7245 records), DS_{recent}^4 (8660 records), and DS_{recent}^5 (10077 records) at five consecutive sampling times $T_1, ..., T_5$. As a result, we got five up-to-moment datasets $DS_{utm}^1, ..., DS_{utm}^5$ with $DS_{utm}^i = DS_{past} \cup DS_{recent}^i$.

Since DS_{recent} is small, computing the recent preferences (the second step of our approach) takes very little time because it is accomplished simply by counting the frequency of each single product in DS_{recent} (see Algorithm 1). It takes even less time to extract the past preferences from EM_{past} (the first step) and do the combination of the past and recent preferences (the third step; see Algorithm 2). Experiments for such claims were conducted by simulating a customer's shopping process in which we assumed that a customer asked for product recommendation at the times T_1, T_2, ... and T_5, respectively.

At time T_1 the customer's shopping cart was empty. Our system computed the up-to-moment preferences and made a recommendation. The customer selected a product from our recommendation and then asked for more at time T_2. The process then went to the next cycle. For each T_i we recorded the response time for the period of the customer presenting the request and the system sending back the recommendation. Figure 2 shows the response time of our approach in contrast to the response time of the *brutal-force* approach that directly applies Apriori to derive the up-to-moment preferences. Clearly, using our approach takes a very short response time (1.38 seconds on average), while using the brutal-force approach spends 86.7 seconds on average.

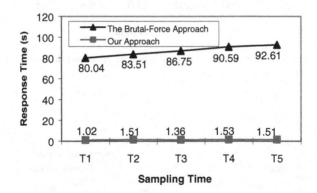

Fig. 2. The response time at five consecutive sampling times.

The above experiments demonstrate that the customer's up-to-moment preferences can be derived in real time by performing the three steps of our approach, provided that the expanded past model is prepared in advance. We now show the efficiency of building an expanded model EM_{past} as compared to building a standard past model SM_{past}. As we mentioned before, this can be measured by showing how close the two models are w.r.t. the number of their rules and the time to build them. In order to see the scalability of our approach, we built from DS_{past} five datasets as past datasets, with DS_1 (269K records) consisting of the data of the first three weeks, DS_2 (372K records) of the first four weeks, ..., and DS_5 (690K records) of the seven weeks. Moreover, by randomly picking up ten days among the seven weeks and counting the sales of each single item in each selected day, we got the number 1217 for the one day's maximum parameter $\mathcal{N}_{max}$.

The comparison of the running time and the number of rules is depicted in Figures 3 and 4, respectively. We see that it does not take much more time to build an expanded past model EM_{past}^i from DS_i than to build a standard past model SM_{past}^i from DS_i (see Figure 3), and that although EM_{past}^1 is about two times bigger than SM_{past}^1, EM_{past}^5 is very close to SM_{past}^5 (see Figure 4). This shows our approach is efficient and has very good scalability. That is, the bigger the past dataset is, the closer the expanded model is to the standard model.

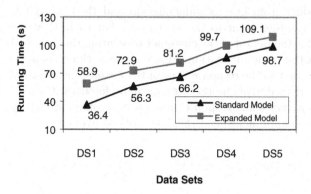

Fig. 3. Comparison of the running time of building an expanded and a standard past preference model.

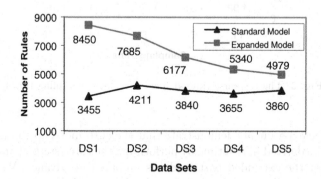

Fig. 4. Comparison of the number of rules of an expanded and a standard past preference model.

6   Conclusions

We have developed an approach to making online e-commerce recommendations based on the up-to-moment data. A key concept of this approach is an expanded past preference model from which the customer's up-to-moment preferences can be derived in real time without the necessity to build an up-to-moment model. Our approach is suitable for applications where knowledge of customer preferences may change frequently but due to the huge size of past datasets the preference models cannot be updated instantly in response to the changes. Therefore, it is an enhancement to existing past data model-based approaches which are suitable only for applications where customer preferences are relatively stable.

Acknowledgment. The authors thank the anonymous referees for their helpful comments. Yi-Dong Shen is supported in part by Chinese National Natural Science Foundation, Trans-Century Training Program Foundation for the Talents by the Chinese Ministry of Education, and Foundations from Chinese Academy of Sciences. Qiang Yang thanks NSERC, IRIS-III program and Hong Kong RGC for their support.

References

1. Agrawal, R., Srikant, R.: Fast algorithm for mining association rules. In: *VLDB* (1994) 487–499.
2. Aumann, Y., Feldman, R., Lipsttat, O., Manilla, B.H.: An efficient algorithm for association generation in dynamic databases. *Journal of Intelligent Information Systems* **12** (1999) 61–73.
3. Basu, C., Hirsh, H., Cohen, W.: Recommendation as classification: using social and content-based information in recommendation. In: *Proceedings of the National Conference on Artificial Intelligence* (1998) 714–720.
4. Breese, J., Heckerman, D., Kadie, C.: Empirical analysis of predictive algorithms for collaborative filtering. In: *Proceedings of the 14th Conference on Uncertainty in Artificial Intelligence* (1998) 43–52.
5. Cheung, D., Han, J., Ng, V., Wong, C.: Maintenance of discovered association rules in large databases: an incremental updating technique. In: *Proc. of 12th Intl. Conf. on Data Engineering* (1996) 106–114.
6. Hidber, C.: Online association rule mining. In: *SIGMOD* (1999) 145–156.
7. Herlocker, J., Konstan, J., Borchers, A., Riedl, J.: An algorithmic framework for performing collaborative filtering. In: *ACM SIGIR* (1999) 230–237.
8. Kohavi, R., Provost, F.: Applications of data mining to electronic commerce. *Journal of Data Mining and Knowledge Discovery* **5** (2001) 5–10.
9. Lawrence, R., Almasi, G., Kotlyar, V., Viveros, M., Duri, S.: Personalization of supermarket product recommendations. *Journal of Data Mining and Knowledge Discovery* **5** (2001) 11–32.
10. Resnick, P., Varian, H.: Recommender systems. *Communications of the ACM* **40** (1997) 56–58.
11. Schafer, J., Konstan, J., Riedl, J.:. Electronic commerce recommender applications. *Journal of Data Mining and Knowledge Discovery* **5** (2001) 115–152.
12. Thomas, S., Bodagala, S., Alsabti, K., Ranka, S.: An efficient algorithm for the incremental updation of association rules in large databases. In: *KDD* (1997) 263–266.

A Graph-Based Optimization Algorithm for Website Topology Using Interesting Association Rules

Edmond H. Wu and Michael K. Ng

Department of Mathematics, The University of Hong Kong
Pokfulam Road, Hong Kong
hcwu@hkusua.hku.hk,mng@maths.hku.hk

Abstract. The Web serves as a global information service center that contains vast amount of data. The Website structure should be designed effectively so that users can efficiently find their information. The main contribution of this paper is to propose a graph-based optimization algorithm to modify Website topology using interesting association rules. The interestingness of an association rule $A \Rightarrow B$ is defined based on the probability measure between two sets of Web pages A and B in the Website. If the probability measure between A and B is low (high), then the association rule $A \Rightarrow B$ has high (low) interest. The hyperlinks in the Website can be modified to adapt user access patterns according to association rules with high interest. We present experimental results and demonstrate that our method is effective.

1 Introduction

Web usage mining refers to mine Web logs to discover user access patterns of Web pages [7]. The effective management of a Website is dependent on awakening the needs of potential and profitable customers, analyzing their behaviors and then improving Web service. Therefore, there is a great demand in developing efficient solutions for Web service, no matter in commerce or industry. A major topic in Web usage mining is mining association rules [11]. An association rule is described by the dependence among Web pages in a Website. The support-confidence framework is well known to determine statistically significant association rules [3,4]. However, there are some drawbacks of this framework while mining association patterns in a Website. For example, when two sets of Web pages A and B are very likely to be visited together, we obtain the association rule: $A \Rightarrow B$ or $B \Rightarrow A$. However, if A and B have already been connected by hyperlinks, then such association rule $A \Rightarrow B$ or $B \Rightarrow A$ does not have a high interest. Moreover, an association rule with a very high support rarely exists in Web log mining. This is because in a complex Website with variety of Web pages, and many paths and hyperlinks, one should not expect that in a given time period, a large number of visitors follow only a few paths or Web pages. Otherwise, it would mean that the structure and content of the Website have a

K.-Y. Whang, J. Jeon, K. Shim, J. Srivatava (Eds.): PAKDD 2003, LNAI 2637, pp. 178–190, 2003.

serious problem. Therefore, in general, there are many association rules of about the same support obtained from the Web logs. It is necessary for us to find an additional interestingness measure to identify really interesting association rules. In this paper, we propose a novel interestingness measure called rule interest that is based on Website topology to analyze association rules.

One of the objectives of Web usage mining is to improve the organization of a Website by learning from Web logs. Previous works [6,8,9,10] have focused on the optimization problem of Website topology from different aspects. For instance, Perkowitz and Etzioni [8,9,10] have used clustering algorithms for indexing page synthesis to create an adaptive Website. Garofalakis et al. [6] have used page popularity to rearrange Website structure to make it more accessible and more effective. In this paper, we employ association rules with high interest to develop a graph-based optimization algorithm to modify the current Website topology that can cater to a large population of visitors. Comparing to the other Website optimization solutions, the most virtue of our method is that we not only use on Web logs to analyze user access patterns, but also use the original Website topology together to determine an effective Website structure.

The rest of this paper is organized as follows. In Section 2, we introduce the concept and the calculation of rule interest. In Section 3, we propose a graph-based optimization method to modify a Website topology according to interesting association rules. Section 4 describes the complexity analysis of our algorithm. Experimental results are presented in Section 5. Finally, some concluding remarks are given in Section 6.

2   Discovery of Interesting Association Rules in Website

In order to discover interesting usage patterns from Web logs, we define a new interestingness measure called rule interest in association rules. By using rule interest, together with rule support and confidence, we can identify association rules which are more meaningful to our analysis of visitors' behaviors.

2.1   Website Topology

First, we regard Website topology as the structure of a Website. The nodes in a Website topology represent the Web pages and the edges among nodes represent the hyperlinks among Web pages. We assume that there is at least one edge to connect every node, that is, every Web page in the Website can be visited through at least one path. Figure 1 shows an example of Website topology. All Web pages are assigned with unique labels.

Table 1. An example of transaction itemsets

TID	Itemset
100	ADCE
200	ABG
300	BFE
400	CH

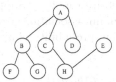

Fig 1. A website topology example

When people visit a Website, the Website server will automatically register the Web access log records including URL requested, the IP address of users and timestamps. Some researchers have carefully survey the behaviors of people while they visit Website [12]. In practical cases, we need to apply some data mining techniques to clean and transform the raw Web logs into transaction itemsets or a data cube [13]. Table 1 presents some transaction itemsets of the given Website topology (see Figure 1). With these preprocessed data sets, we can discover potential association patterns in a Website in the next step.

2.2   Topology Probability Model

We first consider association probability between any two Web pages x_i and x_j in a Website. Given a Website topology G containing n Web pages $X = \{x_1, x_2, \ldots, x_n\}$, we assume the number of hyperlinks of x_k is h_k, $k = 1, \ldots, n$, so a set of the Web pages which have hyperlinks to x_k are $X_k = x'_1, \ldots, x'_{h_k}$. When a user has visited current Web page x_i, he or she may continue to visit other Web pages connected to current Web page or just exit the Website. Therefore, there are $h_i + 1$ choices for the user to select after visiting x_i. We consider the probability of visiting x'_j after having visited x_i is given by: $P(x'_j|x_i) = w_{i,j}/h_{i+1}$, where $w_{i,j}$ is the weighting parameter between x_i and x'_j, $j = 1, \ldots, h_i$ (usually we take $w_{i,j} = 1$). We should note that x'_j must be connected to x_i, that is $x'_j \in X_i$. The distance measure between two Web pages x_i and x'_j is defined by $D(x_i, x'_j) = \log(1/P(x'_j|x_i)) = \log(h_{i+1}/w_{ij})$.

If x_i and x_j do not have a hyperlink between them, we consider the shortest path between x_i and x_j in the Website topology. Since the Website topology is a connected graph, there must be a sequence of nodes connecting x_i and x_j. We can employ classical Floyd algorithm to calculate the shortest paths between any two nodes in a Website topology. Assume the shortest path passes through m Web pages $x_1^*, \ldots, x_m^* \in X$, the length of the shortest path is calculated by $D(x_i, x_j) = D(X_i, x_1^*) + \sum_{k=1}^{m-1} D(x_k^*, x_{k+1}^*) + D(x_m^*, x_j)$. Therefore, the probability of visiting x_j after having visited x_i is given by:

$$P(x_j|x_i) = P(x_1^*|x_i) \prod_{k=1}^{m-1} P(x_{k+1}^*|x_k^*)P(x_j|x_m^*) . \tag{1}$$

Next we discuss the association probability between two sets of Web pages A and B in a Website. Given a Website topology G, $A = \{a_1, \ldots, a_r\}$ and $B = \{b_1, \ldots, b_t\}$ represent two sets of Web pages in the same Website, and $A \cap B = \Phi$, $A, B \subseteq X$. We will construct the topology probability model to calculate the association probability between A and B, which means the chance of A and B will be visited consequently in the Website topology. Suppose the Web page a_i has been visited, we define the minimum association probability of the Web page a_i to Web page b_j as the association probability of a_i to B as follows:

$$P(B|a_i) = \min\{P(b_1, a_i), P(b_2, a_i), \ldots, P(b_t, a_i)\} . \tag{2}$$

Suppose a set of Web pages, $A = \{a_1, \ldots, a_r\}$ have been visited, we define the maximum association probability of the Web page a_i to Web page b_j as the association probability of A to a_j as follows:

$$P(b_j|A) = \max\{P(b_j, a_1), P(b_j, a_2), \ldots, P(b_j, a_r)\} . \tag{3}$$

In (2), we take the minimum because we consider one-to-many possibilities. In (3), we take the maximum because we consider many-to-one possibilities. In our experimental results, we found that the probability model is quite reasonable and effective. However, we remark that other probability models can be used to define these probabilities. From (2) and (3), we can obtain the formula of calculating association probability between any two sets of Web pages in a Website as below:

$$P(B|A) = \max\{\min\{P(b_j|a_i)\}\} , \quad i = 1, \ldots, r , \ j = 1, \ldots, t . \tag{4}$$

Obviously, $0 < P(B|A) < 1$. The association probability represents the extent of relevance of two sets of Web pages in Website topology. Given association rule $A \Rightarrow B$, if the value of this association probability $P(B|A)$ is high, it means that the two sets of Web pages A and B are "close" in the Website. Otherwise, they are "far away" to each other in position of the Website. Therefore, we can use it to determine the user access efficiency from one set of Web pages to another.

Definition 1 A new interestingness measure of the association rules based on Website topology is defined as follows:

$$Interest(B|A) = 1 - P(B|A) . \tag{5}$$

We can use the rule interest to discover interesting user access patterns. For example, we find an association rule $A \Rightarrow B$ which satisfies the minimum rule support and confidence thresholds. If the rule interest of $A \Rightarrow B$ is high, it means the association rule is a frequent access pattern in Website, but it has been ignored, so the Website designer can use this information to improve the access efficiency of the Website topology. Oppositely, if the rule interest of $A \Rightarrow B$ is low, we may conclude that the association rule is a "common sense knowledge" or it has already been considered or known. Therefore, such association rule is an uninteresting pattern. For instance, two association rules were generated, which satisfy $minSupp = 0.2$ and $minConf = 0.8$. These two association rules represent user access patterns in the Website topology (see Figure 1).

$rule1 : \quad x_A \Rightarrow x_B \quad Support = 0.38, Confidence = 0.86, Interest = 0.55$
$rule2 : x_B x_F \Rightarrow x_E \ Support = 0.25, Confidence = 0.82, Interest = 0.97$

In *rule 1*, we find that x_A is the homepage and it already has a hyperlink to connect x_B, so it is an uninteresting association rule for our analysis. But it is not the case in *rule 2*, we find that x_B and x_F are close, but x_E is much far away from x_B and x_F, so we conclude that it is not easy to visit x_E from x_B and x_F. Therefore, *rule 2* attracts more our interest than *rule 1*. If we do not employ rule interest to measure the interestingness of an association rule in Website, we may misunderstand *rule 1* is more interesting just because it has higher rule support and confidence than *rule 2*. In this case, building a hyperlink between x_F and x_E can significantly improve the access efficiency of visitors who follow this access pattern.

2.3   Mining HIARs

We have mentioned that the rule interest can be an effective interestingness to discover High Interest Association Rules (HIARs) in a Website. We summarize the process of mining top k HIARs as follows:

Begin
1. Import raw Web log dataset L, data preprocessing with L
2. Transform into transaction dataset T
3. Set *minSupp* and *minConf* thresholds
4. Generate association rules satisfying thresholds, obtain ruleset R
5. Import the Website topology G
6. Compute association probability matrix P with G
7. Compute the rule interest value of every association rule in R with P
8. Select the top k rules with the highest rule interest values as the top k HIARs

End

---

3   Website Topology Optimization with Top k HIARs

In this section, we propose an algorithm to optimize the Website topology based on the top k HIARs.

3.1   Topology Interest and Variance

Since the HIARs denote the low access efficiency but important user access patterns in a Website, we can use them to redesign the Website topology in order to meet the needs of visitors.

Definition 2 Given the top k HIARs of a Website topology G, which satisfy specified *minSupp* and *minConf*. The corresponding rule interest values of the top k HIARs are $R = \{r_1, r_2, \ldots, r_k\}$. The Website topology interest is defined as follows:

$$TInterest = \frac{\sum_{i=1}^{k} r_i}{k} \tag{6}$$

Clearly, $0 < TInterest < 1$, k is the number of HIARs, which can be specified by a Website analyst.

The topology interest is a statistical measure to evaluate the overall efficiency of accessing a Website. If the topology interest is high, it means current Website topology cannot well adapt the user access patterns. In fact, special Web pages or special group of users may be specified. Therefore more consideration should be paid to them. For example, it could be more suitable to give more weight on association rules related to important users or Web pages which contain clients' advertisement.

However, topology interest should not be the only criterion in deciding which Website topology is more adaptable to user access patterns. For example, if the Website is just convenient for a certain group of users to visit while it is difficult or uninteresting for other users, it will cause the dissatisfaction of a portion of users. To balance the demands for different groups of users, an additional measure named Website topology variance is defined to help determine a good Website topology.

Definition 3 Given top k HIARs of a Website topology G, a new measure representing the extent of difference among rule interest of association rules is defined as follows:

$$TStd = \sqrt{\frac{\sum_{i=1}^{k}(r_i - TInterest)^2}{k}} \ . \tag{7}$$

The structure and the content of a Website should attract visitors as many as possible. Generally, a low topology variance means the Website topology can balance the access efficiency of different groups of users. Therefore, the topology variance can measure the extent on the whole. We will use both topology interest and topology variance to reconstruct the most adaptive Website topology.

3.2   The WTO Algorithm

Since the Web logs are constantly growing and the user access patterns may change from time to time, we propose the WTO (Website Topology Optimization) algorithm to build an adaptive Website that improves itself by learning from user access patterns. But the problem is that the calculation of the Website topology optimization is still tremendous if we try to find the global optimization solution. It is very inefficient to search every possible Website topology structures, especially when the number of Web pages in a Website is very large. However, it is not necessary to search all possible structures to obtain the optimal solution. Some searching strategies are suggested as follows:

Searching strategies: First, considering the simplicity of reconstructing a Website, we solve the optimization problem by constraining the number of hyperlinks which will not exceed too much than the original one. The algorithm can automatically search possible solutions under this constrain, and find the most adaptive Website topology by comparing the topology interest and topology variance.

Second, since we are considering the top k HIARs as low efficiency but important user access patterns, it is reasonable to just rebuild the hyperlink structure of the portion of Web pages related to these HIARs. Obviously, if we build new hyperlinks to connect the Web pages which are highly relevant, it will certainly improve the access efficiency of Website on the whole. On the other hand, we can delete some hyperlinks among the Web pages which are seldom to be visited together in order to reduce the number of hyperlinks in a suitable level. We use topology interest and topology variance as the criteria to measure the overall efficiency of accessing a Website. After reconstructing the Website topology, we recalculate the topology interest and variance, if they are lower than those of the original one, we can conclude that the new Website topology is more adaptable to current user access patterns. We can iterate the process of reconstructing the Website topology until the optimal one is obtained. This iteration process is formalized in the WTO algorithm as follows:

Begin

1. Import top k HIARs $R = \{r_1, \ldots, r_k\}$ and current Website topology G
2. Set current optimal Website topology $OG = G$, calculate $TInterest$ and $TStd$ of OG
3. Orderly add new hyperlinks to connect Web pages related to HIARs
4. Delete some unnecessary hyperlinks by searching other Web pages
5. If the new Website topology is a connected graph, go to Step 6. Otherwise undo the previous two steps' operations and go back to Step 3
6. Update probability matrix P^* of the new topology G^*
7. Recalculate top k HIARs $R^* = \{r_1^*, \ldots, r_k^*\}$ with P^*
8. Calculate $TInterest^*$ and $TStd^*$ of G^*
9. Compare topology interest and topology variance of OG and G^*
10. If $TInterest^* < TInterest$ and $TStd^*$ is acceptable, go to Step 11. Otherwise, go to Step 12
11. Replace current optimal Website topology $OG = G^*$
12. Iterate the searching process. If no better result appears after repeating specified iteration times, stop and output the current optimal Website topology. Otherwise, go to Step 3

End

---

4   Time and Space Complexity Analysis

In our experiment, we use Apriori algorithm to generate association rules. An alternative method is using FP-tree [7], which is suitable to find association rules in large database or mine long pattern association rules. Assume a Website contains E Web pages, H hyperlinks, and K association rules. The process of building the probability matrix requires $O(EH)$ calculations. To compute all the interest values of HIARs needs $O(KUV)$, where U and V denote the average lengths of LHS and RHS in an association rule. Hence, the total time complexity of mining HIARs is $O(EH + KUV)$. In the WTO algorithm, since the searching process mainly focuses on the Web pages related to HIARs, it just takes $O(EHT + KUVT)$ to gain the good result, where T is the iteration times of recalculating the top k HIARs in a new Website topology.

Because our algorithm involves matrix computation, we propose to reside the probability matrix in the main memory. Together with the storage of the association rules, the whole space complexity of the algorithm is $O(MH + R)$, where M is the size of matrix, H is the number of hyperlinks, and R is the number of association rules satisfying thresholds. Even if there are thousands of Web pages, the algorithm is still possible to implement on most PC nowadays.

5   Experiments

In order to verify the performance and efficiency of our algorithms, we have implemented a Web log mining system to discover HIARs and optimize Website topology. All experiments reported below were performed on a PC with a Pentium III CPU of 1.2GB and 384 MB main memory, running on Microsoft Windows 2000 Professional.

5.1   Synthetic Dataset Results

Data Generation: We use a random number generator to generate different synthetic transaction datasets. Table 2 lists the control parameters used to generate these datasets. With these datasets, we can customize variety of Web log datasets and Website topologies for testing.

Table 2. Cotnrol parameters in data generation

Parameter	Definition
Nt	Number of transaction
At	Average transaction size
Ap	Average pattern size
Np	Number of patterns
Nw	Number of Web pages
Nh	Number of hyperlinks

Table 3. Top 5 HIARs

HIARs	Support	Confidence	Interest
$E \Rightarrow F$	0.037	0.81	0.97
$G \Rightarrow H$	0.041	0.93	0.96
$A \Rightarrow E$	0.032	0.62	0.93
$BG \Rightarrow H$	0.031	0.75	0.91
$C \Rightarrow GH$	0.021	0.67	0.86

A Motivation Example: In this section, we try to find the top 5 HIARs and optimize the Website topology (see Figure 1) with a dataset which has 10,000 transaction records. We set $minSupp = 0.02$ and $minConf = 0.5$. If we do not employ rule interest to identify really interesting rules for Website analyst, over 40 rules were found, it will greatly interrupt the effective analysis of access patterns in Website. Therefore the rule interest can prune the number of uninteresting patterns. Table 3 lists the top 5 HIARs. It is significant that the most interesting rules have high rule interest values.

Table 4. Topology interest and variance

Optimization Measure	Original Topology (Figure 1)	Optimal Topology (Figure 3)
Topology Interest	0.752	0.678
Topology Variance	0.117	0.109

Table 4 shows the comparison results of the original Website topology and the optimal Website topology. Obviously, the new Website topology is superior because both the topology interest and topology variance are lower than those of the original Website topology.

We can see from Figure 2 that the relation between topology interest and rule interest. An association rule presents a general user access patterns in Website. But a HIAR can represent an important but not efficient access pattern for certain group of visitors. If we can build hyperlinks among the Web pages related to the HIARs, it will certainly improve the efficiency of the user access patterns. Further, if we can use topology interest and variance which are based on top k HIARs to adjust the Website topology, it will certainly maximize the satisfaction of visitors of Website on the whole. This is exactly the essence of our WTO algorithm.

Figure 3 shows the optimal Website topology in our experiment. Comparing to the original Website topology (see Figure 1), the new Website topology has been added new hyperlinks of AE, EF and GH, and deleted BF and EH. It can be explained as follows, A and E frequently appear together in the original transaction dataset, but the two nodes are "far away" in the original topology. So, the $A \Rightarrow E$ is the HIAR for analysis. If we build a hyperlink between A and E, the new rule interest of $A \Rightarrow E$ in the new Website topology will drop down. It means the access efficiency of $A \Rightarrow E$ has been improved. This is also the same reason why EF and GH are built. However, BF and EH are seldom access patterns of the users, we can delete them to reduce the number of hyperlinks. Obviously, the optimal topology is more accessible than the original one.

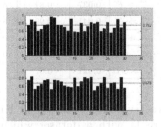   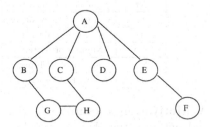

Fig. 2. Comparison of the original and optimal topology The bars denote the rule interest values of top 30 HIARs The horizontal line denotes the value of topology interest

Fig. 3. The optimal Website topology

5.2 Accuracy Studies

We also propose a feedback test in order to check the accuracy of our optimization approach. First, we construct a "best" Website topology that is most efficient to current user access patterns. Second, we randomly change the Website topology by adding and deleting the hyperlinks among the Web pages. Next, we apply the WTO algorithm to optimize the distorted Website topology by using the same Web log dataset. If the new Website topology is the same or quite similar to the original "best" Website topology, then we can guarantee our approach is valid and effective. We use topology accuracy to measure the similarity between two topologies. Topology accuracy is the ratio between L_r and L_o, where L_r is the number of the recovered hyperlinks which are the same with the "best" Website topology, and L_o is the number of the original hyperlinks of the "best" Website topology. The value of accuracy is between 0 and 1.

Hence, a good optimization result is indicated by a high accuracy. However, it is not necessary to recover all the hyperlinks in the original "best" Website topology, especially when the number of Web pages is large. Because our WTO algorithm focus on the important hyperlinks related to the top k HIARs instead of all hyperlinks, some unimportant hyperlinks which mostly include in the unmatched hyperlinks $Lo - Lr$ will not affect the access efficiency of Website topology.

A snowflake like Website topology is proposed in Figure 4(a). Assume node A is the index Web page. Other nodes are sub Web pages which are frequently visited together with A. Therefore the snowflake Website topology is the original "best" Website topology to adapt the specified user access patterns. The generated transaction records are in Table 5. Next, we randomly construct two distorted Website topologies (see Figure 4. (b) and (c)). Then we apply the WTO algorithm to reconstruct the distorted Website topologies, respectively. Although (b) and (c) are quite different, both the optimization results report that Figure 4.(a) is still the optimal Website topology to adapt current user access patterns.

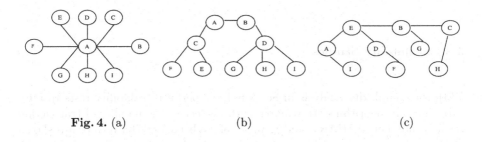

Fig. 4. (a)   (b)   (c)

Table 5. Specified transaction dataset

TID	Itemset
100	AB
200	AC
300	AD
400	AE
...	...

Table 6. Accuracy list using Syn5 with different hyperlinks

Top k HIARs	Rule Interest [min, max]	Original Hyperlinks	Recovered Hyperlinks	Accuracy
100	[0.59,0.98]	1500	1322	88%
150	[0.52,0.98]	2000	1719	86%
200	[0.50,0.99]	3000	2495	83%

Table 7. Accuracy list using different datasets

Datasets	Web Pages	Transaction Records	Top k HIARs	Rule Interest [min, max]	Original Hyperlinks	Recovered Hyperlinks	Accuracy
Syn1	20	1000	10	[0.72,0.93]	26	26	100%
Syn2	65	5000	30	[0.67,0.95]	78	74	95%
Syn3	100	15000	50	[0.65,0.96]	113	106	94%
Syn4	300	50000	100	[0.61,0.97]	367	329	90%
Syn5	1050	105000	150	[0.57,0.98]	1129	1004	89%

In this experiment, we also construct other complex "best" Website topologies and corresponding transaction datasets, and then distort them. However, the distorted Website topologies can still be well recovered to the original ones as we have expected. The accuracy list is shown in Table 6 and Table 7. The rule interest columns in Tables 6 and Table 7 show the interest value range of the top k HIARs. Accuracy ratio denotes the extent how much we can recover the distorted "best" Website topology. We see from the tables that when the number of Web pages and the number of hyperlinks of the original "best" Website topology increase, we must use more high interest association rules in order to recover the hyperlinks and reconstruct the Website topology. Generally, more HIARs used, higher accuracy is expected. The interesting experiment strongly suggests that our WTO algorithm is feasible.

5.3   Scalability Studies

With the complexity analysis in Section 4, we perform scalability tests to indicate that our algorithms are scalable with respect to the number of transaction records, number of HIARs, and number of Web pages. The results are shown in Figure 5, Figure 6 and Figure 7, respectively. As we have expected, Figure 5 shows a linear relationship between the number of transaction records and the execution time. The linear relationship is also expected in increasing interesting association rules (see Figure 6). However it is not the case in Figure 7. Due to the fact that the computation cost for association probability matrix is expensive when the number of Web pages is large, the expected time complexity is $O(EH)$. We note that the time complexity of the global optimization is $O(2^E)$ if we search all the possibilities, but it is unnecessary and impossible especially in a complicated Website with many Web pages and hyperlinks. Because we mainly search those Web pages related to HIARs, this step will significantly reduce the execution time required. Therefore, the total execution time is acceptable in our experiments. According to our experimental results, we see that our approach is feasible to apply in practical applications.

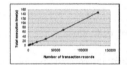 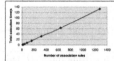 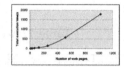

Fig. 5.   Scalability of increasing transaction records

Fig. 6. Scalability of increasing HIARs

Fig. 7.   Scalability of increasing Web pages

5.4   A Real Case Study

In construction of a Website, it is important to know how to organize the Website structure and the Web content properly in order to improve the user service. In this section, we report a real dataset to test our WTO algorithm. The real dataset is the Web logs from a famous Website collected in a month. After data preprocessing, we found that the one-month Web log dataset contains over 500,000 transaction itemsets, and about 1,100 Web pages in the Website have been visited. We set $minSupp$, $minConf$ are 0.01% and 50% respectively. Over 1600 association rules satisfying the thresholds were found. However, we can only use the top 100 HIARs to optimize the original Website topology. The rule interest values of the top 100 HIARs range from 0.78 to 0.98. The original topology interest is 0.87 and the topology variance is 0.27. By applying WTO, we find an optimal Website topology, which topology interest and topology variance has been reduced to 0.79 and 0.24, respectively. The total running time is also satisfying. It just takes about 35 minutes to mine the top 100 HIARs and optimize the Website topology in our machine. The practical experiment demonstrates that our HIARs and WTO algorithm are also effective in practical applications and can be put into online or offline Website service analysis. It provides an easy and flexible way for Website analyst to identify important and potential access patterns, and use the information to improve the Website design and management.

6   Conclusions

In this paper, we have presented a method of mining interesting association rules in Website topology. We adopt a new interestingness measure named rule interest to classify interesting and uninteresting association patterns. Furthermore, Website topology interest and topology variance are introduced to evaluate the access efficiency of a Website. Based on the top k HIARs, we propose the WTO algorithm to optimize the Website topology in order to maximize the visitors' satisfaction on the whole.

The experimental results have shown that the algorithm is effective and scalable. We can get the results in an acceptable period of time. The results can be used to capture and analysis user behaviors and then improve the access efficiency of a Website.

References

1. R. Agrawal and R. Srikant, Fast Algorithm for Mining Association Rules, *Proc. Int'l Conf. Very large Data Bases*, pp.487-499, 1994.
2. R. Agrawal, T. Imielinski, and A. Swami, Database Mining: A Performance Perspective, *IEEE Trans. Knowledge and Data Eng.*, **5**(6) (1993):914-925.
3. R. Agrawal, T. Imielinski, and A. Swami, Mining Association Rules between Sets of Items in Large Databases, In: *Proceedings of the ACM SIGMOD Conference on Management of Data*, 1993: 207-216.

4. R. Agrawal and R. Srikant, Fast algorithms for mining association rules in large databases, In Research Report RJ 9839, *IBM Almaden Research Center*, San Jose, CA, June 1994.

5. R. Agrawal and R. Srikant, Fast algorithms for mining association rules, *In Proc. 1994 Int. Conf. Very Large Data Bases (VLDB'94)*, pages 487-499, Santiago, Chile, Sept.1994.

6. John D. Garofalakis, Panagiotis Kappos, Dimitris Mourloukos: Website Optimization Using Page Popularity. *IEEE Internet Computing* **3**(4): 22-29, 1999.

7. J. Han and M. Kamber, Data Mining: Concepts and Techniques, Morgan Kaufmann, 2000.

8. M. Perkowitz and O. Etzioni, Adaptive Websites: an AI Challenge, *In Proceedings of the Fifteenth International Joint Conference on Artificial Intelligence*, 1997.

9. M. Perkowitz and O. Etzioni, Adaptive Websites: Automatically Synthesizing Web Pages, *In Proceedings of the Fifteenth National Conference on Artificial Intelligence*, 1998.

10. M. Perkowitz and O. Etzioni, Adaptive Websites: Conceptual cluster mining, *In Proc. 16th Joint Int. Conf. on Artificial Intelligence (IJCAI'99)*, pages 264-269, Stockholm, Sweden, 1999.

11. J. Srivastava, R. Cooley, M. Deshpande, and P. N. Tan, Web Usage Mining: Discovery and applications of usage patterns from web data, *SIGKDD Explorations*, 1:12-23, 2000.

12. L. Tauscher and S. Greenberg, How people revisit web pages: Empirical findings and implications for the design of history systems. International Journal of Human Computer Studies, Special issue on World Wide Web Usability, 97-138, 1997.

13. Q. Yang, J. Huang and M. Ng, A data cube model for prediction-based Web prefetching, *Journal of Intelligent Information Systems*, **20**:11-30, 2003.

14. C. Zhang and S. Zhang, Association rule mining: models and algorithms, Springer, 2002.

A Markovian Approach for Web User Profiling and Clustering

Younes Hafri[12], Chabane Djeraba[2], Peter Stanchev[3], and Bruno Bachimont[1]

[1] Institut National de l'Audiovisuel, 4 avenue de l'Europe
94366 Bry-sur-Marne Cedex, France
{yhafri,bbachimont}@ina.fr
[2] Institut de Recherche en Informatique de Nantes, 2 rue de la Houssiniere
43322 Nantes Cedex, France
djeraba@irin.univ-nantes.fr
[3] Kettering University, USA
3, Kettering University, Flint, MI 48504, USA
pstanchev@kettering.edu

Abstract. The objective of this paper is to propose an approach that extracts automatically web user profiling based on user navigation paths. Web user profiling consists of the best representative behaviors, represented by Markov models (MM). To achieve this objective, our approach is articulated around three notions: (1) Applying probabilistic exploration using Markov models. (2) Avoiding the problem of Markov model high-dimensionality and sparsity by clustering web documents, based on their content, before applying the Markov analysis. (3) Clustering Markov models, and extraction of their gravity centers. On the basis of these three notions, the approach makes possible the prediction of future states to be visited in k steps and navigation sessions monitoring, based on both content and traversed paths. The original application of the approach concerns the exploitation of multimedia archives in the perspective of the Copyright Deposit that preserves French's WWW documents. The approach may be the exploitation tool for any web site.

1 Introduction

On the web today, sites are still unable to market each individual user in a way, which truly matches their interests and needs. Sites make broad generalizations based on huge volumes of sales data that don't accurate an individual person. Amazon.com tracks relationships in the buying trends of its users, and makes recommendations based upon that data. While viewing a DVD by Dupont Durand, the site recommends three other Dupont Durant DVDs that are often bought by people who buy first DVD. The agreement calls for the participating web documents to track their users so the advertisements can be precisely aimed at the most likely prospects for web documents. For example, a user who looks up tourist document about Paris might be fed ads for web documents of hotels in Paris. What would be far more useful would be if web sites were able to understand specific user's interests and provide them with the content that was

K.-Y. Whang, J. Jeon, K. Shim, J. Srivatava (Eds.): PAKDD 2003, LNAI 2637, pp. 191–202, 2003.
© Springer-Verlag Berlin Heidelberg 2003

relevant to them. Instead of requiring users to provide their interests, a website could learn the type of content that interests the users and automatically place that information in more prominent locations on the page. Taking a film example, if a web site knew where you lived, it could provide you with information when an actor is touring in the user area. From an advertising standpoint, this technology could be used for better targeting of advertisements to make them more relevant to each user. It is evident that there are strong dependencies between web exploration and different domains and usages. The fundamental requirements that ensure the usability of such systems are: (1) obtaining compressed and exhaustive web representations, and (2) providing different exploration strategies adapted to user requirements. The scope of the paper deals with the second requirement by investigating an exploration based on historical exploration of web and user profiles. This new form of exploration induces the answer to difficult problems. An exploration system should maintain over time the inference schema for user profiling and user-adapted web retrieval. There are two reasons for this. Firstly, the information itself may change. Secondly, the user group is largely unknown from the start, and may change during the usage of exploration processes. To address these problems, the approach, presented in this paper, models profile structures extracted and represented automatically in Markov models in order to consider the dynamic aspect of user behaviors. The main technical contribution of the paper is the notion of probabilistic prediction, path analysis using Markov models, clustering Markov models and dealing with the high dimension matrix of Markov models in clustering algorithm. The paper provides a solution, which efficiently accomplishes such profiling. This solution should enhance the day-to-day web exploration in terms of information filtering and searching.

The paper contains the following sections. Section 2 situates our contribution among state of art approaches. Section 3 describes user-profiling based web exploration. Section 4 highlights the general framework of the system and presented some implementation results. Finally, section 5 concludes the paper.

2   Related Works and Contribution

The analysis of sequential data is without doubts an interesting application area since many real processes show a dynamic behavior. Several examples can be reported, one for all is the analysis of DNA strings for classification of genes, protein family modeling, and sequence alignment. In this paper, the problem of unsupervised classification of temporal data is tackled by using a technique based on Markov Models. MMs can be viewed as stochastic generalizations of finite-state automata, when both transitions between states and generation of output symbols are governed by probability distributions [1]. The basic theory of MMs was developed in the late 1960s, but only in the last decade it has been extensively applied in a large number of problems, as speech recognition [6], handwritten character recognition [2], DNA and protein modeling [3], gesture recognition [4], behavior analysis and synthesis [5], and, more in general, to computer vision problems. Related to sequence clustering, MMs has not been extensively used, and a few papers are present in the literature. Early works were proposed in [6,7], all related to speech recognition. The first interesting approach not directly linked to speech issues was presented by Smyth [8], in which clustering was faced by devising a "distance" measure between sequences using HMMs. Assuming each model structure known, the algorithm trains an HMM for each sequence so that the log-likelihood (LL) of each model, given each sequence, can be computed. This information is used to build a LL distance matrix to be used to cluster the sequences in K groups, using a hierarchical algorithm. Subsequent work, by Li and Biswas [10,11], address the clustering problem focusing on the model selection issue, i.e. the search of the HMM topology best representing data, and the clustering structure issue, i.e. finding the most likely number of clusters. In [10], the former issue is addressed using standard approach, like Bayesian Information Criterion [12], and extending to the continuous case the Bayesian Model Merging approach [13]. Regarding the latter issue, the sequence-to-HMM likelihood measure is used to enforce the within-group similarity criterion. The optimal number of clusters is then determined maximizing the Partition Mutual Information (PMI), which is a measure of the inter-cluster distances. In [11], the same problems are addressed in terms of Bayesian model selection, using the Bayesian Information Criterion (BIC) [12], and the Cheesman-Stutz (CS) approximation [14]. Although not well justified, much heuristics are introduced to alleviate the computational burden, making the problem tractable, despite remaining of elevate complexity. Finally, a model-based clustering method is also proposed in [15], where HMMs are used as cluster prototypes, and Rival Penalized Competitive Learning (RPCL), with state merging is then adopted to find the most likely HMMs modeling data. These approaches are interesting from the theoretical point of view, but they are not tested on real data. Moreover, some of them are very computationally expensive.

Each visitor of a web site leaves a trace in a log file of the pages that he or she visited. Analysis of these click patterns can provide the maintainer of the site with information on how to streamline the site or how to personalize it with respect to a particular visitor type. However, due to the massive amount of data that is generated on large and frequently visited web sites, clickstream analysis is hard to perform 'by hand'. Several attempts have been made to learn the click behaviour of a web surfer, most notably by probabilistic clustering of individuals with mixtures of Markov chains [16,9,17]. Here, the availability of a prior categorization of web pages was assumed; clickstreams are modelled by a transition matrix between page categories. However, manual categorization can be cumbersome for large web sites. Moreover, a crisp assignment of each page to one particular category may not always be feasible. In the following section we introduce the problem and then describe the model for clustering of web surfers. We give the update equations for our algorithm and describe how to incorporate prior knowledge and additional user information. Then we apply the method to logs from a large commercial web site KDDCup (http://www.ecn.purdue.edu/KDDCUP/), discuss both method and results and draw conclusions.

3   User Profiling and Web Exploration

3.1   Mathematical Modeling

Given the main problem "profiling of web exploration", the next step is the selection of an appropriate mathematical model. Numerous time-series prediction problems, such as in [18], supported successfully probabilistic models. In particular, Markov models and Hidden Markov Models have been enormously successful in sequence generation. In this paper, we present the utility of applying such techniques for prediction of web explorations.

A Markov model has many interesting properties. Any real world implementation may statistically estimate it easily. Since the Markov model is also generative, its implementation may derive automatically the exploration predictions. The Markov model can also be adapted on the fly with additional user exploration features . When used in conjunction with a web server, this later may use the model to predict the probability of seeing a scene in the future given a history of accessed scenes. The Markov state-transition matrix represents, basically, "user profile" of the web scene space. In addition, the generation of predicted sequences of states necessitates vector decomposition techniques. The figure 1 shows the graph representing a simple Markov chain of five nodes and their corresponding transitions probabilities. The analogy between the transition probability matrix and the graph is obvious. Each state of the matrix corresponds to a node in the graph and similarly each probability transition in the matrix corresponds to and edge in the graph. A set of three elements defines a discrete Markov model: $< \alpha, \beta, \lambda >$ where α corresponds to the state space. β is a matrix representing transition probabilities from one state to another. λ is the initial probability distribution of the states in α. Each transition contains the identification of the session, the source scene, the target scene, and the dates of accesses. The fundamental property of Markov model is the dependencies of the previous states. If the vector $\alpha(t)$ denotes the probability vector for all the

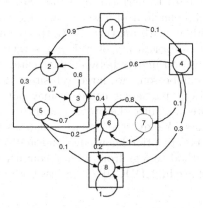

Fig. 1. Transition graph

states at time 't', then:

$$\alpha(t) = \alpha(t-1) \times \beta \tag{1}$$

If there are N states in the Markov model, then the matrix of transition probabilities β is of size N x N. Scene sequence modeling supports the Markov model. In this formulation, a Markov state corresponds to a scene presentation, after a query or a browsing. Many methods estimate the matrix β. Without loss of generality, the maximum likelihood principle is applied in this paper to estimate β and λ. The estimation of each element of the matrix $\beta[v, v']$ respect the following formula:

$$\beta[v, v'] = \phi(v, v')/\phi(v) \tag{2}$$

where $\phi(v, v')$ is the count of the number of times v' follows v in the training data. We utilize the transition matrix to estimate short-term exploration predictions. An element of the matrix state, say $\beta[v, v']$ can be interpreted as the probability of transitioning from state v to v' in one step. The Markovian assumption varies in different ways. In our problem of exploration prediction, we have the user's history available. Answering to which of the previous explorations are *good predictors* for the next exploration creates the probability distribution. Therefore, we propose variants of the Markov process to accommodate weighting of more than one history state. So, each of the previous explorations are used to predict the future explorations, combined in different ways. It is worth noting that rather than compute β and higher powers of the transition matrix, these may be directly estimated using the training data. In practice, the state probability vector $\alpha(t)$ can be normalized and threshold in order to select a list of *probable states* that the user will choose.

3.2   Predictive Analysis

The implementation of Markov models into a web server makes possible four operations directly linked to predictive analysis. In the first one, the server supports Markov models in a predictive mode. Therefore, when the user sends an exploration request to the web server, this later predicts the probabilities of the next exploration requests of the user. This prediction depends of the history of the user requests. The server can also support Markov models in an adaptive mode. Therefore, it updates the transition matrix using the sequence of requests that arrive at the web server. In the second one, prediction relationship, aided by Markov models and statistics of previous visits, suggests to the user a list of possible scenes, of the same or different web bases, that would be of interest to him, and then the user can go to next. The prediction probability influences the order of scenes. In the current framework, the predicted relationship does not strictly have to be a scene present in the current web base. This is because the predicted relationships represent user traversal scenes that could include explicit user jumps between disjointing web bases. In the third one, there is generation of a sequence of states (scenes) using Markov models that predict the sequence of states to visit next. The result returned and displayed to the user consists of a sequence of states. The sequence of states starts at the current scene the user is browsing. We consider default cases, such as, if the sequence of states contains cyclic state, they are marked as "explored" or "unexplored". If multiple states have the same transition probability, a suitable technique chooses the next state. This technique considers the scene with the shortest duration. Finally, when the transition probabilities of all states from the current state are too weak, then the server suggests to the user, the go back to the first state. In the fourth one, we refer to web bases that are often good starting points to find documents, and we refer to web bases that contain many useful documents on a particular topic. The notion of profiled information focuses on specific categories of users, web bases and scenes. The web server iteratively estimate the weights of profiled information based on the Markovian transition matrix.

3.3   Path Analysis and Clustering

To reduce the dimensionality of the Markov transition matrix β, a clustering approach is used. It reduces considerably the number of states by clustering similar states into *similar groups*. The reduction obtained is about log N, where N is the number of scenes before clustering. The clustering algorithm is a variant of k-medoids, inspired of [19]. The particularity of the algorithm (algorithm 1) is the replacement of sampling by heuristics. Sampling consists of finding better clustering by changing one medoid. However, finding the best pair (medoid, item) to swap is very costly $(O(k(nk)2))$. That is why, heuristics have been introduced in [19] to improve the confidence of swap (medoid, data item). To speed up the choice of a pair (medoid, data item), the algorithm sets a maximum number of pairs to test (num_pairs), then choose randomly a pair and compare the dissimilarity. To find the k medoids, our algorithm begins with an arbitrary selection of k objects. Then in each step, a swap between a selected object O_i and a non-selected object O_h is made, as long as such a swap would result in an improvement of the quality of the clustering. In particular, to calculate the effect of such a swap between O_i and O_h , the algorithm computes the cost C_{jih} for all non-selected objects O_j. Combining all possible cases, the total cost of replacing O_i with O_h is given by: $T_{cih} = \sum C_{jih}$. The algorithm 1 is given bellow.

```
Clustering()
  {
    Initialize num_tries and num_pairs;
    min_cost = infinitive;
    for k = 1 to num_tries do
        current=k; randomly selected items in the entire data set.
        L=1;
        repeat
                xi = a; randomly selected item in current
                xh = a; randomly selected item in {entire data set current}.
                if TCih  < 0 then
                    current = currentxi+xh;
                else
                    j = j+1;
```

```
            end if
       until (j <= num_pairs)
       if min_cost < cost(current) then
               best = current;
       end if
  end for
  return   best;
  }
```

Algorithm 1. Clustering function

The algorithm go on choosing pairs until the number of pair chosen reach the maximum fixed. The medoids found are very dependant of the k first medoids selected. So the approach selects k others items and restarts num_tries times (num_tries is fixed by user). The best clustering is kept after the num_tries tries.

4   Approach Implementation

We have implemented the approach, and particularly the clustering method of navigation sessions. Our objective through this implementation is articulated around two points. The first one concerns the study of the different problems that held when we deal with a great number of sessions represented by Markov models. The second one concerns the extraction of the most representative behaviors of the web site. A representative behavior is represented by a Markov model (pages, transitions between pages and the average time in each page) for which the access frequencies to their pages are homogeneous. It means that, the ideal representative user behaviors have not pages accessed frequently and pages rarely accessed, such as presented in the figure presented bellow. Figure 2 represents page frequencies calculated on the basis of the exploration session database, extracted from KDD Cup logs of a commercial web site that will be detailed bellow. In the figure, a curve in two-dimension space is shown In abscise, we have the web page numbers. In the ordinate, we have the frequencies, which measure the number of access to the concerned page. The highest number

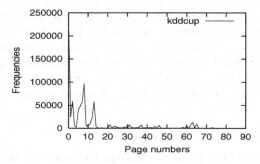

Fig. 2. Classical curve form

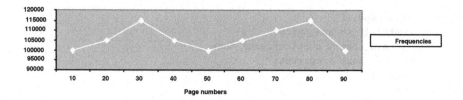

Fig. 3. Ideal curve form

in abscise represents the page numbers of the target web site. In this example, we have 90 pages. The frequency of the page number 0 is between 200000 and 250000. It means that the page number 0 has been accessed more than 200000 times and less than 250000 times. It is the highest frequency in the curve. We can deduce that the page number 0 corresponds to the root page or the main page of the web site. The frequency of the page number 80 is equal to 0. It means that no user accessed to this page. We note that the most frequently accessed pages are between 0 and 15; the less frequently accessed pages are between 15 and 46, and between 60 and 70. The pages between 48 and 58 and between 75 and 90 have never been accessed. Globally, the figure represents a typical state of web site access, where some pages are frequently accessed (ex. page number 0), others are rarely accessed (ex page number 30), and others are not accessed at all (ex. page number 90). In this typical state of web site pages, the cardinality of pages that are frequently accessed is generally less than the cardinality of pages that are rarely or never been accessed. Again, in the ideal result of our approach, we should obtain Markov models composed of a sub set of the target web site pages, and the curve of page frequencies should be homogeneous. It means that the curve should be stable, like in the figure 3. In the figure 3 the page frequencies are between 100000 and 120000 access. So there is a stability of the curve compared to the classical curve form where the page frequencies are between 0 and 250000. This figure is just an example of an ideal curve form where the best behaviors contain pages that have homogeneous page frequencies.

4.1   Data Set

The used data set is provided by KDDCup (www.ecn.purdue.edu/KDDCUP/) which is a yearly competition in data mining that started in 1997. It's objective is to provide data sets in order to test and compare technologies (prediction algorithms, clustering approaches, etc.) for e-commerce, considered as a "killer domain" for data mining because it contains all the ingredients necessary for successful data mining. The ingredients of our data set include many attributes (200 attributes), and many records (232000). Each record corresponds to a session.

4.2   Results

The tests were carried out on a PC Pentium III with 500 MHz and 256 MB of RAM on the Data set of sessions composed of 90785 sessions (individual Markov models). In our tests, we considered different number of classes, iterations and maximum number of neighbors to compute run time and clustering distortion. We will focus our results on run time and distortion obtained when varying the maximum number of neighbors. So we fixed the number of classes and iterations to respectively 4 and 5. For different numbers of classes and iteration, we obtain similar results.

We tested the approach is the data set composed of 90785 markov models, and we supposed that there are 4 typical user behaviors (number of classes equal to 4) and five iteration of the clustering algorithm (number of iteration equal to 5). Previous experiments [19] proved that the distortion is conversely proportional to the number of iterations. That is why we concentrate our experiments on Run time and distortion values on the basis of respectively numbers of classes (clusters) and iterations. On the basis of the figure curve, we can highlight the

Table 1. Run Time (minutes) proposional to the number of neighbors

Run time	Classes	Iterations	Max Neighbors	Medoids	Distortion
15mn	4	5	10	123023, 294044, 328940, 343287	139004.3
15mn	4	5	10	13137, 145762, 387937, 21549	133472.47
15 mn	4	5	10	15368, 59465, 235239, 101451	136211.69
35 mn	4	5	50	101467, 233739, 145567, 5565	130836.16
35 mn	4	5	50	352195, 185969, 246696, 3782	131280.61

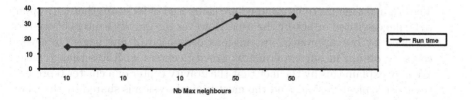

following conclusions. The run time execution is proportional to the maximum number of neighbors. For 10 neighbors, we have 15 minutes run time. For 50 neighbors, we have 35 minutes of execution. We think that the run time will be very high when the number of iteration and classes are high. However the run time is less increasing than the maximum number of neighbors (figure 1). Another remark concerns the distortion (figure 4). The good quality of distortion is proportional to the maximum number of the neighbors. More generally, the results of tests showed some interesting points.

- The first point sub-lined the necessity to clean carefully the data set and to select the useful attribute before any application of the approach. In the data collections, we have a relation table with 200 attributes, and few of them are really useful to achieve our objective. We use only 19 attributes that specify the identification of the web pages, the identifier of the user session, the link relations between web pages and the time spent by the user in each web page.

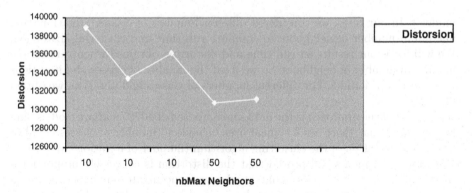

Fig. 4. Distortion conversely proportional to the number of the maximum neighbors

- The second point sub-lined the necessity to create a new data set suitable for our approach. The original data set contain 230000 sessions, and only 90785 sessions are useful for our approach.
- The third point notes that the features of some attributes have been deleted, because they contain confidential information. So we don't know if they are useful or not in the quality of results as we don't know any thing about these attributes.
- The fourth point showed that the gravity centers of clusters are too small. The original sessions to be grouped are composed of 90 states that correspond to 90 pages visited or not by the user. However the gravity centers of clusters, obtained by our approach, are sessions composed of few pages, in several cases we obtain in our experiments gravity centers with less than 5 pages. We may explain this by the fact that the gravity center of a cluster represents the most typical session. And the most typical session is shared by the whole sessions in the cluster. And the shared point is necessary small when we consider a big number of sessions. The different tests showed that higher is the cardinality of the cluster, lesser is the volume of the gravity center. We think that this property is interesting to make accurate decision because the site administrator obtains simple and easy to interpret gravity centers, as they are composed of few states and transitions.
- The fifth point concerns the sparse property of the Markov models of sessions. The original Markov models are high dimensional and too sparse. Each session is represented by a high number of states (90 states) and transitions, however not all state are used. This is the result of the fact that the data set corresponds to a web site composed of many pages, and few number of these pages are used in a session. Our approach is addressed to such voluminous sites. The problem of the high dimension and sparse Markov model matrix is that it needs important resources: too large central memory, powerful processor and a clustering algorithm adapted to this high dimensionality. In our experiment, we considered 90 pages, however many commercial web sites consider hundred pages.

- The sixth point concerns how web site administrators may use the results of our experiments. That is good to obtain the most representative behaviors, but how the representative behaviors (gravity centers of behavior clusters) may be exploited in the real e-commerce environment.

5 Conclusion

The future web sites, particularly web services, will amass detailed records of which uses their web documents and how people use them. However, this "new" future industry highlights the most ambitious effort yet to gather disparate bits of personal information into central databases containing electronic information on potentially every person who surfs the web bases. Profiling explorations are in the interest of the user, providing more customized and directed services through the web bases.

An approach which makes possible prediction of future states to be visited in k steps corresponding to k web pages hyper-linked, based on both content and traversed paths are presented in this paper. To make this prediction possible, three concepts have been highlighted. The first one represents user exploration sessions by Markov models. The second one avoids the problem of Markov model high-dimensionality and sparsely by clustering web documents, based on their content, before applying Markov analysis. The third one extracts the most representative user behaviors (represented by Markov models) by considering a clustering method.

References

1. Rabiner, L.R.: A tutorial on Hidden Markov Models and Selected Applications in Speech Recognition. Proc. of IEEE 77(2) (1989) 257-286
2. Hu, J., Brown, M.K., Turin, W.: HMM based on-line handwriting recognition. IEEE Trans. Pattern Analysis and Machine Intelligence, 18(10) (1996) 1039-1045
3. Hughey, R., Krogh, A.: Hidden Markov Model for sequence analysis: extension and analysis of the basic method. Comp. Appl. in the Biosciences 12 (1996) 95-107
4. Eickeler, S., Kosmala, A., Rigoll, G.: Hidden Markov Model based online gesture recognition. Proc. Int. Conf. on Pattern Recognition (ICPR) (1998) 1755-1757
5. Jebara, T., Pentland, A.: Action Reaction Learning: Automatic Visual Analysis and Synthesis of interactive behavior. In 1st Intl. Conf. on Computer Vision Systems (ICVS'99) (1999)
6. Rabiner, L. R., Lee, C.H., Juang, B. H., Wilpon, J. G.: HMM Clustering for Connected Word Recognition. Proceedings of IEEE ICASSP (1989) 405-408
7. Lee, K. F.: Context-Dependent Phonetic Hidden Markov Models for Speaker Independent Continuous Speech Recognition. IEEE Transactions on Acoustics, Speech and Signal Processing 38(4) (1990) 599-609
8. Smyth, P.: Clustering sequences with HMM, in Advances in Neural Information Processing (M. Mozer, M. Jordan, and T. Petsche, eds.) MIT Press 9 (1997)
9. Smyth, P.: Clustering sequences with hidden markov models. In M.C. Mozer, M.I. Jordan, and T. Petsche, editors, Advances in NIPS 9, (1997)

10. Li, C., Biswas, G.: Clustering Sequence Data using Hidden Markov Model Repre-
    sentation, SPIE'99 Conference on Data Mining and Knowledge Discovery: Theory,
    Tools, and Technology, (1999) 14-21
11. Li, C., Biswas, G.: A Bayesian Approach to Temporal Data Clustering using Hidden
    Markov Models. Intl. Conference on Machine Learning (2000) 543-550
12. Schwarz, G.: Estimating the dimension of a model. The Annals of Statistics, 6(2)
    (1978) 461-464
13. Stolcke, A., Omohundro, S.: Hidden Markov Model Induction by Bayesian Model
    Merging. Hanson, S.J., Cowan, J.D., Giles, C.L. eds. Advances in Neural Informa-
    tion Processing Systems 5 (1993) 11-18
14. Cheeseman, P., Stutz, J.: Bayesian Classification (autoclass): Theory and Results.
    Advances in Knowledge discovery and data mining, (1996) 153-180
15. Law, M.H., Kwok, J.T.: Rival penalized competitive learning for model-based se-
    quence Proceedings Intl Conf. on Pattern Recognition (ICPR) 2, (2000) 195-198
16. Cadez, I., Ganey, S. and Smyth, P.: A general probabilistic framework for clustering
    individuals. Technical report, Univ. Calif., Irvine, March (2000)
17. Smyth, P.: Probabilistic model-based clustering of multivariate and sequential data.
    In Proc. of 7th Int. Workshop AI and Statistics, (1999) 299-304
18. Ni, Z.: Normal orthant probabilities in the equicorrelated case. Jour. Math. Anal-
    ysis and Applications, n° 246, (2000) 280-295
19. Ng, R.T. and Han, J.: CLARANS: A Method for Clustering Objects for Spatial
    Data Mining. TJDE 14(5), (2002) 1003-1016

Extracting User Interests from Bookmarks on the Web

Jason J. Jung[1] and Geun-Sik Jo[1]

School of Computer Engineering, Inha University,
Younghyun-Dong, Nam-Gu,402-751 Incheon, Korea
j2jung@intelligent.pe.kr, gsjo@inha.ac.kr

Abstract. This paper regards bookmarking as the most important information to extract user preferences among user behaviors. Bookmarks are categorized on Bayesian networks by an ontology. Considering the relationships between categories, evidential supports are mutually propagated to improve the coverage of the potential preferences. Consequently, we have attempted to define bookmarking behaviors and apply them to the weight updating on users' preference map. We have measured the causal rate in order to improve accuracy of evidential supports and retrieved relational information between the behavioral patterns and user preferences throught temporally analyzing these patterns. For experiments, we made a dataset organized as 2718 bookmarks and had monitored 12 users' behaviors for 30 days[1].

1 Introduction

As the Internet has become more popular, seeking information is one of the most time-consuming tasks. This problem have been solved by implicit recognizing what a user is interested in. Mining knowledge from users' behavioral patterns during web browsing are essential to personalize web services. A bookmark is stored for revisiting a particular website or memorizing the URL of them. This information along with bookmarks can be evidential to extract user preferences. Nowadays, a number of bookmarks is used and these bookmarks are enough to infer user preferences. In order to recognize what users are interested in, we have first defined bookmarking behaviors. Next, bookmarks are applied to an ontology, a kind of semantic categorizer. Web directory can serve as an ontology by categorizing bookmarks. Moreover, we consider the causality between categories for high coverage of extraction efficiency.

2 Characteristics of Bookmarks and Web Directories

A bookmark indicates URL to revisit a website. Especially, these bookmarks imply that a user is looking for information related to a given topic. However,

[1] This work was supported by the Korea Science and Engineering Foundation(KOSEF) through the Northeast Asia e-Logistics Research Center at University of Incheon.

K.-Y. Whang, J. Jeon, K. Shim, J. Srivatava (Eds.): PAKDD 2003, LNAI 2637, pp. 203–208, 2003.

a bookmark may have more than one kind of content, which is an obstacle to automatically discovering the user's intentions. Web directories are hierarchical taxonomies that classify human knowledge [1]. The causality between categories, however, makes their hierarchical structure more complicated. Different paths of a category and semantically identical categories also cause this kind of problem. In practice, most companies are forced to manage a non-generic tree structure in order to avoid a waste of memory space caused by redundant information [2].

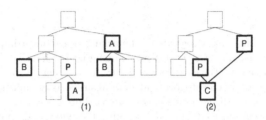

Fig. 1. (1) The multi-attribute of bookmarks; (2) The subordinate relationship between two categories

In brief, the problems with a bookmark and web directory as an ontology are the following:

- The multi-attributes of a bookmark. A bookmark can be involved in more than a category. As shown in Fig. 1 (1), a bookmarks can be included in some other categories, named as 'A' or 'B.'
- The relationship between categories is shown in Fig. 1 (2). The category 'C' can be a subcategory of more than one like 'P.'
  - Redundancy between semantically identical categories
  - Subordination between semantically dependent categories

3   Categorizing Bookmark Based on Ontology

Ontology plays a role of enriching semantic and structural information. Bookmarks can be conceptualized by a web directory functioning as an ontology. Firstly, the bookmark set of a user can be replaced with the category set of that user by referring to the ontology. Next, we can check whether a category is connected to more than one other category or not. Finally, we retrieves the other categories connected to them. As a result, the size of each user's category set become larger than that of his bookmark set because of the incomplete properties of the category structure mentioned in the previous section. Therefore, we have supplemented with a candidate category set. The candidate category set improves the coverage of user preferences. This means that potential preferences can be detected as well.

4   Extracting User Preferences by Using Categorized Bookmarks on Bayesian Networks

Bayesian networks are probabilistic models commonly used for decision tree analysis [3]. Support propagation between mutually linked categories on Bayesian networks can solve this problem [2]. This propagation can extract the most interested categories of a user. In other words, categories can be sorted by the degree of interest. The strength of causal influence between categories is expressed by conditional probabilities [1]. This is how categories reflect their causal relationship on parent nodes. The degree of user preference for the node *parent* is the summation of the evidential supports of the nodes *children* linked to the node *parent*.

$$P(parent, children) = \sum_i P(parent|children(i)) \times P(children(i)) \quad (1)$$

We assume that every category is assigned the *VOP*, the degree of user preference. The *VOP* is caculated by the function *propagate*, the numbers of bookmarks of children categories and their *VOP*'s, recursively.

$$VOP(c) = \sum (propagate[VOP(chidren(c)] \times VOP(children(c))) \quad (2)$$

The following axioms define the rules used to assign the *VOP*'s to every category.

1. The initial *VOP* is equal to the number of times categorized to a proper category by bookmarks.
2. As the number of times a given category found in the user's bookmarks increases, it can be inferred that the user is more interested in this category that the bookmark is included in. This means that the degree of interest for this field is in linear proportion to this value.

$$\text{the number of times categorized} \propto VOP(Category_i)$$

3. Hierarchically, a lower category is more causally influential in deciding user preferences than a higher one. The causal influence propagation from child nodes to their parent node is exponentially decreased as the distance between nodes increases.
4. As the number of subcategories of a node is increased, each subcategory has less influence on that node. This is why user interests are dispersed.
5. The candidate has the same level of influence as the normal one does.
6. All categories' supports are propagated up to the vertex node.
7. After normalizing, categories whose *VOP*'s are over the threshold value can be represented as user preference. The threshold value controls how many categories will be extracted.

The function *propagate* based on axioms 3 and 4 is defined as a logarithm function is given by

$$propagate[VOP(C)] = \frac{\log_k(VOP(C) + 1)}{N} \quad (3)$$

where k is $var(VOP(suncategories)) + 2$ and N is the number of subcategories of a parent category. In the process of normalization, the average VOP's of all categories and the portions of each category are calculated. The categories whose portions are over the threshold value are regarded as the representative categories of user preference.

5   Bookmarking Behaviors on the Web

While web browsing, users perform common actions related to bookmarks. Bookmarking behaviors can be classified into the following patterns:

- *Saving*. He puts an URL of a website in the bookmark repository. This activity is performed as an intention to visit this website again.
- *Reusing*. He prefers clicking a bookmark to use a search engine for the periodic usage patterns of certain bookmarks such as on-line magazines.
- *Deleting*. He deletes a bookmark. This activity is performed as the changes of users' interests.
- *Remembering*. He places a bookmark in his own bookmark set more than once for rearranging, renaming, and making directories.

Each behavior implies how much users are interested or disinterested in a particular category. According to these bookmarking activities, we can update user preferences. We assume that every behavior will be assigned the causal rate, RC, which is the numeric value [-1, 1]. The RC is used for adjusting the propagation of the causal supports among categories. In order to update user preferences adaptively, these rates must be used. A user has the RC's of the four behaviors listed above for every category. More importantly, the RC is applied to propagate the causal supports among categories. This demonstrates the high coverage of the user's potential preferences. The characteristics of a RC are very simple. The summation of all RC's of a category is zero and the absolute value of the RC is related to the association between the preference for the category and the behavioral patterns. Further, the more interested in a field a user is, the larger RC the action taken by him has.

Another important point is that the RC are changeable. The temporal transition of user preferences can be effectively recognized and detected based on hebbian Leaning. The elements are replaced with bookmarking behaviors, the category set of PM and the causal rate RC. To formulate the causal rate updating, the causal rate RC_{kj} with behavioral pattern PB_j and the VOP_k of the category, which the bookmark taken the behavior PB_j is included in, are considered on the PM. The adjustment is applied to the causal rate RC at the time the activities n are performed. It is expressed in the general form

$$\Delta RC_{kj}(n) = \begin{cases} \eta PB_j(n)VOP_k(n) & \text{if } VOP_k(n) > 0; \\ (-1) \times \eta PB_j(n)VOP_k(n) & \text{if } VOP_k(n) = 0; \end{cases} \tag{4}$$

where η is a positive constant that determines the rate of learning. Synaptic enhancement and depression of RC_{kj} depend solely on the relationship between the behavioral pattern and the categories on *PM*. When the category taken a pattern is on *PM* (in other words, the *VOP* is more than zero), the *RC* between them will be enhanced. Otherwise, it will be depressed. To meet the zero-sum condition, the amount of enhancement (or depression) should be charged with the other behaviors.

$$RC_{kh}(n+1) = RC_{kh}(n) - \frac{\Delta RC_{kj}(n)}{3} \qquad \text{if } h \neq j \qquad (5)$$

The causal support between categories is dependent on the causal rate *RC*. The function *rated_propagate* is defined as combining the function *propagate* with the *RC*. The variable k and j are the index of category C on *PM* and the behavioral pattern taken in category C, respectively.

$$rated\_propagate[VOP(C)] \qquad (6)$$
$$= propate[VOP(C)] \times \left[1 + \left(\frac{1}{2 - \exp(-RC_{kj})}\right)\right]$$

As a result, the propagation of causal supports can be reinforced by a user's behaviors. This modification efficiently reflects peculiar user behaviors on their preferences. Sudden changes of user behaviors because of a change in their interests can create a more dramatic dependency between categories than periodical behaviors. As a result, this is related to the sequence of bookmarking behaviors.

6   Experiments and Implementation

A tree structure was created from Yahoo as a test bed. Every time Fifty users found a site related to their own preference, they stored URL's in their bookmark repositories. At last, a total of 2718 bookmarks were collected.

In order to evaluate the performance of how user preferences were extracted, two methods were considered: the causal support propagation and the simple voting method. With the simple voting, the *VOP* was applied to the rate between the number of bookmarks in a category and the total number of bookmarks of a user. The categories extracted in both cases were partially repeated by the categories in the *PM* manually constructed by a user. Looking as recall, the causal support propagation measured higher coverage than the simple voting. Especially, as the threshold increased, the difference between them was more remarkable.

Whereas the normalized threshold of $100*(Threshold/Max.VOP)$ was in the range 24% to 85%, as shown in Fig. 2 (1), the categories acquired by the causal support propagation had almost twice as high coverage as the ones acquired by the voting method. Furthermore, high-ranked categories were retrieved showing that the most interesting category for a particular user can be identified.

In order to evaluate the causal rate updated by bookmarking behaviors, bookmarking log information were recorded. The initial value of the *RC* and the learning rate η were set to $\begin{bmatrix} 0.5 & 0.0 & 0.0 & -0.5 \end{bmatrix}^T$ and 0.01, respectively. To prove

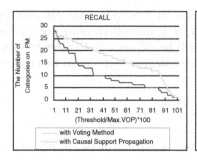 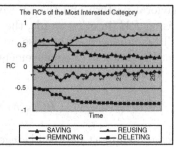

Fig. 2. (1)Evaluating detecting potential preferences; (2)The RC's of the most interested category for behavioral patterns

the following hypothesis *"There is a relationship between bookmarking behavior and user preference,"* the RC's of the most interested categories for each user were measured, as shown in Fig. 2 (2). 'REUSING' and 'SAVING' converged to approximately 0.74 and 0.23, respectively. This shows that these two patterns were used more frequently for the categories users were interested in. Moreover, because the absolute values of the RC's of 'REUSING' and 'DELETING' are larger than those of the other patterns, these patterns have a more critical influence on the category.

7 Conclusion

Bookmarks have been regarded as the most evidential information to help establish user preferences. Behaviors involved in these bookmarks can also be used to adjust users' PM's. With respect to the dependency between categories in the Web directory, the consideration of a causal relationship between categories causes the coverage of the extraction of the user's preference to be higher. Moreover, high-ranked interests can be retrieved for a certain user. Another important point is that user bookmarking behavior can be classified and shown to influence each category according to user preferences.

References

1. Baeza-Yates, R., Ribeiro-Neto, B.: Modern Information Retrieval. Addison-Wesley (1999) 384–387
2. Jung, J.J., Yoon, J.-S., Jo, G.-S.: Collaborative Information Filtering by Using Categorized Bookmarks on the Web. Proceeding of the 14th International Conference on Applications of Prolog (2001) 343–357
3. Pearl, J.: Probabilistic reasoning in intelligent systems. Morgan Kauffman Publisher (1988)

Mining Frequent Instances on Workflows

Gianluigi Greco[1], Antonella Guzzo[1], Giuseppe Manco[2], and Domenico Saccà[1,2]

[1] DEIS, University of Calabria, Via Pietro Bucci, 87036 Rende, Italy
{ggreco,guzzo}@si.deis.unical.it,
[2] ICAR-CNR, National Research Council, Via Pietro Bucci, 87036 Rende, Italy
{manco,sacca}@icar.cnr.it

Abstract. A workflow is a partial or total automation of a business process, in which a collection of *activities* must be executed by humans or machines, according to certain procedural rules. This paper deals with an aspect of workflows which has not so far received much attention: providing facilities for the human system administrator to monitor the actual behavior of the workflow system in order to predict the "most probable" workflow executions. In this context, we develop a data mining algorithm for identifying frequent patterns, i.e., the workflow substructures that have been scheduled more frequently by the system. Several experiments show that our algorithm outperforms the standard approaches adapted to mining frequent instances.

Keywords: workflow management systems, frequent pattern mining.

1   Introduction

Even though the modern enterprisers increasingly use the workflow technology for designing the business process, it is surprising to observe that a little effort has been paid by the research to support workflow administrator's tasks. For instance, only in a recent paper [13], it is formally addressed the problem of finding an appropriate system configuration that guarantees a domain-specific quality of service.

Other attempts have been done in the context of process discovery [1]. The idea is to use the information collected at run-time, in order to derive a structural model of the interactions. Such a model can be used in both the diagnosis phase and the (re)design phase. In this setting the problem encompasses the one of finding sequential patterns; indeed, process graphs are richer in structure as they admit partial ordering among activities and parallel executions (see, e.g, [17,20,6]).

In this paper, we continue on the way of providing facilities for the human system administrator, by investigating the ability of predicting the "most probable" workflow execution. Indeed, in real world-cases, the enterprise must do many choices during workflow execution; some choices may lead to a benefit, others should be avoided in the future. Data mining techniques may, obviously, help the administrator, by looking at all the previous instantiations (collected

K.-Y. Whang, J. Jeon, K. Shim, J. Srivatava (Eds.): PAKDD 2003, LNAI 2637, pp. 209–221, 2003.

into *log* files in any commercial system), in order to extract unexpected and useful knowledge about the process, and in order to take the appropriate decisions in the executions of further coming instances.

A first step consists of identifying the structure of the executions, that have been scheduled more frequently by the workflow system. This problem encompasses well-known techniques, such as frequent itemsets or sequences discovery [2,4], and can be reconduced to more involved pattern mining algorithms, that arise in complex domains.

In this context, we mention the problem of discovering frequent substructures of a given graph [18,19], or the problem of discovering frequent subgraphs in a forest of graphs [23]. It is clear that such approaches could be well-suited to deal with the problem at hand. Indeed, we can model a workflow schema as a graph, and the executions of the workflow as a forest of subgraphs complying with the graph representing the workflow schema. However, many additional features, such as the capability of specifying constraints on the execution of a workflow, make the cited approaches unpractical both from the expressiveness and from the efficiency viewpoint. One could think also at more expressive approaches, such as the multirelational approaches, based on a first-order modeling of the features of a workflow [9]. However, in our opinion, a more effective approach can be obtained by properly formalizing the problem at hand in a suitable way.

Contribution. We investigate the problem of mining frequent workflow instances; despite its particular applicative domain, the problem we consider is of a wider interest because all the proposed techniques, can be profitably exported into other fields. The main reason of this claim is that the approach used in workflow systems is to model the process as a particular directed graph.

Throughout the paper, we propose a quite intuitive and original formalization of the main workflow concepts in terms of graph structures. This gives us the possibility to prove some intractability results in reasoning over workflows. Indeed, the in-depth theoretical analysis we provide strongly motivates the use of Data Mining techniques, thus confirming the validity of the approach. Moreover, the formalization provides a viable mean for developing a levelwise theory which characterizes the problem of mining frequent instances of workflows.

In more detail, we propose a levelwise algorithm for mining frequent sub-graph structures (instances) that conform to a given schema (workflow specification). Several experiments confirm that our approach outperforms pre-existing techniques rearranged to this particular problem.

Organization. The paper is organized as follows. Section 2 provides a formal model of workflows, and many complexity considerations on such a proposed model. A characterization of the problem of mining frequent patterns from workflow schemas is devised in sect. 3, and in sect. 4 a levelwise theory of workflow patterns (and a corresponding algorithm for mining such patterns) is developed. Finally, sect. 5 provides an experimental validation of the approach.

2   The Workflow Abstract Model

A *workflow schema* $\mathcal{WS}$ is a tuple $\langle A, E, A^{\odot}, A^{\vee}, E^!, E^?, E^{\subseteq}, a_0, F \rangle$, where A is a finite set of *activities*, partitioned into two disjoint subsets $A^{\odot}$ and $A^{\vee}$; $E \subseteq (A - F) \times (A - \{a_0\})$ is an acyclic relation of precedences among activities, partitioned into three pairwise disjoint subsets $E^!$, $E^?$, and $E^{\subseteq}$; $a_0 \in A$ is the starting activity, and $F \subseteq A$ is the set of final activities.

For the sake of presentation, whenever it will be clear from the context, a workflow schema $\mathcal{WS} = \langle A, E, A^{\odot}, A^{\vee}, E^!, E^?, E^{\subseteq}, a_0, F \rangle$ will also be denoted by $\langle A, E, a_0, F \rangle$ or even simpler by $\langle A, E \rangle$.

Informally, an activity in $A^{\odot}$ acts as synchronizer (also called a *join* activity in the literature), for it can be executed only after all its predecessors are completed, whereas an activity in $A^{\vee}$ can start as soon as at least one of its predecessors has been completed. Moreover, once finished, an activity a activates all its outgoing $E^!$ arcs, exactly one among the $E^?$ arcs, and any non-empty subset of the $E^{\subseteq}$ arcs.

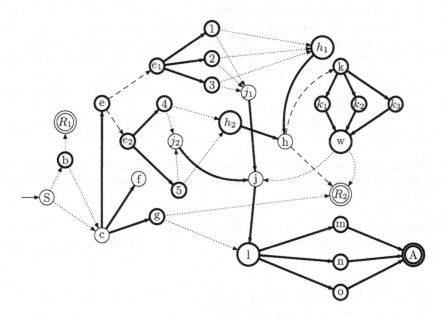

Fig. 1. An example of Workflow Schema

A workflow schema can be represented in a graphical way by means of a directed acyclic graph. We represent arcs in $E^!$, $E^?$ and in $E^{\subseteq}$ with bold, dotted and dashed arcs, respectively; moreover, we draw activities in $A^{\odot}$ and in $A^{\vee}$ with bold and regular circles, respectively. Finally the ending activities are also marked with double circles.

An example of workflow that will be used throughout the paper, is shown in Figure 2. It is easy to notice that the schema actually corresponds to the explosion of the activities of the "sales ordering" process, described in the introduction[1].

A workflow schema may have associated several executions consisting of a sequence of states, where a state S is a tuple $\langle Marked, Ready, Executed \rangle$, with $Ready$ and $Executed$ subsets of activities, and $Marked$ subset of the precedences.

An execution starts with the state $S_0 = \langle \emptyset, \{a_0\}, \emptyset \rangle$. Later on, if the state at the step t is $S_t = \langle Marked_t, Ready_t, Executed_t \rangle$, then the next state S_{t+1} is one of the outcomes of a non-deterministic transition function δ, defined as follows: $\delta(S_t)$ is the set of all states $\langle Marked_t \cup Marked^{!}_{t+1} \cup Marked^{?}_{t+1} \cup Marked^{\subseteq}_{t+1}, Ready^{\vee}_{t+1} \cup Ready^{\odot}_{t+1} - Ready_t, Executed_t \cup Ready_t \rangle$, such that

- $Marked^{!}_{t+1} = \{(a,b)| a \in Ready_t, (a,b) \in E^{!}\}$, i.e., all $E^{!}$ arcs leaving ready activities are marked;
- $Marked^{?}_{t+1}$ is any maximal subset of $\{(a,b)| a \in Ready_t, (a,b) \in E^{?}\}$ s.t. there are no two distinct arcs in it leaving the same node, i.e., exactly one $E^{?}$ arc is marked by a ready activity;
- $Marked^{\subseteq}_{t+1}$ is any subset of $\{(a,b)| a \in Ready_t, (a,b) \in E^{\subseteq}\}$ s.t. for each $a \in Ready_t$ with outgoing $E^{\subseteq}$ arcs, there exists at least one $E^{\subseteq}$ arc in it;
- $Ready^{\vee}_{t+1} = \{a| a \in (A^{\vee} - Executed_t), \exists (c,a) \in Marked_t\}$, i.e., an $A^{\vee}$ activity becomes ready for execution as soon as one of its predecessor activities is completed;
- $Ready^{\odot}_{t+1} = \{a| a \in (A^{\odot} - Executed_t), \not\exists (c,a) \in (E - Marked_t)\}$, i.e., an $A^{\odot}$ activity is ready for execution after all its predecessor activities are completed.

We point out that the above semantics captures the behavior of most commercial workflow products. Now we are in the position to formally define a workflow execution.

Definition 1. *Let $\mathcal{WS}$ be and δ be the transition function. An execution e on a workflow schema $\mathcal{WS} = \langle A, E, a_0, F \rangle$ is a sequence of states $[S_0, ..., S_k]$ such that*

1. $S_0 = \langle \{\emptyset\}, \{a_0\}, \{\emptyset\} \rangle$,
2. $S_{t+1} \in \delta(S_t)$ *for each* $0 < t < k$, *and*
3. $Executed_k \cap F \neq \emptyset$ *or* $Ready_k = \emptyset$.   □

Note that the case $Executed_k \cap F = \emptyset$ and $Ready_k = \emptyset$ (in condition 3) corresponds to an abnormal execution which does not reach a final state. As shown next, checking whether a workflow schema admits a successful execution is intractable – the hardness of the problem depends on the exclusive choices for marking the arcs in $E^{?}$.

---

[1] Due to space limits we do not explain this correspondence here but leave this task to the reader

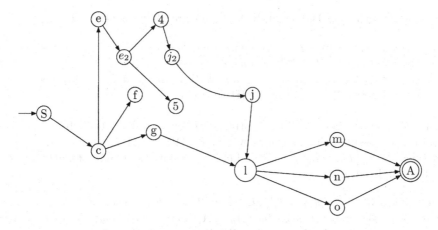

Fig. 2. An instance of the workflow schema in Fig. 2

Proposition 1. *Let* $\mathcal{WS} = \langle A, E, a_0, F \rangle$ *be a workflow schema. Then, deciding whether there exists an execution that reaches a final state (i.e., $Executed_k \cap F \neq \emptyset$) is* **NP**-*complete, but the problem becomes* **P**-*complete if* $E^? = \emptyset$.

Workflow executions can be seen as connected subgraphs of the workflow schema and will be called instances under this graphical perspective.

Definition 2. *Let* $\mathcal{WS} = \langle A, E, a_0, F \rangle$ *be a workflow schema and* $e = [S_0, ..., S_k]$ *be an execution. Then, the* instance *associated to* e *is the graph* $I_e = \langle A_e, E_e \rangle$, *where* $A_e = \cup_{t=1,k} Executed_t$ *and* $E_e = Marked_t$. $\square$

An instance of the workflow of Fig. 2 is reported in Fig. 2. In the following, we denote the set of all instances of a workflow schema $\mathcal{WS}$ by $\mathcal{I}(\mathcal{WS})$.

Deciding whether a subgraph is indeed an instance of $\mathcal{WS}$ is tractable although deciding the existence of an instance (i.e., whether $\mathcal{I}(\mathcal{WS}) \neq \emptyset$) is not because of Proposition 1.

Proposition 2. *Let* $\mathcal{WS} = \langle A, E, a_0, F \rangle$ *be a workflow schema and* I *be a subgraph of* $\mathcal{WS}$. *Then, deciding whether* I *is an instance of* $\mathcal{WS}$ *can be done in polynomial time in the size of* E.

3   Mining Frequent Patterns

In this section we present an algorithm for mining frequent patterns (i.e., subgraphs) in workflow instances. It is important to remark here that the task of discovering frequent patterns is computationally expensive. Indeed, even the problems of deciding whether an arc is included in some instance or whether a pair of nodes always occur together in every instance are not tractable.

Proposition 3. *Let $WS = \langle A, E, a_0, F \rangle$ be a workflow schema. Then,*

1. *given an arc $(a, b) \in E$, deciding whether there exists an instance $I \in \mathcal{I}(WS)$ such that $(a, b) \in I$ is* **NP**-*complete;*
2. *given two nodes a and b, deciding whether, for each $I \in \mathcal{I}(WS)$ a is in I implies b is in I as well is* co-**NP**-*complete.* □

Let us now define the type of patterns we want to discover. Throughout the section, we assume that a workflow schema $WS = \langle A, E, a_0, F \rangle$ and a set of instances $\mathcal{F} = \{I_1, ..., I_n\}$ are given. Let 2^{WS} denote the family of all the subsets of the graph $\langle A, E \rangle$.

Definition 3. *A graph $p = \langle A_p, E_p \rangle \in 2^{WS}$ is a $\mathcal{F}$-pattern (cf. $\mathcal{F} \models p$) if there exists $I = \langle A_I, E_I \rangle \in \mathcal{F}$ such that $A_p \subseteq A_I$ and p is the subgraph of I induced by the nodes in A_p. In the case $\mathcal{F} = \mathcal{I}(WS)$, the subgraph is simply said to be a pattern.* □

Let $supp(p) = |\{I | \{I\} \models p \wedge I \in \mathcal{F}\}| / |\mathcal{F}|$, be the support of a $\mathcal{F}$-pattern p. Then, given a real number $minSupp$, the problem we address consists in finding all the frequent $\mathcal{F}$-patterns, i.e. all the $\mathcal{F}$-patterns whose support is greater than $minSupp$. A frequent $\mathcal{F}$-pattern will be simply called $\mathcal{F}$-frequent.

In order to reduce the number of patterns to generate, throughout the rest of the paper, we shall only consider $\mathcal{F}$-patterns that satisfy two additional constraints: *connectivity* and *deterministic closure*. As we shall see next, these constraints do not actually reduce the generality of our approach.

The first restriction is that the undirected version of an $\mathcal{F}$-pattern must be *connected*. Indeed, the extension of the mining algorithm to cover disconnected $\mathcal{F}$-patterns can be trivially tackled by observing that any $\mathcal{F}$-pattern is frequent if and only if each of its connected components is frequent as well. Hence, the computation of frequent general patterns can be done by simply combining frequent connected components.

The second restriction is formalized next.

Definition 4. *Given a graph $p = \langle A_p, E_p \rangle \in 2^{WS}$, the deterministic closure of p (cf. ws-closure(p)) is inductively defined as the graph $p' = \langle A_{p'}, E_{p'} \rangle$ such that:*

i) $A_p \subseteq A_{p'}$, *and* $E_p \subseteq E_{p'}$ *(basis of induction),*
ii) $a \in A_{p'} \cap A^\odot \wedge (b, a) \in E$, *implies* $(b, a) \in E_{p'}$ *and* $b \in A_{p'}$,
iii) $a \in A_{p'} \wedge (a, b) \in E^!$ *implies* $(a, b) \in E_{p'}$ *and* $b \in A_{p'}$.

A graph p such that $p = ws\text{-}closure(p)$ is said ws-closed. □

The above definition can be used to introduce a third notion of pattern which only depends on the structure of the workflow schema, rather than on the instances $\mathcal{F}$ or $\mathcal{I}(WS)$. The need of this weaker notion will be clear in a while.

Definition 5. *A* weak pattern, *or simply* w-pattern, *is a* ws-closed *graph* $p \in 2^{\mathcal{WS}}$. □

The following proposition characterizes the complexity of recognition for the three notions of pattern; in particular, it states that testing whether a graph is a w-pattern can be done very efficiently in deterministic logarithmic space on the size of the graph $\mathcal{WS}$ rather than on the size of $2^{\mathcal{WS}}$ as it in general happens for $\mathcal{F}$-patterns.

Proposition 4. *Let* $p \in 2^{\mathcal{WS}}$. *Then*

1. *deciding whether* p *is an* $\mathcal{F}$-pattern is feasible in polynomial time in the size of $\mathcal{F}$,
2. *deciding whether* p *is a pattern is* **NP***-complete (w.r.t. the input* $\mathcal{WS}$).
3. *deciding whether* p *is a* w-pattern is in **L** *(w.r.t. the input* $\mathcal{WS}$).

It turns out that the notion of weak pattern is the most appropriate from the computational point of view. Moreover, working with w-patterns rather than $\mathcal{F}$-patterns is not an actual limitation.

Proposition 5. *Let* p *be a frequent* $\mathcal{F}$-pattern. *Then*

1. ws-closure(p) *is both a weak pattern and a frequent* $\mathcal{F}$-pattern;
2. *each weak pattern* $p' \subseteq p$ *is a frequent* $\mathcal{F}$-pattern as well.

We stress that a weak pattern is not necessarily an $\mathcal{F}$-pattern or even a pattern. As shown in the next section, we shall use weak patterns in our mining algorithm to optimize the searching space but we eventually check whether they are frequent $\mathcal{F}$-patterns.

4   The Algorithm for Mining Frequent Patterns

The algorithm we propose for mining frequent $\mathcal{F}$-patterns, uses a levelwise theory, which consist in constructing frequent weak patterns, by starting from frequent "elementary" weak patterns, defined below, and by extending each frequent weak pattern using two basic operations: adding a frequent arc and merging with another frequent elementary weak pattern. As we shall show next, the correctness follows from the results of Proposition 5, and from the observation that the space of all connected weak patterns constitutes a lower semi-lattice, with a particular precedence relation $\prec$, defined next.

The elementary weak patterns, from which we start the construction of frequent patterns, are obtained as the ws-closures of the single nodes.

Definition 6. *Let* $\mathcal{WS} = \langle A, E \rangle$ *be a workflow schema. Then, for each* $a \in A$, *the graph* ws-closure$(\langle \{a\}, \{\} \rangle)$ *is called an* elementary weak pattern *(cf.* ew-pattern*).* □

The set of all *ew*-patterns is denoted by EW. Moreover, let p be a weak pattern, then EW_p denotes the set of the elementary weak patterns contained in p. Note that given an *ew*-pattern e, EW_e is not necessarily a singleton, for it may contain other *ew*-patterns.

Given a set $E' \subseteq \mathrm{EW}$, $Compl(E') = \mathrm{EW} - \bigcup_{e \in E'} \mathrm{EW}_e$ contains all elementary patterns which are neither in E' nor contained in some element of E'.

Let us now introduce the relation of precedence $\prec$ among connected weak patterns. (Observe that the empty graph, denoted by $\perp$, is a connected weak pattern as well.) Given two connected w-patterns, say p and p', $p \prec p'$ if and only if:

a) $A_p = A_{p'}$ and $E_{p'} = E_p \cup \{(a, b)\}$, where $(a, b) \in E^{\subseteq} \cup E^?$ (i.e., p' is obtained from p by adding an arc), or
b) there exists $p'' \in Compl(\mathrm{EW}_p)$ such that $p' = p \cup p'' \cup X$, where X is empty if p and p'' are connected or otherwise $X = \{q\}$ and $q \in E^{\subseteq} \cup E^?$ connects p to p'' (i.e., p' is obtained from p by adding an elementary weak pattern and possibly a connecting arc).

Note that for each $e \in \mathrm{EW}$, we have $\perp \prec e$.

The following result states that all the connected weak patterns of a given workflow schema, can be constructed by means of a chain over the $\prec$ relation.

Lemma 1. *Let p be a connected w-pattern. Then, there exists a chain of connected w-patterns, such that $\perp \prec p_1 \prec \ldots \prec p_n = p$.*

It turns out that the space of all connected weak patterns is a lower semilattice w.r.t. the precedence relation $\prec$. The algorithm w-find, reported in Figure 3, exploits an apriori-like exploration of this lower semi-lattice.

At each stage, the computation of L_{k+1} (steps 5–14) is carried out by extending any pattern p generated at the previous stage ($p \in L_k$), in two ways:

– by adding frequent edges in $E^{\subseteq}$ or $E^?$ (*addFrequentArc* function), and
– by adding an elementary weak patterns (*addEWFrequentPattern* function).

The properties of the algorithm are reported in the following lemma.

Lemma 2. *In the algorithm w-find, reported in Figure 3, the following propositions hold:*

a. *L_0 contains a set of frequent connected $\mathcal{F}$-pattern;*
b. *addFrequentArc add to U connected patterns, which are not necessary $\mathcal{F}$-patterns;*
c. *addFrequentEWPattern add to U connected w-patterns, which are not necessary patterns;*
d. *L_{k+1} contains only frequent connected $\mathcal{F}$-patterns.*   □

Corollary 1. *(Soundness) The set R computed by the algorithm of Figure 3 contains only frequent connected $\mathcal{F}$-patterns.*   □

Proposition 6. *(Completeness) The algorithm of Figure 3 terminates and computes all the frequent connected weak patterns. In particular, let R be the set of $\mathcal{F}$-patterns computed by the algorithm. Then, for each frequent connected weak pattern p, we have $p \in R$.*

Input: A workflow Graph $\mathcal{WS}$, a set $\mathcal{F} = \{I_1, \ldots, I_N\}$ of instances of $\mathcal{WS}$.
Output: A set of frequent $\mathcal{F}$-patterns.
Method: Perform the following steps:

```
 1    L_0 := {e|e ∈ EW, e is frequent w.r.t. F};       //see Lemma 2.a
 2    k := 0, R := L_0;
 3    FrequentArcs := {(a,b)|(a,b) ∈ (E^⊆ ∪ E^?), ⟨{a,b}, {(a,b)}⟩ is frequent w.r.t. F};
 4    E_f^⊆ := E^⊆ ∩ FrequentArcs, E_f^? := E^? ∩ FrequentArcs;
 5    repeat
 6       U := ∅;
 7       forall p ∈ L_k do begin
 8          U := U ∪ addFrequentArc(p);       //see Lemma 2.b
 9          forall e ∈ Compl(EW_p) ∩ L_0 do
10             U := U ∪ addFrequentEWPattern(p,e);   //see Lemma 2.c
11       end
12       L_{k+1} := {p|p ∈ U, p is frequent w.r.t. F};   //see Lemma 2.d
13       R := R ∪ L_{k+1};
14    until L_{k+1} = ∅;
15    return R;
```

Function $addFrequentEWPattern(p = \langle A_p, E_p \rangle, e = \langle A_e, E_e \rangle)$: **w-pattern**;

    $p' := \langle A_p \cup A_e, E_p \cup E_e \rangle$;
    if p' is *connected,* **then return** p' **else return** $addFrequentConnection(p, e)$;

Function $addFrequentConnection(p = \langle A_p, E_p \rangle, e = \langle A_e, E_e \rangle)$: **w-pattern**;

    $S := ∅$
    forall frequent $(a,b) \in (E_f^⊆ \cup E_f^?) - E_p$ s.t. $(a \in A_p, \ b \in A_e) \vee (a \in A_e, \ b \in A_p)$ **do**
    begin
        $p' := \langle A_p, E_p \cup (a,b) \rangle$;
        if $\mathcal{WS} \models p'$ **then** $S := S \cup \{p'\}$;
    end
    return S

Function $addFrequentArc(p = \langle A_p, E_p \rangle)$: **pattern**;

    $S := ∅$
    forall frequent $(a,b) \in (E_f^⊆ \cup E_f^?) - E_p$ s.t. $a \in A_p, b \in A_p$ **do begin**
        $p' := \langle A_p, E_p \cup (a,b) \rangle$
        if $\mathcal{WS} \models p'$ **then** $S := S \cup \{p'\}$;
    end
    return S

Fig. 3. Algorithm w-find$(\mathcal{F}, \mathcal{WS})$

5   Experiments

In this section we study the behavior of the algorithm, by evaluating both its performances and its scalability. As shown in the previous section, the algorithm is sound and complete w.r.t. the set of frequent w-patterns. Nevertheless, the number of candidate w-patterns generated could be prohibitively high, thus making the algorithm unfeasible on complex workflow schemas.

As already mentioned in Section 1, several existing techniques for computing frequent itemsets can be viable solutions to the problem of mining frequent instances in a workflow, provided that suitable representations of instances is exploited. Thus, we compare the w-find algorithm with some of them. In particular, we consider an implementation of the *Apriori* algorithm which computes frequent itemsets in $E^{\subseteq} \cup E^?$, and next connects such arcs with arcs in $E^!$.

A possible further approach to consider is the *WARMR* algorithm devised in [9], that allows an explicit formalization of domain knowledge (like, for example, the connectivity information provided by the workflow schema) which can be directly exploited by the algorithm. However, due to space limitations, we omit the results of such comparison here.

In our experiments, we mainly use synthetic data. Synthetic data generation can be tuned according to: i) the size of $\mathcal{WS}$, ii) the size of $\mathcal{F}$, iii) the size $|L|$ of the frequent weak patterns in $\mathcal{F}$, and iv) the probability $p_{\subseteq}$ of choosing a $\subseteq$-arc. The ideas adopted in building the generator for synthetic data are essentially inspired by [3].

In a first set of experiments, we consider some fixed workflow schemas, and generate synthesized workflow instances, by simulating a number of execution. In particular, for each node a in a ready state, exactly one of the edges $(a, b) \in E^?$ and a subset of the edges $(a, b) \in E^{\subseteq}$ are randomly chosen. The number of $\subseteq$-arcs is chosen by picking from a binomial distribution with mean $p_{\subseteq}$. Frequent instances are forced into the system by replicating some instances (in which some variations were randomly performed) according to $|L|$.

We perform several experiments comparing the performance of *Apriori* and w-find on increasing values of $|\mathcal{F}|$ and *minSupp*. For a dataset of instances generated w.r.t. the workflow schema of Figure 2, the comparison is reported in Figure 5. As expected, w-find outperforms *Apriori* of an order of magnitude. This is mainly due to the fact that, contrarily to w-find, in the *Apriori* implementation arcs in $E^? \cup E^{\subseteq}$ are combined without taking into account the information provided by the workflow schema.

Figure 4(on the right) reports the number of operations (matching of a pattern with an instance), for increasing values of $\mathcal{F}$. The figure shows that the algorithm scales linearly in the size of the input (for different supports).

In a second set of experiments, we randomly generate the workflow schemas to test the efficiency of the approach w.r.t. the structure of the workflow. To this purpose, we fix $|\mathcal{F}|$ and generate workflow instances according to the randomly generated schema. The actual number of nodes and arcs (i.e., each of $|A^{\odot}|$, $|A^{\vee}|$, $|E^!|$, $|E^?|$ and $|E^{\subseteq}|$), is chosen by picking from a Poisson distribution with fixed mean value.

In order to evaluate the contribution of the complexity of workflow schemas, we exploit the factor $f = \frac{|E_?|+|E_{\subseteq}|}{|E_?|+|E_{\subseteq}|+|E_!|}$, which represents the degree of poten-

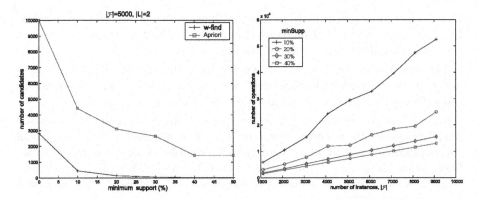

Fig. 4. Left: Number of candidates w.r.t. *minSupp*. Right: Number of matching operations w.r.t. $|\mathcal{F}|$.

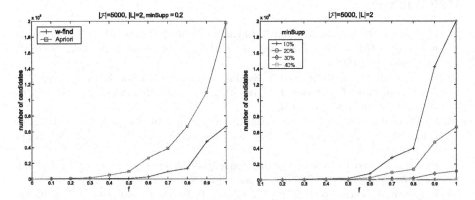

Fig. 5. Number of candidates w.r.t. f, for randomly generated workflow schemas. Left: comparison with a-priori. Right: w-find performances for different *minSupp* values.

tional nondeterminism within a workflow schema. Intuitively, workflow schemas exhibiting $f \simeq 0$ produce instances with a small number of candidate w-patterns. Conversely, workflow schemas exhibiting $f \simeq 1$ produce instances with a huge number of candidate w-patterns. Figure 5 shows the behavior of both *Apriori* and *w-find* when f ranges between 0 (no nondeterminsm) and 1 (full nondeterminism). Again, *Apriori* is outperformed by *w-find*, even though for small values of f both the algorithms produce a small number of candidates.

6   Conclusions

In this paper we presented an efficient algorithm for mining frequent instances of workflow schemas. The main motivation for this work was aimed at providing facilities for the human system administrator to monitor the actual behavior of the workflow system in order to predict the "most probable" workflow executions. In this context, the use of mining techniques is justified by the fact that even "simple" reachability problems become intractable.

The proposed algorithm was shown to efficiently explore the search space; hence, it can be exploited as an effective mean for investigating some inherent properties of the executions of a given workflow schema.

We conclude by mentioning some directions of future research. The proposed model is essentially a *propositional* model, for it assumes a simplification of the workflow schema in which many real-life details are omitted. However, we believe that the model can be easily updated to cope with more complex constraints, such as time constraints, pre-conditions and post-conditions, and rules for exception handling.

References

1. R. Agrawal, D. Gunopulos, and F. Leymann. Mining Process Models from Workflow Logs. In*Proc. 6th Int. Conf. on Extending Database Technology (EDBT)*, pages 469–483, 1998.
2. R. Agrawal, T. Imielinski, and A. Swami. Mining Association Rules between Sets of Items in Large Databases. In *Proc. ACM Conf. on Management of Data (SIGMOD93)*, pages 207–216, 1993.
3. R. Agrawal, and R. Srikant. Fast Algorithms for Mining Association Rules in Large Databases In *Proc. 20th Int. Conf. on Very Large Data Bases(VLDB94)*, pages 487–499,1994.
4. R. Agrawal, and R. Srikant. Mining Sequential Patterns. In *Proc. 11th Int. Conf. on Data Engineering (ICDE95)*, pages 3–14, 1995.
5. E. Cohen, M. Datar, S. Fujiwara, A. Gionis, P. Indyk, R. Motwani, J. D. Ullman, and C. Yang. Finding Interesting Associations without Support Pruning. *IEEE Transactions on Knowledge and Data Engineering*, 13(1), pages 64–78, 2001.
6. J. E. Cook, and A. L. Wolf. Automating Process Discovery Through Event-Data Analysis.In *Proc. 17th Int. Conf. on Software Engineering (ICSE95)*, pages 73–82, 1995.
7. H. Davulcu, M. Kifer, C. R. Ramakrishnan, and I. V. Ramakrishnan. Logic Based Modeling and Analysis of Workflows. In*Proc. 17th ACM SIGACT-SIGMOD-SIGART Symposium on Principles of Database Systems*, pages 25–33, 1998.
8. U. Dayal, M. Hsu, and R. Ladin. Business Process Coordination: State of the Art, Trends, and Open Issues. In*Proc. 27th Int. Conf. on Very Large Data Bases (VLDB01)*, pages 3–13, 2001.
9. L. Dehaspe, and H. Toivonen. Discovery of Frequent DATALOG Patterns. *Data Mining and Knowledge Discovery*, 3(1),pages 7–36, 1999.
10. T. Feder, P. Hell, S. Klein, and R. Motwani. Complexity of Graph Partition Problems. *STOC*, pages 464–472, 1999
11. M. R. Garey, and D. S. Johnson. Computers and Intractability. A Guide to the Theory of **NP**-completeness, Freeman and Comp., NY, USA, 1979.
12. D. Georgakopoulos, M. Hornick, and A. Sheth. An overview of Workflow Management: From Process Modeling to Workflow Automation Infrastructure.*Distributed and Parallel Databases*, 3(2), pages 119–153, 1995.

13. M. Gillmann, W. Wonner, and G. Weikum. Workflow Management with Service Quality Guarantees. In *Proc. ACM Conf. on Management of Data (SIGMOD02)*, 2002.
14. P. Grefen, J. Vonk, and P. M. G. Apers. Global transaction support for workflow management systems: from formal specification to practical implementation. *VLDB Journal* 10(4), pages 316–333, 2001.
15. G. Lee, K. L. Lee, and A. L. P. Chen. Efficient Graph-Based Algorithms for Discovering and Maintaining Association Rules in Large Databases. *Knowledge and Information Systems* 3(3), pages 338–355, 2001.
16. M. Kamath, and K. Ramamritham. Failure handling and coordinated execution of concurrent workflows. In*Proc. 14th Int. Conf. on Data Engineering (ICDE98)*, pages 334–341, 1998.
17. P. Koksal, S. N. Arpinar, and A. Dogac. Workflow History Management. *SIGMOD Record*, 27(1), pages 67–75, 1998.
18. A. Inokuchi, T. Washio, and H. Motoda. An Apriori-Based Algorithm for Mining Frequent Substructures from Graph Data. In *Proc. 4th European Conf. on Principles of Data Mining and Knowledge Discovery*, pages 13–23, 2000.
19. T. Miyahara and others. Discovery of Frequent Tag Tree Patterns in Semistructured Web Documents. In *Proc 6th Pacific-Asia Conf. on Advances in Knowledge Discovery and Data Mining*, pages 356–367, 2002.
20. W. M. P. van der Aalst, and K. M. van Hee. Workflow Management: Models, Methods, and Systems. *MIT Press*, 2002.
21. D. Wodtke, and G. Weikum. A Formal Foundation for Distributed Workflow Execution Based on State Charts. In *Proc. 6th Int. Conf. on Database Theory (ICDT97)*, pages 230–246, 1997.
22. The Workflow Management Coalition, *http://www.wfmc.org/*.
23. M. Zaki. Efficiently Mining Frequent Trees in a Forest. In *Proc. 8th Int. Conf. On Knowledge Discovery and Data Mining (SIGKDD02)*, 2002. to appear.

Real Time Video Data Mining for Surveillance Video Streams

JungHwan Oh, JeongKyu Lee, and Sanjaykumar Kote

Department of Computer Science and Engineering
University of Texas at Arlington
Arlington, TX 76019-0015 U. S. A.
{oh, jelee, kote}@cse.uta.edu

Abstract. We extend our previous work [1] of the general framework for
video data mining to further address the issue such as how to mine video
data using *motions* in video streams. To extract and characterize these
motions, we use an accumulation of quantized pixel differences among all
frames in a video segment. As a result, the accumulated motions of segment are represented as a *two dimensional matrix*. Further, we develop
how to capture the *location* of motions occurring in a segment using the
same matrix generated for the calculation of the amount. We study how
to cluster those segmented pieces using the features (the amount and the
location of motions) we extract by the matrix above. We investigate an
algorithm to find whether a segment has normal or abnormal events by
clustering and modeling normal events, which occur mostly. In addition
to deciding normal or abnormal, the algorithm computes *Degree of Abnormality* of a segment, which represents to what extent a segment is
distant to the existing segments in relation with normal events. Our experimental studies indicate that the proposed techniques are promising.

1 Introduction

There have been some efforts about video data mining for movies [2], and traffic
videos [3,4]. Among them, the developments of complex video surveillance systems [5] and traffic monitoring systems [4] have recently captured the interest
of both research and industrial worlds due to the growing availability of cheap
sensors and processors at reasonable costs, and the increasing safety and security concerns. As mentioned in the literature [3], the common approach in these
works is that the objects (i.e., person, car, airplane, etc.) are extracted from
video sequences, and modeled by the specific domain knowledge, then, the behavior of those objects are monitored (tracked) to find any abnormal situations.
What are missing in these efforts are first, how to index and cluster these unstructured and enormous video data for real-time processing, and second, how
to mine them, in other words, how to extract previously unknown knowledge
and detect interesting patterns.

In this paper, we extend our previous work [1] of the general framework for
video data mining to further address the issues discussed above. In our previous

K.-Y. Whang, J. Jeon, K. Shim, J. Srivatava (Eds.): PAKDD 2003, LNAI 2637, pp. 222–233, 2003.

work, we have developed how to segment the incoming video stream into meaningful pieces, and how to extract and represent the motions for characterizing the segmented pieces. In this work, we investigate how to find the location of the accumulated motion in a segment, how to cluster the segments based on their motions, and how to compute the abnormalities of the segments.

The main contributions of the proposed work can be summarized as follows.

- The proposed technique to compute motions is very cost-effective because an expensive computation (i.e., optical flow) is not necessary. The matrices representing motions are showing not only the amounts but also the exact locations of motions.
- To find the abnormality, our approach uses the normal events which are occurring everyday and easy to obtain. We do not have to model any abnormal event separately. Therefore, unlike the others, our approach can be used for any video surveillance sequences to distinguish normal and abnormal events.

The remainder of this paper is organized as follows. In Section 2, to make the paper self-contained, we describe briefly the video segmentation technique relevant to this paper, which have been proposed in our previous work [1,6]. How to capture the *amount* and the *location* of motions occurring in a segment, how to cluster those segmented pieces, and how to model and detect normal events are discussed in Section 3. The experimental results are discussed in Section 4. Finally, we give our concluding remarks in Section 5.

2   Incoming Video Segmentation

In this section, we briefly discuss the details of the technique in our previous work [1] to group the incoming frames into semantically homogeneous pieces by real time processing (we called these pieces as 'segments' for convenience).

To find segment boundary, instead of comparing two consecutive frames (Figure 1(a)) which is the most common way to detect shot boundary [7,8,9,10,11], we compare each frame with a *background* frame as shown in Figure 1(b). A background frame is defined as a frame with only non-moving components. Since we can assume that the camera remains stationary for our application, a background frame can be a frame of the stationary components in the image. We manually select a background frame using a similar approach as in [3]. The differences are magnified so that segment boundaries can be found more clearly. The algorithm to decompose a video sequence into meaningful pieces (segments) is summarized as follows. The Step.1 is a preprocessing by off-line processing, and the Step.2 through 5 are performed by on-line real time processing. Note that since this segmentation algorithm is generic, the frame comparison can be done by any technique using color histogram, pixel-matching or edge change ratio. We chose a simple color histogram matching technique for illustration purpose.

- Step.1: A background frame is extracted from a given sequence as preprocessing, and its color histogram is computed. In other words, this frame is

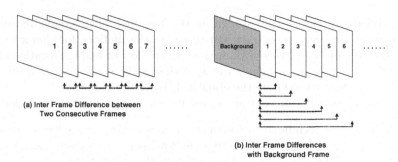

Fig. 1. Frame Comparison Strategies

represented as a *bin* with a certain number (bin size) of quantized colors from the original. As a result, a background frame (F^B) is represented as follows using a *bin* with the size n. Note that P_T is representing the total number of pixels in a background or any other frame.

$$F^B = bin^B = (v_1^B, v_2^B, v_3^B, ..., v_n^B), \quad where \sum_{i=1}^{n} v_i^B = P_T. \tag{1}$$

– Step.2: Each frame (F^k) arriving to the system is represented as follows in the same way, as the background is represented in the previous step.

$$F^k = bin^k = (v_1^k, v_2^k, v_3^k, ..., v_n^k), \quad where \sum_{i=1}^{n} v_i^k = P_T. \tag{2}$$

– Step.3: Compute the difference (D^k) between the background (F^B) and each frame (F^k) as follows. Note that the value of D^k is always between zero and one.

$$D^k = \frac{F^B - F^k}{P_T} = \frac{bin^B - bin^k}{P_T} = \frac{\sum_{i=1}^{n}(v_i^B - v_i^k)}{P_T} \tag{3}$$

– Step.4: Classify D^k into 10 different categories based on its value. Assign a corresponding category number (C_k) to the frame k. We use 10 categories for illustration purpose, but this value can be changed properly according to the contents of video. The classification is stated below.

  - Category 0 : $D^k < 0.1$ - Category 5 : $0.5 \leq D^k < 0.6$
  - Category 1 : $0.1 \leq D^k < 0.2$ - Category 6 : $0.6 \leq D^k < 0.7$
  - Category 2 : $0.2 \leq D^k < 0.3$ - Category 7 : $0.7 \leq D^k < 0.8$
  - Category 3 : $0.3 \leq D^k < 0.4$ - Category 8 : $0.8 \leq D^k < 0.9$
  - Category 4 : $0.4 \leq D^k < 0.5$ - Category 9 : $D^k \geq 0.9$

– Step.5: For real time on-line processing, a temporary table is maintained. To do this, and build a hierarchical structure from a sequence, compare C_k with C_{k-1}. In other words, compare the category number of current frame with the previous frame. We can build a hierarchical structure from a sequence based on these categories which are not independent from each other. We consider that the lower categories contain the higher categories as shown in Figure 2. In our hierarchical segmentation, therefore, finding segment

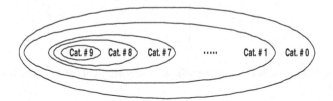

Fig. 2. Relationships (Containments) among Categories

boundaries means finding category boundaries in which we find a starting frame (S_i) and an ending frame (E_i) for each category i.

3   New Proposed Techniques

We propose new techniques about how to capture the *amount* and the *location* of motions occurring in a segment, how to cluster those segmented pieces, and how to model and detect normal events are discussed in this section.

3.1   Motion Feature Extraction

We describe how to extract and represent motions from each segment decomposed from a video sequence as discussed in the previous section. We developed a technique for automatic measurement of the overall motion in not only two consecutive frames but also an entire shot which is a collection of frames in our previous works [6,12]. We extend this technique to extract the motion from a segment, and represent it in a comparable form in this section. We compute *Total Motion Matrix (TMM)* which is considered as the overall motion of a segment, and represented as a *two dimensional matrix*. For comparison purpose among segments with different lengths (in terms of number of frames), we also compute an *Average Motion Matrix (AMM)*, and its corresponding *Total Motion (TM)* and *Average Motion (AM)*.

The TMM, AMM, TM and AM for a segment with n frames is computed using the following algorithm (Step 1 through 5). We assume that the frame size is $c \times r$ pixels.

- **Step.1**: The color space of each frame is quantized (i.e., from 256 to 64 or 32 colors) to reduce unwanted noises (false detection of motion which is not actually motion but detected as motion).

- **Step.2**: An empty two dimensional matrix TMM (its size $(c \times r)$ is same as that of frame) for a segment S is created as follows. All its items are initialized with zeros.

$$
TMM_S = \begin{pmatrix}
t_{11} & t_{12} & t_{13} & \cdots & t_{1c} \\
t_{21} & t_{22} & t_{23} & \cdots & t_{2c} \\
\cdots & \cdots & \cdots & \cdots & \cdots \\
t_{r1} & t_{r2} & t_{r3} & \cdots & t_{rc}
\end{pmatrix} \tag{4}
$$

And AMM_S which is a matrix whose items are averages computed as follows.

$$
AMM_S = \begin{pmatrix}
\frac{t_{11}}{n} & \frac{t_{12}}{n} & \frac{t_{13}}{n} & \cdots & \frac{t_{1c}}{n} \\
\frac{t_{21}}{n} & \frac{t_{22}}{n} & \frac{t_{23}}{n} & \cdots & \frac{t_{2c}}{n} \\
\cdots & \cdots & \cdots & \cdots & \cdots \\
\frac{t_{r1}}{n} & \frac{t_{r2}}{n} & \frac{t_{r3}}{n} & \cdots & \frac{t_{rc}}{n}
\end{pmatrix} \tag{5}
$$

- **Step.3**: Compare all the corresponding quantized pixels in the same position of each and background frames. If they have different colors, increase the matrix value (t_{ij}) in the corresponding position by one (this value may be larger according to the other conditions). Otherwise, it remains the same.
- **Step.4**: Step.3 is repeated until all consecutive pairs of frames are compared.
- **Step.5**: Using the above TMM_S and AMM_S, we compute a motion feature, TM_S, AM_S as follows.

$$
TM_S = \sum_{i=0}^{r} \sum_{j=0}^{c} t_{ij}, \qquad AM_S = \sum_{i=0}^{r} \sum_{j=0}^{c} \frac{t_{ij}}{n} \tag{6}
$$

As seen in these formulae, TM is the sum of all items in TMM and we consider this as total motion in a segment. In other words, TM can indicate an amount of motion in a segment. However, TM is dependent on not only the amount of motions but also the length of a segment. A TM of long segment with little motions can be equivalent to a TM of short segment with a lot of motions. To distinguish these, simply we use AM which is an average of TM.

3.2   Location of Motion

Comparing segments only by the amount of motion (i.e., AM) would not give very accurate results because it ignores the locality such that where the motions occur. We introduce a technique to capture locality information without using partitioning, which is described as follows. In the proposed technique, the locality information of AMM can be captured by two one dimensional matrices which are the summation of column values and the summation of row values in AMM. These two arrays are called as *Summation of Column (SC)* and *Summation of*

Row (SR) to indicate their actual meanings. The following equations show how to compute SC_A and SR_A from AMM_A.

$$SC_A = (\sum_{i=0}^{r} a_{i1} \quad \sum_{i=0}^{r} a_{i2} \quad \cdots \quad \sum_{i=0}^{r} a_{ic}),$$
$$SR_A = (\sum_{j=0}^{c} a_{1j} \quad \sum_{j=0}^{c} a_{2j} \quad \cdots \quad \sum_{j=0}^{c} a_{rj}).$$

To visualize the computed TMM (or AMM), we can convert this TMM (or AMM) to an image which is called *Total Motion Matrix Image (TMMI)* for TMM (*Average Motion Matrix Image (AMMI)* for AMM). Let us convert a TMM with the maximum value, m into a 256 gray scale image as an example. We can convert an AMM using the same way. If m is greater than 256, m and other values are scaled down to fit into 256, otherwise, they are scaled up. But the value zero remains unchanged. An empty image with same size of TMM is created as $TMMI$, and the corresponding value of TMM is assigned as a pixel value. For example, assign white pixel for the matrix value zero which means no motion, and black pixels for the matrix value 256 which means maximum motion in a given shot. Each pixel value for a $TMMI$ can be computed as follows after it is scaled up or down if we assume that $TMMI$ is a 256 gray scale image. *Each Pixel Value* = 256 − *Corresponding Matrix Value*.

Figure 3 shows some visualization examples of $AMMI$, SC and SR such that how these SC and SR can capture where the motions occur. Two SRs in Figure 3 (a) are same, which means that the vertical locations of two motions are same. Similarly, Figure 3 (b) shows that the horizontal locations of two motions are same by SCs. Figure 3 (c) is showing the combination of two, the horizontal and vertical location changes.

3.3 Clustering of Segments

In our clustering, we employ a multi-level hierarchical clustering approach to group segments in terms of category, and motion of segments. The algorithm is implemented in a top-down fashion, where the feature, category is utilized at the top level, in other words, we group segments into k_1 clusters according to the categories. For convenience, we called this feature as *Top Feature*. Each cluster is clustered again into k_2 groups based on the motion (AM) extracted in the previous section accordingly, which are called as *Bottom Feature*. We will consider more features (i.e., SC and SR) for the clustering in the future.

For this multi-level clustering, we adopted K-Mean algorithm and cluster validity method studied by Ngo et. al. [13] since the algorithm is the most frequently used clustering algorithm due to its simplicity and efficiency. It is employed to cluster segments at each level of hierarchy independently. The K-Mean algorithm is implemented as follows.

– **Step.1**: The initial centroids are selected in the following way:
   1. Given v d-dimensional feature vectors, divide the d dimensions to $\rho = \frac{d}{k}$. These subspaces are indexed by $[1, 2, 3, ..., \rho]$, $[\rho, \rho + 1, \rho + 2, ..., 2\rho]$, ..., $[(k-1)\rho + 1, (k-1)\rho + 2, (k-1)\rho + 3, ..., k\rho]$.
   2. In each subspace j of $[(j-1)\rho + 1, ..., j\rho]$ associate a value f_i^j for each feature vector $\mathcal{F}_i$ by $f_i^j = \sum_{m=(j-1)}^{j\rho} \rho \mathcal{F}_i(d)$

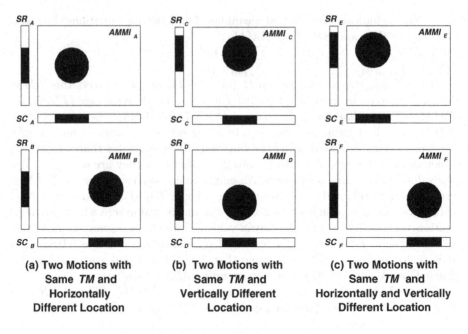

(a) Two Motions with
Same *TM* and
Horizontally
Different Location

(b) Two Motions with
Same *TM* and
Vertically Different
Location

(c) Two Motions with
Same *TM* and
Horizontally and Vertically
Different Location

Fig. 3. Comparisons of Locations of Motions

    3. Choose the initial cluster centroids μ_1, μ_2, ..., μ_k, by $\mu_j = \arg_{\mathcal{F}_i} \max_{1 < i < v} f_i^j$

- **Step.2**: Classify each feature F to the cluster p_s with the smallest distance. $p_s = \arg_{1 \leq j \leq k} \min D(\mathcal{F}, \mu_j)$. This D is a function to measure the distance between two feature vectors and defined as
  $D(\mathcal{F}, \mathcal{F}') = \frac{1}{\mathcal{Z}(\mathcal{F}, \mathcal{F}')} (\sum_{i=1}^{v} |\mathcal{F}(i) - \mathcal{F}'(i)|^k)^{\frac{1}{k}})$,
  where $\mathcal{Z}(\mathcal{F}, \mathcal{F}') = \sum_{i=1}^{v} \mathcal{F}(i) + \sum_{i=1}^{v} \mathcal{F}'(i)$ which is a normalizing function. In this function, $k = 1$ for L_1 norm and $k = 2$ for L_2 norm. The L_1 and L_2 norms are two of the most frequently used distance metrics for comparing two feature vectors. In practice, however, L_1 norm performs better than L_2 norm since it is more robust to outliers [14]. Furthermore, L_1 norm is more computationally efficient and robust. We use L_1 norm for our experiments.

- **Step.3**: Based on the classification, update cluster centroids as $\mu_j = \frac{1}{v_j} \sum_{i=1}^{v_j} \mathcal{F}_i^{(j)}$ where v_j is the number of shots in cluster j, and $\mathcal{F}_i^{(j)}$ is the ith feature vector in cluster j.

- **Step.4**: If any cluster centroid changes the value by Step.3, go to Step.2, otherwise stop.

    The above K-Mean algorithm can be used when the number of clusters k is explicitly specified. To find optimal number (k) clusters, we have employed the cluster validity analysis [15]. The idea is to find clusters that minimize intra-cluster distance while maximize inter-cluster distance. The cluster separation measure $\varphi(k)$ is defined as $\varphi(k) = \frac{1}{k} \sum_{i=1}^{k} \max_{1 \leq v \leq k} \frac{\eta_i + \eta_j}{\xi_{ij}}$ where

$\eta_j = \frac{1}{v_j} \sum_{i=1}^{v_j} D(\mathcal{F}_i^j, \mu_i)$, $\xi_{ij} = D(\mu_i, \mu_j)$. ξ_{ij} is the inter-cluster distance of cluster i and j, while η_j is the intra-cluster distance of cluster j. The optimal number of cluster k' is selected as $k' = \min_{1 \leq k \leq q} \varphi(k)$ In other words, the K-Mean algorithm is tested for $k = 1, 2, ..., q$, and the one which gives the lowest value of $\varphi(k)$ is chosen.

In our multi-level clustering structure, a centroid at the top level represents the category of segments in a cluster, and a centroid at the bottom level represents the general motion characteristics of a sub-cluster.

3.4 Modeling and Detecting of Normal Events

As mentioned in the section 1, to find the abnormal event, we cluster and model the normal events which occur everyday and are easy to obtain. More precisely, the segments with normal events are clustered and modeled using the extracted features about the amount and location of motions. The algorithm can be summarized as follows.

- The existing segments are clustered into k number of clusters using the technique discussed in the section 3.3.
- We compute a *Closeness to Neighbors* (Δ) for a given segment (s_g) as follows,

$$\Delta = \frac{\sum_{i=1}^{m} D(s_g, s_i)}{m} \qquad (7)$$

where $D(s_g, s_i)$ is a distance between s_g and s_i. This Δ is an average value of the distances between a number (m) of closest segments to a given segment s_g in its cluster. We can use the distance function defined in the Step.2 of the previous subsection (3.3) for the computation of $D(s_g, s_i)$. This is possible since a segment can be represented as a feature vector by the features used for the clustering in the above.
- Compute Δ of all existing segments, and an average value of Δs of the segments in each cluster k_1, which is represented as $\bar{\Delta}^{k_1}$.
- If a new segment (S) comes in, then decide which cluster it goes to, its Δ_S. If it goes to the cluster k_1, we can compute whether it is normal or not as follows.

$$If \ \ \Delta^{k_1} \geq \Delta_S, \quad then \ \ S = Normal, \quad Otherwise \ \ S = Abnormal \quad (8)$$

If S is abnormal. then its degree of abnormality (Ψ) can be computed as follows, which is greater than zero.

$$\Psi = |\frac{\bar{\Delta}^{k_1} - \Delta_S}{\bar{\Delta}^{k_1}}| \qquad (9)$$

In addition to determining normal or abnormal, we find to what extent a segment with event or events are distant to the existing segments with normal events. The idea is that if a segment is close enough to the segments with normal events, there are more possibilities in which a given segment can be normal. As seen in the equation (8), if the value of Δ for a new segment is less than or equal to the average of the existing segments in the corresponding cluster, then the new segment can be normal since it is very close to the existing segments as we discussed in the beginning of this subsection. Otherwise, we compute a degree of abnormality using the differences between them.

4   Experimental Results

Our experiments in this paper were designed to assess the following performance issues since we have already examined the issue, "how does the proposed segmentation algorithm work to group incoming frames?" in our previous work [1].

- How do $TM(AM)$, SC and SR capture the amount and the location of motions in a segment?
- How do the proposed algorithms work for clustering and modeling of segments?

Our test video clips were originally digitized in AVI format at 30 frames/second. Their resolution is 160×120 pixels. We used the rates of 5 and 2 frames/second as the incoming frame rates. Our test set has 111 minutes and 51 seconds of raw video taken from a hallway in a building which consist of total 17,635 frames.

4.1   Performance of Capturing Amount and Location of Motions and Clustering

Figure 4 shows some examples of TM, AM, SC and SR for a number of segments in various categories. These features are represented as the images (i.e., $TMMI$ and $AMMI$ as discussed before). As seen in this figure, the amount and the location of motions are well-presented by these features. We will investigate the uniqueness of these SC and SR, and how to compare these in the future. Figure 4 shows a very simple example of clustering segments. As seen in this figure, the segments are clustered by category, and further partitioned using a motion feature, AM. The different categories have the different sizes and/or numbers of object(s), in other words, the segments in the higher categories have relatively larger or more objects. On the other hand, the average motions, represented by AM can distinguish the amount(degree) of motions in different segments. Also, we will consider SC and SR for more detail clustering in the future.

4.2   Performance of Computing Abnormality

A very simple example of computing abnormalities can be seen in Table 1. We consider that these segments in the table are segmented from new incoming stream. The values of Δ_S for the segments (# 130, # 131, # 133 and # 134) are smaller than the values of $\bar{\Delta}^k$ for their corresponding categories. Therefore, the abnormality (Ψ) for those segments can be represented as *normal* as we discussed in the section 3.4. However, since the Δ_{132} (0.15) is larger than $\bar{\Delta}^3$ (0.07), the segment # 132 is considered as an abnormal at the moment, and the actual value of abnormality (Ψ) can be computed as 1.14 as shown in the table. For better illustration, Figure 5 shows that a number of frames in the segment #132 and a typical segment in the category 3 to which the the segment #132 belongs. The new incoming segment # 132 is different from a typical segment in the category 3 in terms of for example, the size and the number of object(s). This difference is captured by our algorithm for computing the abnormality. Eventually, this segment # 132 becomes a normal segment because there is nothing wrong actually. If more number of segments similar to this segment comes, then this kind of segment will be detected as normal at a certain point.

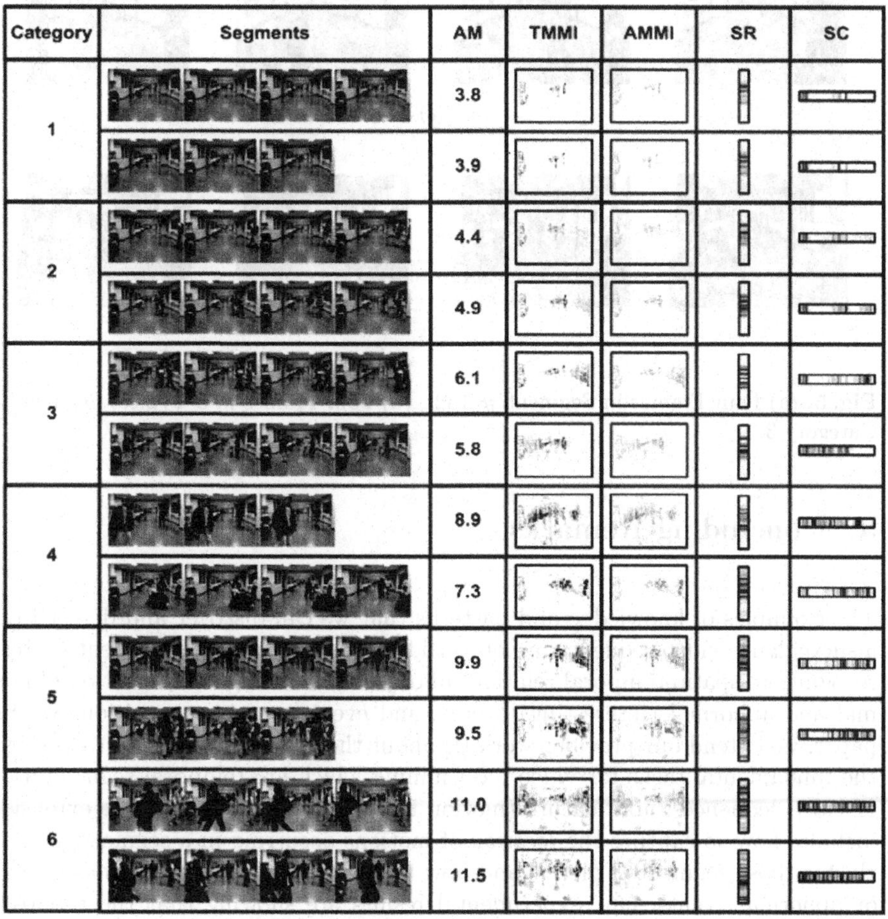

Category	Segments	AM	TMMI	AMMI	SR	SC
1		3.8				
		3.9				
2		4.4				
		4.9				
3		6.1				
		5.8				
4		8.9				
		7.3				
5		9.9				
		9.5				
6		11.0				
		11.5				

Fig. 4. Sample Clustering Results

Table 1. Example of Computing Abnormalities

Segment No.	Number of Frames	Cat. (C_k)	Avg. Motion (AM)	Value of Δ_s	Value of $\overline{\Delta}^k$	Abnormality Ψ
130	23	1	3.5	0.29	0.45	Normal
131	3	2	3.7	0.07	0.09	Normal
132	4	3	4.5	0.15	0.07	1.14
133	68	1	2.7	0.30	0.45	Normal
134	3	2	3.9	0.02	0.09	Normal

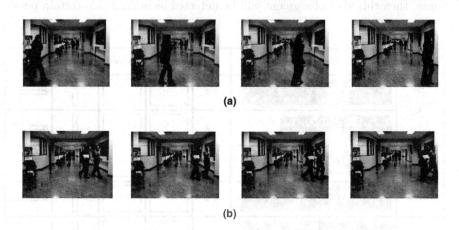

(a)

(b)

Fig. 5. (a) Four Frames in Segment # 132. (b) Four Frames in a Typical Segment in Category 3.

5   Concluding Remarks

The examples of knowledge and patterns that we can discover and detect from a surveillance video sequence are object identification, object movement pattern recognition, spatio-temporal relations of objects, modeling and detection of normal and abnormal (interesting) events, and event pattern recognition. In this paper, we extend our previous work [1] about the general framework to perform the fundamental tasks for video data mining which are temporal segmentation of video sequences, and feature (motion in our case) extraction. The extension includes how to capture the *location* of motions occurring in a segment, how to cluster those segmented pieces, and how to find whether a segment has normal or abnormal events. Our experimental results are showing that the proposed techniques are performing the desired tasks effectively and efficiently. In the future study, we will consider the other features (objects, colors) extracted from segments for more sophisticated clustering and indexing and deal with video files taken by moving camera.

References

1. Oh, J., Bandi, B.: Multimedia data mining framework for raw video sequences. In: Proc. of ACM Third International Workshop on Multimedia Data Mining (MDM/KDD2002), Edmonton, Alberta, Canada (2002)
2. Wijesekera, D., Barbara, D.: Mining cinematic knowledge: Work in progress. In: Proc. of International Workshop on Multimedia Data Mining (MDM/KDD'2000), Boston, MA (2000) 98–103
3. Chen, S., Shyu, M., Zhang, C., Strickrott, J.: Multimedia data mining for traffic video sequences. In: Proc. of International Workshop on Multimedia Data Mining (MDM/KDD'2001), San Francisco, CA (2001) 78–86
4. Cucchiara, R., Piccardi, M., Mello, P.: Image analysis and rule-based reasoning for a traffic monitoring system. IEEE Transactions on Intelligent Transportation Systems 1 (2000) 119–130
5. Pavlidis, I., Morellas, V., Tsiamyrtzis, P., Harp, S.: Urban surveillance systems: From the laboratory to the commercial world. Proceedings of The IEEE 89 (2001) 1478–1497
6. Oh, J., Sankuratri, P.: Automatic distinction of camera and objects motions in video sequences. In: To appear in Proc. of IEEE International Conference on Multimedia and Expo (ICME 2002), Lausanne, Switzerland (2002)
7. Zhao, L., Qi, W., Wang, Y., Yang, S., Zhang, H.: Video shot grouping using best-first model merging. In: Proc. of SPIE conf. on Storage and Retrieval for Media Databases 2001, San Jose, CA (2001) 262–269
8. Han, S., Kweon, I.: Shot detection combining bayesian and structural information. In: Proc. of SPIE conf. on Storage and Retrieval for Media Databases 2001, San Jose, CA (2001) 509–516
9. Oh, J., Hua, K.A.: An efficient and cost-effective technique for browsing and indexing large video databases. In: Proc. of 2000 ACM SIGMOD Intl. Conf. on Management of Data, Dallas, TX (2000) 415–426
10. Oh, J., Hua, K.A., Liang, N.: A content-based scene change detection and classification technique using background tracking. In: SPIE Conf. on Multimedia Computing and Networking 2000, San Jose, CA (2000) 254–265
11. Hua, K.A., Oh, J.: Detecting video shot boundaries up to 16 times faster. In: The 8th ACM International Multimedia Conference (ACM Multimedia 2000), LA, CA (2000) 385–387
12. Oh, J., Chowdary, T.: An efficient technique for measuring of various motions in video sequences. In: Proc. of The 2002 International Conference on Imaging Science, System, and technology (CISST'02), Las Vegas, NV (2002)
13. Ngo, C., Pong, T., Zhang, H.: On clustering and retrieval of video shots. In: Proc. of ACM Multimedia 2001, Ottawa, Canada (2001) 51–60
14. Rousseeuw, P., Leroy, A.M.: Robust Regression and Outlier Detection. John Wiley and Sons (1987)
15. Jain, A.K.: Algorithm for Clustering Data. Prentice Hall (1988)

Distinguishing Causal and Acausal Temporal Relations

Kamran Karimi and Howard J. Hamilton

Department of Computer Science
University of Regina
Regina, Saskatchewan
Canada S4S 0A2
{karimi,hamilton}@cs.uregina.ca

Abstract. In this paper we propose a solution to the problem of distinguishing between causal and acausal temporal sets of rules. The method, called the Temporal Investigation Method for Enregistered Record Sequences (TIMERS), is explained and introduced formally. The input to TIMERS consists of a sequence of records, where each record is observed at regular intervals. Sets of rules are generated from the input data using different window sizes and directions of time. The set of rules may describe an instantaneous relationship, where the decision attribute depends on condition attributes seen at the same time instant. We investigate the temporal characteristics of the system by changing the direction of time when generating temporal rules to see whether a set of rules is causal or acausal. The results are used to declare a verdict as to the nature of the system: instantaneous, causal, or acausal.

1 Introduction

In this paper we introduce the TIMERS method (Temporal Investigation Method for Enregistered Record Sequences). This method is based on a temporal order among the observed values of the attributes. Suitable input consists of a chronologically ordered set of records, where each record contains the values of the attributes, all observed at the same time. An example of such a record is: $<x = 1, y = 2, z = 3>$. These records are obtained at regular intervals. Here we are not concerned about the relation among individual attributes, such as "x and y are causes of z." Instead, TIMERS judges a *set of temporal rules* that involves the values of x and y to predict the value of z, as being either causal or acausal. An example rule that could belong to this rule set is: if $\{(x = 1)$ **and** $(y = 2)\}$ **then** $(z = 3)$. x and y are considered to be condition attributes, while z is the decision attribute. We cannot tell if this rule, considered by itself, represents a causal relation or not, i.e., do x and y cause z to have a certain value, or do they just happen to be seen together, with all their values caused by some hidden variable(s)? The TimeSleuth software [2] implements the TIMERS method and tries to answer this question. This method is especially appropriate when we have access to many attributes of a system, because the more attributes we have, the better the chance of finding a meaningful relationship among them.

A popular method for assessing the causality of a relationship is to use the concept of conditional independence to determine how two attributes influence each other [8]. In previous work, we looked at other methods of discovering causality, such as TETRAD [9] and CaMML [6]. One important property that differentiates the method we will present in this paper from these is that the presented method deals with data originating from the same source over time, while others deal with data

K.-Y. Whang, J. Jeon, K. Shim, J. Srivatava (Eds.): PAKDD 2003, LNAI 2637, pp. 234–240, 2003.
© Springer-Verlag Berlin Heidelberg 2003

generated from different sources with no special temporal ordering. Our previous work [3,4] was concerned with discovering temporal rules, with no consideration of causality and acausality. Here we discuss an extension to our method to aid a domain expert in making that distinction.

The rest of the paper is organized as follows. Section 2 defines the two directions for time, forward and backward, describes an operation called flattening, and formally defines causality and acausality in the context of the TIMERS method. The distinction among temporal and atemporal rules is also made clear. Section 3 explains how TIMERS determines the nature of a set of rules. Section 4 presents the results of experiments performed with the TimeSleuth software using real and synthetic data sets. Section 5 concludes the paper.

2 Forward and Backward Directions of Time

We consider a set of rules to define a relationship among the condition attributes and the decision attribute. A temporal rule is one that involves variables from times different than the decision attribute's time of observation. An example temporal rule is:

If $\{$(At time T_{-3}: $x = 2$) **and** (At time T_{-1}: $y > 1, x = 2$)$\}$ **then** (At time T: $x = 5$).   (Rule 1).

This rule indicates that the current value of x (at time T) depends on the value of x, 3 time steps ago, and also on the value of x and y, 1 time step ago. We use a preprocessing technique called flattening [3] to change the input data into a suitable form for extracting temporal rules with tools that are not based on an explicit representation of time. With flattening, data from consecutive time steps are put into the same record, so if in two consecutive time steps we have observed the values of x and y as: Time n: $<x = 1, y = 2>$, Time $n + 1$: $<x = 3, y = 2>$, then we can flatten these two records to obtain $<$Time $T - 1$: $x_1 = 1, y_1 = 2$, Time T: $x_2 = 3, y_2 = 2>$. The "Time $<$number$>$" keywords are implied, and do not appear in the records. The initial temporal order of the records is lost in the flattened records, and time always starts from $(T - w - 1)$ inside each flattened record, and goes on until T. Time T signifies the "current time" which is relative to the start of each record. Such a record can be used to predict the value of either x_2 or y_2 using the other attributes. Since we refrain from using any condition attribute from the current time, we modify the previous record by omitting either x_2 or y_2.

In the previous example we used *forward flattening*, because the data is flattened in the same direction as the forward flow of time. We used the previous observations to predict the value of the decision attribute. The other way to flatten the data is *backward flattening*, which goes against the natural flow of time. Given the two previous example records, the result of a backward flattening would be $<$ Time T: $y_1 = 2$, Time $T + 1$: $x_2 = 3, y_2 = 2>$. Inside the record, time starts at T, and ends at $(T + w - 1)$. This record could be used to predict the value of y_1 based on the other attributes. x_1 is omitted because it appears at the same time as the decision attribute y_1. In the backward direction, *future* observations are used to predict the value of the decision attribute.

Given a set of N temporally ordered observed records $D = \{rec_1, ..., rec_N\}$, the problem is to find a set of rules, as described in more detail below. Each record $rec_t = <c_{t1}, ..., c_{tm}>$ gives the values of a set of variables $V = \{v_1, ..., v_m\}$ observed at time step t. . The *forward window set* $P_f(w, t) = \{d_t, c_{ki} \mid (w \le t)$ & $t - w + 1 \le k < t, 1 \le i \le m\}$ represents all observations in the window of size w, between time $(t - w + 1)$ and time

t, inclusive, where t is the current time. Time flows forward, in the sense that the decision attribute appears at the end (time t). d_t is the decision attribute at time t. The backward window set $P_b(w, t) = \{ d_t, c_{ki} \mid (t \leq |D| - w + 1) \& t < k \leq (t + w - 1), 1 \leq i \leq m \}$ represents all observations in the window between time t and time $(t + w - 1)$, inclusive. Time flows backward, in the sense that the decision attribute appears at the beginning (time t). At the time step containing the decision attribute, condition attributes do not appear. In other words, d_t is the only variable at current time t.

Formally, the flattening operator $F(w, D, direction, d)$ takes as input a window size w, the input records D, a time direction $direction$, and the decision attribute d, and outputs flattened records according to the algorithm in Figure 1.

for ($t = 1$ to $|D|$)
    if (($direction = forward$) and ($t \geq w$))
        output ($z = <z_{ki} \mid d_{pi} \in P_f(w, t) \& k = w\text{-}1\text{-}t+p \& z_{ki} = d_{pi}>$)
    else if (($direction = backward$) and ($|D| - w +1 \geq t$))
        output ($z = <z_{ki} \mid d_{pi} \in P_b(w, t) \& k = p - t \& z_{ki} = d_{pi}>$)

Fig. 1. The flattening operation. The decision attribute d is used by the $P_f()$ and $P_b()$ sets.

The flattened record contains the neighboring w records in the appropriate direction of time. The F_w operator renames the time index values so that in each record, time is measured relative to the start of that record. In each flattened record, the time index ranges from 0 to $w\text{-}1$. The flattened records are thus independent of the time variable t.

Each rule r, generated from these flattened records, is a pair. The first member of a rule is a set of tests. The other member of the rule is the value that is predicted for the decision variable at time 0 or $w\text{-}1$. $r = (Tests_r, d_{val})$, where $Tests_r = \{ Test = (a, x, Cond) \}$, where $a \in V$, x is the time in which the variable a appears, and $Cond$ represents the condition under which $Test$ succeeds. One example is: $a_x > 5$. d_{val} is the value predicted for d_t (the decision attribute at time t). The CONDITION operator yields the set of variables that appear in the condition side of a rule, i.e., $CONDITION(r) = \{a_x \mid (a, x, Cond) \in Tests_r\}$. Similarly, we define $DECISION(r) = \{d_{\{0, w\text{-}1\}}\}$.

There is no consensus on the definitions of terms like causality or acausality. For this reason, we provide our own definitions here. In previous research, we detected sets of temporal rules and assigned the task of whether such a relationship is causal to a domain expert [4]. Here we provide a way to make such distinction. Even though TIMERS provides an algorithmic method for making a decision through a set of metrics, a domain expert must still make the final decision.

2.1   Instantaneous

An *instantaneous* set of rules is one in which the current value of the decision attribute in each rule is determined solely by the current values of the condition attributes in each rule [10]. An instantaneous set of rules is an *atemporal* one. Another name for an instantaneous set of rules is a (atemporal) *co-occurrence*, where the values of the decision attribute are associated with the values of the condition attributes.

Instantaneous definition: For any rule r in rule set R, if the decision attribute d appears at time T, then all condition attributes should also appear at time T, i.e., R is instantaneous iff ($\forall\, r \in R$, if $d_T = \text{DECISION}(r)$, then $\forall\, a_t \in \text{CONDITION}(r)$, $t = T$).

2.2   Temporal

A *temporal* set of rules is one that involves attributes from different time steps. A temporal set of rules can be causal or acausal.

Temporal definition: For any rule r in the rule set R, if the decision attribute appears at time T, then all condition attributes should appear at time $t \neq T$, i.e., R is temporal iff ($\forall\, r \in R$, if $d_T = \text{DECISION}(r)$, then $\forall\, a_t \in \text{CONDITION}(r)$, $t \neq T$).

We now define the two possible types of a temporal rule:

2.2.1 Causal

In a *causal* set of rules, the current value of the decision attribute relies only on the previous values of the condition attributes in each rule [10].

Causal definition: For any rule r in the rule set R, if the decision attribute d appears at time T, then all condition attributes should appear at time $t < T$, i.e., R is causal iff ($\forall\, r \in R$, if $d_T = \text{DECISION}(r)$, then $\forall\, a_t \in \text{CONDITION}(r)$, $t < T$).

2.2.2 Acausal

In an *acausal* set of rules, the current value of the decision attribute relies only on the future values of the condition attributes in each rule [7].

Acausal definition: For any rule r in the rule set R, if the decision attribute d appears at time T, then all condition attributes should appear at time $t > T$. i.e., R is acausal iff ($\forall\, r \in R$, if $d_T = \text{DECISION}(r)$, then $\forall\, a_t \in \text{CONDITION}(r)$, $t > T$).

All rules in a causal rule set have the same direction of time, and there are no attributes from the same time as the decision attribute. This property is guaranteed simply by not using condition attributes from the same time step as the decision attribute, and also by sorting the condition attributes in an increasing temporal order, until we get to the decision attribute. The same property holds for acausal rule sets, where time flows backward in all rules till we get to the decision attribute. Complementarily, in an instantaneous rule set, no condition attribute from other times can ever appear. The TIMERS methodology guarantees that all the rules in the rule set inherit the property of the rule set in being causal, acausal, or instantaneous.

3   The TIMERS Method

The TIMERS method is based on finding classification rules to predict the value of a decision attribute using a number of condition attributes that may have been observed at different times. We extract different sets of rules to predict the value of a condition attribute based on different window sizes and different directions for the flow of time.

The quality of the set of rules determines whether the window size and time direction are appropriate. We choose either the training accuracy or the predictive accuracy of the set of rules as the metric for the quality of the rules, and the appropriateness of the window size that was used to generate the rules. TIMERS is presented in Figure 2.

We use ε subscripts in the comparison operators to allow TIMERS to ignore small differences. We define $a >_\varepsilon b$ as: $a > b + \varepsilon$, and $a \geq_\varepsilon b$ as $a \geq b + \varepsilon$. The value of ε is determined by the domain expert.

Input: A sequence of temporally ordered data records D, minimum and maximum flattening window sizes α and β, where $\alpha \leq \beta$, a minimum accuracy threshold Ac_{th}, a tolerance value ε, and a decision attribute d. The attribute d can be set to any of the observable attributes in the system, or the algorithm can be tried on all attributes in turn.
Output: A verdict as to whether the system behaves in a instantaneous, causal or acausal manner when predicting the value of a specified decision attribute.
RuleGenerator() is a function that receives input records, generates decision rules, and returns the training or predictive accuracy of the rules.

TIMERS(D, α, β, Ac_{th}, ε, d)
ac_i = RuleGenerator(D, d); // instantaneous accuracy. window size = 1
for ($w = \alpha$ to β)
    ac_{fw} = RuleGenerator($F(w, D, forward, d)$, d)   // causality test
    ac_{bw} = RuleGenerator($F(w, D, backward, d)$, d) // acausality test
end for

ac_f = max($ac_{f\alpha}, ..., ac_{f\beta}$)   // choose the best value
ac_b = max($ac_{b\alpha}, ..., ac_{b\beta}$)

if $((ac_i < Ac_{th}) \wedge (ac_f < Ac_{th}) \wedge (ac_b < Ac_{th}))$ then discard results and stop. // not enough info

if $((ac_i >_\varepsilon ac_f) \wedge (ac_i >_\varepsilon ac_b))$ then verdict = "the system is instantaneous"
else if $(ac_f \geq_\varepsilon ac_b)$ then verdict = "the system is acausal"
else verdict = "the system is causal"

if $w_1 = w_2$ and verdict $\neq$ "the system is instantaneous" then
     verdict = verdict + "at window size" + w_1

verdict = "for attribute d, " + verdict
return verdict.

Fig. 2. The TIMERS algorithm, performed for the decision attribute d.

4 Experimental Results

We use three data sets: synthetic artificial life data from a simulated world, a Louisiana weather database, with seven relevant condition attributes, and a Helgoland weather database with three relevant condition attributes.

Series 1: The first series of experiments used data set from an artificial life program called URAL [12]. In a two-dimensional world, a robot moves around randomly, and records its current location plus the action that will take it to the next

(x, y) position. It also records the presence of food at each location. We first set the position along the x axis as the decision attribute. In this world, the current x value depends on the previous x value and the previous movement direction, and is perfectly predictable. The results appear in Table 1(a). The system is not instantaneous, because a window size of 1 (current time) gives relatively poor results. Rather, the system is causal, because the forward test gives the best results. The same conclusions are obtained for the y values.

Next we set food as the decision attribute. In this simple world, the presence of food is associated with its position. Because the location of the food does not change, the presence of food can be predicted using only the robot's previous position at a neighboring location and the robot's previous action. As shown in Table 1(b), the system is instantaneous, because a window size of 1 gives relatively good results. We expect about the same results with any other value for the window size.

Table 1. URAL data.(a) Decision attribute is x. (b) Decision attribute is presence of food

Window	Causality	Acausality
1	46.0%	
2	100%	70.6%
3	100%	71.7%
4	100%	72.8%
5	100%	**74.5%**
6	100%	73.7%
7	100%	73.3%
8	100%	73.6%
9	100%	72.6%
10	100%	73.9%

(a)

Window	Causality	Acausality
1	99.5%	
2	99.3%	99.1%
3	99.3%	99.3%
4	99.3%	99.4%
5	99.4%	99.5%
6	99.4%	**99.6%**
7	99.4%	99.5%
8	**99.5%**	99.5%
9	99.3%	99.6%
10	99.3%	99.6%

(b)

Series 2: The second experiment was done on a real-world data set, comprising hourly Louisiana weather observations [11]. The observed attributes consist of air temperature, amount of rain, maximum wind speed, average wind speed, wind direction, humidity, solar radiation, and soil temperature. We used 343 consecutive observations to predict the value of the soil temperature attribute. The results are given in Table 2(a). The system is not instantaneous, because a window size of 1 gives poor results. The system is acausal, since the acausality tests gives results as good as or better than those of the causality tests.

Table 2. (a) Louisiana data, $d =$ Soil temperature. (b) Helgoland data, $d =$ Wind speed.

Window	Causality	Acausality
1	27.7%	
2	82.7%	75.1%
3	**86.8%**	**87.1%**
4	84.4%	84.7%
5	86.7%	82.9%
6	77.5%	81.4%
7	79.5%	79.8%
8	80.7%	79.8%
9	77.9%	77.3%
10	79.2%	74.0%

(a)

Window	Causality	Acausality
1	18.9%	
2	**17.7%**	**20.7%**
3	14.7%	17.2%
4	14.2%	16.9%
5	13.9%	14.5%
6	14.0%	15.2%
7	13.4%	15.0%
8	13.2%	14.9%
9	12.2%	13.9%
10	12.0%	14.7%

(b)

Series 3: The next test was done on the Helgoland weather data set [1], which consists of hourly observations of the year, month, day, hour, air pressure, wind direction, and wind speed attributes. Wind speed was selected as the decision attribute. The results from 3000 hours of consecutive observations are given in Table 2(b). With an accuracy threshold of 21% or higher, the data are insufficient to make a judgement. The records are not rich enough to allow TimeSleuth to create rules that can reliably predict the value of the decision attribute. We expect that additional records would not change this.

5 Concluding Remarks

We introduced the TIMERS method for distinguishing causal and acausal sets of temporal rules. We implemented TIMERS in the TimeSleuth software and applied it to three data sets. TimeSleuth correctly categorized two rules sets as causal and another as acausal for an Artificial Life data set, categorized a rule set for the Louisiana weather data set as acausal, and said data were insufficient to make a conclusion about the Helgoland dataset. TimeSleuth is available at http://www.cs.uregina.ca/~karimi/downloads.html.

References

1. Hoeppner, F., Discovery of Temporal Patterns: Learning Rules about the Qualitative Behaviour of Time Series, *Principles of Data Mining and Knowledge Discovery (PKDD'2001)*, 2001.
2. Karimi, K., and Hamilton, H.J., TimeSleuth: A Tool for Discovering Causal and Temporal Rules, *14th IEEE International Conference On Tools with Artificial Intelligence (ICTAI'2002)*, Washington DC, USA, November 2002, pp. 375–380.
3. Karimi K., and Hamilton H.J., Learning With C4.5 in a Situation Calculus Domain, *The Twentieth SGES International Conference on Knowledge Based Systems and Applied Artificial Intelligence (ES2000)*, Cambridge, UK, December 2000, pp. 73–85.
4. Karimi K., and Hamilton H.J., Discovering Temporal Rules from Temporally Ordered Data, *The Third International Conference on Intelligent Data Engineering and Automated Learning (IDEAL'2002)*, Manchester, UK, August 2002, pp. 334–338.
6. Kennett, R.J., Korb, K.B., and Nicholson, A.E., Seabreeze Prediction Using Bayesian Networks: A Case Study, *Proc. Fifth Pacific-Asia Conference on Knowledge Discovery and Data Mining (PAKDD'01)*. Hong Kong, April 2001.
7. Krener, A. J. Acausal Realization Theory, Part I; Linear Deterministic Systems. *SIAM Journal on Control and Optimization*. 1987. Vol 25, No 3, pp. 499–525.
8. Pearl, J., *Causality: Models, Reasoning, and Inference*, Cambridge University Press. 2000.
9. Scheines, R., Spirtes, P., Glymour, C. and Meek, C., *Tetrad II: Tools for Causal Modeling*, Lawrence Erlbaum Associates, Hillsdale, NJ, 1994.
10. Schwarz, R. J. and Friedland B., *Linear Systems*. McGraw-Hill, New York. 1965.
11. http://typhoon.bae.lsu.edu/datatabl/current/sugcurrh.html. Content varies.
12. http://www.cs.uregina.ca/~karimi/downloads.html/URAL.java

Online Bayes Point Machines

Edward Harrington[1], Ralf Herbrich[2], Jyrki Kivinen[1], John Platt[3], and
Robert C. Williamson[1]

[1] Research School of Information Sciences and Engineering
The Australian National University
Canberra, ACT 0200
{edward.harrington, jyrki.kivinen, bob.williamson}@anu.edu.au
[2] Microsoft Research
7 J J Thomson Avenue
Cambridge, CB3 0FB, UK
rherb@microsoft.com
[3] Microsoft Research
1 Microsoft Way
Redmond, WA 98052, USA
jplatt@microsoft.com

Abstract. We present a new and simple algorithm for learning large
margin classifiers that works in a truly online manner. The algorithm
generates a linear classifier by averaging the weights associated with several perceptron-like algorithms run in parallel in order to approximate
the Bayes point. A random subsample of the incoming data stream is
used to ensure diversity in the perceptron solutions. We experimentally
study the algorithm's performance on online and batch learning settings.
The online experiments showed that our algorithm produces a low prediction error on the training sequence and tracks the presence of concept
drift. On the batch problems its performance is comparable to the maximum margin algorithm which explicitly maximises the margin.

1 Introduction

An online classifier tries to give the best prediction based on the example sequence seen at time t in contrast to a batch classifier which waits for the whole
sequence of T examples. There have been a number of recent attempts [1,2,3]
in the online setting to achieve a approximation to the maximum margin solution of batch learners such as SVMs. In this paper we present a truly online
algorithm for linear classifiers which achieves a large margin by estimating the
centre-of-mass (the so-called Bayes point) by a randomisation trick.

It is well accepted that a (fixed) large margin classifier provides a degree of
immunity to attribute noise and concept drift. One advantage of the algorithm
presented here is the ability to also track concept drift.

The primary appeal of the algorithm is its simplicity: one merely needs to
run a number of perceptrons in parallel on different (random) subsamples of the
training sequence. Hence, it is well suited to very large data sets. We illustrate

K.-Y. Whang, J. Jeon, K. Shim, J. Srivatava (Eds.): PAKDD 2003, LNAI 2637, pp. 241–252, 2003.

the effectiveness of the algorithm to track concept drift with an experiment based on the MNIST OCR database.

We define a sequence of training examples seen by the online learning algorithm in the form of a sequence of T training examples, $\boldsymbol{z} = (z_1, \ldots, z_T) :=$ $((\mathbf{x}_1, y_1), \ldots, (\mathbf{x}_T, y_T))$, comprising an instance $\mathbf{x}$ in the instance space $\mathcal{X} \subseteq \mathbb{R}^d$ and $y \in \{-1, +1\}$ the corresponding binary label. The weight vector $\mathbf{w} \in \mathbb{R}^d$ defines the d-dimensional hyperplane[1] $\{\mathbf{x} \colon (\mathbf{w} \cdot \mathbf{x}) = 0\}$ which is the decision boundary associated with the hypothesis $\mathrm{sgn}(\mathbf{w} \cdot \mathbf{x})$ of the perceptron algorithm [4], where $\mathrm{sgn}(v) = 1$ if $v \geq 0$ and $\mathrm{sgn}(v) = -1$ if $v < 0$. We often assume that the target labels y_i are defined by $y_i = \mathrm{sgn}(\mathbf{u} \cdot \mathbf{x}_i)$; in other words, there exists a linear classifier which can correctly classify the data. The *margin* $\gamma_{\boldsymbol{z}}(\mathbf{w})$ plays a key role in understanding the performance and behaviour of linear classifier learning algorithms. It is the minimum Euclidean distance between an instance vector $\mathbf{x}$ and the separating hyperplane produced by $\mathbf{w}$. Formally, the margin of a weight vector on an example z_i is $\gamma_{z_i}(\mathbf{w}) := y_i(\mathbf{w} \cdot \mathbf{x}_i)/\|\mathbf{w}\|_2$, and the margin with respect to the whole training sequence is $\gamma_{\boldsymbol{z}}(\mathbf{w}) := \min\{\gamma_{z_i}(\mathbf{w}) \colon i \in \{1, \ldots, T\}\}$. The *version space* $V(\boldsymbol{z})$ is the set of hypotheses consistent with the training sample. Since we only consider linear classifiers, we consider $V(\boldsymbol{z})$ as a subset of the weight space:

$$V(\boldsymbol{z}) := \{\mathbf{w} \in \mathbb{R}^d \colon \mathrm{sgn}(\mathbf{w} \cdot x_i) = y_i \text{ for all } (\mathbf{x}_i, y_i) \in \boldsymbol{z}\}. \tag{1}$$

Clearly, the target function $\mathbf{u}$ is in $V(\boldsymbol{z})$ for any $\boldsymbol{z}$. A learning algorithm that attains zero training error finds an hypothesis in the version space.

2   (Large Margin) Perceptron Learning Algorithms

We first consider the generalised perceptron (see Algorithm 1). The algorithm differs from the standard perceptron algorithm in two ways: 1) an update is made on example z_k if $\gamma_{z_k}(\mathbf{w}_k) \leq \rho_k$, where ρ_k is the *update margin* ($\rho_k = 0$ for all k in the standard perceptron), and 2) the update step size is η_k instead of always being fixed to 1. When $\rho_k = 0$, and η_k is fixed it is known by a result of Novikoff [5] that the algorithm makes no more than $(\beta/\gamma_{\boldsymbol{z}})^2$ mistakes where $\beta := \max_{i=1,\ldots,T} \|\mathbf{x}_i\|_2$ (we assume that the instances throughout the paper are normalised so $\beta = 1$) and $\gamma_{\boldsymbol{z}} := \max\{\gamma_{\boldsymbol{z}}(\mathbf{w}) : \mathbf{w} \in \mathbb{R}^d\}$ is the margin attained by the maximum margin hyperplane. This perceptron convergence theorem only guarantees that the perceptron's hypothesis $\mathbf{w}$ is in $V(\boldsymbol{z})$; *i.e.* only that $\gamma_{\boldsymbol{z}}(\mathbf{w}_{\mathrm{perc}}) > 0$ where $\mathbf{w}_{\mathrm{perc}}$ is the weight vector the perceptron algorithm converges to. In the marginalised perceptron [4], ρ_k is a positive constant and the final margin, if run until separation of the training sequence in the batch setting, can be guaranteed to be $\geq \rho_k$. The perceptron algorithm has several key advantages: it is extremely simple, on-line, and it can be readily kernelised since it only uses inner products between input vectors [6].

---

[1] Note that in the case of a weight vector with bias term (also referred to as the threshold) an extra attribute of $+1$ is added.

Algorithm 1 Generalised Perceptron algorithm

Require: A linear separable training sequence $\boldsymbol{z} = ((\mathbf{x}_1, y_1), (\mathbf{x}_2, y_2), \dots)$.
Require: Margin parameter sequence ρ_k and learning rate η_k.

1: $t = 1, \ k = 1$.
2: **repeat**
3:    **if** $y_t (\mathbf{w}_k \cdot \mathbf{x}_t) / \|\mathbf{w}_k\|_2 \le \rho_k$ **then**
4:       $\mathbf{w}_{k+1} = \mathbf{w}_k + \eta_k y_t \mathbf{x}_t$
5:       $k = k + 1$
6:    **end if**
7: **until** No more updates made

The motivation for setting $\rho_k = \rho > 0$ in order to guarantee $\gamma_{\boldsymbol{z}}(\mathbf{w}) > \rho$ for the final solution $\mathbf{w}$ is that a larger margin hyperplane has a better guarantee of generalisation performance than a smaller margin one [6]. It should be noted that the maximum margin hyperplane is not unique in this regard; indeed, generically it is not the hyperplane with the best generalisation performance (the so-called Bayes point) [7], but historically much effort has been expended on maximising the margin.

Whilst such analyses and guarantees are proven in the batch setting, the idea of seeking a large margin hyperplane with an online algorithm still makes sense for two reasons. First, one can always use an online algorithm to learn in the batch setting by repeatedly iterating over the sample $\boldsymbol{z}$. Second, even in a truly online setting, a large margin solution provides some immunity to attribute noise and concept drift.

There have been a few recent attempts to develop further online algorithms that achieve an approximation to the maximum margin. Kivinen *et al.* [3] studied the marginalised perceptron (and issues arising when it is kernelised). Li and Long [2] studied an algorithm they called ROMMA where if there is a mistake at the tth trial then $\mathbf{w}_{t+1}$ is the smallest element of the constrained set of $\left\{\mathbf{w} : \mathbf{w}_t \cdot \mathbf{w} \ge \|\mathbf{w}_t\|_2^2\right\} \cap \{\mathbf{w} : y_t(\mathbf{w} \cdot \mathbf{x}_t \ge 1)\}$, else $\mathbf{w}_{t+1} = \mathbf{w}_t$. It has a similar mistake bound to the perceptron and is computationally slightly more costly than the perceptron. Gentile [1] has presented an Approximately Large Margin Algorithm (ALMA)[2] which he analysed in a batch setting showing that for any $\delta \in (0, 1)$ it can achieve a margin of at least $(1 - \delta)\gamma_{\boldsymbol{z}}$ (in the linearly separable case), requiring $O(1/(\delta^2 \gamma_{\boldsymbol{z}}^2))$ updates. Thus, ALMA is obtained from the generalised perceptron by setting $\rho_k = (1-\delta)B/\sqrt{k}$ and $\eta_k = C/\sqrt{k}$ and, if necessary, re-normalising $\mathbf{w}_k$ so that $\|\mathbf{w}_k\|_2 \le 1$.

3 Online Bayes Point Machines

Whilst the variants on the classical perceptron discussed above can guarantee convergence to a hyperplane with a large margin, there is a price to pay. The

---

[2] We only consider the 2-norm case of ALMA. ALMA is a more general algorithm for p-norm classifiers.

Algorithm 2 OBPM algorithm

Require: A training sample $z = ((\mathbf{x}_1, y_1), \ldots, (\mathbf{x}_T, y_T))$.
Require: A online learning algorithm with update rule $\mathcal{L}$ for linear discrimination and associated step-size update rule $\mathcal{S}$.
Require: A subroutine Bernoulli(p) which returns independent Bernoulli random variables with probability p of taking the value 1.
Require: Parameters $N \in \mathbb{N}$ and $\tau \in [0, 1]$.

1: Initialise step sizes $\eta_{j,1}$, for all $j \in \{1, \ldots, N\}$.
2: Initialise weights $\mathbf{w}_{j,1} = \mathbf{0}$, for all $j \in \{1, \ldots, N\}$.
3: **for** $t = 1$ to T **do**
4:     $\tilde{\mathbf{w}}'_t = \mathbf{0}$
5:     **for** $j = 1$ to N **do**
6:         $b_{j,t} = \text{Bernoulli}(\tau)$
7:         **if** $b_{j,t} = 1$ **then**
8:             $\mathbf{w}_{j,t+1} = \mathcal{L}(\mathbf{w}_{j,t}, \eta_{j,t}, \mathbf{x}_t, y_t)$
9:         **else**
10:             $\mathbf{w}_{j,t+1} = \mathbf{w}_{j,t}$
11:         **end if**
12:         $\eta_{j,t+1} = \mathcal{S}(t, \eta_{j,t}, \ldots)$
13:         $\tilde{\mathbf{w}}'_t = \tilde{\mathbf{w}}'_t + \mathbf{w}_{j,t+1}/N$
14:     **end for**
15:     $\tilde{\mathbf{w}}_t = \tilde{\mathbf{w}}'_t / \max\{1, \|\tilde{\mathbf{w}}'_t\|_2\}$
16: **end for**
17: **return** $\tilde{\mathbf{w}}_T$

---

closer one desires the algorithm's hypothesis margin to be to the maximal possible margin, the slower the convergence. In addition, there are several extra parameters one has to choose. This suggests it is worthwhile exploring alternate variants on classical perceptrons.

Our starting point for such variants is the observation that (1) defines a *convex* set $V(z)$. Thus if one found a number of weight vectors $\mathbf{w}_1, \ldots, \mathbf{w}_N$ in $V(z)$, one could optimise over the set of convex combinations $C^N :=$ $C^N(\mathbf{w}_1, \ldots, \mathbf{w}_N) := \{\sum_i \alpha_i \mathbf{w}_i : \|\boldsymbol{\alpha}\|_1 = 1, \boldsymbol{\alpha} \geq \mathbf{0}\}$ of them. Intuitively one would expect that the more "diverse" $\mathbf{w}_1, \ldots, \mathbf{w}_N$ are, the greater the proportion of $V(z)$ could be covered by C^N. The centre of C^N provides an rough approximation of the Bayes point and thus potentially better performance (dependent on the closeness to the true Bayes point). This convex combination of *weight vectors* $\mathbf{w}_j$ is different to hypothesis aggregation methods such as boosting which form convex combinations of classifier *hypotheses* $\mathbf{x} \mapsto \text{sgn}(\mathbf{w}_j \cdot \mathbf{x})$. The sampling presented in this paper can be mistaken for an online version of Bagging [8] since the sample is drawn randomly with replacement from a fixed set. The difference with bagging (and boosting [9]) is that we combine the classifier weights not the hypotheses.

If $\mathbf{w}_1, \ldots, \mathbf{w}_N$ are all identical, C^N is a singleton and nothing is gained. A well known technique for achieving diversity is to generate $\mathbf{w}_1, \ldots, \mathbf{w}_N$ by running (a suitable variant of) the perceptron algorithm on different permutations $\pi(z) :=$

$(z_{\pi(1)}, \ldots, z_{\pi(T)})$ of the training sample z [7], where $\pi : \{1, \ldots, T\} \to \{1, \ldots, T\}$ is a permutation. Although an elegant and effective trick, it is not an online algorithm — one needs the entire training sample z before starting.

This motivates the algorithm we study in this paper: the Online Bayes Point Machine (OBPM) (Algorithm 2). Given a training sequence z, we run N Perceptrons "in parallel" and ensure diversity of their final solutions by randomly choosing to present a given sample z_k to perceptron j only if $b_{jk} = 1$, where b_{jk}, $j = 1, \ldots, N$, $k = 1, 2, \ldots$ are independent Bernoulli random variables with $\Pr\{b_{jk} = 1\} = \tau$.

Although there are theoretical reasons for expecting that refined optimisation of the α_i would lead to better solutions than naively setting them all equal to $1/N$, we have found experimentally this expectation not to hold. Furthermore, such optimisation can only occur in the batch setting: one cannot determine the margin of a candidate weight vector without the whole training sample. One method tried in optimizing the choice of α_i was to maximize the following lower bound in the batch setting (see Appendix A for proof):

$$\gamma_{\mathbf{z}}(\tilde{\mathbf{w}}) \geq \frac{\sum_{j=1}^{N} \alpha_j \gamma_{\mathbf{z}}(\mathbf{w}_j)}{\sqrt{\sum_{i=1}^{N} \sum_{j=1}^{N} \alpha_i \alpha_j (\mathbf{w}_i \cdot \mathbf{w}_j)}}. \tag{2}$$

It was found by experimentation that maximizing the bound of (2) had no performance gain compared to setting $\alpha_i = 1/N$. Hence in the rest of the paper we will only consider the situation where $\alpha_i = 1/N$ for all i. Step 15 of algorithm 2 is optional as it only bounds $\|\tilde{\mathbf{w}}\|$ by one and given the instances are normalized, this results in $\rho \leq 1$ in the generalised Perceptron algorithm 1.

On average each perceptron j sees only a fraction τ of all the examples in z. The smaller τ, the more diverse the solutions $\mathbf{w}_j$ become, although the algorithm takes longer to reach these solutions.

In this paper we take $\mathcal{L}$ the base learner to be the perceptron update rule and fix $\eta_{j,t} = 1$. This leaves OBPM two parameters required for tuning: τ and N.

We see that the number of arithmetic operations for OBPM on average is $O(\tau N dT)$, for both ALMA and the perceptron it is $O(dT)$. OBPM is a factor τN more expensive computationally than ALMA. A point worth noting is that each perceptron used in OBPM is independent and so OBPM can be readily implemented using parallel computing.

In the case of OBPM, the implementation may be made more efficient by noting that the kernel values $K(\mathbf{x}, \mathbf{x}_t)$ are the same for all N perceptrons. For a truly online algorithm we need to bound the number of instances $\mathbf{x}_t$ needed to be stored for the kernel expansion [3].

4   Experiments

We used a toy problem to analyse the effect of τ and N on approximating the maximum margin. We show that for this example we can do better than other

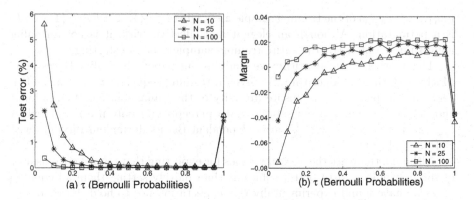

Fig. 1. Artificial data set averages from 30 Monte Carlo simulations training OBPM with no label noise ($e = 0.0$) for a single epoch: (a) test error versus τ, (b) margin $\gamma_z (\tilde{\mathbf{w}}_T)$ versus τ.

online methods but there is a trade-off with computational cost. We demonstrate that OBPM has the ability to handle concept drift by applying it to a problem generated using the MNIST OCR data set. Last, we show the performance of OBPM on some real-world data sets using the online prediction error analysis and batch test errors.

The tuning parameters used in the experiments for OBPM were τ (ranging from 0.01 to 0.5 in non-uniform steps) and $N \in \{100, 200\}$. For ALMA, B was adjusted from 0.5 up to 20 in non-uniform steps, and $C \in \{\sqrt{2}/4, \sqrt{2}/2, \sqrt{2}, 2\sqrt{2}\}$[3]. Thoughout the experiments the final reported parameters were chosen based on the best training error for that experiment. The SVM results were produced with a linear kernel and soft margin using an exact optimisation method. Training and test errors and margins for the algorithms were measured using the last weights as the hypothesis. The results were produced using C code on a 1GHz DEC alpha.

4.1   Parameter Selection

A target $\mathbf{u} \in \{-1, 0, +1\}^{100}$ and T instances $\mathbf{x}_t$ were generated randomly with a margin, $\gamma_z (\mathbf{u}) \geq m = 0.05$. The associated label y_t was allocated $+1$ if $(\mathbf{u} \cdot \mathbf{x}_t) / \|\mathbf{u}\|_2 \geq m$ and -1 if $(\mathbf{u} \cdot \mathbf{x}_t) / \|\mathbf{u}\|_2 \leq -m$. This process ensured that the training sample has a margin greater than m; we generated the test sample identically to the training sample. This problem is very similar to the artificial data experiment of Gentile [1]. We use a different data set size of T, and we use the same margin in the training and test samples. We reduced T from 10000 to 1000, m from 1 to 0.05. This gives a mistake bound of 400 for the classic

---

[3] This was after initial experimentation to determine the best B and C for all the problems considered here.

Table 1. Table of test error averages from 30 Monte Carlo simulations of a single epoch of the parameter selection experiment, with 95 percent confidence interval for Student's t-distribution.

NOISE	PERCEPTRON	ALMA	OBPM
0.0	2.03 $\pm$0.32	0.07$\pm$0.04(B=1.11,C=0.71)	0.00$\pm$0.00($\tau = 0.35$,N=100)
0.01	3.35$\pm$0.44	0.14$\pm$0.07(B=0.83,C=0.71)	0.10$\pm$0.06($\tau = 0.50$,N=100)
0.1	12.96$\pm$1.0	0.76$\pm$0.13(B=1.25,C=0.71)	0.96$\pm$0.13($\tau = 0.15$,N=100)

Perceptron, making it highly unlikely that the perceptron could learn this in a single epoch. Simulations with label noise on the training sample (not the test sample) were done by flipping each label y_t with probability e.

There is a trade-off between computational cost and $\tilde{\mathbf{w}}$ achieving an accurate estimate of the maximum margin. It is indicated in Figure 1 (b) that by increasing N, the margin was increased. As the margin increased, the total number of perceptron updates was increased. For example when $\tau = 0.3$ in the noise free case, the total number of updates of all the perceptrons went from 1334 to 5306, as a result of the total number of perceptrons going from 25 to 100. The relationship between τ and N is shown by Figure 1; if τ is too small then the number of N must be increased to ensure $\tilde{\mathbf{w}}$ is in the version space. From Figure 1 (b) OBPM achieved a margin of 0.02, whereas ALMA achieved 0.006. Table 1 shows with label noise that the performance of ALMA and OBPM were similar, except in the noise-free case. We also noted that ALMA's test error performance was more sensitive to the choice of parameter settings compared to OBPM.

4.2 Drifting Concept

In order to demonstrate OBPM's ability to learn a drifting concept we designed a drifting experiment using the well known MNIST OCR data set[4]. To simulate the drift we set the positive class to a single label which varied over time. The negative class was set to be the remaining 9 labels. We took 1000 examples. The labels where swapped in two phases gradually using a linear relationship in the number of examples seen so far. The mixing of the labels in each trial was random and we averaged results over 10 trials, therefore the transition boundaries in Figure 2 are not obvious. The following psuedo code shows how we allocated labels $l = 1, \ldots, 3$ to the positive class of $z_t = z_t^+$ (i.e. $y = 1$) according to the uniform random variable $r_t \in [0, 1]$ at time t, where $z_t^l = (\mathbf{x}_t^l, y^l = 1)$:

$$z_t^+ = z_t^1$$
$$\text{if } r_t > (1 - \tfrac{t}{T/2}) \text{ then } z_t^+ = z_t^2$$
$$\text{if } r_t > (1 - \tfrac{t}{1.11T}) \text{ then } z_t^+ = z_t^3$$

---

[4] Available at http://yann.lecun.com/exdb/mnist/index.html.

Selected the labels this way we ensure by $t > T/2$ all the positive class examples are with label 2 and by 900 they are all label 3. Replacing the labels randomly in the above way also gives a smoother and more gradual transition between labels.

Figure 2 presents the results of the MNIST drift experiment for 10 trials with permutations. We see that OBPM made fewer prediction errors compared to several other online algorithms. One of the online algorithms compared with OBPM was LMS (Least Mean Squared [12]), which is a regression algorithm where we take the sign of the predictions to convert to the classification setting. One reason why ALMA did not perform as well as OBPM is that ALMA assuming a stationary target in order to adjust the learning rate η_t and margin parameter ρ_k. Note that this is not an exhaustive study on drifting concepts but it appears the large margin of OBPM allows for the improvement over the perceptron (the generalised Perceptron $\rho = 0$).

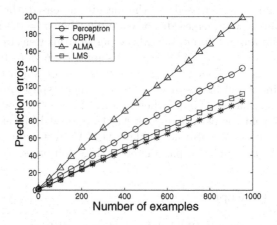

Fig. 2. Prediction errors from drifting concept experiment with the MNIST OCR data set comparing LMS (step size 0.5), ALMA (B=2, C=$\sqrt{2}$), Perceptron ($\rho = 0$) and OBPM (τ=0.4, N=100).

4.3   Real-World Data Sets

We consider three different data sets. The first data set was derived from a genome of the nematode Ceanorhabditis Elegans (C. Elegans). The classification task was to locate the specific splice sites (for more details see [10]). The C. Elegans training sample had 6125 dimensions and 100 000 examples, with a separate test set of 10 000 examples. The other two data sets were the UCI Adult the Web data sets [5]. Each input consists of 14 mixed categorical and

[5] Available at http://www.research.microsoft.com/~jplatt/smo.html

continuous attributes. We increase the input dimensionality to 123 by quantising the continuous attributes and by representing the categorical inputs with a 1-of-N code. The training sample size of Adult is 32562 and the test set size is 16282. The prediction task of the adult data set is to determine whether a person earns over $50k a year. The Web task consists of predicting whether a web page belongs to a topic or not, based on the presence of 300 words. The Web task has a training sample size of 49 749 and a test set size of 21 489. To study the

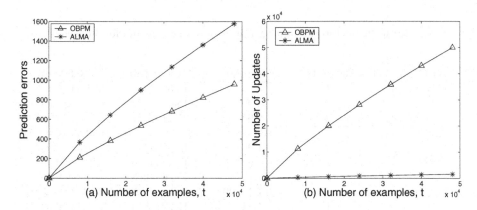

Fig. 3. (a) Prediction errors for OBPMs $\tilde{\mathbf{w}}_t$ (τ=0.3, N=100) and ALMAs weight solutions and (b) total number updates for the Web data after training example t.

online behaviour of this algorithm we measured the number of prediction errors made at t examples and compared this to ALMA. The prediction errors made on the Web data in Figure 3(a) demonstrate a lower prediction error compared to ALMA. The prediction error was consistently lower than ALMA with all the experiments in this paper. The lower prediction error comes at a cost, as shown in Figure 3(b): the total number of updates made by OBPM is larger than the number made by ALMA.

The batch setting results benchmarking OBPM with other algorithms are given in Table 2. The UCI data results of Table 2 report the average test error for 30 trials of permutations of the training sequence with their 95 percent confidence interval using the Student's t-distribution. The maximum margin benchmark, the soft margin linear SVM gave test errors of 15.05% and 1.25% for the Adult and Web respectively. The C. Elegans data set OBPM with $\tau = 0.05, N = 100$, was able to achieve a test error of 1.59% after 20 epochs which was comparable to the SVM (soft margin) result 1.58%, ALMA achieved 1.72%. The Cochran test statistic for C. Elegans after 20 epochs comparing ALMA, OBPM and SVM was 3.77 which indicates no difference at a significance level of 95 percent. A general comment of the results of Table 2 was that in the batch setting OBPM performed better than the standard perceptron whereas ALMA and OBPM were comparable. For these experiments it was not obvious how to tune ALMA's parameters easily, whereas with OBPM good results were achieved by simply adjusting τ over a small range 0.1 to 0.4.

The time taken to process the C. Elegans results were: OBPM with CPU time 19 minutes (real time 21 minutes) and ALMA CPU time of 3 minutes (real time 13 minutes). The CPU time is not prohibitive, especially given that the SVM[6] optimisation took approximately 300 hours of real time. Of course, on a parallel machine, OBPM could be sped up further.

Table 2. Real world data set test errors in percent run for 3 epochs.

ALGORITHM	EPOCHs	C. Elegans	UCI Adult	UCI Web
Perceptron	1	3.33	19.59±1.54	1.84±0.12
	2	2.99	18.81±0.96	2.12±1.03
	3	2.93	21.21±2.21	1.72±0.18
ALMA(B,C)		(B=3.33,C=0.70)	(B=20,C=0.70)	(B=0.625,C=1.41)
	1	1.95	15.88±0.19	1.36±0.03
	2	1.92	15.35±0.13	1.29±0.03
	3	1.78	15.26±0.13	1.26±0.03
OBPM(τ,N)		(τ=0.2,N=200)	(τ=0.01,N=200)	(τ=0.3,N=200)
	1	1.94	15.29±0.15	1.42±0.04
	2	1.86	15.24±0.19	1.36±0.04
	3	1.79	15.17±0.14	1.31±0.03

5   Conclusions

We have presented OBPM, which is a simple meta-algorithm for the online training of linear classifiers with large margin. OBPM trains N linear classifiers in parallel on random subsets of the data. We have shown experimentally that compared to online algorithms such as the standard perceptron and ALMA, OBPM was able to achieve a lower prediction error and track concept drift. We were able to demonstrate that OBPM is a truly online algorithm with a large margin which is simple to implement with the potential for a parallel architecture. OBPM's CPU times were not prohibitive and its performance compared favourably with SVM and ALMA when trained in the batch setting on three real datasets.

Acknowledgements. This work was supported by the Australian Research Council. We thank Gunnar Rätsch, Alex Smola and Sören Sonnenburg for their

---

[6] We used the fast interior point optimization package SVlab of Alex Smola which has been shown to have comparable speed to SVM light.

help with the C. Elegans data and for providing their SVM results for C. Elegans. Edward's research was funded by the Defence Science and Technology
Organisation, Australia.

A   Lower Bound Proof of (2)

Proof. Recall that the convex hull of weights $\mathbf{w}_1, \ldots, \mathbf{w}_N$ is defined by $C^N :=$
$C^N(\mathbf{w}_1, \ldots, \mathbf{w}_N) := \{\sum_i \alpha_i \mathbf{w}_i : \|\boldsymbol{\alpha}\|_1 = 1, \boldsymbol{\alpha} \geq \mathbf{0}\}$. From the definition of C^N,
and $\tilde{\mathbf{w}} = \sum_{i=1}^N \alpha_i \mathbf{w}_i$, we find that $\|\tilde{\mathbf{w}}\|_2$ is equal to

$$\sqrt{\left(\sum_{i=1}^N \alpha_i \mathbf{w}_i \cdot \sum_{j=1}^N \alpha_j \mathbf{w}_j\right)} = \sqrt{\sum_{i=1}^N \sum_{j=1}^N \alpha_i \alpha_j (\mathbf{w}_i \cdot \mathbf{w}_j)}. \tag{A.1}$$

For any example $z_i := (\mathbf{x}_i, y_i)$ from $\mathbf{z} = (z_1, \ldots, z_t)$ and exploiting the normal-
ization $\|\mathbf{w}_i\|_2 = 1$ for $i = 1, \ldots, N$, we conclude that

$$y_i (\tilde{\mathbf{w}} \cdot \mathbf{x}_i) = y_i \left(\sum_{j=1}^N \alpha_j \mathbf{w}_i \cdot \mathbf{x}_i\right) = \sum_{j=1}^N \alpha_j y_i \frac{(\mathbf{w}_i \cdot \mathbf{x}_i)}{\|\mathbf{w}_i\|_2}. \tag{A.2}$$

Using the definition for the margin for a particular weight solution $\mathbf{w}$ over the
sequence $\mathbf{z}$, $\gamma_{\mathbf{z}}(\mathbf{w}) := \min\{y_i (\mathbf{w} \cdot \mathbf{x}_i) / \|\mathbf{w}\|_2 : i \in \{1, \ldots, T\}\}$ with the result of
(A.2), we obtain

$$y_i (\tilde{\mathbf{w}} \cdot \mathbf{x}_i) \geq \sum_{j=1}^N \alpha_j \gamma_{\mathbf{z}}(\mathbf{w}_j). \tag{A.3}$$

Combining (A.1) and (A.3) completes the proof that

$$\gamma_{\mathbf{z}}(\tilde{\mathbf{w}}) \geq \frac{\sum_{j=1}^N \alpha_j \gamma_{\mathbf{z}}(\mathbf{w}_j)}{\sqrt{\sum_{i=1}^N \sum_{j=1}^N \alpha_i \alpha_j (\mathbf{w}_i \cdot \mathbf{w}_j)}}.$$

References

1. C. Gentile, (2001) A new approximate maximal margin classification algorithm. *Journal of Machine Learning Research*, **2**:213-242.
2. Y. Li. & P. Long, (2002) The relaxed online maximum margin algorithm. *Machine Learning*, **46(1-3)**:361-387.
3. J. Kivinen, A. Smola, and R. C. Williamson, (2002) Online Learning with kernels. *Advances in Neural Information Processing Systems 14*, Cambridge, MA: MIT Press (pp. 785-793).
4. R.O. Duda & P.E. Hart & D.G. Stork, (2000) *Pattern Classification And Scene Analysis 2nd Edition*. John Wiley.

5. A.B.J. Novikoff, (1962) On convergence proofs on perceptrons. *In Proceedings of the Symposium on Mathematical Theory of Automata*, vol. XII, (pp. 615-622).

6. R. Herbrich, (2002) *Learning Kernel Classifiers*, Cambridge, MA: MIT Press.

7. R. Herbrich, T. Graepel & C. Campbell, (2001) Bayes Point Machines. *Journal of Machine Learning Research*, 1:245-279.

8. L. Breiman, (1996) Bagging predictors. *Machine Learning*, **24(2)**:120-140.

9. R.E. Schapire, (1990) The strength of weak learnability. *Machine Learning*, **5**:197-227.

10. S. Sonnenburg, (2002) New Methods for Splice Site Recognition. Master's thesis, Humbold University.

11. W.J. Conover, (1980) *Practical nonparametric statistics, 2nd Edition.* John Wiley.

12. B. Widrow and M. E. Hoff, (1960) Adaptive switching circuits. *1960 IRE WESCON Convention Record*, pt. 4, pp. 96-104.

Exploiting Hierarchical Domain Values for Bayesian Learning*

Yiqiu Han and Wai Lam

Department of Systems Engineering and Engineering Management
The Chinese University of Hong Kong
Shatin, Hong Kong
{yqhan, wlam}@se.cuhk.edu.hk

Abstract. This paper proposes a framework for exploiting hierarchical structures of feature domain values in order to improve classification performance under Bayesian learning framework. Inspired by the statistical technique called shrinkage, we investigate the variances in the estimation of parameters for Bayesian learning. We develop two algorithms by maintaining a balance between precision and robustness to improve the estimation. We have evaluated our methods using two real-world data sets, namely, a weather data set and a yeast gene data set. The results demonstrate that our models benefit from exploring the hierarchical structures.

1   Introduction

Hierarchical structures are commonly found in different kinds of data sets. Previous methods mainly focused on utilizing the hierarchy of class labels. In these methods, the classification task is divided into a sequence of sub-problems where each sub-problem is concerned with a classification task for a certain node in the hierarchy. Apart from class hierarchy, some data sets may contain features in which the domain values rather than classes are organized in a hierarchical structure. For example, the attributes handled by "drill down" or "roll up" operations in OLAP are basically hierarchical features. Little attention was paid to this kind of hierarchical features in classification problems despite their presence in real-world problems.

In this paper, we refer *hierarchical features* to those features with categorical domain values organized in a hierarchical structure. The hierarchical structure is basically a tree and each node in this tree corresponds to a distinct value in the domain of the feature. The value represented by a parent node denotes a more general concept with respect to the value represented by a child node. Therefore concept nodes in such a hierarchy become more abstract if we navigate the hierarchy from bottom-up. Data instances can attain domain values corresponding to any node in the hierarchy including leaf nodes and intermediate nodes.

---

* The work described in this paper was partially supported by grants from the Research Grant Council of the Hong Kong Special Administrative Region, China (Project Nos: CUHK 4385/99E and CUHK 4187/01E)

K.-Y. Whang, J. Jeon, K. Shim, J. Srivatava (Eds.): PAKDD 2003, LNAI 2637, pp. 253–264, 2003.
© Springer-Verlag Berlin Heidelberg 2003

Hierarchical features may bring forth the problem of data sparseness especially when the number of domain values is large. Traditional model parameter estimation for Bayesian learning becomes very unreliable since increased number of feature domain values decreases the average number of training instances associated with a particular domain value for estimating the posterior probability. This situation becomes worse when the data set itself is already sparse.

A clue for solving this problem may lie in the hierarchy itself. A hierarchy in fact reflects existing knowledge about feature values, revealing their interrelationships and similarities in different levels. If utilized properly, this kind of background knowledge can help deal with problems related to hierarchical features including data sparseness.

McCallum et al. [6] applied shrinkage [5,10] technique in a hierarchy of classes and showed some improvements on text classification. An improved estimate for each leaf node is derived by "shrinking" its maximum likelihood estimate towards the estimates of its ancestors. The hierarchy they used is composed of classes and their model mainly utilizes ancestor information in the hierarchy. Segal and Koller [9] introduced a general probabilistic framework for clustering data into a hierarchy. Global optimization is used to construct the class hierarchy over the biological data. Their model is not designed for handling hierarchical features considered in this paper. Exploiting the hierarchical structure of classes has been investigated in classification problems, but little work has been done for hierarchical features.

Domingos and Pazzani [2] explained why good predictive performance can be achieved by Naive Bayesian learning on classification problems even it requires independence assumption which may be invalid in the data set. Under Maximum A Posteriori Principle, the independence assumption does not seriously degrade the classification.

This paper investigates Bayesian learning model in classification problems with hierarchical features. Inspired by the statistical technique called *shrinkage*, we investigate the variance of posterior probabilities in the domain value hierarchy. We develop a model which attempts to offset these variances by maintaining a balance between precision and robustness and hence improving the parameter estimation for Bayesian learning. We introduce two algorithms designed to exploit the hierarchical structure of domain values.

We have conducted experiments using two real-world data sets. The first one is a weather data set concerned with rainfall measurements of different geographical sites. The second one is a yeast gene data set obtained from KDD Cup 2002 Task 2. The results demonstrate that our proposed methods can improve the classification performance under Bayesian learning model.

2   Simple Bayesian Learning

We investigate the learning problem by considering a parametric model. The training data is used to estimate the model parameters. After training, the model equipped with the model parameters can be used for predicting the class label or calculating the membership likelihood for a certain class. One issue we need to deal with is the handling of hierarchical features. As mentioned above, we refer hierarchical features to those features with categorical domain values organized in a hierarchical structure. To facilitate the discussion, we first provide some definitions.

Definition. We define *raw frequency*, denoted by $R(n)$, of a node n in a hierarchy as the number of training instances having the same feature value represented by the node n. ◇

Definition. Let C denote a class. We define *raw hit frequency*, denoted by $R(C, n)$, of a node n in a hierarchy as the number of training instances having class label C and the same feature value represented by the node n. ◇

Suppose we are given a new instance and we wish to estimate the membership likelihood of a particular class. Let the new instance be represented by a set of features $F = (f_1, \ldots, f_m)$ where f_i denotes the values of the feature F_i for this instance. For a hierarchical feature, the value corresponding to f_i's is associated with a node in the hierarchy. To obtain the likelihood of the new instance belonging to the class C, we can make use of Bayes' theorem as follows:

$$P(C|F) = \frac{P(C)P(F|C)}{P(F)} \tag{1}$$

Given the assumption that the features are conditionally independent of each other, Equation 1 can be expressed as:

$$P(C|F) = \frac{P(C)\prod_i^m P(F_i = f_i|C)}{\prod_i^m P(F_i = f_i)} \tag{2}$$

Given a set of training instances, Equation 2 can be estimated by:

$$\hat{P}(C|F) = N^{-1}N(C)^{1-m}\frac{\prod_i^m \Theta(C, f_i)}{\prod_i^m \Theta(f_i)} \tag{3}$$

where N is the total number of training instances; $N(C)$ is the total number of training instances with class C; $\Theta(C, f_i)$ and $\Theta(f_i)$ are two model parameters estimated from the training instances. A common technique is to employ the maximum likelihood estimates. Then, $\Theta(C, f_i)$ and $\Theta(f_i)$ are given by $R(C, f_i)$ and $R(f_i)$ respectively.

Suppose we are given a set of new instances. If we are interested in a ranked list of the likelihood of these instances belonging to the class C, we can compute the likelihood, $L(C|F)$, by deriving from Equation 3:

$$L(C|F) = \frac{\prod_i^m R(C, f_i)}{\prod_i^m R(f_i)} \tag{4}$$

3   Exploiting the Hierarchical Structure

Our model considers subtrees in a hierarchy as representation of hierarchical concepts. The motivation is to tackle problems where an ordinary concept-node hierarchy is not applicable. For example, the feature value of an instance may not only be defined on a particular leaf node, it may contain values corresponding to more abstract concepts consisting of both leaf nodes and internal nodes. One reason is due to the fact that the information carried by a feature value may be insufficient or unclear. Hence the value can only be represented as a more abstract concept. Another reason is that some feature values do not comfortably fit into any known category and can only be associated with the closest concept in the hierarchy.

Suppose we are going to conduct learning for a binary classification model for the class C given a training data set. Consider a certain feature H with its domain values organized in a hierarchy. For each node n in the hierarchy, let T_n denote the subtree rooted by the node n. We define *tree frequency* and *tree hit frequency* as follows:

Definition. We define *tree frequency*, denoted by $R(T_n)$, of a node n in a hierarchy as the number of training instances having feature values represented by the node n or any node in the sub-tree rooted by n. The relationship of $R(T_n)$ and $R(n)$ can be expressed as follows:

$$R(T_n) = \sum_{i \in T_n} R(i) \tag{5}$$

where $R(i)$ is defined in Definition 2. ◇

Definition. We define *tree hit frequency*, denoted by $R(C, T_n)$, of a node n in a hierarchy as the number of training instances having class label C and the same feature value represented by the node n or any node in the sub-tree rooted by n. The relationship of $R(C, T_n)$ and $R(C, n)$ can be expressed as follows:

$$R(C, T_n) = \sum_{i \in T_n} R(C, i) \tag{6}$$

where $R(C, i)$ is defined in Definition 2. ◇

There is a parameter associated with the subtree T_n, namely, the posterior probability $P(C|H = T_n)$ where $H = T_n$ denotes the fact that the domain value can be any element found in T_n. We wish to estimate this probability from the training data set. Let $\Theta_c(T_n)$ be the observation for estimating $P(C|H = T_n)$, expressed as follows:

$$\Theta_c(T_n) = \frac{R(C, T_n)}{R(T_n)} \tag{7}$$

The instances in the training data set satisfying $H = T_n$ can be regarded as $R(T_n)$ independent identical distributional Bernoulli trials with success probability $P(C|H = T_n)$. Therefore, the observed value of $\Theta_c(T_n)$ distributes around $P(C|H = T_n)$.

Consider the expectation and variance of $\Theta_c(T_n)$

$$E(\Theta_c(T_n)) = P(C|H = T_n) \tag{8}$$

$$\text{Var}(\Theta_c(T_n)) = \frac{P(C|H = T_n)(1 - P(C|H = T_n))}{R(T_n)} \tag{9}$$

According to Chebyshev theorem,

$$P(|\Theta_c(T_n) - P(C|H = T_n)| \geq \epsilon) \leq \frac{P(C|H = T_n)(1 - P(C|H = T_n))}{R(T_n)\epsilon^2} \tag{10}$$

In the traditional Bayesian learning model as described in Section 2, the feature value is limited to the raw frequency of the instances associated with the node n. Equation 10 shows that if we use $\Theta_c(n)$ as the estimator of $P(C|H = n)$, the variance decreases with the increasing of $R(n)$. When $R(n)$ and $P(C|H = n)$ are very small, the variance of $\Theta_c(n)$ almost reaches $\frac{1}{R(n)}$ of $P(C|H = n)$. Hence Bayesian learning becomes very unreliable especially when the data set is sparse. Furthermore, hierarchical structures tend to exhibit the data sparseness problem due to the increase in the number of domain values.

4   Shrinkage-Based Approach

To cope with the problems commonly found in hierarchical features, we propose a technique for conducting Bayesian learning by exploiting the domain value hierarchical structure. The main idea is inspired by a statistical technique called *shrinkage* which provides a way for improving parameter estimation coping with limited amount of training data

The parent-child relationship of feature values should indicate some mean-variance relationships in the distribution of class labels. Let m be the parent node of the node n. We assume the posterior probability $P(C|H = T_n)$ of subtree T_n follows a distribution with mean $P(C|H = T_m)$ and variance σ_m^2. Then, we have the upper bound of the error using the parent's posterior probability to estimate that of the child subtree.

$$P(\, |P(C|H = T_n) - P(C|H = T_m)| \, \geq \, \epsilon \,) \leq \frac{\sigma_m^2}{\epsilon^2} \tag{11}$$

Here σ_m^2 is independent of the variance of $\Theta_c(T_n)$ or $\Theta_c(T_m)$. It is only determined by the position of T_m in the hierarchy as well as how T_m is partitioned into different subtrees.

Let ψ be the error of using the observed posterior probability $\Theta_c(T_m)$ to estimate the $P(C|H = T_n)$ and it is defined as:

$$\psi = \Theta_c(T_m) - P(C|H = T_n)$$
$$= (\Theta_c(T_m) - P(C|H = T_m)) + (P(C|H = T_m) - P(C|H = T_n)) \tag{12}$$

Consider the expectation and variance of ψ

$$E(\psi) = 0 \tag{13}$$

$$\text{Var}(\psi) = \sigma_m^2 + \frac{P(C|H = T_m)(1 - P(C|H = T_m))}{R(T_m)} \tag{14}$$

So we have

$$P(|\Theta_c(T_m) - P(C|H = T_n)| \geq \epsilon) \leq \frac{\sigma_m^2 + \frac{P(C|H=T_m)(1-P(C|H=T_m))}{R(T_m)}}{\epsilon^2} \tag{15}$$

Recall from Equation 10 that $\Theta_c(T_m)$ should be chosen to estimate $P(C|H = T_n)$ instead of $\Theta_c(T_n)$ if the following equation holds:

$$\sigma_m^2 + \frac{P(C|H = T_m)(1 - P(C|H = T_m))}{R(T_m)} < \frac{P(C|H = T_n)(1 - P(C|H = T_n))}{R(T_n)} \tag{16}$$

Since a node x corresponding to a feature value can be treated as a subtree containing only one node, one can iteratively use the observed posterior probability of "ancestor" subtrees to estimate the original $P(C|H = x)$ until Equation 16 does not hold.

Based on the above idea, we develop a parameter estimation algorithm called SB1 which can handle hierarchical domain values for Bayesian learning. This algorithm looks for a subtree, whose tree frequency should be large enough to produce a robust estimation and offset the variances between its posterior probability and that of the target node. The details of SB1 is shown in Figure 1. Suppose the average number of shrinkage is N and the average fan-out number of a node is M. Then the computational cost of SB1 for an instance is $O(NM)$. Generally both N and M are not too large, so the cost is low.

5   Subtree Clustering Approach

By utilizing the observed posterior probability of other concept nodes, the SB1 algorithm trades the precision of estimation for robustness. Equation 16 gives the upper bound of the estimation error. However it also limits the possibility of exploiting the feature hierarchy since SB1 only considers a particular path in the hierarchy. In fact, the potential candidate sets of nodes for improving estimation may com e from neighboring nodes. It have been observed that some hierarchical structures in real world problems is basically "flat" ones. These kinds of "flat" hierarchical structures are characterized by a large average fan-out nodes and small number of hierarchy levels. These characteristics make Equation 16 difficult to meet and hence degrading the performance of SB1 to simple Bayesian learning.

We treat the "flat" hierarchy as a kind of reliable but "coarse" categorization of the domain values. It may cause the average variance of posterior probabilities between the parent subtree and children subtrees to be too large to handle. However, it could still be useful for there may be a potential set of children subtrees which can meet the requirements stated in Equation 16. Based on this

```
        // Statement 1 to 7 estimate the variance of the
        // posterior probability distribution P(C|H = Tj)
1   For every node j in the hierarchy of H
2       If j is leaf node
3           σ²j = 0
4       Else
            // Maximum likelihood estimation of the variance
5           σ̂²j = Avg((Θc(Ti) − Θc(Tj))²) for all i satisfying Parent(i) = j
        // Statement 8 to 19 estimate the posterior probability P(C|H = Tj)
6   For every node j in the hierarchy of H
7       If (σ̂²j + Θc(Tj)(1−Θc(Tj))/R(Tj)  >  Θc(j))(1−Θc(j))/R(j))
            // Use direct estimation when conditions are not met
8           Θ(C, j) = R(C, j);
9           Θ(j) = R(j);
10      Else
            // Recursively improve the estimation by shrinkage
11          x = j;
12          y = Parent(j);
13          While(Root(H) ≠ x and σ̂²y + Θc(Ty)(1−Θc(Ty))/R(Ty)  >  Θc(Tx)(1−Θc(Tx))/R(Tx))
14              x = y;
15              y = Parent(x);
16          Θ(C, j) = R(C, Tx);
17          Θ(j) = R(Tx);
```

Fig. 1. The Shrinkage-Based Bayesian Learning (SB1) Algorithm

idea, we employ an agglomerative clustering technique to search for such a cluster of children subtrees and generate virtual parent subtrees.

We modify SB1 to incorporate this idea of subtree clustering and develop an algorithm called SB2. The design of SB2 is depicted in Figure 2. First we use the target subtree T_n as the initial cluster V, V should be viewed as a parent subtree containing only T_n. Then we consider all possible sibling subtrees T_k of T_n, V' is produced by adding T_k as a child of V. The distance metric used for agglomerative clustering is defined as follows:

$$Dis(V, T_k) = \frac{\sigma^2_{V'} + \frac{P(C|H=V')(1-P(C|H=V'))}{R(V')}}{\sigma^2_V + \frac{P(C|H=V)(1-P(C|H=V))}{R(V)}} \tag{17}$$

This distance metric calculates the change of error bound by adding a new subtree. If it is less than 1, it implies that adding this new subtree is helpful in controlling the error. If it is greater than 1, it indicates that adding this new subtree is not helpful. The subtree T_k with the smallest $Dis(V, T_k)$ will be added to the V as a child subtree iteratively until $Dis(V, T_k)$ becomes more than 1. Then the value of $P(C|H = V)$ is returned as the estimation of $P(C|H = T_n)$.

Compared with SB1, the additional computational cost of SB2 is $O(M^3)$ where M is the average fan-out number of a node. If M is not large, the computational cost of SB2 is acceptable.

```
1   For every node j in the hierarchy of H
2       If j is leaf node
3           σ²_j = 0
4       Else
5           σ̂²_j = Avg((Θ_c(T_i) − Θ_c(T_j))²) for all i satisfying Parent(i) = j
6   For every node j in the hierarchy of H
7       If (σ̂²_j + [Θ_c(T_j)(1−Θ_c(T_j))]/R(T_j)  >  [Θ_c(j))(1−Θ_c(j))]/R(j))
8           Θ(C, j) = R(C, j);
9           Θ(j) = R(j);
10      Else
11          x = j;
12          y = Parent(j);
13          While(Root(H) ≠ x and σ̂²_y + [Θ_c(T_y)(1−Θ_c(T_y))]/R(T_y)  >  [Θ_c(T_x)(1−Θ_c(T_x))]/R(T_x))
14              x = y;
15              y = Parent(x);
16          V = T_x;
17          Loop
18              For all T_k satisfying Parent(k) = Parent(x)
19                  V' = V + T_k;
20                  If Dis(V, M) = Min_{T_k}(Dis(V, T_k)) and Dis(V, M) < 1
21                      V = V + M;
22          Until (Dis(V, M) ≥ 1)
23          Θ(C, j) = R(C, V);
24          Θ(j) = R(V);
```

Fig. 2. The Subtree Clustering (SB2) Algorithm

6   Experimental Results

6.1   Experiments on the Weather Data Set

Data Set and Experiment Setup. We have also conducted some experiments on a weather data set to investigate our models dealing with hierarchical features. The data set was collected from the measurements of the weather of 206 geographical sites across China from January to August in 2002 [1]. It contains noises since the data is collected from Global Telecommunication System and some records may be missing.

The geographical locations of the sites can be organized into a hierarchy. We focus on the average probability of raining for a geographical site as the posterior probability $P(Rain|Site)$ to be estimated. The root of the hierarchy represents the country-wide parameters. The first level of the hierarchy represents the "district" level. The second level represents the "province" level and the last level represents different sites in different provinces.

There are 40,000 data points in the original data set. To control the degree of data sparseness in the empirical investigation, we randomly selected a fraction of the original data set records for learning in every trial. Data records for the nonleaf nodes in the hierarchy were generated in the following way. In every trial we

---

[1] This data set can be download from http://cdc.cma.gov.cn/realtime.asp

randomly transferred some instances to the ancestor nodes of their associated nodes so that every node in the hierarchy has roughly equal number of the corresponding data records. The average probability of rainfall of every node in the geographical hierarchy is calculated for evaluation. We applied our proposed methods to estimate the posterior probability of rainfall for 206 nodes in the hierarchy. The direct estimation without using the hierarchy is also computed to serve as a baseline. The performance of the model is measured by the sum of squared error of 2,000 random trials.

Results and Discussions. Figure 3 depicts the estimation error under different degrees of data sparseness. We can see that SB1 reduced the error of estimation significantly. And the improvement increases with the data sparsness. On the other hand, SB2 improves the performance compared with SB1. The improvement over the direct estimation increases with the degree of data sparseness.

The improvement of SB2 over SB1 also increases with the data sparseness. It demonstrates the effect of using the hierarchy for increasing robustness and using clustering to handle "flat" structures.

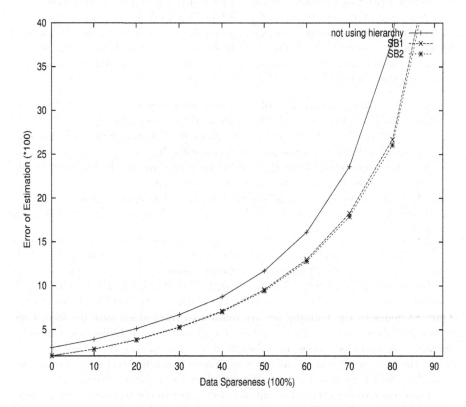

Fig. 3. Empirical investigation of SB1 and SB2 using the weather data set under different degrees of data sparseness

Table 1. Estimation error of Bayesian learning for different algorithms on the weather data set

data sparseness	direct estimation	SB1	SB2
0%	2.95	2.61	2.56
10%	3.85	3.35	3.32
20%	5.06	4.42	4.40
30%	6.72	5.89	5.86
40%	8.83	7.78	7.70
50%	11.74	10.37	10.19
60%	16.13	13.85	13.41
70%	23.44	19.21	18.45
80%	38.26	27.91	26.79
90%	81.42	47.92	45.95

6.2 Experiments on the Gene Data Set

Data Set and Experiment Setup. The first data set used in our experiment is a yeast gene data set obtained from KDD Cup 2002 Task 2 [2]. Every instance in the data set corresponds to a gene (gene-coded protein) and the class label is based on the discrete measurement of how active one (hidden) system is when the gene is knocked out (disabled). Each instance is represented by a set of features listed as follows:

- The sub-cellular *localization* of the gene-coded protein. This is a hierarchical feature whose domain values are organized in a hierarchy.
- The *protein class* of the gene-coded protein. This is a hierarchical feature whose domain values are organized in a hierarchy.
- The *functional class* of the gene-coded protein. This is a hierarchical feature whose domain values are organized in a hierarchy.
- *Abstracts* drawn from the MEDLINE articles in free text form.
- The *interaction* of the gene-coded protein. This feature is organized as a pair list.

The data set is very sparse and noisy, reflecting existing knowledge of the yeast genes. There are missing feature values, aliases, and typographical errors.

There are in total 4,507 instances with class label in the data set. 3,018 labeled ones are for training use and 1,489 instances are held aside for testing. Only 84 genes in the training set are related to the hidden system. 38 of these genes are labeled with "change" which are of interests to biologists. The other instances in those 84 genes are labeled with "control" which means that they affect the hidden system in indirect way. The remaining 2,934 genes are labeled with "nc" which means that they do not affect the hidden system.

The objective is to return a ranked list of genes in the testing set in the order of their likelihood of belonging to a particular category. The first category is

[2] The detail can be found at http://www.biostat.wisc.edu/~ craven/kddcup/faqs.html

called "narrow" classification whose aim is to predict the class "change". The second category is called "broad" classification whose aim is to predict the union of the classes "change" and "control".

To focus on the investigation on effectiveness of exploiting hierarchical structures, we only utilize three hierarchical features, namely, *localization*, *protein class*, and *functional class* for conducting Bayesian learning.

The evaluation metric used to measure the classification performance on the gene data set is based on a score metric offered by the KDD Cup 2002 Task 2. This metric is called Receiver Operating Characteristic (ROC) score. An ROC curve is a plot of the true positive rate (TP rate) against the false positive rate (FP rate) for different classification parameter thresholds. Under a particular threshold, TP rate is the fraction of positive instances for which the system predicts "positive". The FP rate is the fraction of negative instances for which the system erroneously predicts "positive". The area under the plot is the ROC score.

Results and Discussions. Tables 2 and 3 depict the classification performance of different models under 3-fold cross validation and under the same training/testing data split as specified in KDD Cup 2002 Task 2 respectively. The results show that the simple Bayesian learning cannot handle the data sparseness problem in the gene data set. The ROC score of the simple Bayesian method is similar to random prediction. The SB1 algorithm shows a significant improvement on simple Bayesian learning.

Table 2. Classification performance measured by ROC score for different algorithms on the gene data set under 3-fold cross validation

	Simple Bayesian	SB1	SB2
Broad	0.512	0.640	0.646
Narrow	0.520	0.644	0.646

Table 3. Classification performance measured by ROC score for different algorithms on the gene data set using the same training/testing data split as specified in KDD Cup 2002 Task 2

	Simple Bayesian	SB1	SB2
Broad	0.506	0.633	0.637
Narrow	0.537	0.656	0.660

7   Conclusions and Future Work

We have developed a framework exploiting the domain value hierarchy of a feature for improving classification performance under Bayesian learning. The data sparseness and "flat" structures are tackled by our algorithms which employ shrinkage and subtree clustering techniques. We have conducted experiments on two real-world data sets. They are a weather data set and a yeast gene data set. The results demonstrate the effectiveness of our framework.

We intend to explore several future research directions. One direction is to investigate different kinds of hierarchies, especially those with large depth. Another promising direction is to explore "global" methods of exploiting the hierarchy instead of local operations which focus on the neighborhood and ancestors. At this stage, our framework mainly handles the nodes given by the hierarchy. The improved performance of SB2 suggests that it may be worthwhile to consider beyond the limit of predefined tree structure. We plan to use grid structure to exploit inter-relationships of domain values.

References

1. L. D. Baker and A. K. McCallum, *Distributional Clustering of Words for Text Classification*, Proceedings of the 21st Annual International ACM SIGIR Conference on Research and Development in Information Retrieval, pp. 96–102, 1998.
2. P. Domingos and M. Pazzani, *Beyond Independence: Conditions for the Optimality of the Simple Bayesian Classifier*, Machine Learning 29, pp. 103–130, 1997.
3. S. Dumis and H. Chen, *Hierarchical Classification of Web Content*, Proceedings of the 23rd Annual International ACM SIGIR Conference on Research and Development in Information Retrieval, pp. 256–263, 2000.
4. D. Freitag and A. McCallum. *Information Extraction with HMMs and Shrinkage.* Proceedings of the AAAI-99 Workshop on Machine Learning for Information Extraction. 1999
5. W. James and C. Stein, *Estimation with Quadratic Loss.*, Proceedings of the Fourth Berkeley Symposim on Mathematical Statistics and Probability 1, pp. 361–379, 1961.
6. A. McCallum, R. Rosenfeld, T. Mitchell and A. Y. Ng, *Improving Text Classification by Shrinkage in a Hierarchy of Classes*, Proceedings of the International Conference on Machine Learning (ICML), pp. 359–367, 1998.
7. A. McCallum and K. Nigam, *Text Classification by Bootstrapping with Keywords, EM and Shrinkage*, ACL Workshop for Unsupervised Learning in Natural Language Processing, 1999
8. K. Qu, L. Nan, L. Yanmei and D. G. Payan, *Multidimensional Data Integration and Relationship Inference*, IEEE Intelligent Systems 17, pp. 21–27, 2002.
9. E. Segal and D. Koller, *Probabilistic Hierarchical Clustering for Biological Data*, Annual Conference on Research in Computational Molecular Biology, pp. 273–280, 2002.
10. C. Stein, *Inadmissibility of the Usual Estimator for the Mean of a Multivariate Normal Distribution*, Proceedings of the Third Berkeley Symposim on Mathematical Statistics and Probability 1, pp. 197–206, 1955.
11. F. Takech and E. Suzuki, *Finding an Optimal Gain-Ratio Subset-Split Test for a Set-Valued Attribute in Decision Tree Induction*, Proceedings of the International Conference on Machine Learning (ICML), pp. 618–625, 2002

A New Restricted Bayesian Network Classifier

Hongbo Shi, Zhihai Wang, Geoffrey I. Webb, and Houkuan Huang

[1] School of Computer and Information Technology
Northern Jiaotong University, Beijing, 100044, China
[2] School of Computer Science and Software Engineering
Monash University, Clayton, Victoria, 3800, Australia

Abstract. On the basis of examining the existing restricted Bayesian network classifiers, a new Bayes-theorem-based and more strictly restricted Bayesian-network-based classification model $DLBAN$ is proposed, which can be viewed as a double-level Bayesian network augmented naive Bayes classification. The experimental results show that the $DLBAN$ classifier is better than the TAN classifier in the most cases.

Keywords: Naive Bayes, Bayesian Network, Classification

1 Introduction

Many approaches and techniques have been developed to create a classification model. The naive Bayesian classifier is one of the most widely used in interactive applications due to its computational efficiency, competitive accuracy, direct theoretical base, and its ability to integrate the prior information with data sample information. However its attribute independence assumption rarely holds in real world problems. Previous research has shown that semi-naive techniques [1] and Bayesian networks [2,3] that explicitly adjust the naive strategy to allow for violations of the independence assumption, can improve upon the prediction accuracy of the naive Bayesian classifier in many domains.

A Bayesian network classifier is a probability classification method, which can describe the probability distributions over the training data and show better classification performance in some domains. However, learning unrestricted Bayesian network is very time consuming and quickly becomes intractable as the number of attributes increases [3,4]. Therefore, restricting the structure of Bayesian networks has become an active research area. TAN (tree augmented naive Bayes) is a tree-like Bayesian networks classifier [3,5]. BAN (Bayesian network augmented naive Bayes) extends the structure of TAN by allowing the attributes to form an arbitrary graph, rather than just a tree [3,6], which tends to search the whole arc space in a complete directed graph in order to select the best arc set.

2 Restricted Bayesian Network Classifiers

Bayes theorem is the theoretical basis of Bayesian network learning method, which associates the prior probabilities with posterior probabilities. Let $A_1, A_2,$

K.-Y. Whang, J. Jeon, K. Shim, J. Srivatava (Eds.): PAKDD 2003, LNAI 2637, pp. 265–270, 2003.

$\cdots, A_n$ be attribute variables, C be the class label (or variable). Bayes theorem can be expressed as follows.

$$P(C|A_1, A_2, \cdots, A_n) = \frac{P(C) \cdot P(A_1, A_2, \cdots, A_n|C)}{P(A_1, A_2, \cdots, A_n)} \tag{1}$$

$$= \alpha \cdot P(C) \cdot P(A_1, A_2, \cdots, A_n|C) \tag{2}$$

$$= \alpha \cdot P(C) \cdot \prod_{i=1}^{n} P(A_i|A_1, A_2, \cdots, A_{i-1}, C) \tag{3}$$

where α is a normalization factor. Therefore, the key issue of building a Bayesian classification model is how to estimate $P(A_i|A_1, A_2, \cdots, A_{i-1}, C)$.

The main difference among Bayesian classification models is about the different way to calculate $P(A_i|A_1, A_2, \cdots, A_{i-1}, C)$. The simplest restricted Bayesian network classifier is the naive Bayesian classifier. Each attribute node A_i in the network is just dependent to the class node C, and $P(A_i|A_1, A_2, \cdots, A_{i-1}, C)$ in equation 3 can be simplified as $P(A_i|C)$.

TAN is another restricted Bayesian network classification model. In TAN classification model, the class node is the root and has no parents, i.e. $\prod_C = \emptyset$ ($\prod_C$ represents the set of parents of C). The class variable is a parent of each attribute variables, i. e. $C \in \prod_{A_i}$ ($\prod_{A_i}$ represents the set of parents of A_i, $i = 1, 2, \cdots, n$). And except for the class node, each attribute variable node has at most one other attribute variable node as its parent, i.e. $|\prod_{A_i}| \leq 2$. Therefore $P(A_i|A_1, A_2, \cdots, A_{i-1}, C)$ in equation 3 can be simplified as $P(A_i|C)$ or $P(A_i|A_j, C)$, where $A_j \in \{A_1, A_2, \cdots, A_{i-1}\}$. In principle, there is no any restriction on the number of parents for any attribute node in the BAN classification model and general Bayesian network classification model. According to the criterion for selecting the dependence, the attribute node A_i might associate any other attribute nodes from $\{A_1, A_2, \cdots, A_{i-1}\}$. Each attribute node A_i may have more than two parents. In the following section, we try to describe a more strictly restricted Bayesian network classification model, which shows better performance than the TAN classifier in the most cases.

3   DLBAN: A Restricted Double Level Bayesian Network

Let $\{G_1, G_2\}$ be a partition of the attribute set $\{A_1, A_2, \cdots, A_n\}$, a variant of Bayes theorem can be written as follows.

$$P(C|G_1, G_2) = \frac{P(C|G_1)P(G_2|C, G_1)}{P(G_2|G_1)} \tag{4}$$

$$= \beta \cdot P(C|G_1) \cdot P(G_2|C, G_1) \tag{5}$$

where β is a normalization factor. Assume that $G_1 = \{A_{k_1}, A_{k_2}, \cdots, A_{k_m}\}$ and $G_2 = \{A_{l_1}, A_{l_2}, \cdots, A_{l_{n-m}}\}$, if the class label C and G_1 are given, and each attribute in the subset G_2 is conditionally independent of any other attribute in the subset G_2, then a naive-Bayes-like simplifying independence assumption can be applied to the above formula.

$$P(C|G_1, G_2) = \beta \cdot P(C|A_{k_1}, A_{k_2}, \cdots, A_{k_m}) \cdot \prod_{i=1}^{n-m} P(A_{l_i}|C, A_{k_1}, A_{k_2}, \cdots, A_{k_m})$$
(6)

Zheng and Webb [7] proposed the lazy Bayesian rule (LBR) learning technique, which can be viewed as a lazy approach to classification using this variant of Bayes theorem. The more attributes in subset G_1 the weaker the assumption required. However, a counter-balancing disadvantage of adding attribute values to G_1 is that the numbers of training instances from which the required conditional probabilities are estimated decrease and hence the accuracy of estimation can be expected to also decrease. In this paper, we restrict the number of attributes belonging to the subset G_1 is less than a fixed number. If all the attributes in the subset G_1 could be found from the attribute set $\{A_1, A_2, \cdots, A_n\}$, the other attributes in the subset G_2 are dependent on them. In the Bayesian network, all the attributes in the subset G_1 would be the common parents of the other attributes in the subset G_2. Therefore, $P(A_i|A_1, A_2, \cdots, A_{i-1}, C)$ in equation 2 can be simplified as $P(A_i|K_{A_i}, C)$, where K_{A_i} is the set of parents of node A_i, then the equation 6 can be written as:

$$\gamma \cdot P(C) \cdot \prod_{i=1}^{n} P(A_i|K_{A_i}, C)$$
(7)

where γ is a normalization factor.

Given G_1 and C, any attribute in G_2 is conditionally independent of other attributes in G_2. The class variable C is a parent of each attribute in A. Each attribute in G_1 may be the parents of each attribute in G_2. If a Bayesian network model satisfies these conditions, it is called a $DLBAN$ model.

There might be a certain dependence between any two attributes in $\{A_1, A_2, \cdots, A_n\}$, and the degree of dependence is different from each other between two attributes for two different categories. The mutual information can measure the degree of providing information between two attributes. In this paper, we use the conditional mutual information to represent dependence between attribute A_i and attribute A_j. Given the class C, the conditional mutual information of attribute A_i and attribute A_j is written as below.

$$I(A_i, A_j|C) = \sum_{A_i, A_j, C} P(A_i, A_j|C) log \frac{P(A_i, A_j|C)}{P(A_i|C) \cdot P(A_j|C)}$$
(8)

The attributes in G_1 are called stronger attributes, and the attributes in G_2 are called weaker attributes.

The learning algorithm of a $DLBAN$ model is described as follows.

1) The set of stronger attributes $G_1 = \emptyset$, the set of weaker attributes $G_2 = \{A_1, A_2, \cdots, A_n\}$, the threshold ϵ is a smaller real and the number of stronger attributes at most is k;

2) Evaluate classification performance of the current classifier, and save the evaluation result $OldAccuracy$;

3) Let $i = 1$, and $ImpAccuracy[i] = 0$;

4) If A_i does not belong to G_2, go to step 7), else continue;

5) Remove A_i from G_2 into G_1, assign A_i to parents of every weaker attributes in G_2, and evaluate classification performance of the current classifier, then save the evaluation result $ImpAccuracy[i]$;

6) Remove A_i from G_1 into G_2;

7) $i = i + 1$, if $i \leq n$, go to step 4), else go to step 8);

8) For $i = 1, 2, \cdots, n$, if $ImpAccuracy[i] - OldAccuracy < 0$, go to step 10); else choose A_i as stronger attribute, where $ImpAccuracy[i] - OldAccuracy$ is the maximum for all i, and remove A_i from G_2 in to G_1;

9) If the number of stronger attributes less than k, go to step 3), else go to step 10);

10) Compute the conditional mutual information $I(A_i, A_j|C)$ according to equation 8. If $I(A_i, A_j|C) > \epsilon$, return the arc from to A_i to A_j, else remove this arc.

Table 1. Descriptions of Data

	Domain	Size♯	Classes♯	Attributes♯	Missing Value
1	Car	1728	4	6	No
2	Contact-Lenses	24	3	5	No
3	Flare-C	1389	8	10	No
4	House-Votes-84	435	2	16	No
5	Iris Classification	150	3	4	Yes
6	Chess	3196	2	36	No
7	LED	1000	10	7	No
8	Lung Cancer	32	3	56	Yes
9	Mushroom	8124	2	22	Yes
10	Nursery	12960	5	8	No
11	Post-Operative	90	3	8	Yes
12	Promoter Gene Sequences	106	2	57	No
13	Soybean Large	683	19	35	Yes
14	Tic-Tac-Toe End Game	958	2	9	No
15	Zoology	101	7	16	No

4   Experimental Methodology and Results

We chose fifteen data sets (Table 1)from the UCI machine learning repository for our experiments. In data sets with missing value, we regarded missing value as a single value besides $Post - Operative$. For $Post - Operative$ data set, we simply removed 3 instances with missing values from the data set. Our experiments have compared *DLBAN* classifier with the naive Bayes classifier and a *TAN* classifier by the classification accuracy. The classification performance was evaluated by ten-folds cross-validation for all the experiments on each data set.All the experiments were performed in the *Weka* system [9], which provides a workbench that includes full and working implementations of many popular learning schemes that can be used for practical data mining or for research.

Table 2 shows the classification accuracies. Boldface font indicates that the accuracy of *DLBAN* is higher than that of *TAN* at a significance level better than 0.05 using a two-tailed pairwise t-test on the results of the 20 trials in a domain. From Table 2, the significant advantage of *DLBAN* over *TAN* in terms of higher accuracy can be clearly seen. On average over the 15 domains, *DLBAN* increases the accuracy of *TAN* by 2%. In 12 out of these fifteen domains, *DLBAN* achieves significantly higher accuracy than *TAN*.

Table 2. Descriptions of Data

	Domain	Naive Bayes	TAN	DLBAN
1	Car	85.5757±0.32	91.6001±0.22	**94.3302±0.38**
2	Contact-Lenses	**72.7667±3.33**	65.8333±4.68	**72.7667±3.33**
3	Flare-C	79.0137±0.23	83.1209±0.31	**83.8408±0.19**
4	House-Votes-84	90.0690±0.14	93.1954±0.32	**94.1724±0.34**
5	Iris Classification	93.1667±0.73	91.7500±1.47	**93.4000±1.07**
6	Chess	87.8989±0.12	**93.4473±0.12**	87.8989±0.12
7	Led	73.8874±0.34	**73.9600±0.24**	73.8874±0.34
8	Lung Cancer	**53.2157±3.06**	46.0938±3.55	51.4999±3.69
9	Mushroom	95.7680±0.03	99.4090±0.03	**99.6147±0.05**
10	Nursery	90.2847±0.05	92.5319±0.23	**95.5128±0.16**
11	Post-Operative	**68.8889±0.86**	66.0000±1.63	**68.8890±0.86**
12	Promoter Gene Sequences	**91.2735±1.76**	82.9717±3.26	**91.2735±1.76**
13	Soybean Large	**92.7306±0.13**	87.3585±0.36	92.6574±0.21
14	Tic-Tac-Toe End Game	69.7390±0.32	**74.4394±1.17**	72.8497±0.80
15	Zoology	94.0325±1.04	95.5270±0.84	96.0230±0.29

On the data sets *Chess* and *Tic−Tac−ToeEndGame*, the *DLBAN* classifier was inferior to the *TAN*. In particular, the accuracy of the *DLBAN* classifier is lower than the *TAN* classifier by 6% on *Chess*. Why comes this situation? In our experiments, the most number of stronger attributes is limited to three in order to avoid making the probability estimates of the attributes unreliable. However, whether three stronger attributes is enough for higher dimension attributes is worthy researching. Debugging the learning process on *Chess*, we found that the classification accuracy is increased from 84.6996% to 87.8989% as the number of stronger attributes is added from 1 to 3. If the number of stronger attributes will continue increasing, the classification accuracy maybe will continue increasing.

5   Conclusions

The naive Bayes classifier is a simple and efficient classification algorithm, but its independence assumption makes it unable to express the real dependence among attributes in the practical data. At present, many methods and techniques are brought to improve the performance of the naive Bayes classifier. In this paper, we present a new Bayesian model $DLBAN$, which can determine the dependence relationship among attributes by selecting some suitable attributes. It could not only extend the attributes number on which one attribute depends, but also determine the dependence relationship among attributes by searching the attribute space. The experimental results show that a $DLBAN$ classifier has a bit higher classification performance than the TAN classifier. In the process of learning $DLBAN$, it is very important to select the stronger attributes. The method we use is to select attributes according to their conditional mutual information value. Additionally, the maximum number of the stronger attributes is defined to be three in our experiments. In fact, different data sets might have its maximum suitable number of the stronger attributes.

References

1. Kononenko, I.: Semi-Naive Bayesian Classifier. In: Proceedings of European Conference on Artificial Intelligence, (1991) 206-219
2. Pearl J.: Probabilistic Reasoning in Intelligent Systems: Networks of Plausible Inference. San Francisco, CA: Morgan Kaufman Publishers. 1988
3. Friedman, N., Geiger, D., Goldszmidt, M.: Bayesian Network Classifiers. Machine Learning, 29 (1997) 131–163
4. Chickering D. M.: Learning Bayesian networks is NP-Hard. Technical Report MSR-TR-94-17, Microsoft Research Advanced Technology Division, Microsoft Corporation, (1994)
5. Keogh, E. J., Pazzani, M. J.: Learning Augmented Bayesian Classifiers: A Comparison of Distribution-Based and Classification-Based Approaches. In: Proceedings of the Seventh International Workshop on Artificial Intelligence and Statistics. (1999) 225-230
6. Cheng J., Greiner R.: Comparing Bayesian Network Classifiers. In: Proceedings of the Fifteenth Conference on Uncertainty in Artificial Intelligence (Laskey K. B. and Prade H. Eds.). San Franscico, CA: Morgan Kaufmann Publishers.(1999): 101-108.
7. Zheng, Z., Webb, G. I.: Lazy learning of Bayesian Rules. Machine Learning. Boston: Kluwer Academic Publishers.(2000) 1-35
8. Cheng J., Bell D. A., Liu W.: Learning Belief Networks from Data: An Informaition Theory Based Approach. In: Proceedings of the Sixth ACM International Conference on Information and Knowledge Management, (1997)
9. Witten, I. H., Frank, E.: Data Mining: Practical Machine Learning Tools and Techniques with Java Implementations. Seattle, WA: Morgan Kaufmann Publishers. (2000)

AGRID: An Efficient Algorithm for Clustering Large High-Dimensional Datasets [*]

Zhao Yanchang[1] and Song Junde[2]

Electronic Engineering School, Beijing University of Posts and Telecommunications,
Beijing 100876, China
[1]zhaoyanchang@hotmail.com, [2]jdsong@bupt.edu.cn

Abstract. The clustering algorithm GDILC relies on density-based clustering with grid and is designed to discover clusters of arbitrary shapes and eliminate noises. However, it is not scalable to large high-dimensional datasets. In this paper, we improved this algorithm in five important directions. Through these improvements, AGRID is of high scalability and can process large high-dimensional datasets. It can discover clusters of various shapes and eliminate noises effectively. Besides, it is insensitive to the order of input and is a non-parametric algorithm. The high speed and accuracy of the AGRID clustering algorithm was shown in our experiments.

Keywords. Data mining, clustering, grid, iso-density line, dimensionality

1 Introduction

Because of the fast technological progress, the amount of data which is stored in databases increases very fast. For example, there are about 3,000,000 billing records of international telephone calls in China Telecom everyday. Meanwhile, in many applications, the databases consist of data objects with several tens to a few hundreds of dimensions. Clustering large high-dimensional datasets is a great challenge.

Many studies have been conducted in clustering, and most clustering algorithms fall into four categories: partitioning clustering, hierarchal clustering, density-based clustering and grid-based clustering. The idea of partitioning clustering is to partition the dataset into k clusters which are represented by the centroid of the cluster (K-Means) or one representative object of the cluster (K-Medoids). It uses an iterative relocation technique that attempts to improve the partitioning by moving objects from one group to another. Well-known partitioning algorithms are K-Means [ARS98], K-Medoids [Hua98], and CLARANS [NH94]. Hierarchical clustering creates a hierarchical decomposition of the dataset in bottom-up approach (agglomerative) or in top-down approach (divisive). A major problem of hierarchical methods is that they cannot correct erroneous decisions. Famous hierarchical algorithms are AGENS, DIANA, BIRCH [ZRL96], CURE [GRS98], ROCK [GRS99], and Chameleon [KHK99]. The general idea of density-based clustering is to continue growing the given cluster as long as the density (number of objects) in the neighborhood exceeds some threshold. Such a method can be used to filter out noise and discover clusters of arbitrary shapes. Typical density-based methods are DBSCAN [EKSX96], OPTICS

---

[*] The participation of the conference is supported by NOKIA.

K.-Y. Whang, J. Jeon, K. Shim, J. Srivatava (Eds.): PAKDD 2003, LNAI 2637, pp. 271–282, 2003.
© Springer-Verlag Berlin Heidelberg 2003

[ABKS99], and DENCLUE [HK98]. Grid-based algorithms quantize the data space into a finite number of cells that form a grid structure and all of the clustering operations are performed on the grid structure. The main advantage of this approach is its fast processing time. However, it does not work effectively and efficiently in high-dimensional space due to the so-called "curse of dimensionality". Well-known grid-based methods are STING [WYM97], WaveCluster [SCZ98], and CLIQUE [AGGR98].

GDILC is first presented in [ZS01] to cluster low-dimensional datasets using iso-density line and grid. It uses the same notion of density-based cluster as DBSCAN, but a concept of iso-density line is put forward and a different approach is employed to build clusters. Besides, a grid structure is employed to reduce the complexity of computing density. GDILC can detect clusters of arbitrary shapes and various sizes, and can eliminate noise effectively. Nevertheless, its performance degrades with the increasing of the size and the dimensionality of dataset.

In this paper, we present the algorithm AGRID (Advanced GRid-based Iso-Density line clustering), an advanced version of GDILC, for clustering large high-dimensional datasets. This new algorithm is improved in five important directions, which are the new definition of neighbor, the new data structure for storing cells, a better method for partitioning, a heuristic to get parameters without human intervention, and the feature of multi-resolution.

AGRID, as a density-based algorithm, is similar to DBSCAN and GDBSCAN [SEKX98] in the definition of density and the criterion of combination. However, they are much different in three aspects. First, the clustering procedures of them are different. DBSCAN selects one seed randomly and expand the cluster of the seed as large as possible. Then it uses another seed and repeats the selecting and expanding procedure. However, AGRID clusters by combining eligible pairs of objects. Second, AGRID employs an automatic method to compute thresholds, while DBSCAN uses a sorted k-distance plot to get parameters and requires a user to recognize the first valley in the graphical representation. If the densities of the clusters differ not much from the density of noise, the first valley will be less clear and difficult to find. Finally, they employ different methods of reducing complexity. DBSCAN uses R*-tree, while AGRID employs grid and neighbor. These differences between them lead to different complexities. The time complexity of AGRID is nearly $O(n)$ while DBSCAN $O(n\log n)$ where n is the size of a dataset.

The rest of this paper is organized as follows. Part 2 and 3 introduce respectively the idea of clustering with iso-density line and the idea of grid and neighbor. The AGRID algorithm for large high-dimensional datasets is described in part 4. We analyze the time complexity and space complexity of AGRID in part 5. Part 6 shows you the results of experiments. A summary is presented in the last section.

2   Idea of Iso-Density Line Clustering

The idea of clustering with iso-density line comes from contour figures in geography. With a contour figure, it's easy to tell where mountains are located. Moreover, one can find those mountains higher than a threshold (say, 3000 meters). In the same way, when clustering, iso-density line figure (i.e. the contour figure of density) obtained from the distribution densities of objects, can be used to discover clusters. One can choose iso-density lines of a certain value, and the regions circumscribed by the chosen isolines are clusters. Clusters of different densities can be obtained by selecting different isolines. For example, in Fig.1, if he choose the iso-density lines of 5, two regions, A and E, are circumscribed by the lines. Thus, two clusters are obtained. If isolines of 3 are chosen, four clusters, A, B, C and E, will be discovered.

Since the densities of those objects which are located in the area circumscribed by iso-density lines of a certain value are all no less than the value of the lines, we can simply combine those objects which are of higher density and near to each other into a group, and the borders of the group are precisely the iso-density lines. To cluster with iso-density line, the distance between each pair of objects should be calculated. For each object, count the numbers of objects within a

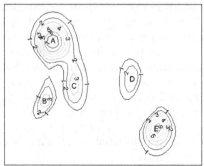

Fig. 1. Iso-density lines

distance threshold as its density. From the densities of objects, a density threshold is obtained. Then, for each pair of objects, say object α and β, if the densities of both of them are greater than the density threshold and the distance between them is less then the distance threshold, the two clusters which object α or β belong to respectively are combined into one cluster. Continue combining in such a way, until all eligible pairs of object have been checked. In the end, those clusters, whose sizes are less than a size threshold, are removed as noise clusters and those objects in these clusters are eliminated as noises (or outliers).

One key point of clustering with iso-density line is how to compute density. In our algorithm, the following definition of density is used.

Definition 1 (Density): For an object, its density is the count of those objects lying in the neighboring region of it.

$$Den(\alpha) = \left|\{\beta \mid f(\beta,\alpha) \le T\}\right| \tag{1}$$

In the above formula, α and β are objects. $Den(\alpha)$ is the density of object α. $|A|$ stands for the size of object set A. $f(\beta,\alpha)$ is a function to measure the dissimilarity between object α and β. T is a given threshold. Euclidean distance is used as the dissimilarity function in our algorithm. Besides, we assume that all attributes are numerals and are normalized and scaled into the range of [0, 1]. For categorical attributes, the similarity can be calculated using the method given in [Huang97].

3   Idea of Grid and Neighbor

The idea of grid and neighbor, which is first put forward in [ZS01], is employed in our algorithm to reduce the complexity of density computing and combining. In our previous procedure of clustering with iso-density line, we have to compute the distance between each pair of objects. That is to say, there are C_n^2 (i.e., $n(n-1)/2$) times of distance computing. In fact, in order to calculate the density of an object, those objects that are very far away can be ignored because they do not contribute to the density of it. Consequently, for a large amount of object pairs, because the distances between them are much farther than the radius threshold RT, the distance between them need not be calculated. From this point, we use a grid-based method to reduce the number of object pairs to be considered in our algorithm.

Before computing distances, each dimension is divided into several intervals. Thus the data space is partitioned into many hyper-rectangular cells. For all objects in a cell, we merely consider those objects in the neighboring cells of it. For each object, those objects that are not in the neighboring cells of it are very far away, and their distances away from it are much greater than our threshold RT. The distance between them need not be calculated because they donot contribute to the densities of each other and there is little chance for them to be qualified for the criteria of combining. Much time can be saved in such a way.

First we give the definition of neighbor used in the former version of our algorithm in [ZS01] (Fig.2).

Definition 2 (Neighboring cells): Cell $C_{i_1 i_2 \cdots i_d}$ and $C_{j_1 j_2 \cdots j_d}$ are neighboring cells when the following condition is meet.

$$\forall p, \ (1 \le p \le d), \ \left| i_p - j_p \right| \le 1$$

Here, $i_1 i_2 \ldots i_d$ and $j_1 j_2 \ldots j_d$ are sequences of interval IDs of cell $C_{i_1 i_2 \cdots i_d}$ and $C_{j_1 j_2 \cdots j_d}$ respectively. Generally speaking, there are 3^d neighbors for each cell in a d-dimensional data space.

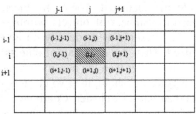

Fig. 2. 3^d Neighboring Cells

4   AGRID for Clustering Large High-Dimensional Datasets

This section will introduce our techniques to process high-dimensional datasets, which includes the new definition of neighbor, the new technique of partitioning data space, the method of storing cells, and a heuristic to get parameters.

4.1   Definition of Neighbor for High-Dimensional Datasets

With the definition of neighbors in definition 2, each cell has 3^d neighbors, which would be too large when the dimensionality is high. In AGRID, a new definition of neighbor is employed to solve the problem (Fig.3).

Definition 3 (Neighboring cells): Cell $C_{i_1 i_2 \cdots i_d}$ and $C_{j_1 j_2 \cdots j_d}$ are neighboring cells when the following conditions are meet.

$$\begin{cases} \left| i_p - j_p \right| \le 1, \ p = l, \ (1 \le l \le d) \\ i_p = j_p, \ p = 1, 2, \ldots, l-1, l+1, \ldots, d \end{cases}$$

In the formula above, l is an integer between 1 and d. $i_1 i_2 \ldots i_d$ and $j_1 j_2 \ldots j_d$ are respectively the sequences of interval IDs of cell $C_{i_1 i_2 \ldots i_d}$ and $C_{j_1 j_2 \ldots j_d}$. Generally speaking, in d-dimensional data space, each cell has $(2d+1)$ neighbors instead of 3^d neighbors. By the way, the neighbors of a cell lying in corners are less than $(2d+1)$.

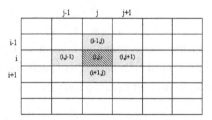

Fig. 3. $(2d+1)$ Neighboring Cells

4.2   Partitioning Data Space

In some papers, all dimensions are partitioned into the same number of intervals (say, m intervals), and there are m^d cells in the data space. This method of partitioning is not suitable for high-dimensional datasets because the number of cells increases exponentially with the dimensionality. Our new method of partitioning high-dimensional data space is presented in the following.

The Technique of Partitioning
The performance of AGRID is largely dependent on the partition of the data space. In low-dimensional data space, each dimension can be partitioned into equal number of intervals (say m), so the number of cells is m^d. This method of partitioning is effective when the dimension is very low. But it is not the case in high-dimensional data space where the number of cells would be too large a number. In addition, when the dimension is high, it is very difficult to choose an appropriate value for m, and a little change of it can lead to a great variance of the number of cells. Therefore, it is difficult to choose an appropriate granularity when all dimensions are partitioned in the same way. To resolve the above problem, a technique of dividing different dimension into different number of intervals is employed to partition the whole data space.

In our algorithm, different interval numbers are used for different dimensions. For the first p dimensions, each dimension is divided into m intervals, while $(m-1)$ intervals for each of the remaining $(d-p)$ dimensions. With such a technique of partitioning, the total number of cells is $m^p (m-1)^{d-p}$ and the number of cells can vary smoothly by adjusting the values of m and p. Let's assume that the percentage of non-empty cells is ω and the number of objects is n. The average number of objects contained in each non-empty cell is $\dfrac{n}{m^p (m-1)^{d-p} \omega}$. Assume that the average number of neighboring cells of a non-empty cell is n_{nc}. Then, the total number of distance computing is $\dfrac{n}{m^p (m-1)^{d-p} \omega} \times n_{nc} \times n$, that is, $\dfrac{n^2 n_{nc}}{m^p (m-1)^{d-p} \omega}$. We set the time complexity to be linear with both n and d, and get the following equation,

$$\frac{n^2 n_{nc}}{m^p (m-1)^{d-p} \omega} = nd \text{ , namely, } \frac{nn_{nc}}{m^p (m-1)^{d-p} \omega} = d \qquad (2)$$

In the above formula, both m and p are positive integers. When m is a large number, most cells have $(2d+1)$ neighbors (i.e., $n_{nc} \approx 2d+1$), and the values of m and p can be derived from the formula (2).

Average Number of Neighbors per Cell

With definition 3, most cells have $(2d+1)$ neighbors if m is no less than 4. However, when the dimensionality is high, m may be no greater than 3 and the neighbors of the majority of cells would be less than $(2d+1)$. In the following, a theorem is presented to compute the average number of neighbors of each cell. The values of m and p can be computed according to Equation (2) and Theorem 1.

Theorem 1: When the data space is partitioned using our method, the average number of neighbors of each cell is

$$N_{nc} = 1 + \frac{2(m-1)}{m} p + \frac{2(m-2)}{m-1}(d-p), \quad m \geq 2 \qquad (3)$$

The proof of the above theorem is not given here for the sake of saving space.

4.3 Storage of Cells

In high-dimensional data space, the number of cells can be too large a number and it is impossible to store all cells in memory. Fortunately, not all of these cells contain objects. Especially when the dimensionality is high, the objects are very sparse and the majority of cells are empty. Therefore, in our algorithm, only those cells that are not empty are stored in a hash table (Fig.4). The field "headObject" and "tailObject" are used to retrieve the objects located in the cell. Since the conflict of hash is unavoidable, a conflict chain ("conflictCell") is used to store the conflicting cells for each node in the hash table. Besides, there is a pointer named "nextCell" to keep track of all non-empty cells.

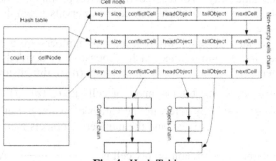

Fig. 4. Hash Table

Because each non-empty cell contains no less than one object, the number of non-empty cells is no more than that of objects. In order to lower the probability of hash conflict, the size of the hash table is set to the minimal prime number that is greater than 1.5 times of the number of objects.Our experiments showed that the number of conflicts is nearly 20 percent of the number of non-empty cells averagely.

4.4 Choice of *RT* and *DT*

While it is very easy to discover clusters from the iso-density line figure, it is not so easy to obtain an appropriate iso-density line figure. In fact, it depends on the choice of distance threshold *RT*. As Fig.5 shows, when *RT* is great enough (e.g. $RT_1 = r_3$) that all the objects in all the neighbors of a cell are within the range of *RT*, AGRID will become somewhat like grid-based clustering. The densities of all the objects in a cell will be the same, and can be gotten by simply counting those objects in all its neighboring cells. But they are different because AGRID uses the neighbor definition that a cell has *(2d+1)* instead of 3^d neighbors. Let RT_1 be the minimum of such kind of

radius threshold, then $RT_1^2 = \left(2 \times Len_h\right)^2 + \sum\limits_{i=1,\dots,h-1,h+1,\dots,d} \left(Len_i\right)^2$. Len_i is the interval

length in the *i*-th dimension, and the *h*-th dimension has the largest interval length. On the other hand, if *RT* is less than the lengths of all edges of the hyper-rectangular cell (e.g. $RT_2 = r_1$), AGRID will become somewhat like density-based clustering. In this case, the density of an object is largely decided by the number of objects circumscribed by *RT*. However, the partition of data space into cells helps reducing the number of distances to be computed and making AGRID much faster than density-based clustering. Let RT_2 be the maximum of such kind of radius threshold, then we can get

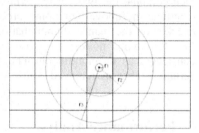

$RT_2^2 = \min\limits_{i=1,\dots,d}\left(\dfrac{1}{2} \times Len_i\right)^2$. If we choose a

radius, say r_2, which lies between the two extremes, our algorithm can be effective and accurate.

Fig. 5. Choice of *RT*

From the above analysis, we use the following formula to compute the distance threshold *RT*.

$$RT = (1 - \lambda) \times RT_1 + \lambda \times RT_2, \ \ 0 \le \lambda \le 1 \tag{4}$$

In the above formula, λ is an adjustable coefficient. When the clustering result is not satisfying, the coefficient can be adjusted to get a better result. Our experiments showed that good results can be obtained when λ is set to 0.5.

The final result of clustering is decided by the value of density threshold *DT*. If it is too small, neighboring clusters will be combined into one cluster. If too large, those objects that should be grouped into a cluster would be separated into several clusters and many objects would be mistaken as outliers. In AGRID algorithm, we calculate *DT* dynamically according to the mean of densities according to the following formula.

$$DT = \frac{1}{\theta} \times \frac{\sum\limits_{i=1}^{n} Density(i)}{n} \tag{5}$$

θ is an adjustable coefficient which be modified to get clustering results of different resolution levels. It is set to 16 in our algorithm and good clustering results can be obtained in most conditions.

4.5 Procedure of AGRID

There are six steps in the AGRID algorithm, which is shown in the following.
1)   Computing m and p according to Equation (2) and Theorem 1.
2)   Partitioning: The whole data space is partitioned into many cells according to m and p calculated in the first step. Each object is assigned to a cell according to its coordinates and non-empty cells are inserted into the hash table shown in Fig.4.
3)   Computing RT: The distance threshold RT is computed according to Formula (4) with the interval lengths of every dimension.
4)   Calculating DT: For each object, compute its density with the help of grid and neighbor. The average density is calculated and then the density threshold DT is computed. From the density vector, the iso-density line figure can be derived. In fact, the figure need not explicitly exist during the procedure of clustering.
5)   Clustering automatically: At first, we take each object whose density is greater than the density threshold DT as a cluster. Then, for each object α, check each object in the neighboring cells of $C\alpha$, whether its density is greater than the density threshold DT and whether its distance from object α is less than the distance threshold RT. If so, then combine the two clusters which those two objects are located in respectively. Continue the above combining procedure until all eligible object pairs have been checked for the chance of combination.
6)   Removing noises: In those clusters obtained, many are too small to be considered as meaningful clusters. So those clusters whose sizes are less than a certain number are removed. Thus, AGRID can eliminate outliers effectively.

5 Performance Analysis

The performance of AGRID depends on the values of n (the size of dataset), m (the number of intervals in each dimension) and d (the dimensionality of dataset).

The time complexity is set to be linear with n and d in Equation (2). But the time complexity we computed above is under the ideal condition that the number of objects in every cell is equal to one another. In nearly all cases, the number of objects varies from cell to cell. So the time complexity is dependent on the distribution of objects in a dataset. Our experiments show that the time complexity is nearly linear both with the dataset size and with the dimensionality.

In the hash table, each hash node is five bytes long and each cell node is of $(2d+20)$ bytes. The size of the hash table is about $1.5n$ and the number of non-empty cells is $cellNum$, which is no more than n, the object number, then the space occupied by hash table is

$$1.5n*5B + cellNum*(2d+20)B \leqslant 1.5n*5B + n*(2d+20)B = (27.5+2d)n\ bytes.$$

Besides, the spaces used to store densities and clusters in memory are $4n$ and $16n$ bytes respectively, so the total space complexity is less than $(47.5+2d)*n$ bytes. So the space complexity is linear both with the object number (n) and with the dimensionality (d) of datasets.

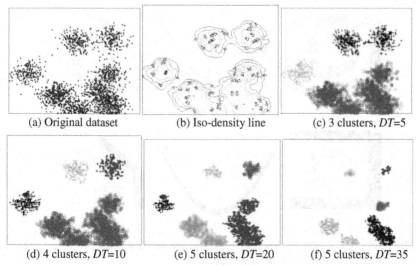

| (a) Original dataset | (b) Iso-density line | (c) 3 clusters, DT=5 |
| (d) 4 clusters, DT=10 | (e) 5 clusters, DT=20 | (f) 5 clusters, DT=35 |

Fig. 6. Multi-resolution Clustering

6 Experimental Results

Experiments were performed on a PC with 256MB RAM and an AMD Athlon K7 700MHz CPU and running Windows 2000 Professional. Experiments have been made on datasets of different sizes, of different dimensionalities and of different distribution, and the results are given in the following.

6.1 Results of Multi-resolution Clustering

Multi-resolution property of AGRID can help to detect clusters at different levels, which is shown in Fig.6. Although the multi-resolution property of AGRID is somewhat like that of WaveCluster, they are much different in that AGRID achieves by adjusting the value of density threshold while WaveCluster by "increasing the size of a cell's neighborhood". Fig.6a is the original dataset of 2,000 objects and Fig.6b is the iso-density lines derived from the densities of the objects. The other four figures are clustering results with different DT. The values of density threshold are 5, 10, 20 and 35 in Fig.6c, Fig.6d, Fig.6e and Fig.6f respectively. Generally speaking, the greater the density threshold is, the smaller the clusters are, and the more is the number of objects that would be treated as outliers. When DT, the density threshold, is a very small number, the number of clusters is small and just a few objects are treated as outliers. With the increasing of DT, some clusters break into more and smaller clusters. A hierarchical clustering tree can be built with selecting a series of DT and the appropriate resolution level for choosing clusters can be decided on the needs of users.

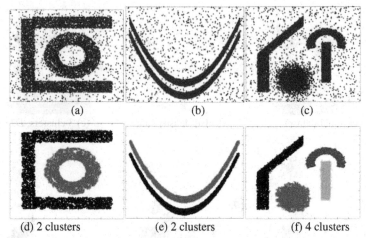

(d) 2 clusters   (e) 2 clusters   (f) 4 clusters

Fig. 7. Results of Complex Shapes (a), (b) and (c) are original datasets; (d), (e) and (f) are the results of clustering the three datasets respectively.

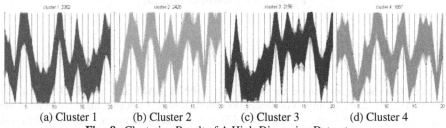

(a) Cluster 1   (b) Cluster 2   (c) Cluster 3   (d) Cluster 4

Fig. 8. Clustering Result of A High-Dimension Dataset

6.2 Results of Complex Shapes

Here, three datasets (Fig.7) were used to test whether AGRID can find clusters of complex shapes. In the following, the first two datasets were used by Gholam Sheikholeslami in WaveCluster [SCZ98] and the third is a synthetic dataset generated by ourselves. The sizes of the three datasets are all 10,000 and the dimensionalities are two. The result shows that AGRID can not only find clusters of sphere shape and line shape, but also can find clusters of non-convex shape, such as the curve shape and semi-circle shape. Moreover, it can discriminate clusters that are very near to each other. From the clustering results, we can see that AGRID is very powerful in handling any type of sophisticated patterns. Besides, noises are removed effectively.

6.3 Results of Clustering High-Dimensional Datasets

The method of parallel coordinates used in [SCZ98] is employed here to visualize the clusters in high-dimensional space. There are d parallel vertical lines, which stand for dimensions. Each object is shown as a zigzag line in the figure (Fig.8).

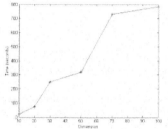

Fig. 9. Performance of clustering datasets of different dimensionalities (n =300,000).

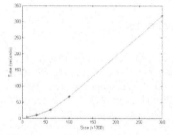

Fig. 10. Performance of clustering datasets of different sizes ($d = 50$)

The dataset used in our experiment is 20-dimensional and is of 10,000 objects, and contains four clusters. Besides, 10 percent of the objects are noises, which are uniformly distributed. The clustering result is shown in Fig.8 and the size of each cluster is shown on the top of it. All four clusters were successfully discovered correctly by AGRID and about 11.4 percent of objects were eliminated as noises. The experiments clearly show that AGRID can handle high-dimension datasets efficiently.

6.4 Scalability with Dataset Size and Dimensionality

Experiments haven been conducted with datasets of different sizes and of different dimensionalities. The AGRID algorithm is very fast, and a 100-dimensional dataset with the size of 300,000 can be processed in about 13 minutes. The performance of clustering datasets with size of 300,000 and with dimensionalities ranging from 10 to 100 is shown in Fig.9. Fig.10 gives the performance of clustering datasets with dimensionality of 50 and with sizes ranging from 10,000 to 300,000. As the figures show, the time complexity is nearly linear both with the size and with the dimensionality of datasets.

7 Conclusion

Clustering large high-dimensional datasets is an important problem in data mining. Density-based clustering algorithms are well known for their ability to discover clusters of various shapes and eliminate noises, while grid-based methods are famous for high speed. However, grid-based clustering is not scalable to high-dimensional datasets because the counts both of cells and of neighbors would grow exponentially with the dimensionality. By combining density-based and grid-based clustering together, AGRID can discover clusters of complex shapes and eliminate noises with high speed. Through the new definition of neighbor and the new technique of partitioning, it can be scalable to large high-dimensional datasets. In addition, with a heuristic to get important parameters, it need no input and therefore does not require any domain knowledge. Nevertheless, the performance of AGRID is dependent on the distribution of datasets.

AGRID has been tested only with synthetic datasets generated by us. In future research, experiments will be performed with real datasets in our future work.

References

[ABKS99] M. Ankerst, M. Breunig, et al, "OPTICS: Ordering points to identify the clustering structure", In *Proc. 1999 ACM-SIGMOD Int. Conf. Management of Data (SIGMOD'99)*, pp. 49–60, Philadelphia, PA, June 1999.

[AGGR98] R. Agrawal, J. Gehrke, et al, "Automatic subspace clustering of high dimensional data for data mining aplications", In *Proc. 1998 ACM-SIGMOD Int. Conf. Management of Data (SIGMOD'98)*, pp. 94–105, Seattle, WA, June 1998.

[ARS98] K. Alsabti, S. Ranka, V. Singh, "An Efficient K-Means Clustering Algorithm," *Proc. the First Workshop on High Performance Data Mining*, Orlando, Florida, 1998.

[BM00] P.S. Bradley, O.L. Mangasarian, "K-Plane Clustering," *Journal of Global Optimization* 16, Number 1, 2000, pp. 23–32.

[EKSX96] M. Ester, H.-P. Kriegel et al, "A Density-Based Algorithm for Discovering Clusters in Large Spatial Databases with Noise," In *Proc.1996 Int. Conf. On Knowledge Discovery and Data Mining (KDD'96)*, pp. 226–231, Portland, Oregon, Aug. 1996.

[GRS98] S. Guha, R. Rastogi, and K. Shim, "Cure: An efficient clustering algorithm for large databases", In *Proc. 1998 ACM-SIGMOD Int. Conf. Management of Data (SIGMOD'98)*, pp. 73–84, Seattle, WA, June 1998.

[GRS99] S. Guha, R. Rastogi, and K. Shim, "Rock: A robust clustering algorithm for categorical attributes", In *Proc. 1999 Int. Conf. Data Engineering (ICDE'99)*, pp. 512–521, Sydney, Australia, Mar. 1999.

[HK98] A. Hinneburg and D.A. Keim, "An efficient approach to clustering in large multimedia databases with noise", In *Proc. 1998 Int. Conf. Knowledge Discovery and Data Mining (KDD'98)*, pp. 58–65, New York, Aug. 1998.

[HK01] Jiawei Han, Micheline Kamber, "Data Mining: Concepts and Techniques," Higher Education Press, Morgan Kaufmann Publishers, 2001.

[Hua98] Z. Huang, "Extensions to the k-means algorithm for clustering large data sets with categorical values", *Data Mining and Knowledge Discovery*, 2:283–304,1998.

[Huang97] Zhexue Huang, "A Fast Clustering Algorithm to Cluster Very Large Categorical Data Sets in Data Mining," *In SIGMOD Workshop on Research Issues on Data Mining and Knowledge Discovery* (SIGMOD-DMKD'97), Tucson, Arizona, May 1997.

[KHK99] G. Karypis, E.-H. Han, and V. Kumar, "CHAMELEON: A hierarchical clustering algorithm using dynamic modeling", *IEEE Computer, Special Issue on Data Analysis and Mining*, Vol. 32, No. 8, August 1999, pp. 68–75.

[NH94] R. Ng and J. Han, "Efficient and effective clustering method for spatial data mining", In *Proc. 1994 Int. Conf. Very Large Data Bases (VLDB'94)*, pp. 144–155, Santiago, Chile, Sept. 1994.

[SCZ98] G. Sheikholeslami, S. Chatterjee, and A. Zhang, "WaveCluster: A multi-resolution clustering approach for very large spatial databases", In *Proc. 1998 Int. Conf. Very Large Data Bases (VLDB'98)*, pp. 428–429, New York, Aug. 1998.

[SEKX98] Jörg Sander, Martin Ester, Hans-Peter Kriegel, Xiaowei Xu: "Density-Based Clustering in Spatial Databases: The Algorithm GDBSCAN and Its Applications", *Data Mining and Knowledge Discovery*, Vol. 2, No 2, June 1998.

[WYM97] W. Wang, J. Yang, and R. Muntz, "STING: A statistical information grid approach to spatial data mining", In *Proc. 1997 Int. Conf. Very Large Data Bases (VLDB'97)*, pp. 186–195, Athens, Greece, Aug. 1997.

[ZRL96] T. Zhang, R. Ramakrishnan, and M. Livny, "BIRCH: An efficient data clustering method for very large databases", In *Proc. 1996 ACM-SIGMOD Int. Conf. Management of Data (SIGMOD'96)*, pp. 103–114, Montreal, Canada, June 1996.

[ZS01] Zhao Yanchang, Song Junde, "GDILC: A Grid-based Density Iso-line Clustering Algorithm," *Proc. Int. Conf. Info-tech and Info-net* (ICII 2001), Beijing, China, Oct. 2001.

Multi-level Clustering and Reasoning about Its Clusters Using Region Connection Calculus

Ickjai Lee and Mary-Anne Williams

Innovation and Technology Research Laboratory, Faculty of Information Technology,
University of Technology Sydney, NSW 2007, Australia
ickjai@cafe.newcastle.edu.au, maryanne@it.uts.edu.au

Abstract. Spatial clustering provides answers for "where?" and "when?" and evokes "why?" for further explorations. In this paper, we propose a divisive multi-level clustering method that requires $O(n \log n)$ time. It reveals a cluster hierarchy for the "where?" and "when?" queries. Experimental results demonstrate that it identifies quality multi-level clusters. In addition, we present a solid framework for reasoning about multi-level clusters using Region Connection Calculus for the "why?" query. In this framework, we can derive their possible causes and positive associations between them with ease.

1 Introduction

Due to the developments in data gathering processes, Geographic Information Systems (GIS) have shifted from a data-poor and computation-poor to a data-rich and computation-rich environment [15]. The sheer volume of digital spatial data stowed in GIS turns spatial analysis into a complex task that demands intelligent data mining techniques. Clustering is one of the most popular and frequently used spatial data mining techniques. It suggests global concentrations and localized excesses in spatial and temporal dimensions. The hierarchical nature of spatial data often demands multi-level (hierarchical) clustering structure that can reveal comprehensive hierarchical spatial aggregations. Several hierarchical clustering methods [10,17,18] have been studied in the spatial data mining community. However, the use of global parameters in these methods contradicts the peculiar spatial characteristics (spatial dependency and spatial heterogeneity [1]) that make geo-space special. Their main goal is to detect global concentrations and thus easily miss localized excesses that are also of significant interest in many spatial settings. This paper proposes a multi-level clustering method based on Short-Long clustering [8] that incorporates these spatial peculiarities. Further, this paper provides a robust framework for reasoning about multi-level clusters using Region Connection Calculus (RCC) [3]. This framework offers an inference mechanism that detects causal relationships between multi-level clusters and suggests their positive associations.

The rest of paper is structured as follows. Section 2 outlines the working principle of Short-Long clustering. Section 3 discusses the details of multi-level Short-Long clustering. We provide algorithms, examine the performance with

K.-Y. Whang, J. Jeon, K. Shim, J. Srivatava (Eds.): PAKDD 2003, LNAI 2637, pp. 283–294, 2003.

real datasets and analyze the time complexity. Section 4 introduces a framework for reasoning about multi-level clusters using RCC and Section 5 explains cluster reasoning with indeterminate cluster regions. Section 6 offers final remarks.

2   Short-Long Clustering

Short-Long clustering [8] is a graph-based clustering method especially designed for spatial data mining. It attempts to locate cluster boundaries where sharp changes in point density occur in order to detect spatial aggregations. Short-Long clustering identifies sharp alterations using local neighborhood relationships defined by the Delaunay Diagram (removing ambiguities from the Delaunay Triangulation, the dual of Voronoi Diagram, when co-circularity occurs) of a set P of points, denoted by $DD(P)$, and detects globally significant cluster boundaries using all local neighborhood relationships within the study region R. Different from most graph-based clustering, Short-Long clustering is a node-centric approach. It is based on three node-centric statistics. Note that, these statistics convey more information than edges and clustering is a process of grouping nodes not edges. The three node-centric statistics are as follows.

$$LocalMean_k(p_i) = \sum_{e \in N_k(p_i)} |e|/\|N_k(p_i)\|, \qquad (1)$$

$$LocalStDev_k(p_i) = \sqrt{\sum_{j=1}^{\|N_k(p_i)\|} (LocalMean_k(p_i) - |e_j|)^2 /\|N_k(p_i)\|,} \qquad (2)$$

$$MeanStDev(P) = \sum_{i=1}^{n} LocalStDev_1(p_i)/n, \qquad (3)$$

where $|e|$ denotes the length of e and $N_k(p_i)$ denotes the k-neighborhood of p_i, namely $N_k(p_i) = \{e_j \in DD(P)|e_j$ is a path of k or less edges starting at $p_i\}$.

$LocalMean_k(p_i)$ and $LocalStDev_k(p_i)$ represent local neighborhood strength and variation respectively while $MeanStDev(P)$ represents the global degree of variation. Short-Long clustering classifies $N_k(p_i)$ for each p_i into three groups: *short* edge (denoted by $SEdges_k(p_i)$), *long* edge (denoted by $LEdges_k(p_i)$) and *other* edge (denoted by $OEdges_k(p_i)$). These are defined as follows.

$$SEdges_k(p_i) = \{e_j \in N_k(p_i) \mid |e_j| < LocalMean_k(p_i) - m \cdot MeanStDev(P)\}, (4)$$
$$LEdges_k(p_i) = \{e_j \in N_k(p_i) \mid |e_j| > LocalMean_k(p_i) + m \cdot MeanStDev(P)\}, (5)$$
$$OEdges_k(p_i) = N_k(p_i) - (SEdges_k(p_i) \cup LEdges_k(p_i)), (6)$$

where m is a control value.

Delaunay edges in $LEdges_k(p_i)$ represent exceptionally long edges in the k-neighborhood of p_i while Delaunay edges in $SEdges_k(p_i)$ represent exceptionally short edges. Short-Long clustering removes exceptionally long Delaunay edges in the 1-neighborhood ($LEdges_1$) to extract rough cluster boundaries. And then, it eliminates some short Delaunay edges in the 1-neighborhood ($SEdges_1$) to overcome the chain effect [10]. Short-Long clustering extends local neighborhood to the 2-neighborhood to remove inconsistently long edges ($LEdges_2$).

The control value m serves as an exploratory tool. Larger values of m result in clusters with more heterogeneous lengths of edges. The user may increase the value of m to explore which clusters are growing fast in terms of their sizes or which clusters are going to merge. Conversely, smaller values of m result in clusters with more homogeneous lengths of edges. The user may decrease the value of m to find breakable or vulnerable regions in clusters where they are about to split. The control value m is set to 1 as default [8]. Readers may refer to the original paper [8] for further details.

3   Multi-level Short-Long Clustering

One of the unique characteristics of spatial data is its generic hierarchical complexity. Thus, partitioning an initial set of points into mutually exclusive and collectively exhaustive clusters is insufficient to understand its complex structure. Instead, multi-level clustering is more informative in some spatial settings. This chapter extends Short-Long clustering to multi-level Short-Long clustering.

3.1   Intuition and Algorithms

In graph-based clustering, connected components denoted by $CC[p_i]$ represent clusters. These connected components are further decomposed into subclusters in hierarchical clustering methods unless they are homogeneous. In traditional hierarchical clustering [10,17,18], user-supplied global arguments determine termination or merge conditions. However, these arguments are not easily determined without prior knowledge which may hinder exploration [7]. Multi-level Short-Long clustering overcomes this difficulty. Note that, Short-Long clustering does not detect any clusters for most Complete Spatial Randomness (CSR) datasets with $m = 1$, but rather assigns P to the same cluster [8]. Thus, the termination condition is met in multi-level Short-Long clustering when a set of points does not exhibit any clusters, but rather exhibits point-centric statistics similar to CSR. That is, multi-level Short-Long clustering does not require a termination condition from users, but automatically derives it from the data. In this paper, we propose two approaches for multi-level Short-Long clustering.

Reconstruction based approach
This approach first applies Short-Long clustering to the whole dataset P to extract globally significant clusters. And then, it recomputes the Delaunay Diagram for each cluster and recursively applies Short-Long clustering to the cluster. That is, P is decomposed into subsets (C_1, C_2,..., C_k) and noise after the initial clustering. For each subset C_i ($1 \leq i \leq k$), this approach reconstructs $DD(C_i)$ and applies Short-Long clustering to $DD(C_i)$. Although this approach provides reasonable grounds for multi-level clustering, it has several drawbacks. First, it requires extra time for recomputing the Delaunay Diagram for each cluster. Second, recomputing $DD(C_i)$ is not affected by the

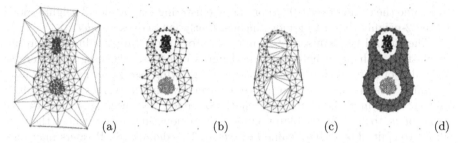

(a)   (b)   (c)   (d)

Fig. 1. Effect of two approaches to delineate multi-level clusters: reconstruction *vs* recuperation ($\|P\| = 250$): (a) A dataset with $DD(P)$; (b) After Short-Long clustering (3 clusters); (c) Reconstruction of the Delaunay Diagram of a "8-like" sparse cluster; (d) Recuperation of the Delaunay edges of the "8-like" sparse cluster.

distribution of points $p \in P \setminus C_i$. Since the cluster C_i is a subset of P, it is correlated to every point in P. Third, recomputing independent $DD(C_i)$ may cause some non-informative interactions (edges) that do not exist in $DD(P)$. Especially, this is the case when the cluster C_i contains other clusters. Some edges in $DD(C_i)$ may go over the surrounded clusters. This is illustrated in Fig. 1. Let C_i be the "8-like" sparse cluster. Fig. 1(a) depicts $DD(P)$. Fig. 1(b) shows a subgraph of $DD(P)$ after the initial clustering. $DD(C_i)$ illustrated in Fig. 1(c) contains Delaunay edges running across the dense clusters. These edges are non-informative in $DD(P)$. In order to overcome these drawbacks, we propose another approach that is based on valid Delaunay edges recuperation.

Recuperation based approach

Similar to the previous approach, it initially applies Short-Long clustering to P to extract significant clusters in the context of R. Since each edge $e_{i,j} = (p_i, p_j)$ represents a valid interaction when $CC[p_i] = CC[p_j]$ after the clustering, this edge must be recuperated for the next level of clustering. Thus, this approach recuperates all Delaunay edges in $DD(P)$ when their two end-points belong to the same cluster after the clustering. Once these intra-cluster edges are recuperated, then Short-Long clustering is applied to each connected components. Fig. 1(d) depicts a recuperated subgraph of $DD(P)$ after the clustering. To summarize these algorithms, we provide multi-level Short-Long clustering in pseudo-code.

Algorithm **Multi-level Short-Long clustering**
    Input: A graph G of P and the control value m;
    Output: Multi-level clusters;
    1) **begin**
    2) WriteCluster(G);
    3) *MeanStDev(P)* $\Leftarrow$ ComputeMeanStDev(G);
    4) *outG* $\Leftarrow$ ShortLongClustering(G, m, *MeanStDev(P)*);
    5) ComputeConnectedComponents(*outG*);
    6) **for** each connected component CC

7)   **if**$(\frac{\|CC\|}{\|P\|} \geq NoiseRatio$ && $\|CC\| \neq \|G\|)$ **then**

8) $\begin{cases} [\text{v1}] \ subG \Leftarrow \texttt{BuildDelaunayDiagram}(CC); \\ [\text{v2}] \ subG \Leftarrow \texttt{RecuperateDelaunayDiagram}(CC); \end{cases}$

9)   Call **Multi-level Short-Long clustering** with the $subG$;
10)   **else**
11)   Return;
12) **end for**
13) **end**

The algorithm for multi-level Short-Long clustering is based on recursion. Initial inputs are $DD(P)$ and m. The process labeled `WriteCluster`(G) reports all points in G as an initial cluster. The process labeled `ShortLongClustering` is Short-Long clustering. The process labeled `ComputeConnectedComponents`$(outG)$ (Step 5) computes connected components of the output graph $outG$. The label $NoiseRatio$ (Step 7) represents a noise ratio defined by the user. Step 8 has two versions with a version identifier. One or the other version is to be used. The reconstruction based approach uses Step 8.v1 while the recuperation based approach uses Step 8.v2. The process labeled `BuildDelaunayDiagram`(CC) builds a new Delaunay Diagram with points in connected component CC while the process `RecuperateDelaunayDiagram`(CC) recuperates intra-cluster edges in CC. Since the recuperation based approach is faster and more informative as discussed earlier in this section, the recuperation based approach is set as default and subsequent sections are based on this approach.

3.2   Experimental Results

Fig. 2 and Fig. 3 demonstrate experimental results of multi-level Short-Long clustering with real-world datasets recorded in 1997 at 217 suburbs around Brisbane in Australia. These datasets explain the complex and hierarchical nature of real-world data. That is, spatial aggregations are tiered and not easily identifiable. Fig. 2(a) and Fig. 3(a) show locations of reserves and locations of robbery incidents within R, respectively. Their corresponding Voronoi diagrams are also shown. Fig. 2(b) and Fig. 3(b) depict first-level clusters. These clusters demonstrate the effectiveness of multi-level clustering. Clusters are in arbitrary shapes, different sizes and heterogeneous densities. Shaded regions represent cluster regions computed by the cluster-to-region transformation [13]. These first-level clusters are significant spatial concentrations in the context of R.

The increase of $NoiseRatio$ will remove smaller clusters such as a first-level cluster D in Fig. 2(b) while the decrease of $NoiseRatio$ detects several more clusters including two concentrations in the East of R and one concentration in the North of R in Fig. 3(b). For illustration purposes, we set $NoiseRatio$ to 3% for both datasets. For first-level clusters exhibiting heterogeneous distributions (unlike CSR) are further split into subclusters. For instance, a first-level cluster B in Fig. 2(b) is further split into four subclusters (BA,BB,BC and BD) as

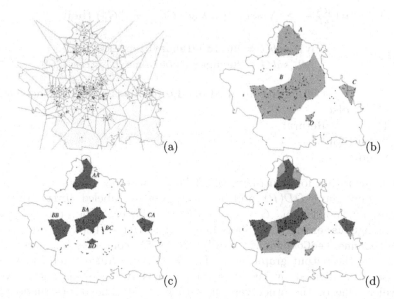

Fig. 2. Reserves recorded in the year of 1997 at 217 suburbs around Brisbane in Australia ($\|P\| = 248$, $NoiseRatio = 3\%$): (a) Reserves with $VD(P)$; (b) First-level clusters; (c) Second-level clusters; (d) Overlay of the first-level and the second-level clusters.

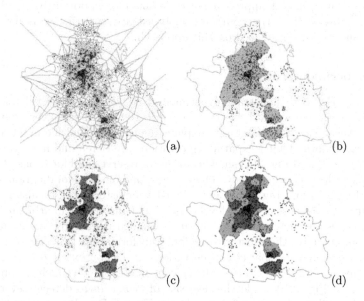

Fig. 3. Robbery incidents recorded in the year of 1997 at 217 suburbs around Brisbane in Australia ($\|P\| = 1328$, $NoiseRatio = 3\%$): (a) Robbery incidents with $VD(P)$; (b) First-level clusters; (c) Second-level clusters; (d) Overlay of the first-level and the second-level clusters.

shown in Fig. 2(c). On the other hand, clusters exhibiting their point-centric statistics similar to CSR are not partitioned. For instance, a first-level cluster D in Fig. 2(b) does not produce subclusters, but terminates.

3.3 Time Complexity Analysis

First note that, Short-Long clustering requires $O(n \log n)$ time to construct $DD(P)$. Once this proximity graph is available, it performs the remaining processes in linear time. For the recuperation approach, the preprocessing to create the proximity graph (dominating Short-Long clustering time) is not replicated at each level. Multi-level Short-Long clustering repeatedly calls itself until the termination condition is met. For time complexity analysis, we examine the total time complexity at each level of the hierarchy. At the root level of the clustering hierarchy, the input number of edges is $O(n)$. At each level of hierarchy, we have a remaining proximity graph that is bounded by $O(n)$. Thus, multi-level Short-Long clustering requires $O(n)$ time at each level. If there are t levels in the hierarchy, then it needs $O(tn)$ time. The structure of the splitting process of multi-level Short-Long clustering is similar to that of binary search, removing a constant fraction of the edges at each level. Thus, the expected depth (t) is likely to be $O(\log n)$. Consequently, the time complexity of the proposed clustering is $O(n \log n)$ expected time. This compares extremely favorably with other clustering methods that require $O(n^2)$ time [16].

4   Reasoning about Multi-level Clusters

Despite the importance of cluster reasoning, it has attracted less attention than clustering itself in the spatial data mining community. The main reason for the lack of cluster reasoning arises from the fact that points constituting clusters are dimensionless. Thus, spatial analysis (for instance, *intersection*) becomes difficult tasks with points [2]. This section provides a robust framework for reasoning about multi-level clusters by transforming them into well-defined regions which can be used by RCC, a well-developed region-based reasoning theory.

4.1   Cluster-to-Region Transformation and RCC

Cluster-to-region transformation identifies shapes of clusters. The Minimum Bounding Rectangle (MBR) and the Convex Hull (CH) are the two simplest ways of representing the shape of a cluster $C_i \subset P$. They are unique for C_i, thus they are natural choices. However, they are too crude to capture details of the shape of C_i. The α-shape [6] (this is a generalization of the CH) overcomes the crudeness of $CH(C_i)$. The set of α values leads to a whole family of shapes capturing crude $(CH(C_i))$ to fine shapes of C_i (ultimately points themselves). However, it is a difficult task to determine the best value α that produces a neither too crude nor too fine shape [11] and several trial-and-error steps are necessary for tuning the value of α. If R consists of l layers and each layer contains s clusters, then we need to tune α for as many as ls times in order to find the best shape of each cluster. Thus, the α-shape is not well-suited for data-rich environments. Recently, Lee and Estivill-Castro [13] proposed a linear time cluster-to-region transformation based on the Delaunay edge recuperation discussed in the previous section. This approach minimizes the need for parameter-tuning, instead it automatically derives the shape of a cluster C_i from both the spread of points in C_i and the spread of points in $P \backslash C_i$. In spatial settings, clusters are truly the result of the peculiarities in the entire distribution sampled by P. That is, although a point in $P \backslash C_i$ does not belong to a cluster C_i, it has some effect on the shape of C_i. Since this approach is more effective than the MBR and CH, and more efficient than the α-shape, it is well-suited for reasoning about multi-level clusters in data-rich environments. Fig. 1(d) shows a transformed region of the "8-like" cluster (shaded region) derived by this approach. The region is much finer than the corresponding MBR and CH generated regions.

RCC is a well founded qualitative spatial reasoning method based on regions. It uses a connection relation $C(X, Y)$ that represents "region X connects with region Y". RCC comprises several families of binary topological relations. One family, RCC8 uses a set of eight pairwise base relations: DC(DisCrete), EC(Externally Connected), PO(Partial Overlap), EQ(EQual), TPP(Tangential Proper Part), NTPP(Non-Tangential Proper Part), TPP^{-1}(Inverse of TPP) and NTPP^{-1}(Inverse of NTPP). Fig. 4 illustrates the eight base relations of RCC8. The cluster reasoning in the remainder of this paper is based on RCC8.

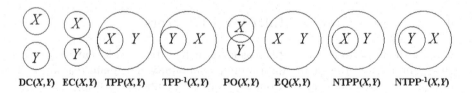

Fig. 4. The eight base relations of RCC8.

4.2   Intra-layer Cluster Reasoning

The popular way of modeling real-world spatial data within GIS is the McHarg's multi-layer view of the world [14]. In this view, each geographical layer captures something unique to it. To begin with analyzing multi-layer structure within spatial databases, it is essential to find spatial aggregations within each layer and reason about clusters within a layer (intra-layer clusters). *Intra-layer cluster reasoning* involves detecting relationships between X and Y, where both X and Y are cluster regions belonging to the same layer within GIS. Once clusters are transformed into corresponding regions, then it is straightforward to derive spatial relationships using RCC8. Some spatial relationships derived from the reserve dataset in Fig. 2 are as follows.

- DC(A,B), DC(A,C), DC(B,C), DC(C,D), ...
- TPP^{-1}(A,AA), TPP^{-1}(B,BA), TPP^{-1}(C,CA), ...
- DC(AA,B), DC(A,BA), DC(AA,BA), DC(AA,CA), DC(BA,CA), ...

However, comprehensive search is not necessary for detecting these relationships. We can derive several properties that limit the search space of possible relationships of intra-layer clusters. Note that, Short-Long clustering is a crisp clustering method that assigns each point $p_i \in P$ to a single cluster. And, the cluster-to-region transformation underlying multi-level Short-Long clustering derives cluster boundaries from $DD(P)$ that is a planar graph. Namely, for two distinct cluster regions X and Y, their boundaries $\partial(X)$ and $\partial(Y)$ are subsets of $DD(P)$. Thus, $\partial(X) \cap \partial(Y) \subset P$. Based on these discussions, we can derive the following properties. Here, we let L_k denote the set of kth-level cluster regions and $L = \{L_1, L_2, \ldots, L_k\}$.

Property 1. If $X,Y \in L$ and $X \neq Y$, then EC(X,Y), PO(X,Y) and EQ(X,Y) do not hold.

Property 2. If $X_i,Y_i \in L_i$ ($1 \leq i \leq k$) and $X_i \neq Y_i$, then only DC(X_i,Y_i) holds.

Property 3. If $X \in L$ and X^d denotes a descendant cluster of X, then either TTP^{-1}(X,X^d) or NTTP^{-1}(X,X^d) holds.

Property 4. If $X,Y \in L$, X^c is a child cluster of X, Y^c is a child cluster of Y and DC(X,Y), then DC(X^c,Y), DC(X,Y^c) and DC(X^c,Y^c) hold.

These properties are similar to the progressive refinement approach to spatial data mining [12]. That is, the characteristics of a parent (ancestor) cluster limit the space of possible relationships of its child (descendant) clusters. Thus, a top-down approach can prune the search space. From these properties, now we can derive the intra-layer cluster relationships derived from Fig. 2 more efficiently.

4.3   Inter-layer Cluster Reasoning

In contrast to intra-layer cluster reasoning, all the eight base relations are possible in *inter-layer cluster reasoning*. It involves in detecting relationships between X and Y, where X and Y are cluster regions belonging to different layers. Similar to intra-layer cluster reasoning, we can perform inter-layer cluster reasoning using RCC8 once clusters are transformed into corresponding regions. Some inter-layer spatial relationships from the datasets in Fig. 2 and Fig. 3 are as follows.

- DC(Reserve(A),Robbery(C)), DC(Reserve(AA),Robbery(C)), ...
- PO(Reserve(B),Robbery(A)), PO(Reserve(B),Robbery(AA)), ...

Table 1. Invalid relationships between inter-layer clusters.

Current relationship	Invalid relationships
$DC(X, Y)$	$\forall \gamma \in \Gamma - \{DC\}$: $\gamma(X^d, Y)$, $\gamma(X, Y^d)$, $\gamma(X^d, Y^d)$
$EC(X, Y)$	$\forall \gamma \in \Gamma - \{DC, EC\}$: $\gamma(X^d, Y)$, $\gamma(X, Y^d)$, $\gamma(X^d, Y^d)$
$TPP(X, Y)$	$DC(X^d, Y)$, $EC(X^d, Y)$, $TPP^{-1}(X^d, Y)$, $PO(X^d, Y)$, $EQ(X^d, Y)$, $NTPP^{-1}(X^d, Y)$, $NTPP(X, Y^d)$
$TPP^{-1}(X, Y)$	$NTPP^{-1}(X^d, Y)$, $DC(X, Y^d)$, $EC(X, Y^d)$, $TPP(X, Y^d)$, $PO(X, Y^d)$, $EQ(X, Y^d)$, $NTTP(X, Y^d)$
$PO(X, Y)$	$TPP^{-1}(X^d, Y)$, $EQ(X^d, Y)$, $NTPP^{-1}(X^d, Y)$, $TPP(X, Y^d)$, $EQ(X, Y^d)$, $NTPP(X, Y^d)$
$EQ(X, Y)$	$DC(X^d, Y)$, $EC(X^d, Y)$, $TPP^{-1}(X^d, Y)$, $PO(X^d, Y)$, $EQ(X^d, Y)$, $NTPP^{-1}(X^d, Y)$, $DC(X, Y^d)$, $EC(X, Y^d)$, $TPP(X, Y^d)$, $PO(X, Y^d)$, $EQ(X, Y^d)$, $NTPP(X, Y^d)$
$NTPP(X, Y)$	$DC(X^d, Y)$, $EC(X^d, Y)$, $TPP(X^d, Y)$, $TPP^{-1}(X^d, Y)$, $PO(X^d, Y)$, $EQ(X^d, Y)$, $NTPP^{-1}(X^d, Y)$
$NTPP^{-1}(X, Y)$	$DC(X, Y^d)$, $EC(X, Y^d)$, $TPP(X, Y^d)$, $TPP^{-1}(X, Y^d)$, $PO(X, Y^d)$, $EQ(X, Y^d)$, $NTPP(X, Y^d)$

Since inter-layer cluster reasoning involves more clusters and layers, it is much more complicated than intra-layer cluster reasoning. However, the space of possible relationships between inter-layer clusters can be reduced similar to the discussion made in the previous subsection. Let Γ be the eight base relations of RCC8, L^h be a set of cluster regions in the h-th layer, X^d be a descendant cluster of X and Y^d be a descendant cluster of Y. For cluster regions $X \in L^i$ and $Y \in L^j$ where $i \neq j$, invalid relationships are outlined in Table 1.

Invalid relationships in Table 1 can be used to avoid the comprehensive search and thus for efficiently deriving inter-layer cluster relationships. For instance, $DC(\text{Reserve}(A), \text{Robbery}(C))$ implies three relationships according to the table above: $DC(\text{Reserve}(AA), \text{Robbery}(CA))$, $DC(\text{Reserve}(A), \text{Robbery}(CA))$ and $DC(\text{Reserve}(AA), \text{Robbery}(C))$.

5   Blurring Multi-level Clusters

Cluster reasoning in the previous section is based on crisp regions of clusters. However, in many spatial settings regions of clusters may not be determinate but rather vague. This section defines the upper and lower bounds of a cluster, and uses the vagueness of a cluster region for reasoning clusters. Fig. 5 illustrates reasoning with the upper bounds. Fig. 5(b) depicts crisp regions of three clusters. Since these crisp regions are the finest regions derived by the cluster-to-region transformation [13], we define them as the lower bounds in this paper. The Voronoi Diagram represents the region of influence of each point. Thus, the upper bound of a cluster can be intuitively defined by the union of Voronoi regions of points in the cluster. A region in light grey in Fig. 5(c) is the upper bound of the "8-like" cluster. Fig. 5(d) displays the upper bounds of the three

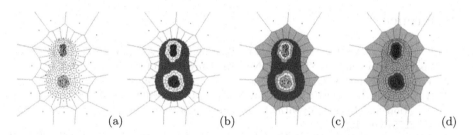

Fig. 5. The upper and lower bounds (P is the same as in Fig. 1): (a) $VD(P)$; (b) Crisp regions; (c) The upper bound of the "8-like" cluster; (d) Reasoning with the upper bounds.

clusters. Now, the upper bound of the "8-like" cluster is externally connected to the upper bounds of the "ball-like" clusters. The indeterminacy of a cluster discussed in this section is similar to the Egg-York theory [4]. Thus, we can use the extended RCC that includes an irreflexive, asymmetric and transitive binary relation $X < Y$ read as "X is a crisping of Y" or "Y is a blurring of X". For instance, we can have $X <$ [the upper bound of X].

6   Final Remarks

We proposed an effective and efficient divisive multi-level clustering method and a robust framework for reasoning about its clusters using RCC8. The reasoning framework is well-suited for multi-layered GIS. However, it can be also used for concept management [9]. The multi-level clustering method minimizes user-oriented bias and constraints hindering explorations and maximizes user-friendliness. It does not require a termination condition, but rather derives it from the data. If a subset exhibits point-centric statistics similar to CSR, then the multi-level clustering method leaves it as a single cluster and terminates. Its effectiveness includes clusters of arbitrary shapes, clusters of different densities, clusters of heterogeneous sizes, robustness to noise and insensitiveness to bridges. It requires $O(n \log n)$ time which compares extremely favorably with other hierarchical clustering approaches typically requiring $O(n^2)$ time. In addition, we demonstrated how reasoning about clusters can be performed using RCC8. The reasoning framework proposed in this paper first transforms clusters to regions and uses RCC8 to perform intra-layer reasoning and inter-layer reasoning. This framework also considers the indeterminacy of a cluster region defined by the upper and lower bounds. The crisping and blurring relation ($<$) together with RCC8 can serve for reasoning these vague cluster regions. This vagueness is especially useful for interactive spatial clustering and cluster process analysis. The upper bound of a cluster is a set of Voronoi regions of influence of points belonging to the cluster. Thus, new points added in this blurring can be assigned to the cluster. Also, the cluster can grow and shrink within the upper and lower bounds.

References

1. L. Anselin. What is Special About Spatial Data? Alternative Perspectives on Spatial Data Analysis. TR 89-4, NCGIA, University of California, 1989.
2. T. C. Bailey and A. C. Gatrell. *Interactive Spatial Analysis*. 1995.
3. A. G. Cohn, B. Bennett, J. Gooday, and N. M. Gotts. Qualitative Spatial Representation and Reasoning with the Region Connection Calculus. *GeoInformatica*, 1(3):275–316, 1997.
4. A. G. Cohn and N. M. Gotts. Representing Spatial Vagueness: A Mereological Approach. In *Proc. of the 5th Int. Conf. on Principles of Knowledge Representation and Reasoning*, pages 230–241, Massachusetts, 1996. Morgan Kaufmann.
5. P. J. Diggle. *Statistical Analysis of Spatial Point Patterns*. 1983.
6. H. Edelsbrunner, D. Kirkpatrick, and R. Seidel. On the Shape of a Set of Points in the Plane. *IEEE Transactions on Information Theory*, 29(4):551–559, 1983.
7. M. Ester, H. P. Kriegel, J. Sander, and X. Xu. A Density-Based Algorithm for Discovering Clusters in Large Spatial Databases with Noise. In *Proc. of the 2nd Int. Conf. on KDDM*, pages 226–231, Portland, Oregon, 1996. AAAI Press.
8. V. Estivill-Castro and I. Lee. Argument Free Clustering via Boundary Extraction for Massive Point-data Sets. *Computers, Environments and Urban Systems*, 26(4):315–334, 2002.
9. P. Gärdenfors, and M-A. Williams. Reasoning about Categories in Conceptual Spaces. In *Proc. of the IJCAI*, pages 385–392, Seattle, Washington, 1998. Morgan Kaufmann.
10. S. Guha, R. Rastogi, and K. Shim. CURE: An Efficient Clustering Algorithm for Large Databases. In *Proc. of the ACM SIGMOD'98*, pages 73–84, Seattle, Washington, 1998. ACM Press.
11. E. M. Knorr, R. T. Ng, and D. L. Shilvock. Finding Boundary Shape Matching Relationships in Spatial Data. In *Proc. of the 5th Int. Symposium on Spatial Databases*, LNCS 1262, pages 29–46, Berlin, Germany, 1997. Springer.
12. K. Koperski. *A Progressive Refinement Approach to Spatial Data Mining*. PhD thesis, Computing Science, Simon Fraser University, B.C., Canada, 1999.
13. I. Lee and V. Estivill-Castro. Polygonization of Point Clusters through Cluster Boundary Extraction for Geographical Data Mining. In *Proc of the 10th Int Symposium on Spatial Data Handling*, pages 27–40, Ottawa, Canada, 2002. Springer.
14. I. L. McHarg. *Design with Nature*. Natural History Press, New York, 1969.
15. H. J. Miller and J. Han. *Geographic Data Mining and Knowledge Discovery: An Overview*. Cambridge University Press, Cambridge, UK, 2001.
16. F. Murtagh. Comments on Parallel Algorithms for Hierarchical Clustering and Cluster Validity. *IEEE Transactions on Pattern Analysis and Machine Intelligence*, 14(10):1056–1057, 1992.
17. W. Wang, J. Yang, and R. R. Muntz. STING: A Statistical Information Grid Approach to Spatial Data Mining. In *Proc. of the 23rd Int. Conf. on Very Large Data Bases*, pages 186–195, Athens, Greece, 1997. Morgan Kaufmann.
18. T. Zhang, R. Ramakrishnan, and M. Livny. BIRCH: An Efficient Data Clustering Method for Very Large Databases. In *Proc. of the ACM SIGMOD'96*, pages 103–114, Montreal, Canada, 1996. ACM Press.

An Efficient Cell-Based Clustering Method for Handling Large, High-Dimensional Data

Jae-Woo Chang

Dept. of Computer Engineering
Research Center of Industrial Technology, Engineering Research Institute
Chonbuk National University, Chonju, Chonbuk 561-756, South Korea
jwchang@dblab.chonbuk.ac.kr

Abstract. In this paper, we propose an efficient cell-based clustering method for handling a large of amount of high-dimensional data. Our clustering method provides an efficient cell creation algorithm using a space-partitioning technique and a cell insertion algorithm to construct clusters as cells with more density than a given threshold. To achieve good retrieval performance on clusters, we also propose a new filtering-based index structure using an approximation technique. In addition, we compare the performance of our cell-based clustering method with the CLIQUE method in terms of cluster construction time, precision, and retrieval time. The experimental results show that our clustering method achieves better performance on cluster construction time and retrieval time.

1 Introduction

Data mining is concerned with the extraction of interesting knowledge from a large amount data, i.e. rules, regularities, patterns, constraints. Among the typical research topics of data mining, one of the most important topics is clustering. Clustering is the process of grouping data into classes or clusters so that objects within a cluster can have high similarity in comparison to one another but can be very dissimilar to objects in other clusters [1]. In data mining applications, there have been several existing clustering methods, such as BIRCH(Balanced Iterative Reducing and Clustering using Hierarchies) [2], DBSCAN(Density Based Spatial Clustering of Applications with Noise) [3], STING(STatistical INformation Grid) [4], and CLIQUE(CLustering In QUEst) [5]. The conventional clustering methods have a critical drawback that they are not suitable for handling a large data set with more than millions of data because the data set is restricted to be resident in a main memory. Besides, they do not work well for clustering high-dimensional data because their retrieval performance is generally degraded as the number of dimension increases.

In this paper, we propose an efficient cell-based clustering method for dealing with a large amount of high-dimensional data. Our clustering method provides an efficient cell creation algorithm which makes cells by splitting each dimension into a set of partitions using a split index. It also provides a cell insertion algorithm to construct

K.-Y. Whang, J. Jeon, K. Shim, J. Srivatava (Eds.): PAKDD 2003, LNAI 2637, pp. 295–300, 2003.

clusters as cells with more density than a given threshold and insert the clusters into an index structure. By using an approximation technique, we also propose a new filtering-based index structure to achieve good retrieval performance on clusters.

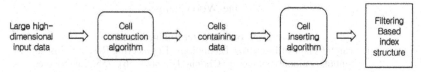

Fig. 1. Overall architecture of our cell-based clustering method

2   An Efficient Cell-Based Clustering Method

Our cell-based clustering method provides both a cell creation and a cell insertion algorithm and it makes use of a filtering-based index structure. The figure 1 shows the overall architecture of our cell-based clustering method.

2.1 Cell Creation Algorithm

Our cell creation algorithm makes cells by splitting each dimension into a group of sections using a split index. The split index based on density is used for creating split sections and is efficient for splitting multi-group data. Our cell creation algorithm first finds the optimal split section by repeatedly examining a value between the maximum and the minimum in each dimension. That is, it find the optimal one while the difference between the maximum and the minimum is greater than one and the value of a split index after splitting is greater than the previous value. The split index value is calculated by Eq. (1) before splitting and Eq. (2) after splitting.

$$Split(S) = 1 - \sum_{j=1}^{0} P_j^{\ 2} \qquad ----- \ Eq.(1)$$

$$Split(S) = \frac{n_1}{n} Split(S_1) + \frac{n_2}{n} Split(S_2) \qquad ----- \ Eq.(2)$$

By Eq. (1), we can determine the split index value for a data set S in three steps: (i) divide S into C classes, (ii) calculate the square value of the relative density of each class, and (iii) add the square value of all the density of C classes. By Eq. (2), we compute the split index value for S after S is divided into S_1 and S_2. If the split index value is larger than the previous value before splitting, we actually divide S into S_1 and S_2 and otherwise we stop splitting. Secondly, our cell creation algorithm creats cells being made by the optimal n split sections for n-dimensional data. As a result, our cell creation algorithm makes much less cells than the existing clustering methods using equivalent interval. If a data set has n dimensions and the number of the initial split sections in each dimension is m, the conventional cell creation algorithm make m*n, but our algorithm makes only K1*K2*...*Kn cells ($1 \bullet \leq$ K1, k2, ...,K $\leq$ n•m).

2.2 Cell Insertion Algorithm

For storing the created cells, we make clusters as cells with more density than a given cell threshold and store them into a cluster information file. In case the number of the created cells is large, it may take much time to answer users' queries for a large amount of high-dimensional data. To improve an overall retrieval performance, we adopt a filtering technique by using an approximation information file. Therefore, we store into an approximation information file all the sections with more density than a given section threshold. As a result, the insertion algorithm to store data is as follows. First, we calculate the frequency of a section of all dimensions whose frequency is greater than a given section threshold. Secondly, we store one bit set by '1' in an approximation information file for sections whose frequencies are greater than the threshold and we store one bit set by '0' for the remainder sections. Thirdly, we calculate the frequency of data in a cell. Finally, we store a cell id and its frequency into the cluster information file for a cell whose frequency is greater than a given cell threshold. The cell threshold and the section threshold are shown in Eq. (3).

$$Section\ threshold = \begin{cases} \lambda = \dfrac{NR \times F}{NI} \\ NI : \text{the number of input data} \\ NR : \text{the number of sections per dimension} \\ F : \text{minimum section frequency being regarded as '1'} \end{cases} \quad -\text{-- } Eq.(3)$$

$Cell\ threshold(\tau)$: positive integer

2.3 Filtering-Based Index Structure

In order to reduce the number of I/O accesses to a cluster information file which is generally large in size, it is possible to construct a new filtering-based index structure using an approximation information file. Figure 2 shows a two-level filtering-based index structure containing both an approximation information file and a cluster information file. When a query is given in our two-level index structure, we first obtain the corresponding section of each dimension. In case the corresponding sections in all the dimensions have one as their value in an approximation information file, we calculate a cell number and access a cluster information file using it. Otherwise, we can improve retrieval performance by not accessing the approximation information file. An increase in dimensionality may cause high probability that at least one dimension in the approximation information file has zero.

Figure 2 also shows a procedure to answer a user query in our two-level index structure when a cell threshold and a section threshold are 1, respectively. For a query Q1, we determine 0.6 in X axis as the third section and 0.8 in Y axis as the fourth section. In the approximation-information file, the value for the third section in X axis has '1' and the one for the 4-th section in Y axis has '0'. If there is one or more sections with '0' in an approximation-information file, a query is discarded and the query processing is terminated without searching the corresponding cluster information file. So, Q1 is discarded at the first phase. For a query Q2, the value of 0.55 in X axis and the value of 0.7 in Y axis belong to the third section, respectively. In the approximation information file, the third bit for X axis and the third bit for Y axis have '1', we can calculate a cell number and obtain its cell frequency by accessing the cluster information file. As a result, in case of Q1, it is proved that there is no resulting set. In case of Q2, we obtain the cell number of 11 and its frequency of 3 in the cluster information file.

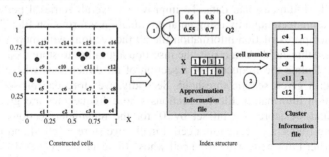

Fig. 2. Two-level filtering-based index structure

3   Performance Analysis

For our performance analysis, we implement our cell-based clustering method under Linux server with 650 MHz dual processor and 512 MB main memory. In addition, we make use of one million 16-dimensional data which are made by Synthetic Data Generation Code for Classification in IBM Quest Data Mining Project [6]. A record for the data is composed of numeric attributes, like salary, commission, age, hvalue, hyears, loan, tax, interest, cyear, and balance, as well as categorical attributes, like elevel, zipcode, area, children, ctype, and job. The measures of our performance analysis are cluster construction time, precision, and retrieval time. The cluster construction time means the addition of cell creation time and cell insertion time. We compare our cell-based clustering method with the CLIQUE method, which is one of the most efficient conventional clustering method for handling high-dimensional data. In addition, the CLIQUE+A is the method which is made by applying our approximation information file to the CLIQUE method for comparison.

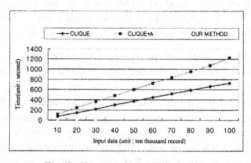

Fig. 3. Cluster construction time

For the performance analysis of cluster construction time, we set threshold values to 0 without considering section thresholds for the three methods. Figure 3 shows the performance of cluster construction time. It is shown that the cluster construction time is increased linearly in proportion to the amount of data. This result is acceptable for a large amount of data. The experimental result shows that the CLIQUE and the CLIQUE+A need about 730 and 1,200 seconds, respectively. However, our cell-based clustering method needs only 100 seconds. Because our clustering method makes much less cells than the CLIQUE, our method leads to 85% decrease in cluster construction time. Because the CLIQUE+A uses an approximation information file, it causes about 160% increase in cluster construction time, compared to the CLIQUE.

Figure 4 shows average retrieval time for a given user query. The CLIQUE, the CLIQUE+A, and our clustering method need 31, 8, and 1 second, respectively. It is shown that our cell-based clustering method is about thirty times better on retrieval performance than the CLIQUE and is about ten times better than the CLIQUE+A. This is because our clustering method creates less cells by using our cell creation algorithm and achieves good filtering effect by using the approximation information file. Figure 5 shows the performance of precision. The precision is calculated by dividing the number of objects retrieved by the total number of objects relevant for a given query. The result shows that our cell-based clustering method and the CLIQUE+A achieve over 90% precision while the CLIQUE achieves about 94% precision. In the viewpoint of the precision, it is shown that the performance of our clustering method is slightly worse than the CLIQUE. This is because our clustering method makes much less cells, compared with the CLIQUE.

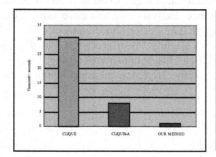
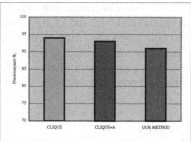

Fig. 4. Retrieval time   **Fig. 5.** Precision

Because two major performance measures, i.e., retrieval time and precision, have a trade-off relationship, we estimate a system efficiency measure to combine retrieval time with precision. We define the system efficiency measure as Eq. (5). Here E_{MD} means the system efficiency of MD(method) and W_p and W_t mean the weight of precision and that of retrieval time, respectively. P_{MD} and T_{MD} mean the precision and the retrieval time of the three methods. P_{MAX} and T_{MIN} mean the maximum precision and the minimum retrieval time, respectively, for all methods. We assume that the weight of precision and that of retrieval time are 0.5, respectively. Figure 6 depicts the performance results of their system efficiency. It is shown that our clustering method achieves about 1.0 on the performance of system efficiency while the CLIQUE and the CLIQUE+A achieve about 0.5 and 0.6, respectively. Conclusively, our clustering method achieves the best performance in term of the system efficiency.

$$E_{MD} = W_p \bullet \frac{P_{MD}}{P_{MAX}} + W_t \bullet \frac{1}{\frac{T_{MD}}{T_{MIN}}} \qquad ----- Eq.(5)$$

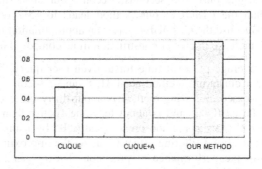

Fig. 6. System efficiency

4   Conclusion

We proposed a new cell-based clustering method with two. The first one allows us to create the small number of cells for a large amount of high-dimensional data. The second one allows us to apply an approximation technique to our cell-based clustering method for fast clustering. For performance analysis, we compare our cell-based clustering method with the CLIQUE method. The performance analysis results show that our cell-based clustering method shows slightly lower precision, but it achieves good performance on retrieval time as well as cluster construction time. Finally, our method shows good performance on a system efficiency measure which is one to combine both precision and retrieval time.

References

[1]   Jiawei Han, Micheline Kamber, "Data Mining : Concepts and Techniques", Morgan Kaufmann, 2000.
[2]   Zhang T., Ramakrishnan R., Linvy M., "BIRCH : An Efficient Data Clustering Method for Very Large Databases", Proc. ACM Int. Conf. on Management of Data, ACM Press, 1996, pp. 103–114.
[3]   Ester M., Kriegel H.-P., Sander J., Xu X., "A Density Based Algorithm for Discovering Clusters in Large Spatial Databases with Noise", Proc. 2nd Int. Conf. on Knowledge Discovery and Data Mining, AAAI Press, 1996.
[4]   Wang W., Yang J., Muntz R., "STING : A Statistical Information Grid Approach to Spatial Data Mining", Proc. 23rd Int. Conf. on Very Large Data Bases, Morgan Kaufmann, 1997.
[5]   Rakesh Agrawal, Johannes Gehrke, Dimitrios Gunopulos, Prabhakar Raghavan, "Automatic Subspace Clustering of High Dimensional Data Mining Applications", Proc. ACM SIGMOD Int. Conf. on Management of Data, 1998, pp. 94–105.
[6]   http://www.almaden.ibm.com/cs/quest.

Enhancing SWF for Incremental Association Mining by Itemset Maintenance

Chia-Hui Chang and Shi-Hsan Yang

Department of Computer Science and Information Engineering,
National Central University, Chung-Li, Taiwan 320
chia@csie.ncu.edu.tw, carry@db.csie.ncu.edu.tw

Abstract. Incremental association mining refers to the maintenance and utilization of the knowledge discovered in the previous mining operations for later association mining. Sliding window filtering (SWF) is a technique proposed to filter false candidate 2-itemsets by segmenting a transaction database into partitions. In this paper, we extend SWF by incorporating previously discovered information and propose two algorithms to boost the performance for incremental mining. The first algorithm FLSWF (SWF with Frequent Itemset) reuses the frequent itemsets of previous mining task to reduce the number of new candidate itemsets that have to be checked. The second algorithm CLSWF (SWF with Candidate Itemset) reuses the candidate itemsets from the previous mining task. Experiments show that the new proposed algorithms are significantly faster than SWF.

1 Introduction

Association rule mining from transaction database is an important task in data mining. The problem, first proposed by [2], is to discover associations between items from a (static) set of transactions. Many interesting works have been published to address this problem [9,8,11,1,6,7,10]. However, in real life applications, the transaction databases are changed with time, where updates take particular way: delete the oldest stuff at one end and add new stuff at the other end. For example, Web log records, stock market data, and grocery sales data, etc. In these dynamic databases, the knowledge acquired previously can facilitate successive discovery processes. How to maintain and reuse such knowledge is called incremental association mining.

Incremental association mining was first introduced in [4] where the problem is to update the association rules in a database when new transactions are added to the database. The proposed algorithm, FUP, stores the counts of all the frequent itemsets found in a previous mining phrase and iteratively discover $(k + 1)$-itemsets as the Apriori algorithm [2]. An extension of the work is addressed in [5] where the authors propose the FUP2 algorithm for updating the existing association rules when transactions can be deleted from the database. In essence, FUP2 is equivalent to FUP for the case of insertion, and is, however, a complementary algorithm of FUP for the case of deletion.

K.-Y. Whang, J. Jeon, K. Shim, J. Srivatava (Eds.): PAKDD 2003, LNAI 2637, pp. 301–312, 2003.

In the recent years, several algorithms, including the MAAP algorithm [13] and the PELICAN algorithm [12], are proposed to solve the problem of incremental mining. Both MAAP and PELICAN reuse the previous results as FUP2 does, but they only focus on how to maintain maximum frequent itemsets when the database is updated. In other words, they do not care non-maximum frequent itemsets, therefore, the counts of non-maximum frequent itemsets can not be calculated. The difference of these two algorithms is that MAAP calculates maximum frequent itemsets by Apriori-based framework while PELICAN calculates maximum frequents itemsets based on vertical database format and lattice decomposition.

Generally speaking, Apriori-like algorithms tend to suffer from two inherent problems, namely (1) the occurrence of a potentially huge set of candidate itemsets, and (2) the need of multiple scans of databases. To remedy these problems, Lee et al. proposed a sliding-window-filtering (SWF) technique for incremental association mining [3]. By partitioning a transaction database into partitions, SWF filters unnecessary candidate itemsets by employing a local minimum support in each partition such that the information in the prior partition is selectively carried over toward the generation of candidate itemsets in the subsequent partitions. It is shown that SWF maintains a set of candidate itemsets that is very close to the set of frequent itemsets and outperforms FUP2 in a great deal. However, SWF does not utilize previously discovered frequent itemsets which are used to improve incremental mining in FUP2. In this paper, we show two simple extensions of SWF which significantly improve the performance by incorporate previously discovered information such as the frequent itemsets or candidate itemsets from the old transaction database.

The rest of this paper is organized as follows. Preliminaries on association mining and incremental mining are illustrated in Section 2. The idea of extending SWF with previously discovered information is introduced in Section 3. Experiments are presented in Section 4. Finally, this paper concludes with Section 5.

2   Preliminaries

Let $I=\{i_1, i_2, ..., i_m\}$ be a set of literals, called items. Let D be a set of transactions, where each transaction T is a set of items such that $T \subseteq I$. Note that the quantities of items bought in a transaction are not considered, meaning that each item is a binary variable representing if an item was bought. Each transaction is associated with an identifier, called TID. Let X be a set of items. A transaction T is said to contain X if and only if $X \subseteq T$. An association rule is an implication of the form $X \to Y$, where $X \subset I, Y \subset I$ and $X \cap Y = \emptyset$. The rule $X \to Y$ holds in the transaction set D with confidence c if $c\%$ of the transactions in D that contain X also contain Y. The rule $X \to Y$ has support s in the transaction set D if $s\%$ of transactions in D contain $X \cup Y$. For a given pair of confidence and support thresholds, the problem of association rule mining is to find out all the association rules that have confidence and support greater than the corresponding thresholds. This problem can be reduced to the problem of finding all frequent itemsets for the same support threshold [2].

For a dynamic database, old transactions ($\triangle^-$) are deleted from the database D and new transactions ($\triangle^+$) are added. Naturally, $\triangle^- \subseteq D$. Denote the updated database by D', therefore $D' = (D - \triangle^-) \cup \triangle^+$. We also denote the set of unchanged transactions by $D^- = D - \triangle^-$.

2.1 The SWF Algorithm

Sliding-window-filtering was proposed by Lee, et al in [3]. The key idea of SWF algorithm is to compute a set of candidate 2-itemsets as close to frequent 2-itemset L_2 such that the time for database scanning is minimized. Suppose the database is divided into n partitions $P_1, P_2, \ldots, P_n$, and processed one by one. For each frequent itemset I, if I is not frequent from P_1 to current partition, there must exist some partition P_k such that I is frequent from partition P_k to P_n. Therefore, if we know a candidate 2-itemset is not frequent from the first partition where it presents frequent to current partition, we can delete this itemset even though we don't know wheter it is frequent or not. If somehow this itemset is indeed frequent, it must be frequent from some partition P_k, where we can add it to the candidate set again and check it for the successive partitions.

The SWF algorithm maintains a list of 2-itemsets CF that is frequent for the lasted partitions as described below. For each partition P_i, SWF adds frequent 2-itemsets (together with its starting partition P_i and supports) that is not in CF and checks if the present 2-itemsets are continually frequent from its' start partition to the current partitions. If a 2-itemset is no longer frequent, SWF deletes this itemset from CF. If an itemset becomes frequent at partition P_j ($j > i$), the algorithm put this itemset into CF and check it's frequency from P_j to each successive partition P'_j ($j' > j$). It was shown that the number of the reserved candidate 2-itemsets will be close to the number of the frequent 2-itemsets.

For a moderate number of candidate 2-itemsets, scan reduction technique [9] can be applied to generate all candidate k-itemsets from candidate itemsets (instead of frequent itemsets). Therefore, one database scan is enough to calculate all candidate itemsets' supports and determine the frequent ones. In such cases, the total number of database scans is kept as small as 2.

2.2 An Example

In the following, we use an example to demonstrate the SWF algorithm and introduce how to improve the SWF algorithm. The example is the same as [3]. Figure 1(a) shows the original transaction database where it contains 9 transactions, tagged $t_1, \ldots, t_9$, and the database is segmented into three partition, P_1, P_2, P_3 (denoted by $db^{1,3}$). After the database is updated (delete P_1 and add P_4), the updated database contains partition P_2, P_3, P_4 (denoted by $db^{2,4}$) as shown in Figure 1(b). The minimum support is assumed to be $s = 40\%$. The incremental mining problem is decomposed into two procedures:

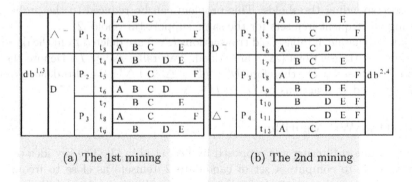

(a) The 1st mining   (b) The 2nd mining

Fig. 1. Example

1. **Preprocessing procedure** which deals with mining in the original transaction database.
2. **Incremental procedure** which deals with the update of the frequent itemsets for an ongoing time-variant transaction database.

Preprocessing Procedure. In the preprocessing procedure, each partition is scanned sequentially for the generation of candidate 2-itemsets. After scanning the first partition of 3 transactions, candidate 2-itemsets {AB, AC, AE, AF, BC, BE, CE} are generated, and frequent itemsets {AB, AC, BC} are added to CF_2 with two attributes *start* partition P_1 and occurrence *count*. Note that SWF begins from 2-itemset instead of 1-itemset.

Similarly, after scanning partition P_2, the occurrence counts of candidate 2-itemsets in the present partition are recorded. For itemsets that are already in CF_2, add the occurrence counts to the corresponding itemsets in CF_2. For itemsets that are not in CF_2, record the counts and its *start* partition from P_2. Then, check if the itemsets are frequent from its start partition to current partition and delete those non-frequent itemsets from CF_2. In this example, the counts of the three itemsets {AB, AC, BC} in P_2, which start from partition P_1, are added to their corresponding counts in CF_2. The local support (or called the *filtering threshold*) of these 2-itemsets is given by $\lceil (3 + 3) * 0.4 \rceil = 3$. The other candidate 2-itemsets of P_2, {AD, AE, BD, BE, CD, CF, DE}, has filtering threshold $\lceil 3 * 0.4 \rceil = 2$. Only itemsets with counts greater than its filtering threshold are kept in $CF_2 = \{AB, AC, BC, BD\}$, others are deleted. Finally, partition P_3 is processed by algorithm SWF. The resulting CF_2={AB, AC, BC, BD, BE}.

After all partitions are processed, the scan reduction technique is employed to generate C'_3 from CF_2. Iteratively, C'_{k+1} can be generated from C'_k for $k = 3, \ldots, n$, as long as all candidate itemsets can been stored in memory. Finally, we can find $\cup_k L_k$ from $\cup_k C_k$ with only one database scan.

Incremental Procedure. Consider Figure 1(b), when the database is updated form $db^{1,3}$ to $db^{2,4}$ where transaction, t_1 t_2 t_3, are deleted from the database and t_{10}, t_{11}, t_{12} are added to the database. The incremental step can be divided into three sub-steps: (1) deleting $\triangle^-$ and generating C_2 in $D^- = db^{2,3}$ (2) adding $\triangle^+$ and generating C_2 in $db^{2,4} = D^- + \triangle^+$ (3) scanning the database $db^{2,4}$ for the generation of all frequent itemsets L_k. In the first sub-step, we load CF_2 and check if the counts of the itemsets in CF_2 will be influenced by deleting P_1. We decrease the influenced itemsets' counts that are supported by P_1 and set the *start* position of influenced itemsets to 2. Then, we check the itemsets of CF_2 from the start position of the itemsets to partition P_3, and delete non-frequent itemsets of CF_2. It can be seen that itemsets {AB, AC, BC} were removed. Now we have CF_2 corresponding to $db^{2,3}$.

The second sub-step is similar to the operation in the preprocessing step. We add the counts of the 2-itemset in P_4 to the counts of the same itemsets of CF_2, and reserve the frequent itemsets of CF_2. New frequent 2-itemsets that are not in CF_2 will be added with its *start* position P_4 and counts. Finally, at the third sub-step, SWF uses scan reduction technique to generate all candidate itemsets from CF_2={BD, BE, DE, DF, EF} and scan database $db^{2,4}$ once to find frequent itemsets. This is how SWF works for incremental mining.

As we can see, SWF does not utilize previously discovered frequent itemsets as applied in FUP2. In fact, for most frequent itemsets, the supports can be updated by simply scan the changed partitions of the database. It doesn't need to scan the whole database for these old frequent itemsets. Only those newly generated candidate itemsets have to be checked by scanning the total database. In such cases, we can speed up not only candidate generation but also database scanning. The algorithm will be described in the next section.

3   The Proposed Algorithm

The proposed algorithm combines the idea of $FUP2$ with SWF. Consider the candidate itemsets generated in the preprocessing procedure ({A, B, C, D, E, F, AB, AC, BC, BD, BE, ABC}) and the incremental procedure ({A, B, C, D, E, F, BD, BE, DE, DF, EF, BDE, DEF}) in SWF, we find that there are eight candidate itemsets in common and five of them are frequent itemsets in $db^{1,3}$. If we incorporate previous mining result, i.e. the counts of the frequent itemsets, we can get new counts by only scanning the changed portions of database instead of scanning the whole database in the incremental procedure.

The proposed algorithm divides the incremental procedure into six sub-steps: the first two sub-steps update CF_2 as those for the incremental procedure in SWF. Let us denote the set of previous mining result as KI (known itemsets). Consider the example where KI ={A(6), B(6), C(6), E(4), AB(4), AC(4), BC(4), BE(4)}. In the third and fourth sub-step, we scan the changed portion of the database, P_1 and P_4 to update the counts of itemsets in KI. Denote the set of verified frequent itemsets from KI as FI (frequent itemsets). In this example, FI={A(4), B(5), C(5), E(5), BE(4)}.

In the fifth sub-step, we generate candidate itemsets that need to be checked by scanning the whole new database $db^{2,3}$. Denote those 2-itemsets that are not in KI as new candidate 2-itemsets, NC_2 ($= CF_2 - KI$). The set NC_2 contains itemsets that have no count information for $db^{2,4}$ and really need to scan the whole database to get the counts. By scan reduction, we use NC_2 to generate NC_3 by three JOIN operations. The first JOIN operation is $NC_2 \star NC_2$. The second JOIN operation is $NC_2 \star FI_2$ (2-itemsets of FI). The third JOIN operation is $FI_2 \star FI_2$. However, result from the third JOIN can be deleted if the itemsets are present in KI. The reason for the third JOIN operation is that NC_2 may generate itemsets which are not frequent in previous mining task because one of the 2-itemset are not frequent. While new itemsets from NC_2 sometimes makes such 3-itemsets satisfy the anti-monotone Apriori heuristic in present mining task. From the union of the three JOIN operations, we get NC_3. Similarly, we can get NC_4 by three JOIN operations from $NC_3 \star NC_3$, $NC_3 \star FI_3$ (3-itemsets of FI), and $FI_3 \star FI_3$. Again, result from the third JOIN can be deleted if the itemsets are present in KI. The union of the three JOINs results in NC_4. The process goes on until no itemsets can be generated. Finally, at the sixth sub-step, we scan the database to find frequent itemsets from $\cup_k NC_k$ and add these frequent itemsets to FI.

In this example, NC_2 is {BD, DE, DF, EF} ($CF_2 - KI$). The first JOIN operation, $NC_2 \star NC_2$, generates {DEF}. The second JOIN operation, $NC_2 \star FI_2$, generates {BDE}. The third JOIN operation, $FI_2 \star FI_2$, generates nothing. So NC_3 is {BDE, DEF}. Again we use NC_3 to generate NC_4 by the three JOIN operations. Because the three JOIN operations generate nothing for NC_4, we stop the generation. At last, we scan the database and find new frequent itemsets ({BE, DE}) from the union of NC_2 and NC_3. Compare the original SWF algorithm with the extension version, we can find that the original SWF algorithm needs to scan database for thirteen itemsets, {A, B, C, D, E, F, BD, BE, DE, DF, EF, BDE, DEF}, but the proposed algorithm only needs to scan database for eight itemsets, {D, F, BD, DE, DF, EF, BDE, DEF}.

In addition to reuse previous large itemsets $\cup_k F_k$ as KI, when the size of candidate itemsets $\cup_k C'_k$ from $db^{1,3}$ is close to the size of frequent itemsets $\cup_k F_k$ from $db^{1,3}$, KI can be set to candidate itemsets instead of frequent itemsets. This way, it can further speed up the runtime of the algorithm.

3.1   SWF with Known Itemsets

The proposed algorithm, KI_SWF (SWF with known itemsets), employs the preprocessing procedure of SWF (omitted here) and extend the incremental procedure as Figure 2. The change portion of the algorithm is presented in bold face. For ease exposition, the meanings of various symbols used are given in Table 1. At beginning, we load the cumulative filter CF_2 generated in preprocessing procedure (Step 7). From $\triangle^-$, we decrease the counts of itemsets in CF_2 and delete non-frequent itemsets from CF_2 with the filtering threshold $\lceil s * D^- \rceil$ (Step

Incremental Procedure of **KI_SWF**

1. Original databases = $db^{m,n}$;
2. New database= $db^{i,j}$;
3. Database removed $\triangle^- = \sum_{k=m,i-1} P_k$;
4. Database added $\triangle^+ = \sum_{k=n+1,j} P_k$;
5. $D^- = \sum_{k=i,n} P_k$;
6. $db^{i,j} = db^{m,n} - \triangle^- + \triangle^+$;
7. loading $C_2^{m,n}$ of $db^{m,n}$ into CF_2;
8. **begin for** $k = m$ **to** $i - 1$
9.   **begin for each** 2-itemset $I \in P_k$;
10.   **if**($I \in CF_2$ and $I.start \leq k$)
11. $I.count = I.count - N_{P_k}(I)$;
12. $I.start = k + 1$;
13.   **if** $(I.count < \lceil s * \sum_{t=i}^{n} |P_t| \rceil)$
14. $CF_2 = CF_2 - I$;
15.   **end**
16. **end**
17. **begin for** $k = n + 1$ **to** j
18.   **begin for each** 2-itemset $I \in P_k$
19.   **if**($I \notin CF_2$)
20. $I.count = N_{P_k}(I)$;
21. $I.start = k$;
22.   **if** $(I.count \geq \lceil s * |P_k| \rceil)$
23. $CF_2 = CF_2 \cup I$;
24.   **if** $(I \in CF_2)$
25. $I.count = I.count + N_{P_k}(I)$;
26.   **if** $(I.count \geq \lceil s * \sum_{t=i}^{k} |P_t| \rceil)$
27. $CF_2 = CF_2 - I$;
28.   **end**
29. **end**
30. **loading** $L_{2,h}^{m,n}$ of $db^{m,n}$ into KI;
31. **begin for** $k = m$ **to** $i - 1$
32.   **begin for each itemset** $I \in P_k$
33.   **if** $(I \in KI)$
34. $I.count = I.count - N_{P_k}(I)$;
35.   **end**
36. **end**
37. **begin for** $k = n+1$ **to** j
38.   **begin for each itemset** $I \in P_k$
39.   **if** $(I \in KI)$
40. $I.count = I.count + N_{P_k}(I)$;
41.   **end**
42. **end**;

43. **set** $FI = \emptyset$;
44. **begin for each itemset** $I \in KI$
45.   **if** $(I.count \geq \lceil s * |db^{i,j}| \rceil)$
46. $FI = FI \cup I$;
47. **end**
48. **set** $NC_2 = CF_2 - KI$;
49. $h = 2$;
50. **begin while**($NC_h \neq \emptyset$)
51. $NC_{h+1} = $**New_Candidate**
             (NC_h, FI_h, KI_h);
52. $h = h + 1$;
53. **end**
54. **refresh** $I.count = 0$ **where** $I \in NC_{2,h}$;
55. **begin for** $k = i$ **to** j
56.   **for each itemset** $I \in NC_{2,h}$;
57. $I.count = I.count + N_{P_k}(I)$;
58. **end**

59. **begin for each itemset** $I \in NC_{2,h}$;
60.   **if** $(I.count \geq \lceil s * |db^{i,j}| \rceil)$
61. $FI = FI \cup I$;
62. **end**
63. **return** FI;

Procedure of **New_Candidate**(P_h, Q_h, KI_h)
1. set $R_h = $JOIN$(P_k, Q_h, P_h \cup Q_h)$;
2. $R_h = R_h \cup$JOIN$(P_h, P_h, P_h \cup Q_h)$;
3. $R_h = R_h \cup ($JOIN$(Q_h, Q_h, P_h \cup Q_h) - KI_h)$;
4. **return** R_h;

Procedure of **JOIN**(P_h, Q_h, C_h)
1. set $R_h = \emptyset$;
2. for any two itemsets I_1 of P_h and I_2 of Q_h,
   where $I_1 = \{a_1, a_2, \ldots, a_h\}$ $I_2 = \{b_1, b_2, \ldots, b_h\}$
4. if $(a_1 = b_1 \wedge a_2 = b_2 \wedge \ldots \wedge a_{h-1} = b_{h-1} \wedge a_h > b_h)$
5.   set $I_3 = \{a_1, a_2, \ldots, a_h, b_h\}$;
6.   if (all h-itemsets of I_3 are in C_h)
7. $R_h = R_h \cup I_3$;
8. endfor
9. **return** R_h;

Fig. 2. Incremental procedure of Algorithm KI_SWF

8 to 16). Similarly, we increase the counts of itemsets in CF_2 by scanning $\triangle^+$ and delete non-frequent itemsets with filtering threshold $\lceil s * (D^- + \triangle^+) \rceil$ (Step 17 to 29). The resulting CF_2 is the set of candidate 2-itemsets corresponding to database $db^{i,j}$.

Continually, we load the frequent itemsets that was generated in previous mining task into KI. We check the frequency of every itemset from KI by scanning $\triangle^-$ and $\triangle^+$, and put those remaining frequent into FI (Step 30 to 47). Let NC_2 denotes the itemsets from $CF_2 - KI$, i.e. new 2-itemsets that are not frequent in the previous mining procedure and need to be checked by scanning $db^{i,j}$. From NC_2, we can generate new candidate 3-itemsets (NC_3) by JOIN 2-

Table 1. Meanings of symbols used in Figure 2

$db^{i,j}$	Database D from partition P_i to P_j
s	Minimum support threshold
$|P_k|$	Number of transactions in partition P_k
$N_{p_k}(I)$	Number of transactions in P_k that cantain itemset I
$C_k^{i,j}$	The candidate k-itemsets corresponding to $db^{i,j}$
$C_{2,h}^{i,j}$	The union of candidate k-itemset $C_k^{i,j}$ $(k = 2, \ldots, h)$ from $db^{i,j}$
$L_{2,h}^{i,j}$	The union of frequent k-itemset $L_k^{i,j}$ $(k = 2, \ldots, h)$ from $db^{i,j}$
FI_j	The j-itemsets of FI (frequent itemsets)
KI_j	The j-itemsets of KI (known itemsets)
NC_i	The i-itemsets of NC (new candidate itemsets)

itemsets from NC_2, FI_2, and KI_2. The function New_Candidate performs three JOIN operations: $NC_2 \star NC_2$, $NC_2 \star FI_2$, and $FI_2 \star FI_2$. However, if the itemsets are present in KI, they can be deleted since they have already been checked. The third operation is necessary since new candidates may contain supporting $(k - 1)$-itemsets for a k-itemset to become candidate. We unite the results of the three JOIN operations and get NC_3. Iteratively (Step 48 to 53), NC_4 can be obtained from NC_3, FI_3, and KI_3. Finally, we scan the database $db^{i,j}$ and count supports for itemsets in $\cup_{i=2,h} NC_i$ and put the frequent ones to FI (Step 54 to 63).

Sometimes, the size of candidate itemsets in the previous mining procedure is very close to the size of frequent itemsets. If we kept the support counts for all candidate itemsets, and load them to KI at Step 30 in the incremental procedure:

30. **Loading** $C_{2,h}^{m,n}$ **of** $db^{m,n}$ **into** KI;

The performance, including candidate generation and scanning, can further be speed up. In the next section, we will conduct experiment to compare the original SWF algorithm with the KI_SWF algorithms. If the itemsets of KI are the candidate itemsets of the previous mining task, we call this extension CI_SWF. If the itemsets of KI are the frequent itemsets of previous mining task, we call this extension FI_SWF.

4   Experiments

To evaluate the performance of the FI_SWF algorithm and the CI_SWF algorithm, we preformed several experiments on a computer with a CPU clock rate of 800 MHz and 768 MB of main memory. The transaction data resides in the FAT32 system and is stored on a 30GB IDE 3.5 drive. The simulation program was coded in C++. The method to generate synthetic transactions is similar to the ones used in [4,9,3]. The performance comparison of SWF, FI_SWF, and

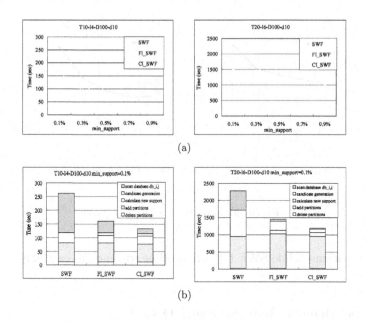

(a)

(b)

Fig. 3. Execution time vs. minimum support

CI_SWF is presented in Section 4.1. Section 4.2 shows the memory required and I/O cost for the three algorithms. Results on scaleup experiments are presented in Section 4.3.

4.1   Running Time

First, we test the speed of the three algorithms, FI_SWF, CI_SWF, and SWF. We use two different datasets, $T10-I4-D100-d10$ and $T20-I6-D100-d10$, and set N to 1000 and $|L|$ to 2000. Figure 3(a) shows the running time of the three algorithms as the minimum support threshold decreased from 1% to 0.1%. When the support threshold is high, the running times of three algorithms are similar, because there are only a limited number of frequent itemsets produced. When the support threshold is low, the difference in the running time becomes large. This is because when the support threshold is low (0.1%), there are a lot number of candidate itemsets produced. SWF scanned the total databases $db^{i,j}$ for all candidate itemsets. FI_SWF scanned $db^{i,j}$ for the candidate itemsets that are not frequent or not present in previous mining tasks. CI_SWF scanned $db^{i,j}$ only for the candidate itemsets that are not present in previous mining tasks. The details of the running time of three algorithms are shown in Figure 3(b). For SWF, the incremental procedure includes four phase: delete partitions, add partitions, generate candidates and scan database $db^{i,j}$. For FI_SWF and CI_SWF, they take one more phase to count new supports for known itemsets by scanning $\triangle^-$ and $\triangle^+$. The time for deleting or add partitions is about the same for

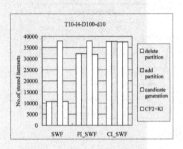 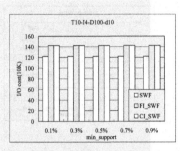

(a) Memory required at each phase   (b) I/O cost for three algorithms

Fig. 4. Memory required and I/O cost

three algorithms, while SWF takes more time to generate candidates and scan database.

4.2   The Memory Required and I/O Cost

Intuitively, FI_SWF and CI_SWF require more memory space than SWF because they need to load the itemsets of the previous mining task and CI_SWF needs more memory space than FI_SWF. However, the maximum memory required for these three algorithms are in fact the same. We record the number of itemsets kept in memory at running time as shown in Figure 4(a). The dataset used was $T10 - I4 - D100 - d10$ and the minimum support was set to 0.1%. At beginning, the memory space used by SWF algorithm was less than the others, but the memory space grew significantly at the candidate generation phase. The largest memory space required at candidate generation is the same as that used by CI_SWF and FI_SWF. In other words, SWF still needs memory space as large as CI_SWF and FI_SWF. The disk space required is also shown in Figure 4(a), where "$CF_2 + KI$" denotes the number of itemsets saved for later mining task. From this figure, CI_SWF require triple disk space of SWF. However, frequent itemsets are indeed by-product of previous mining task.

To evaluate the corresponding cost of disk I/O, we assume that each read of a transaction consumes one unit of I/O cost. We use $T10 - I4 - D100 - d10$ as the dataset in this experiment. In Figure 4(b), we can see that the I/O costs of FI_SWF and CI_SWF were a little higher than SWF. This is because that FI_SWF and CI_SWF must scan the changed portions of the database to update the known itemsets' supports.

4.3   Scaleup Test

Continuously, we test the scaleup ability of FI_SWF and CI_SWF with large databases. We conduct two experiments. In the first experiment, we use $T10 - I4 - DX - d(X * 0.1)$ as the dataset, where X represents the size of the databases

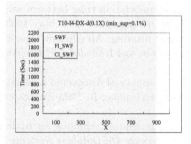 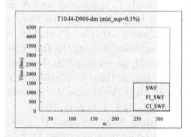

(a) Execution time vs. the database size (b) Execution time vs. the changed portions

Fig. 5. Scalable test

and the changed portions are 10 percents of the original databases. We set $X = 300, 600, 900$ with min_support=0.1%. The result of the experiment is shown in Figure 5(a). We find the the response time of these two algorithms were linear to the size of the databases. In the second experiment, we use $T10 - I4 - D900 - dm$ as the dataset and set min_support to 0.1%. The value of m is changed from 50 to 300. The result of the this experiment is shown in Figure 5(b). As $\triangle^+$ and $\triangle^-$ became larger, it requires more time to scan the changed partitions of the dataset to generate CF_2 and calculate the new supports. From this two experiments, we can say that FI_SWF and CI_SWF have the ability of dealing with large databases.

5   Conclusion

In this paper, we extend the sliding-window-filtering technique and focus on how to use the result of the previous mining task to improve the response time. Two algorithms are proposed for incremental mining of frequent itemsets from updated transaction database. The first algorithm FI_SWF (SWF with Frequent Itemset) reuse the frequent itemsets (and the counts) of previous mining task as FUP2 to reduce the number of new candidate itemsets that have to be checked. The second algorithm CI_SWF (SWF with Candidate Itemset) reuses the candidate itemsets (and the counts) from the previously mining task. From the experiments, it shows that the new incremental algorithm is significantly faster than SWF. In addition, the need for more disk space to store the previously discovered knowledge does not increase the maximum memory required during the execution time.

References

1. R. Agarwal, C. Aggarwal, and V.V.V Prasad. A tree projecction algorithm for generation of frequent itemsets. *Journal of Parallel and Distributed Computing (Special Issue on High Performance Data Mining)*, 2000.

2. R. Agrawal, T. Imielinski, and A. Swami. Mining association rules between sets of items in large databases. In *Proc. of ACMSIGMOD*, pages 207–216, May 1993.
3. M.-S. Chen C.-H. Lee, C.-R. Lin. Sliding-window filtering: An efficient algorithm for incremental mining. In *Intl. Conf. on Information and Knowledge Management (CIKM01)*, pages 263–270, November 2001.
4. D. Cheung, J. Han, V. Ng, and C.Y. Wong. Maintenance of discovered association rules in large databases: An incremental updating technique. In *Data Engineering*, pages 106–114, February 1996.
5. D. Cheung, S.D. Lee, and B. Kao. A general incremental technique for updaing discovered association rules. In *Intercational Conference On Database Systems For Advanced Applications*, April 1997.
6. J. Han and J. Pei. Mining frequent patterns by pattern-growth: Methodology and implications. In *ACM SIGKDD Explorations (Special Issue on Scaleble Data Mining Algorithms)*, December 2000.
7. J. Han, J. Pei, B. Mortazavi-Asl, Q. Chen, U. Dayal, and M.-C. Hsu. Freespan: Frequent parrten-projected sequential pattern mining. In *Knowledge Discovery and DAta Mining*, pages 355–359, August 2000.
8. J.-L. Lin and M.H. Dumham. Mining association rules: Anti-skew algorithms. In *Proc. of 1998 Int'l Conf. on Data Engineering*, pages 486–493, 1998.
9. J.-S Park, M.-S Chen, and P.S Yu. Using a hash-based method with transaction trimming for minng association rules. In *IEEE Transactions on Knowledge and Data Engineering*, volume 9, pages 813–825, October 1997.
10. J. Pei, J. Han, and L.V.S. Lakshmanan. Mining frequent itemsets with convertible constraints. In *Proc. of 2001 Int. Conf. on Data Engineering*, 2001.
11. A. Savasere, E. Omiecinski, and S. Navathe. An efficient algorithm for mining association rules in large databases. In *Proc. of the 21th International Conference on Very Large Data Bases*, pages 432–444, September 1995.
12. A. Veloso, B. Possas, W. Meira Jr., and M. B. de Carvalho. Knowledge management in association rule mining. In *Integrating Data Mining and Knowledge Management*, 2001.
13. Z. Zhou and C.I. Ezeife. A low-scan incremental association rule maintenance method. In *Proc. of the 14th Conadian Conference on Artificial Intelligence*, June 2001.

Reducing Rule Covers with Deterministic Error Bounds*

Vikram Pudi and Jayant R. Haritsa

Database Systems Lab, SERC
Indian Institute of Science
Bangalore 560012, India
{vikram, haritsa}@dsl.serc.iisc.ernet.in

Abstract. The output of boolean association rule mining algorithms is often too large for manual examination. For dense datasets, it is often impractical to even generate all frequent itemsets. The closed itemset approach handles this information overload by pruning "uninteresting" rules following the observation that most rules can be derived from other rules. In this paper, we propose a new framework, namely, the generalized closed (or g-closed) itemset framework. By allowing for a small tolerance in the accuracy of itemset supports, we show that the number of such redundant rules is far more than what was previously estimated. Our scheme can be integrated into both levelwise algorithms (Apriori) and two-pass algorithms (ARMOR). We evaluate its performance by measuring the reduction in output size as well as in response time. Our experiments show that incorporating g-closed itemsets provides significant performance improvements on a variety of databases.

1 Introduction

The output of boolean association rule mining algorithms is often too large for manual examination. For dense datasets, it is often impractical to even generate all frequent itemsets. Among recent approaches [16,15,9,6,8,5,4] to manage this gigantic output, the *closed itemset* approach [16,15] is attractive in that both the identities and supports of all frequent itemsets can be derived *completely* from the frequent closed itemsets. However, the usefulness of this approach critically depends on the presence of frequent itemsets that have supersets with *exactly the same support*. This causes the closed itemset approach to be sensitive to *noise* in the database – even minor changes can result in a significant increase in the number of frequent closed itemsets. For example, adding a select 5% transactions to the mushroom dataset (from the UC Irvine Repository) caused the number of closed frequent itemsets at a support threshold of 20% to increase from 1,390 to 15,541 – a factor of 11 times! The selection of transactions was made so as to break exact equalities in itemset supports.

In order to overcome this limitation, we propose in this paper the *generalized closed (or g-closed) itemset framework*, which is more robust to the database contents. In our scheme, although we do not output the *exact* supports of frequent itemsets, we estimate the supports of frequent itemsets within a *deterministic*, user-specified "tolerance" factor.

---

* A poster of this paper appeared in Proc. of IEEE Intl. Conf. on Data Engineering (ICDE), March 2003, Bangalore, India.

K.-Y. Whang, J. Jeon, K. Shim, J. Srivatava (Eds.): PAKDD 2003, LNAI 2637, pp. 313–324, 2003.

A side-effect of allowing for a tolerance in itemset supports is that the supports of some "borderline" infrequent itemsets may be over-estimated causing them to be incorrectly identified as frequent. Since our typical tolerance factors are much less than the minimum support threshold, this is not a major issue. Further, an extra (quick) database pass can always be made to check these borderline cases.

We provide theoretical arguments to show why the g-closed itemset scheme works and substantiate these observations with experimental evidence. Our experiments were run on a variety of databases, both real and synthetic, as well as sparse and dense. Our experimental results show that even for very small tolerances, we produce *exponentially fewer rules* for most datasets and support specifications than the closed itemsets, which are themselves much fewer than the total number of frequent itemsets.

Our scheme can be used in one of two ways: (1) as a post-processing step of the mining process (like in [9,6]), or (2) as an integrated solution (like in [17,11]). We show that our scheme can be integrated into both levelwise algorithms as well as the more recent two-pass mining algorithms. We chose the classical Apriori algorithm [2] as a representative of the levelwise algorithms and the recently-proposed ARMOR [14], as a representative of the class of two-pass mining algorithms. Integration into Apriori yields a new algorithm, g-**Apriori** and into ARMOR, yields g-**ARMOR**. Our experimental results show that these integrations often result in a significant reduction in response-time, especially for dense datasets.

We note that integration of our scheme into two-pass mining algorithms is a novel and important contribution because two-pass algorithms have several advantages over Apriori-like levelwise algorithms. These include: (1) significantly less I/O cost, (2) significantly better overall performance as shown in [12,14], and (3) the ability to provide approximate supports of frequent itemsets at the end of the first pass itself, as in [7,14]. This ability is an essential requirement for mining *data streams* [10] as it is infeasible to perform more than one pass over the complete stream.

2   Generalized Closed Itemsets

In addition to the standard boolean association rule mining inputs ($\mathcal{I}$, the set of database columns, $\mathcal{D}$, the set of database rows and $minsup$, the minimum support), the frequent g-closed itemset mining problem also takes as input ϵ, the user-specified tolerance factor. It outputs a set of itemsets (that we refer to as the frequent g-closed itemsets) and their supports. The frequent g-closed itemsets are required to satisfy the following properties: (1) The supports of all frequent itemsets can be derived from the output within an error of ϵ. (2) If $\epsilon = 0$, the output is precisely the frequent closed itemsets.

Note that the set of g-closed itemsets may not be unique for a given database and mining parameters. We are interested in obtaining any set of itemsets that satisfies the above properties since that would ensure that the frequent itemsets and their supports can be estimated sufficiently accurately.

2.1 ϵ-Equal Support Pruning

The key concept in the g-closed itemset framework lies in the *generalized openness propagation property*, which is stated as a corollary to the following theorem[1]. Here, the supports of itemsets X and Y are said to be approximately equal or ϵ-equal (denoted as $support(X) \approx support(Y)$) iff $|support(X) - support(Y)| \leq \epsilon$. Also, we refer to the allowable error in itemset counts as *tolerance count*. The term "tolerance" is reserved for the allowable error in itemset supports and is equal to the tolerance count normalized by the database size.

Theorem 1. *If Y and Z are supersets of itemset X, then $support(Z) - support(Y \cup Z) \leq support(X) - support(Y)$.*

Corollary 1 *If X and Y are itemsets such that $Y \supseteq X$ and $support(X) \approx support(Y)$, then for every itemset $Z : Z \supseteq X$, $support(Z) \approx support(Y \cup Z)$.*

This result suggests a general technique to incorporate into mining algorithms, which we refer to as ϵ-*equal support pruning*: If an itemset X has an immediate superset Y, with ϵ-equal support, then prune Y and avoid generating any candidates that are supersets of Y. The support of any of these pruned itemsets, say W, will be ϵ-equal to one of its subsets, $(W - Y) \cup X$. The remaining unpruned itemsets are referred to as *generators*. Note that this is a generalization of the notion of generators proposed in [11] and would reduce to it when $\epsilon = 0$.

2.2 Generating g-Closed Itemsets When $\epsilon = 0$

We now present a simple technique to generate closed itemsets from their generators. This technique does not involve an additional database scan as is required in the A-Close algorithm [11] for generating frequent closed itemsets. Also, the technique will directly carry over to the g-closed itemset case. For any closed itemset Y with generator X, the following theorem enables us to determine $Y - X$ using information that could be gathered while performing ϵ-equal support pruning. We refer to $Y - X$ as $X.pruned$.

Theorem 2. *Let Y be a closed itemset and $X \subseteq Y$ be the generator of Y. Then for each item A in $Y - X$, $X \cup \{A\}$ would be pruned by the equal support pruning technique. No other immediate supersets of X would be pruned by the same technique.*

Therefore, in order to generate a closed itemset Y from its generator X, it is sufficient to compute $X.pruned$ while performing ϵ-equal support pruning. Note that if some subset W of X had a proper subset V with equal support, then W would be pruned using the ϵ-equal support pruning technique. It would then be necessary to include all the items in $V.pruned$ in $X.pruned$. That is, the *pruned* value of any itemset needs to be propagated to all its supersets.

---

[1] Proofs of theorems are available in [13]

2.3   Generating g-Closed Itemsets When $\epsilon > 0$

The technique outlined above for $\epsilon = 0$ will not produce correct results when $epsilon >$ 0 – the supports of all frequent itemsets will not be derivable even approximately from the output. This is because Corollary 1 considers for any itemset X, only *one* superset Y with ϵ-equal support. If X has more than one superset (say $Y_1, Y_2, \ldots, Y_n$) with ϵ-equal support then *approximation error accumulates*. A naive interpretation of the generalized openness propagation property would seem to indicate the following: Every itemset $Z : Z \supset X \bigwedge Z \not\supseteq Y_k, k = 1 \ldots n$, also has a proper superset $\bigcup_{k=1}^{n} Y_k \cup Z$ with ϵ-equal support. Although this is valid when $\epsilon = 0$, in the general case, it is not necessarily true. However, the following theorem reveals an upper bound on the difference between the supports of $\bigcup_{k=1}^{n} Y_k \cup Z$ and Z.

Theorem 3. *If* $Y_1, Y_2, \ldots, Y_n, Z$ *are supersets of itemset* X, *then*
$$support(Z) - support(\bigcup_{k=1}^{n} Y_k \cup Z) \leq \sum_{k=1}^{n}(support(X) - support(Y_k)).$$

In our approach we solve the problem of approximation error accumulation by ensuring that an itemset is pruned using the ϵ-equal support pruning technique only if the *maximum* possible cumulative error in approximation does not exceed ϵ. Whenever an itemset X, having more than one immediate superset $Y_1, Y_2, \ldots, Y_n$, with ϵ-equal support is encountered, we prune each superset Y_k only as long as the sum of the differences between the supports of each pruned superset and X is within tolerance.

While performing the above procedure, at any stage, the sum of the differences between the support counts of each pruned superset and X is denoted by $X.debt$. Recall from Section 2.2 that these pruned supersets are included in $X.pruned$. Since $X.pruned$ needs to be propagated to all unpruned supersets of X, it becomes necessary to propagate $X.debt$ as well.

For any itemset X, $X \cup X.pruned$ is referred to as its corresponding g-closed itemset and will have ϵ-equal support. This is a natural extension of the closed itemset concept because when $\epsilon = 0$, $X \cup X.pruned$ is the closed itemset corresponding to X.

2.4   Rule Generation

Given the frequent g-closed itemsets and their associated supports, it is possible to generate association rules with approximate supports and confidences. This is stated in the following theorem:

Theorem 4. *Given the* g-*closed itemsets and their associated supports, let* $\hat{c}$ *and* $\hat{s}$ *be the estimated confidence and support of a rule* $X_1 \longrightarrow X_2$, *and* c *and* s *be its actual confidence and support. Then,* $\hat{s} - \epsilon \leq s \leq \hat{s}$; *and* $\hat{c} \times \lambda \leq c \leq \hat{c}/\lambda$ *where* $\lambda = (1 - \epsilon/minsup)$.

Further, it has been shown earlier [15,16] that it suffices to consider rules among adjacent frequent closed itemsets in the itemset lattice since other rules can be inferred by transitivity. This result carries over to frequent g-closed itemsets.

3 Incorporation in Levelwise Algorithms

In the previous sections we presented the g-closed itemset framework and the theory supporting it. In this section we show that the framework can be integrated into levelwise algorithms. We chose the classical Apriori algorithm as a representative of the levelwise mining algorithms. Integration of our scheme into Apriori yields g-Apriori, an algorithm for mining frequent g-closed itemsets.

The g-Apriori algorithm is obtained by combining the ϵ-equal support pruning technique described in Section 2.1 with the subset-based pruning of Apriori. The pseudo-code of the g-Apriori algorithm is shown in Figure 1 and works as follows: The code between lines 1–9 of the algorithm, excluding lines 6 and 7, consists of the classical Apriori algorithm. The SupportCount function (line 4) takes a set of itemsets as input and determines their counts over the database by making one scan over it.

Every itemset X in C_k (the set of candidate k-itemsets), G_k (frequent k-generators) and G (the frequent generators produced so far) has an associated counter, $X.count$, to store its support count during algorithm execution. Every itemset X in G has two fields in addition to its counter: (1) $X.pruned$ (described in Section 2.2). (2) $X.debt$: an integer value to check the accumulation of approximation error in itemset supports.

The Prune function is applied on G_k (line 6) before the $(k+1)$-candidates C_{k+1} are generated from it using AprioriGen (line 8). Its responsibility is to perform ϵ-equal support pruning while ensuring that approximation error in the supports of itemsets is not accumulated. The pseudo-code for this function is shown in Figure 2 and it performs the following task: it removes any itemset X from G_k if X has a subset Y with ϵ-equal count, provided $Y.debt$ remains within tolerance.

The code in lines 1–9, excluding line 7, is analogous to the A-Close algorithm [11] for generating frequent closed itemsets. At the beginning of line 10, G would contain the equivalent of the "generators" of the A-Close algorithm.

The PropagatePruned function is applied on G_k (line 7) and it ensures that the $pruned$ value of each itemset X in G_k is appended with the $pruned$ values of each immediate subset of X. The pseudo-code for this function is shown in Figure 3. The necessity for performing this function was explained in Section 2.2, where we showed that the $pruned$ value of an itemset should be propagated to all its supersets.

Finally, in lines 10–11, the g-closed itemsets are output.

4 Incorporation in Two Pass Algorithms

In this section we show that the g-closed framework can be incorporated into two-pass mining algorithms. As mentioned in the Introduction, this is a novel and important contribution because two pass algorithms are typically much faster than level-wise algorithms and also because they can be tweaked to work on data streams [10]. We selected the recently-proposed ARMOR [14] as a representative of the class of two-pass mining algorithms. Integration of the g-closed framework into ARMOR yields g-ARMOR, a two-pass algorithm for mining frequent g-closed itemsets.

g-Apriori $(\mathcal{D}, I, minsup, tol)$
Input: Database $\mathcal{D}$, Set of Items I, Minimum Support $minsup$, Tolerance Count tol
Output: Generalized Closed Itemsets
1. C_1 = set of all 1-itemsets;
2. $G = \phi$;
3. **for** $(k = 1; |C_k| > 0; k + +)$
4.   SupportCount$(C_k, \mathcal{D})$; // Count supports of C_k over $\mathcal{D}$
5. G_k = Frequent itemsets in C_k
6.   Prune(G_k, G, tol);
7.   PropagatePruned(G_k, G, tol);
8. C_{k+1} = AprioriGen(G_k);
9. $G = G \cup G_k$;
10. **for each** itemset X in G
11.   **Output** $(X \cup X.pruned, X.count)$;

Fig. 1. The g-Apriori Algorithm

Prune (G_k, G, tol)
Input: Frequent k-itemsets G_k, Generators G, Tolerance Count tol
Output: Remove non-generators from G_k
1. **for each** itemset X in G_k
2.   **for each** $(|X| - 1)$-subset Y of X, in G
3. $debt = Y.count - X.count$;
4.   **if** $(debt + Y.debt \leq tol)$
5. $G_k = G_k - \{X\}$
6. $Y.pruned = Y.pruned \cup (X - Y)$
7. $Y.debt \mathrel{+}= debt$

Fig. 2. Pruning Non-generators from G_k

PropagatePruned (G_k, G, tol)
Input: Frequent k-itemsets G_k, Generators G, Tolerance Count tol
Output: Propagate pruned value to generators in G_k
1. **for each** itemset X in G_k
2.   **for each** $(|X| - 1)$-subset Y of X, in G
3.   **if** $(X.debt + Y.debt \leq tol)$
4. $X.pruned = X.pruned \cup Y.pruned$
5. $X.debt\mathrel{+} = Y.debt$

Fig. 3. Propagate Pruned Value to Supersets

4.1   The ARMOR Algorithm

We first review the overall structure of the ARMOR algorithm – its details are available in [14]. In this algorithm, the database is conceptually partitioned into disjoint blocks and data is read from disk and processed partition by partition.

In the first pass, the algorithm starts with the set of all 1-itemsets as candidates. After processing each partition, the set of candidates (denoted as C) is updated – new candidates may be inserted and existing ones removed. The algorithm ensures that at any stage, if d is the database scanned so far, then the frequent itemsets within d (also called d-frequent itemsets) are available. The algorithm also maintains the *partial counts* of these itemsets – the partial count of an itemset is its count within the database scanned so far from the point it has been inserted into C.

In the second pass, complete counts of the candidates obtained at the end of the first pass are determined. During this pass, there are no new insertions into C. However, candidates that can no longer become frequent are removed at each stage.

4.2 Details of Incorporation

In ARMOR, the complete supports of candidate itemsets are not available during the first pass. However, for each candidate X, its *partial support* is available representing its support over the portion of the database that has been processed so far starting from the point where the candidate was inserted into C. The rule that we follow in the integrating the g-closed itemset framework is simple[2]: While processing a partition during the first pass, if we find the partial support of an itemset X to be ϵ-equal to that of its superset Y, then prune every proper superset of Y from C while ensuring that the approximation error does not accumulate beyond the tolerance limit. That is, whenever an itemset X, that has more than one immediate superset $Y_1, Y_2, \ldots, Y_n$, with ϵ-equal partial support is encountered, we prune each superset Y_k only as long as the sum of the differences between the partial supports of each pruned superset and X is within ϵ.

Next, if after processing a few more partitions, the supports of X and Y are no longer ϵ-equal, then regenerate the pruned supersets of Y as follows: For every d-frequent itemset Z in C such that $Z \supset X \wedge Z \not\supseteq Y$, insert a new candidate $Y \cup Z$ into C. The partial support of the new candidate should be set equal to that of Z.

5   Performance Study

In the previous sections, we have described the g-closed itemset framework along with the g-Apriori and g-ARMOR algorithms. We have conducted a detailed study to assess the utility of the framework in reducing both the output size and the response time of mining operations. Our experiments cover a range of databases and mining workloads including the real datasets from the UC Irvine Machine Learning Database Repository, the synthetic datasets from the IBM Almaden generator, and the real dataset, **BMS-WebView-1** from Blue Martini Software. Due to lack of space, we show only representative samples of our results. The complete results are available in [13].

All the algorithms were coded in C++ and the experiments were conducted on a 700-MHz Pentium III workstation running Red Hat Linux 6.2, configured with 512 MB main memory and a local 18 GB SCSI 10000 rpm disk. The same data-structures (hashtrees [2]) and optimizations (using arrays to store itemset counters in the first two database passes) were used in both g-Apriori and Apriori to ensure fairness. We chose tolerance count values ranging from zero (corresponding to the exact closed itemset case)

---

[2] The resulting pseudo-code is not shown here due to lack of space, but is available in [13]

to 1000. While higher values of tolerance are uninteresting, their inclusion is useful in studying the effect of increasing tolerance on the output size.

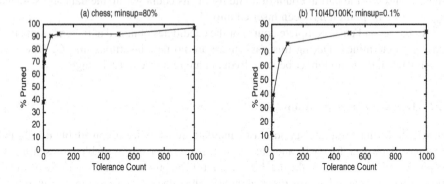

Fig. 4. Output Size Reduction

5.1 Experiment 1: Output Size Reduction

We now report on our experimental results. In our first experiment, we measure the output size reduction obtained with the g-closed itemset framework as the percentage of frequent itemsets pruned to result in frequent g-closed itemsets. The results of this experiment are shown in Figures 4a–b. The x-axis in these graphs represents the tolerance count values, while the y-axis represents the percentage of frequent itemsets pruned.

In these graphs, we first see that the pruning achieved is significant in most cases. For example, on the chess dataset (Figure 4a) for a minimum support of 80%, the percentage of pruned itemsets is only 38% at zero tolerance (closed itemset case). For the same example, at a tolerance count of 50 (corresponding to a maximum error of 1.5% in itemset supports), *the percentage of pruned itemsets increases to 90%!*

The pruning achieved is significant even on the sparse datasets generated by the IBM Almaden generator. For example, on the T10I4D100K dataset (Figure 4b) for a minimum support of 0.1%, the percentage of pruned itemsets is only 12% at zero tolerance, whereas it increases to 41.5% at a tolerance count of 10 (corresponding to a maximum error of 0.01% in itemset supports).

An interesting trend that we notice in all cases is that the percentage of pruned itemsets increases dramatically at low tolerances and then plateaus as the tolerance is increased further. This trend is significant as it indicates that the maximum benefit attainable using the g-closed itemset framework is obtained at low tolerances. The reason for this trend is as follows:

As the length of an itemset X increases, its subsets increase exponentially in number. This means that the chance of one of the subsets being ϵ-equal to one of its subsets becomes exponentially high. Hence most of the long generators get pruned at low tolerances itself. This accounts for the initial steep rise in the curve. Shorter generators get pruned at a slower pace with regard to the increase in tolerance. This accounts for the gradual upward slope in the curve after the initial exponential increase.

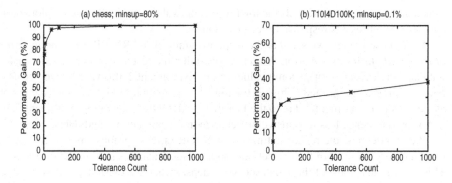

Fig. 5. Response Time Reduction

5.2   Experiment 2: Response Time Reduction

In our second experiment, we measure the performance gain obtained from the g-closed itemset framework. This is measured as the percentage reduction in response time of g-Apriori over Apriori. The results of this experiment are shown in Figures 5a–b. The x-axis in these graphs represents the tolerance count values, while the y-axis represents the performance gain of g-Apriori over Apriori.

In all these graphs, we see that the performance gain of g-Apriori over Apriori is significant. In fact, the curves follow the same trend as in Experiment 1. This is expected because the bottleneck in Apriori (and other frequent itemsets mining algorithms) lies in the counting of the supports of candidates. Hence any improvement in pruning would result in a corresponding reduction in response-time.

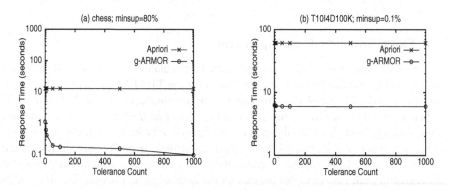

Fig. 6. Response Times of g-ARMOR

5.3   Experiment 3: Response Times of g-ARMOR

In our third experiment, we measure the response times of g-ARMOR and compare them against those of Apriori. The results of this experiment are shown in Figures 6a–b. The x-axis in these graphs represents the tolerance count values, while the y-axis (plotted on a **log-scale**) represents response times in seconds.

In all these graphs, we see that the response times of g-ARMOR are over an *order of magnitude* faster than Apriori. We also notice that the response times become faster with an increase in tolerance count values. As in Experiment 2, this is expected because more candidates are pruned at higher tolerances. The reduction in response time is not as steep as in Experiment 2 due to the fact that g-ARMOR is much more efficient than g-Apriori and hence less responsive to a change in the number of candidates.

We do not show the response times of ARMOR in these graphs since it ran out of main memory for most of the datasets and support specifications used in our evaluation. This was because most of these datasets were dense, whereas ARMOR, as described in [14], is designed only for sparse datasets and is memory intensive.

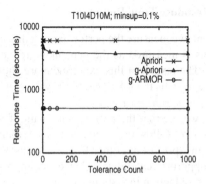

Fig. 7. Scale-up Experiment

5.4   Experiment 4: Scale-up Experiment

In our fourth (and final) experiment, we studied the scalability of g-ARMOR, g-Apriori and Apriori by measuring their response times for the T10I4D10M database having 10 million records. The results of this experiment are shown in Figure 7. These results (when analyzed along with Figures 5b and 6b) show that the performances of all three algorithms are linear w.r.t. database size. This behaviour is due to the following reasons: (1) The number of database passes in these algorithms depends only on the pattern density and not on the number of transactions. (2) The rate at which transactions are processed in each pass does not depend on the number of transactions but only on the distribution from which the transactions are derived, the number of candidate itemsets and on the efficiency of the data-structure holding the counters of candidates.

6   Related Work

A number of *post-mining schemes* to discover "redundancy" in association rules have been proposed – [1,3,6,9]. These schemes are to be applied *after* frequent itemsets have been mined. They are therefore inefficient and sometimes even infeasible because the number of frequent itemsets could be very large, especially for dense databases.

Techniques for pruning uninteresting rules during mining have been previously presented in [16,15,9,6,8,5,4]. In most of these studies (other than those following the closed itemset approach), it is sufficient for a rule to be considered uninteresting or redundant if it has no additional predictive power over another rule with fewer items. Techniques based on the closed itemset approach [16,15], on the other hand, have a tighter requirement for a rule to be considered redundant: A rule is redundant only if its identity and support can be derived from another "non-redundant" rule.

In this paper, we follow the tighter approach. However, as mentioned in the Introduction, we relax the requirement of deriving exact supports – instead, it is sufficient if the supports can be estimated within a deterministic user-specified tolerance factor. This strategy of relaxing the requirement of deriving exact supports has also been considered in [5,4]. In [5], the authors develop the notion of *freesets* along with an algorithm called *MINEX* to mine them. The bound on approximation error in the freesets approach increases linearly with itemset length in contrast to the constant bound featured in our approach. In [4], the authors do not provide any bounds on approximation error. Further, the focus in [5,4] is only on highly correlated, i.e. "dense" data sets, whereas we show that our techniques can be profitably applied even on sparse data sets. Finally, there was no attempt in [5,4] to incorporate their scheme into two-pass mining algorithms, which as mentioned earlier is essential for mining data streams.

The A-Close algorithm [11] for mining frequent closed itemsets is a levelwise algorithm based on Apriori. g-Apriori significantly differs from A-Close (even for the zero tolerance case) in that it does not require an additional database scan to mine closed itemsets from their respective generators. This is achieved by utilizing the technique described in Section 2.2. Our technique also bypasses the additional processing that is required in the MINEX [5] algorithm to test for "freeness".

7   Conclusions

In this paper we proposed the generalized closed itemset framework (or g-closed itemset framework) in order to manage the information overload produced as the output of frequent itemset mining algorithms. This framework provides an order of magnitude improvement over the earlier closed itemset concept. This is achieved by relaxing the requirement for exact equality between the supports of itemsets and their supersets. Instead, our framework accepts that the supports of two itemsets are equal if the difference between their supports is within a user-specified tolerance factor. We also presented two algorithms – g-Apriori (based on the classical levelwise Apriori algorithm) and g-ARMOR (based on a recent two-pass mining algorithm) for mining the frequent g-closed itemsets. g-Apriori is shown to perform significantly better than Apriori solely because the frequent g-closed itemsets are much fewer than the frequent itemsets. Finally, g-ARMOR was shown to perform over an order of magnitude better than Apriori over all workloads used in our experimental evaluation.

Acknowledgements. This work was supported in part by a Swarnajayanti Fellowship from the Dept. of Science and Technology, Govt. of India, and a research grant from the Dept. of Bio-technology, Govt. of India.

References

1. C. Aggarwal and P. Yu. Online generation of association rules. In *Intl. Conf. on Data Engineering (ICDE)*, February 1998.
2. R. Agrawal and R. Srikant. Fast algorithms for mining association rules. In *Proc. of Intl. Conf. on Very Large Databases (VLDB)*, September 1994.
3. R. Bayardo, R. Agrawal, and D. Gunopulos. Constraint-based rule mining in large, dense databases. In *Intl. Conf. on Data Engineering (ICDE)*, February 1999.
4. J-F. Boulicaut and A. Bykowski. Frequent closures as a concise representation for binary data mining. In *Pacific-Asia Conference on Knowledge Discovery and Data Mining (PAKDD)*, April 2000.
5. J-F. Boulicaut, A. Bykowski, and C. Rigotti. Approximation of frequency queries by means of free-sets. In *European Conference on Principles and Practice of Knowledge Discovery in Databases (PKDD)*, September 2000.
6. G. Dong and J. Li. Interestingness of discovered association rules in terms of neighborhood-based unexpectedness. In *Pacific-Asia Conference on Knowledge Discovery and Data Mining (PAKDD)*, 1998.
7. C. Hidber. Online association rule mining. In *Proc. of ACM SIGMOD Intl. Conf. on Management of Data*, June 1999.
8. M. Klemettinen, H. Mannila, P. Ronkainen, H. Toivonen, and A.I. Verkamo. Finding interesting rules from large sets of discovered association rules. In *Intl. Conf. on Information and Knowledge Management (CIKM)*, November 1994.
9. B. Liu, W. Hsu, and Y. Ma. Pruning and summarizing the discovered association rules. In *Intl. Conf. on Knowledge Discovery and Data Mining (KDD)*, August 1999.
10. G. Manku and R. Motwani. Approximate frequency counts over streaming data. In *Proc. of Intl. Conf. on Very Large Databases (VLDB)*, August 2002.
11. N. Pasquier, Y. Bastide, R. Taouil, and L. Lakhal. Discovering frequent closed itemsets for association rules. In *Proc. of Intl. Conference on Database Theory (ICDT)*, January 1999.
12. J. Pei et al. H-mine: Hyper-structure mining of frequent patterns in large databases. In *Intl. Conf. on Data Mining (ICDM)*, December 2001.
13. V. Pudi and J. Haritsa. Generalized closed itemsets: Improving the conciseness of rule covers. Technical Report TR-2002-02, DSL, Indian Institute of Science, 2002.
14. V. Pudi and J. Haritsa. On the efficiency of association-rule mining algorithms. In *Pacific-Asia Conference on Knowledge Discovery and Data Mining (PAKDD)*, May 2002.
15. R. Taouil, N. Pasquier, Y. Bastide, and L. Lakhal. Mining basis for association rules using closed sets. In *Intl. Conf. on Data Engineering (ICDE)*, February 2000.
16. M. J. Zaki. Generating non-redundant association rules. In *Intl. Conf. on Knowledge Discovery and Data Mining (KDD)*, August 2000.
17. M. J. Zaki and C. Hsiao. Charm: An efficient algorithm for closed itemset mining. In *SIAM International Conference on Data Mining*, 2002.

Evolutionary Approach for Mining Association Rules on Dynamic Databases

P. Deepa Shenoy[1], K.G. Srinivasa[1], K.R. Venugopal[1], and L.M. Patnaik[2]

[1] University Visvesvaraya College of Engineering, Bangalore-560001.
{shenoypd@yahoo.com, vkrajuk@vsnl.com}
http://www.venugopalkr.com
[2] Microprocessor Application Laboratory, Indian Institute of Science ,
Bangalore-560012, India.
{lalit@micro.iisc.ernet.in}
http://www.geosites.com//ResearchTriangle/campus/8937/Lmp.html

Abstract. A large volume of transaction data is generated everyday in a number of applications. These dynamic data sets have immense potential for reflecting changes in customer behaviour patterns. One of the strategies of data mining is association rule discovery, which correlates the occurrence of certain attributes in the database leading to the identification of large data itemsets. This paper seeks to generate large itemsets in a dynamic transaction database using the principles of Genetic Algorithms. Intra transactions, Inter transactions and distributed transactions are considered for mining association rules. Further, we analyze the time complexities of single scan DMARG(Dynamic Mining of Association Rules using Genetic Algorithms), with Fast UPdate (FUP) algorithm for intra transactions and E-Apriori for inter transactions. Our study shows that DMARG outperforms both FUP and E-Apriori in terms of execution time and scalability, without compromising the quality or completeness of rules generated. The problem of mining association rules in the distributed environment is explored in DDMARG(Distributed and Dynamic Mining of Association Rules using Genetic Algorithms).

1 Introduction

Data mining attempts to extract valuable and hidden information from large databases ranging from gigabytes to terabytes. It finds wide applicability in the fields of market strategies, financial forecast and decision support. The tasks performed in data mining can be broadly categorized into clustering, discovery of association rules, classification rules and decision rules. Mining association rules is the process of discovering expressions of the form $X \Rightarrow Y$. For example, customers buy bread (Y) along with butter (X). These rules provide valuable insights to customer buying behaviour, vital to business analysis. New association rules, which reflect the changes in the customer buying pattern, are generated by mining the updations in the database and is called Dynamic mining [1].

K.-Y. Whang, J. Jeon, K. Shim, J. Srivatava (Eds.): PAKDD 2003, LNAI 2637, pp. 325–336, 2003.

1.1   Genetic Algorithms

Genetic algorithms(GA) is highly nonlinear, multifaceted search process which uses simple operators like mutation, crossover and reproduction. Recent trends indicate the application of GA in diverse fields ranging from character script recognition to cockpit ambience control [3]. It is an iterative technique which works with the solution set, called population, directly. The population of solutions undergo selection using operators of reproduction, crossover and mutation. The *fitness function,* a property that we want to maximise, is used to evaluate quality of solutions [4]. The genetic algorithm used in this paper is summarised below.

1. Create an initial population of schema *(M(0)),* where each member contains a single one in its pattern. For example, for transactions of length four, the initial population would contain 1*** , *1** , **1* , ***1.
2. Evaluate the fitness of each individual with respect to the schema. Fitness of each individual, *m,* in the current population *M(t)* is denoted by *u(m),* where
$$u(m) = \frac{number\ of\ matching\ bits - number\ of\ nonmatching\ bits}{Total\ number\ of\ fields}$$
3. Enter the schema into the Roulette wheel. The selection probabilities *p(m)* for each *m* in *M(t)* determines the size of each slot in the Roulette wheel.
4. Generate *M(t+1)* by probabilistically selecting individuals from *M(t)* to produce offspring via crossover. The offspring is a part of the next generation.
5. Repeat step (1) until no new schema are generated. Steps (2), (3) and (4) create a generation which constitutes a solution set.

1.2   Dynamic and Distributed Mining

Dynamic mining algorithms are proposed to handle updation of rules when increments or decrements to database occur. It should be done in a manner which is cost effective, without involving the database already mined and perimitting reuse of the knowledge mined earlier. The two major operations involved are (i) Additions: Increase in the support of appropriate itemsets and discovery of new itemsets. (ii) Deletions: Decrease in the support of existing large itemsets leading to the formation new large itemsets. The process of addition and deletion may result in the invalidation of certain existing rules.

With the unprecedented growth of data and inspite of the advancements in hard disk technology, storage in a centralized location is becoming a rare phenomenon. In some cases the database itself is inherently distributed. This demands computation in a distributed environment. Implementation of data mining in a high-performance parallel and distributed computing paradigm is thus becoming crucial. The proliferation of data in the recent years has made it impossible to store it in a single global server. Emerging technologies seek to resolve this problem by using a distributed system where several inter-connected hosts share the load. The problem of data mining now reduces to mining local databases and combining the results to form global knowledge. The concept of distributed data mining ensures system and database scalability as data continues to grow inexorably in size and complexity.

1.3 Association Rule Mining

Let $I = \{i_1, i_2... i_m\}$ be a set of literals called items. Let DB denote a set of transactions where each transaction T is a set of items such that T is a subset of I. Associated with each transaction there is a unique Transaction IDentifier (TID). A transaction T contains X, a set of some items in I, if $X \subseteq T$. An association rule is an implication of the form $X \Rightarrow Y$ where $X \subseteq I$, $Y \subseteq I$ and $X \cap Y = \phi$. The rule $X \Rightarrow Y$ holds in a transaction set DB with confidence c, if $c\%$ of transactions in DB that contain X also contain Y. The rule $X \Rightarrow Y$ has support s in the transaction set DB if $s\%$ of transactions contain $X \cup Y$. The process of discovering association rules can be split into two domains,

- Finding all itemsets with appreciable support(large itemsets).
- Generate the desired rules.

Rule $X \Rightarrow Y$ is generated if, support $(X \cup Y)$ / support $(X) \geq$ minimum confidence.

An inter transaction association describes the association among the items of different transactions, such as "if company A's stock price goes up on day 1, B's stock price will go down on day 2, but go up on day 4". Intra transactions can be treated as a special case of inter transaction association. Mining inter transaction associations poses more challenges than intra transaction associations as the number of potential association rules becomes extremely large. A frequent inter-transaction itemsets must be made up of frequent intra-transaction itemsets. It has wide applications in predicting stock market price movements.

Let $I = \{i_1, i_2... i_s\}$ be a set of literals called items. A transaction database DB is a set of transactions $\{t_1, t_2... i_n\}$, where t_i ($1 \leq i \leq m$) is a subset of I. Let $x_i = <v>$ and $x_j = <u>$ be two points in the one-dimensional space, then the relative distance between x_i and x_j is defined as $\Delta(x_i, x_j) = <u-v>$. A single dimensional space is represented by one dimensional attribute, whose domain is a finite subset of non-negetive integers. A single dimensional inter transaction association rule is an implication of the form $X \Rightarrow Y$, where $X \subset I_e$, $Y \subset I_e$ and $X \cap Y = \phi$. A single dimensional inter-transaction association rule can be expressed as $\Delta_0(a)$, $\Delta_1(c) \Rightarrow \Delta_3(e)$, which means if the price of stock a goes up today, and price of stock b increases next day, then most probably the price of stock e will increase on the fourth day.

The paper is organized into the following sections. Section 2 presents a review of related work. Section 3 addresses the problem of mining association rules using algorithms DMARG and DDMARG on dynamic and distributed databases. An example illustrating the algorithms is given in Section 4. Results are analysed in Section 5. Section 6 presents conclusions.

2   Related Work

The incremental mining of association rules using genetic algorithms was introduced in [1]. This algorithm was extended to mine the association rules on distributed databases [2]. Dynamic subspace clustering on high dimensional databases was discussed in[6]. Mining optimized association rules with respect to categorical and numerical attributes was introduced in [5]. The technique for discovering localized associations in segments of the data using clustering was discussed in [7]. Algorithm for generation of significant association rules between items in a large database is given in [8]. A graph based approach for association rule mining is introduced in [9]. The incremental updating technique presented in [10] is similar to Apriori algorithm. It operates iteratively and makes a complete scan of the current database in each iteration. At the end of kth iteration all the frequent itemsets of size-k are derived. Mining inter-transaction association rules was first introduced in [11], in which extended aprori and extended hash apriori were proposed to deal with single and multidimensional inter transaction association rules. Various parallel and distributed data mining techniques like count distribution, data distribution and candidate distribution algorithm, which effectively uses aggregate main memory to reduce synchronisation between the processors and has load balancing built into it, was discussed in [12].

3   Algorithms

Conventional association rule mining has focussed primarily on intra transaction association rules that are limited in scope. This has led to the development of the more powerful inter-transactional association rules. Inter transaction association rule offer insight into co-relations among items belonging to different transaction records. The classical association rules express associations among items purchased by a customer in a transaction i.e. associations among the items within the same transaction record. Inter transaction rules express the relations between items from different transactional records.

We consider a sliding window of some specific size (size is the number of transactions to be handled at a time) to generate inter-transaction association rules. Mining intra transaction association rules, is the special case of inter transaction association rule mining when the window size is one. The emergence of network based distributed computing environments such as the internet, private intranet and wireless networks coupled with the development of high performance data mining systems, has created a need for scalable techniques for harnessing data patterns from very large databases.

3.1   Problem Definition

Consider a large database DB with horizontal layout consisting of a finite set of transactions $\{t_1, t_2 \ldots t_m\}$. Each transaction is represented by a binary string of fixed length n, where n is the total number of items present. Let an incremental set of new transactions be added to DB. The objective is to generate all the large inter transaction association rules in $DB \cup db$ without involving DB using scan only once technique through an algorithm DMARG.

3.2 Assumptions

- A transaction is represented by a binary coded string.
- The items in the transaction are ordered.
- The transactions are generated by synthetic data generator.

3.3 Algorithm *DMARG*

1. Initialize database *DB*.
2. Read_Window_Size(Window_Size)
3. Prepare_Database(DB, Window_Size)
4. Initialize itemsets database *I*.
5. Read an itemset *i* from the incremental database *db*. Update those Itemsets in I which share one or more common transactions with i.
6. [*Discover new itemsets*]
   Repeat step 4.
   a) Apply Genetic Algorithm to members of *I*.
      i. Crossover parents with high fitness values.
      ii. Mutate certain itemsets to obtain a new itemset, if complete set not found.
   b) Update itemsets until complete set found.
7. If end of database *db* has not been reached, goto Step 3.
8. Identify large itemsets above minimum support.
9. Generate all rules above the minimum confidence.
10. If the incremental operation is to be repeated, goto Step 1.

Given n hosts $\{h_1, h_2, \ldots h_n\}$ where each h_i $(1 \leq i \leq n)$, contains a local database db_i, we mine db_i to discover large itemsets local to h_i and generate local association rules. The local large itemsets are combined to generate global rules. The local databases are horizontally partitioned. Databases are statically distributed across all the nodes of the homogeneous network.

3.4 Algorithm *DDMARG*

1. The server waits for the request from the clients.
2. Clients register themselves within the allotted time frame.
3. Server registers the clients which participate in mining process.
4. Server requests the clients to check for the integrity of the executables required for distributed mining.
5. Clients return status of association rules mined from the local database db_i.
6. Server waits for clients to perform local mining on the database (db_i) and places it in WAIT state.
7. Clients interrupt server when the local mining on db_i is completed and generates the local association rules using the algorithm *DMARG*.
8. Server requests the clients to send the entire rule chromosome mined.

9. Server builds the universal database of rule chromosomes after receiving the complete set of association rules mined in all the registered clients.
10. Server generates the global association rules from the result obtained from step 9.
11. Server closes connections with all the clients.
12. End.

4   Example

4.1   Intra Transaction Association Rule Mining

Consider an example transaction database given in Fig. 1. The first column represents the transaction identifier TID. The various items purchased in a particular transaction is given in column 2. The third column consists of the encoded binary string. As the number of items considered in the database is five($n = 5$), the length of the encoded transaction is also five.

TID	Transaction	Encoded Trans.
1	AD	10010
2	ABE	11001

Fig. 1. A transaction database

The algorithm scans the database one transaction at a time and tries to approximate or classify the data by the nearest itemset if not already present and update its support. Assuming, this is the first run of the code and the mode of operation is incremental, the initial population contains the following strings with individual supports as zero. The initial population, the itemsets and corresponding supports are shown in the Fig. 2.

String	Itemset	Supprot
1****	{A}	0
*1***	{B}	0
1	{C}	0
***1*	{D}	0
****1	{E}	0

Fig. 2. The Strings of intial population

When the first transaction record is read, it is compared with the existing strings in the itemset list and the fitness value of the string is computed with respect to the encoded transaction TID 1 as follows

$$\text{fitness} = \frac{No.\ of\ matching\ bits - No.\ of\ non\ matching\ bits}{No.\ of\ items\ in\ the\ transaction(n)}$$

If the fitness is one, then the support of the string in question is incremented. In Fig. 3 the fitness of the item A is one, therefore its support is incremented to 1.

All strings with fitness 1, are placed in a dynamiaclly created array which serves as a Roulette wheel. In the case of TID 1, the Roulette wheel would contain strings 1**** and ***1*(as shown in Fig. 3). A random crossover point (position3) is chosen and crossover operation is applied on the two parents strings 1**** and ***1* to obtain chidren 1**1* and *****. The null string is discarded as it serves no purpose. Since a perfect match(number of ones in the encoded transaction 10010 matching the number ones in the generated string 1**1*) has been reached, the process stops and the updated table with the string 1**1*. The procedure is repeated till all the transactions are covered in the database.

String	Itemset	Fitness	Support	No. of Matching bits = 5 No. of non–matching bits=0
1****	{A}	1	1	Fitness(10010,1****)=1 Fitness(10010,*1***)=0.6
*1***	{B}	0.6	0	Roulette Wheel(10010)
1	{C}	0.6	0	1****, ***1* Parent1 = 1**|**
***1*	{D}	1	1	Parent2 = ***|1*
****1	{E}	0.6	0	Child1 = 1**1* Child2 = *****

Fig. 3. Updations and reproduction by the transaction TID 1

4.2   Inter Transaction Association Rule Mining

When the window size is greater than one, then the algorithm generates inter transaction association rules. The database is prepared for mining inter transaction association rules as shown in Fig. 4. The major task in this phase is to organize the transactions. We consider equal length transactions and the window size is taken as three. The output of the *Prepare_Database*() is the augmented transaction T'. Now we run the algorithm DMARG on this augmented transaction T'. This data preparation step is not necessary in the case of intra transaction association rule mining as the window size is one.

TID	Transaction	Binary Encoded String	Augmented Trans T'
T_0	AB	110	110@101@111 $(T_0 \cup T_1 \cup T_2)$
T_1	AC	101	101@111@110 $(T_1 \cup T_2 \cup T_3)$
T_2	ABC	111	111@110@011 $(T_2 \cup T_3 \cup T_4)$
T_3	AB	110	
T_4	BC	011	

Fig. 4. A transaction database and corresponding augmented transactions

Since the rules generated on the augmented transaction database T' is explosive, only few rules of interest are considered in the Fig. 5. As the string length is three and the window size considered is also three, the first three bits in the augmented transaction T' belong to TID T_0, similarly second three bits belong TID T_1 and the last three bits belong to TID T_2 in the first row of Fig. 4.

Example Rules	Support	Rules generated
1** *** ***	3	A[0]
1 *** ***	2	B[0]
*** 1** ***	3	A[1]
11* *** ***	2	A[0]B[0]
1** 1** ***	3	A[0]A[1]
1 1** ***	2	B[0]A[1]
11* 1** ***	2	A[0]B[0] A[1]

Fig. 5. The inter transaction association rules

Considering the itemsets, A[0]B[0]A[1] and A[0]B[0] with support 2 in Fig. 5., we can conclude a valid inter transaction association rule i.e., A[0]B[0] $\Rightarrow$ A[1].

4.3   Distributed Mining of Association Rules

Consider the example of a transaction database present in two client nodes H_1 and H_2 as shown in Fig. 6. The genetic operators are applied on the database of each node to generate local large itemsets as shown in the Fig. 7.

	Node H_1			Node H_2	
TID	Transaction	Encoded Trans.	TID	Transaction	Encoded Trans.
100	AD	10010	103	BE	01001
101	ABE	11001	104	ABD	11010

Fig. 6. A transaction database in two nodes H_1 and H_2

	Node H_1			Node H_2	
List	Rule String	Sup	List	Rule String	Sup
1	1****	2	1	1****	1
2	*1***	1	2	*1***	2
3	***1*	1	3	***1*	1
4	****1	1	4	****1	1
5	1**1*	1	5	*1**1	1
6	1***1	1	6	11***	1
7	11***	1	7	1**1*	1
8	11**1	1	8	*1*1*	1
			9	11*1*	1

Fig. 7. Local rules generated at nodes H_1 and H_2

Given the local itemsets of H_1 and H_2 in Fig. 7, the next step is to obtain global list which is the aggregate of rule strings produced in all the hosts. This is the responsibility of the server. The server polls each of the clients on a round robin basis. Node H_1 is first prompted to send its rule string to the server. In reply, H_1 sends the first string in its list (i.e., 1****). The server now sends a copy of this rule string to all clients. These nodes are instructed to send the local count or support of the string, if it is present in their database. Thus, node H_1 will send the value 2 and node H_2,will send the value 1.

List	Rule String	Global Count	List	Rule String	Global Count
1	1****	3	7	1**1*	2
2	*1***	3	8	*1*1*	1
3	***1*	2	9	*1**1	2
4	****1	2	10	11**1	1
5	1***1	1	11	11*1*	1
6	11***	2			

Fig. 8. The global rules generated at server

The process is repeated for all the strings in the database of node H_1. Once node H_1 exhausts its list of strings, the server will now move to the next node, i.e. node H_2. Node H_2 will now search for strings from its database that have not been considered so far, for example, *1**1. It then sends the string to the server. The server will now gather the global support of this string by surveying the other nodes. The above algorithm is repeated until every node in the network has been scanned for local rules. The result is shown in Fig. 8.

5   Performance Analysis

Simulations were performed on a system of physical memory capacity 256 MB and a PentiumIII processor of speed 750 MHz. Transaction data, in the form of binary strings of desired length, was generated using a Synthetic Data Generator. The number of items in a transaction was 1000, the average size of each, 10, the maximum number of items, 25 and the average size of maximal potentially large itemsets was 4. DMARG was compared with a standard algorithm, FUP (Fast UPdate), which works on the principles of Apriori.

Fig. 9. shows the plot of the execution time of DMARG versus the number of transactions. The transaction databases were varied in the range of [100k-300k]. The *support* considered was 1. This low value has been explicitly chosen to exhaustively generate all the possible combinations of itemsets. The execution time varies from 2 minutes to 31minutes. To avoid losing any small itemsets which may become large in future, pruning is not employed. It is evident from Fig.9., that DMARG can handle large databases without compromising performance and has excellent scalability.

Fig. 10. compares the execution times of DMARG with the standard algorithm for incremental mining, FUP. The initial database *DB* was of size 100k.

The incremental database *db* was varied from 1% to 100% of *DB*. The execution time for DMARG was observed to vary from 2.6 seconds to 269.5 seconds. Under identical conditions, the execution time of FUP varied from 20.7 seconds to 861.4 seconds. It is obvious from Fig. 10, DMARG is faster than FUP. This difference in the time complexity between DMARG and FUP is due to the single scan feature of DMARG.

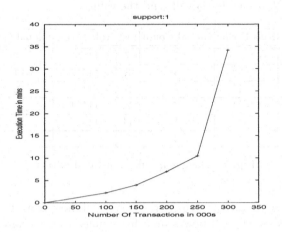

Fig. 9. Execution time of DMARG vs No. of transactioons (Intra transaction)

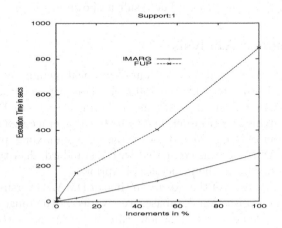

Fig. 10. Execution time vs Increment database (Intra transaction)

The variation of the execution time of DMARG for various decrements of *DB* is also studied. The initial database *DB* considered is of size 100k. The decrements were varied from 1% to 50% of *DB*. DMARG is versatile in handling both increments and decrements to *DB*. For inter transaction association rules the database of size of 1,00,000 transactions with 300 items and the window size of three is taken. The execution time of DMARG is compared with Extended-Apriori.

In case of distributed mining, simulations were performed on an Ethernet Local Area Network. Database is distributed uniformly amongst all the processors. Fig. 11. shows the variation of execution time in minutes versus the number processors. The execution time decreases as the increase in the number of processors. The time taken for the synchronization is negligible. Instead of synchronizing at every iteration, the synchronization is done at the final stage without pruning away redundant rules, hence the DDMARG consumes minimal time for synchronization. In our experiments, the database was kept constant and the number of processors were varied.

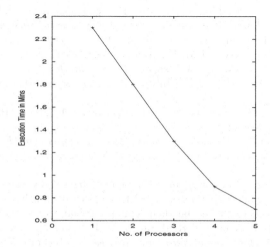

Fig. 11. Execution time vs No. of processors (Distributed mining)

6   Conclusions

We examine in this paper, the issue of mining intra and inter transaction association rules for dynamic databases. The algorithm proposed is effective in generating large itemsets and in addressing updations in the database. The rules generated using DMARG, are qualitatively sound. The single scan feature reduces computational costs significantly. The algorithm generates inter transaction association rules using the augmented transactions. The performance of DMARG is studied and compared with that of FUP(in case of intra transactions) and E-Apriori(in case of inter transactions), the results show that DMARG is faster and more efficient than FUP and E-Apriori. DMARG also takes care of decremental transactions. DMARG handles identical and skewed databases. The algorithms DMARG and DDMARG are efficient and effective; Efficient, because of the inherent speed and scalability of genetic algorithms in performing data mining and a low communication overhead achieved when building global knowledge. It is effective because, the quality of the rule chromosomes are not compromised in the process. Further, pruning is not done on the local string chromosomes set. This ensures that no rule is lost until the final filtering process on the server side. In DMARG, a single scan technique, the support is required only before the association rule generation, to build stronger and relevant rules. DDMARG at present, works on a homogeneous database. With certain modifications, i.e., the exchange of details regarding the properties of the database before the mining process, the code can cater to a heterogeneous database as well.

References

1. P. Deepa Shenoy, Srividya M, Kavitha K, Venugopal K. R and L. M Patnaik, *"Incremental Mining of Association rules using Genetic algorithm"*, Proc. Intl. Conf on High Performance Computing(Asia), Dec. 2002.
2. P. Deepa Shenoy, Srividya M, Kavitha K, Srinivasa K. G, Venugopal K. R and L. M. Patnaik, *"NetMiner: A Distributed Algorithm for Association Rule Mining using Genetic Algorithm"*, Proc. Intl. Conf on Agile Manufacturing(ICAM), India, Dec. 2002.
3. M. Srinivas and L.M Patnaik, *"Genetic Algorithms: A Survey"*, IEEE Computers, Vol. 27, No. 6, 1994, pp. 17–26 .
4. M. Srinivas and L.M Patnaik, *"Genetic Search, Analysis using Fitness moments"*, IEEE Transactions on Knowledge and Data Engineering, Vol. 8, No. 1, Feb 1996, pp. 120–133.
5. Rajiv Rastogi and Kyuseok Shim, *"Mining Optimized Association Rules with Categorical and Numeric Attributes"* , IEEE Transactions on Knowledge and Data Engineering, Vol. 14, No. 1, Jan/Feb 2002, pp. 29–50.
6. P. Deepa Shenoy, Venugopal K. R and L. M. Patnaik, *"Dynamic Subspace Clustering for High-Dimensional Very Large Databases"*, Technical Report, University Visvesvaraya College of Engineering, Bangalore University, June 2002.
7. Charu C. Aggarwal, Cecilia Procopiuc and Philip S, Yu, *"Finding Localized Associations in Market Basket Data"* , IEEE Transactions on Knowledge and Data Engineering, Vol. 14, No. 1, Jan/Feb 2002, pp. 51–62.
8. Rakesh Agrawal, Tomasz Imielinski and Arun Swami, *"Mining Association Rules between Sets of Items in Large Databases"*, Proc. of ACM SIGMOD Intl. Conf. on Management of Data, 1993, pp. 207–216.
9. Show-jane Yen and Arbee L. P Chen, *"A Graph-Based Approach for Discovering Various Types of Association Rules"*, IEEE Transactions on Knowledge and Data Engineering, Vol. 13, No. 5, Sep/Oct 2001, pp. 839–845.
10. David W Cheung, Jiawei Han, Vincent T. and Wong C Y. *"Maintenance of Discovered Association Rules in Large Databases: An Incremental Updating Technique"*. Proc. of the International Conference On Data Engineering, Feb. 1996, pp. 06–114.
11. Hongjun Lu, Ling Feng and Jaiwei Han, *"Beyond Intra-Transaction Association Analysis: Mining Multi-Dimensional Inter-Transaction Association Rules"*, ACM Transactions on Information Systems, 18(4), 2000, pp. 423–454.
12. M. J. Zaki. *"Parallel and distributed association mining: A survey"*, IEEE Concurrency: Special Issue on Parallel Mechanism for Data Mining. Vol 7, No. 4, 1999, pp. 14–25.

Position Coded Pre-order Linked WAP-Tree for Web Log Sequential Pattern Mining

Yi Lu and C.I. Ezeife*

School of Computer Science, University of Windsor,
Windsor, Ontario, Canada N9B 3P4
cezeife@uwindsor.ca

Abstract. Web access pattern tree algorithm mines web log access sequences by first storing the original web access sequence database on a prefix tree (WAP-tree). WAP-tree algorithm then mines frequent sequences from the WAP-tree by recursively re-constructing intermediate WAP-trees, starting with their suffix subsequences.

This paper proposes an efficient approach for using the preorder linked WAP-trees with binary position codes assigned to each node, to mine frequent sequences, which eliminates the need to engage in numerous re-construction of intermediate WAP-trees during mining. Experiments show huge performance advantages for sequential mining using prefix linked WAP-tree technique.

Keywords: Sequential Mining, Web Usage Mining, WAP-tree Mining, Pre-Order Linkage, Position Codes

1 Introduction

In sequential pattern mining, an order exists between the items (events) making up an itemset, called a sequence, and an item may re-occur in the same sequence. An example sequence of events that could represent a sequential pattern is the order that customers in a video rental store may follow in renting movie videos, such as: rent "star wars", then "Empire strikes back", and then "Return of Jedi". The measures of support and confidence, used in basic association rule mining for deciding frequent itemsets, are also used in sequential pattern mining to determine frequent sequences and strong rules generated from them.

Sequential pattern mining technique is useful for finding frequent web access patterns [8,2,5]. For example, the access information in a simplified web log format could stand for access web site represented as events {a, b, c, d, e, f}. The web log events are first pre-processed, to group them into sets of access sequences for each user identifier and to create web access sequences in the form of a transaction database for mining. The web log sequences in a transaction

---

* This research was supported by the Natural Science and Engineering Research Council (NSERC) of Canada under an operating grant (OGP-0194134) and a University of Windsor grant.

K.-Y. Whang, J. Jeon, K. Shim, J. Srivatava (Eds.): PAKDD 2003, LNAI 2637, pp. 337–349, 2003.

database, obtained after pre-processing the web log has each tuple consisting of a transaction ID and the sequence of this transaction's web accesses as shown in Table 1. Thus, for example, user ID 100, from the table has accessed sites, *a* then *b*, *d*, *a*, and *c*. The problem of mining sequential patterns from web logs are now

Table 1. Sample Web Access Sequence Database

TID	Web access Sequences
100	abdac
200	eaebcac
300	babfaec
400	afbacfc

based on the database of Table 1. Given a set of events E, the access sequence S can be represented as $e_1 e_2 \ldots e_n$, where $e_i \in E$ $(1 \leq i \leq n)$. Considering the sequences S' = ab, S = babcd, we can say that S' is a subsequence of S. We can also say that ac is a subsequence of S. A frequent pattern is an access sequence to be discovered during the mining process and it should have a support that is higher than minimum support. In sequence eaebcac, eae is a prefix of bcac, while the sequence, bcac is a suffix of eae. The support of pattern S in web access sequence database, WASD is defined as the number of sequences S_i, which contain the subsequence S, divided by number of transactions in the database WASD. Although events can be repeated in an access sequence, a pattern gets at most one support count contribution from one access sequence. The problem of web usage mining is that of finding all patterns which have supports greater than λ, given the web access sequence database WASD and a minimum support threshold λ.

1.1 Related Work

Sequential mining was proposed in [1], using the main idea of association rule mining presented in Apriori algorithm. Later work on mining sequential patterns in web log include the GSP [9], the PSP [4], and the graph traversal [5] algorithms. Pei et al. [6] proposed an algorithm using WAP-tree, which stands for web access pattern tree. This approach is quite different from the Apriori-like algorithms. The main steps involved in this technique are: the WAP-tree algorithm first scans the web log once to find all frequent individual events. Secondly, it scans the web log again to construct a WAP-tree over the set of frequent individual events of each transaction. Thirdly, it finds the conditional suffix patterns. In the fourth step, it constructs the intermediate conditional WAP-tree using the pattern found in previous step. Finally, it goes back to repeat steps 3 and 4 until the constructed conditional WAP-tree has only one branch or is empty.

Although WAP-tree technique improves on mining efficiency, it recursively re-constructs large numbers of intermediate WAP-trees during mining and this entails expensive operations of storing intermediate patterns.

1.2 Contributions

This paper proposes the Pre-order Linked WAP tree algorithm, which stores the sequential data in a Pre-order Linked WAP tree. Each of this tree's nodes has a binary position code assigned for directly mining the sequential patterns without re-constructing the intermediate WAP trees improving a lot on the efficiency of the WAP tree technique. This paper also contributes the technique for assigning a binary position code to nodes of any general tree, which, can be used to quickly define the ancestors and descendants of any node.

1.3 Outline of the Paper

Section 2 presents the proposed Pre-Order Linked WAP-Tree Mining (PLWAP) algorithm with the Tree Binary Code Assignment rule. Section 3 presents an example sequential mining of a web log database with the PLWAP algorithm. Section 4 discusses experimental performance analysis, while section 5 presents conclusions and future work.

2 Pre-order Linked WAP-Tree Mining Algorithm

PLWAP algorithm is able to quickly determine the suffix trees or forests of any frequent pattern prefix event under consideration by comparing the assigned binary position codes of nodes of the tree. Binary Code Assignment (TreBCA) technique is defined for assigning unique binary position codes to nodes of any general tree, by first transforming the tree to its binary tree equivalent and using a rule similar to that used in Huffman coding [7], to define a unique code for each node.

2.1 Important Concepts for PLWAP-Tree Based Mining

From the root to any node in the tree defines a sequence. Ancestor nodes of a node, e_i in PLWAP-tree form its prefix sequence, while its suffix sequence are the descendant nodes. The count of this node e_i is called the count of the prefix sequence. Each branch from a child of node e_i, to a leaf node represents a suffix sequence of e_i, and these suffix branches of e_i are called the suffix trees (forest) of e_i. The suffix trees of a node, e_i are rooted at several nodes that are children of e_i, called the suffix root set of e_i. The suffix root sets are used to virtually represent the suffix forests without the need to physically store each forest. For example, in Figure 1(a), the suffix trees of node, Root are rooted at nodes (a:3:1) and (b:1:10), while the suffix trees of node (a:3:1) are rooted at (b:3:11). To avoid reconstructing trees, the idea of PLWAP is to use the suffix trees of the last frequent event in an *m-prefix sequence* to recursively extend the subsequence to m+1 sequence by adding a frequent event that occurred in the suffix sequence. Thus, binary *position codes* are introduced for identifying the position of every node in the PLWAP tree.

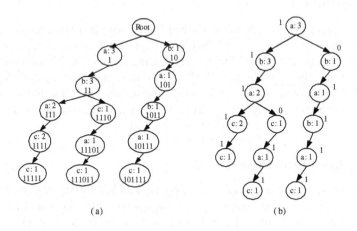

(a)   (b)

Fig. 1. Position Code Assignment Using Single Codes in its Binary Tree

2.2   Tree Binary Position Code Assignment Rule

A temporary position code of 1 is assigned to each left child of a node in the binary tree transformation of a given PLWAP-tree, while a temporary code of 0 is assigned to each right child. The actual position code of a node on the PLWAP tree is, then, defined as the concatenation of all temporary position codes of its ancestors from the root to the node itself (inclusive) in the transformed binary tree equivalent of the tree. For example, given a general tree shown as Figure 1(a), it can be transformed into its binary tree equivalent shown in Figure 1(b). In Figure 1(a), the actual position code of the rightmost leaf node (c:1) is obtained by concatenating all temporary position codes from path (a:3) to (c:1) of the rightmost branch of Figure 1(b) to obtain 101111. The transformation to binary tree equivalent is mainly used to come up with a technique for defining and assigning position codes (presented below as Rule 21).

Rule 21 *Given a PLWAP-tree with some nodes, the position code of each node can simply be assigned following the rule that the root has null position code, and the leftmost child of the root has a code of 1, but the code of any other node is derived by appending 1 to the position code of its parent, if this node is the leftmost child, or appending 10 to the position code of the parent if this node is the second leftmost child, the third leftmost child has 100 appended, etc. In general, for the nth leftmost child, the position code is obtained by appending the binary number for 2^{n-1} to the parent's code.* ∎

Property 21 *A node α is an ancestor of another node β if and only if the position code of α with "1" appended to its end, equals the first x number of bits in the position code of β, where x is the ((number of bits in the position code of α) + 1).* ∎

For example, in Figure 1(a), (c:1:1110) is an ancestor of (c:1:111011) because after appending 1 to the position code of (c:1:1110), which is 1110, we get 11101. This code 11101 is equal to the first 5 (i.e, length of c +1) bits of code from (c:1:111011). Position codes are also used to decide if one node belongs to the right-tree or left-tree of another node.

2.3   The PLWAP Algorithm

The PLWAP algorithm is based on the following properties.

Property 22 *If there is a node in an e_i current suffix subtree, which is also labeled e_i, the support count of the first e_i is the one that contributes to the total support count of e_i, while the count of any other e_i node in this same subtree is ignored.* ∎

From Figure 1(a), property 22 allows us to obtain the support count of event a on the left suffix subtree of Root as 3, and that on the right subtree as 1.

Property 23 *The support count of a node e_i, in the current e_i suffix trees (also called current conditional PLWAP-trees), ready to be mined is the sum of all first e_i nodes in all suffix trees of e_i, or in the suffix trees of the Root if the first event of a frequent pattern is being mined.* ∎

Applying property 23 to Figure 1(a), the support count of node a in the Root suffix trees is 4 (the sum of a:3:1 and a:1:101). The support count of a in subsequent a suffix trees (that is subtrees rooted at b:3:11 and b:1:1011) is 4, from the sum of a:2:1111, a:1:11101 and a:1:10111. The support count of event b in this same a suffix trees rooted at (b:3:11 and b:1:1011) is 4.

Property 24 *e_i is the next frequent event in the mined prefix subsequence if the node e_i in the current suffix tree of e_i has a support count greater than or equal to the minimum support threshold.* ∎

Property 25 *For any frequent event e_i, all frequent subsequences containing e_i can be visited by following the e_i linkage starting from the last visited e_i record in the PLWAP-tree being mined.* ∎

The sequence of steps involved in the PLWAP-tree algorithm is:
Step 1: The PLWAP algorithm scans the access sequence database first time to obtain the support of all events in the event set, E. All events that have a support of greater than or equal to the minimum support are frequent. Each node in a PLWAP-tree registers three pieces of information: node label, node

count and node position code, denoted as label: count: position. The root of the tree is a special virtual node with an empty label and count 0.

Step 2: The PLWAP scans the database a second time to obtain the frequent sequences in each transaction. The non-frequent events in each transaction are deleted from the sequence. PLWAP algorithm also builds a prefix tree data structure, called PLWAP tree, by inserting the frequent sequence of each transaction in the tree the same way the WAP-tree algorithm would insert them. The insertion of frequent subsequence is started from the root of the PLWAP-tree. Considering the first event, denoted as e, increment the count of a child node with label e by 1 if there exists this node, otherwise create a child labeled by e and set the count to 1, and set its position code by applying Rule 21 above. Once the frequent sequence of the last database transaction is inserted in the PLWAP-tree, the tree is traversed pre-order fashion (by visiting the root first, the left subtree next and the right subtree finally), to build the frequent header node linkages. All the nodes in the tree with the same label are linked by shared-label linkages into a queue, called event-node queue.

Step 3: Then, PLWAP algorithm recursively mines the PLWAP-tree using prefix conditional sequence search to find all web frequent access patterns. Starting with an event, e_i on the header list, it finds the next prefix frequent event to be appended to an already computed m-sequence frequent subsequence, by applying property 24 , which confirms an e_n node in the suffix root set of e_i, frequent only if the count of e_n in the current suffix trees of e_i is frequent. It continues the search for each next prefix event along the path, using subsequent suffix trees of some e_n (a frequent 1-event in the header table), until there are no more suffix trees to search. For example, the mining process would start with an e_i event a, and given the PLWAP-tree, it first mines the first a event in the frequent pattern by obtaining the sum of the counts of the first e_n (i.e, a) nodes in the suffix subtrees of the Root. This event is confirmed frequent if this count is greater than or equal to minimum support. Now, the discovered frequent pattern list is {a}. To find frequent 2-sequences that start with this event a, the next suffix trees of last known frequent event in the prefix subsequence, e_i (i.e., a) are mined for events a, b, c (1-frequent events in the header node), in turn to possibly obtain frequent 2-sequences aa, ab, ac respectively if support thresholds are met. Frequent 3-sequences are computed using frequent 2-sequences and the appropriate suffix subtrees. All frequent events in the header list are searched for, in each round of mining in each suffix tree set. The formal algorithm for PLWAP mining is presented as Figure 2. An example, showing the construction and mining of the PLWAP-tree is given in section 3.

3   An Example – Constructing and Mining PLWAP-Tree for Frequent Patterns

The same web access sequence database used in section 1 for introducing WAP tree mining is used here for showing how the PLWAP algorithm constructs a

Algorithm 21 *(PLWAP-Mine - Mining the Pre-Order Linked WAP Tree)*

Algorithm PLWAP-Mine()
Input: PLWAP tree T, header linkage table L,
                 minimum support λ ($0 < \lambda \leq 1$), Frequent m-sequence F).
                 suffix tree roots set R (R is root and F is empty first time)
                 Extendible set L (is frequent 1-sequence set the first time)
Output: Frequent (m+1)-sequence, F'.
Other Variables: S stores whether node is ancestor of the following nodes
                    in the queue, C stores the total number of events e_i in the suffix trees.
begin
(1) If R is empty, or the summation of R's children is less than λ, return
(2) For each event, e_i in L, find the suffix tree of e_i in T (i.e,$e_e|suffixtree$), do
       (2a) Save first event in e_i-queue to S.
       (2b) Following the e_i-queue
           If event e_i is the descendant of any event in R, and is not descendantof S,
               Insert it into suffix-tree-header set R'
               Add count of e_i to C.
               Replace the S with e_i.
       (2c) If C is greater than λ
               Append e_i after F to F' and output F'
               Call Algorithm PLWAP-Mine of Figure 2 passing R' and F'.
       Else Remove e_i from extendible set L
end // of PLWAP-Mine //

Fig. 2. The PLWAP-Tree Mining for Frequent Patterns Algorithm

PLWAP tree and how it mines the tree to obtain frequent sequences with a
minimum support of 75%.

3.1   Example Construction of PLWAP Tree

The PLWAP algorithm scans the WASD once to obtain the supports of the
events in the event set E = {a, b, c, d, e, f} and stores the frequent 1-events in
the header List, L as {a:4, b:4, c:4}. The non-frequent events are d:1, e:2, f:2.
The PLWAP algorithm next uses only the frequent events in each transaction
to construct the PLWAP tree and pre-order linkages of the header nodes for
frequent events a, b, and c. Thus, the PLWAP algorithm inserts the first frequent
sequence abac in the PLWAP tree with only the root by checking if an *a* child
exists for the root, inserting an *a* as the leftmost child of the root as shown in
Figure 3(a). The count of this node is 1 and its position code is 1 since it is
the leftmost child of the root. Next, the *b* node is inserted as the leftmost child
of the *a* node with a count of 1 and postion code of 11 from Rule 21. Then,
the third event in this sequence, *a*, is inserted as a child of node *b* with count
1 and position code 111. Finally, the last event *c* is inserted with count 1 and
position code 1111. All inserted nodes on the tree have information recorded as
(node label: count : position code). Next, the second frequent sequence, abcac is
inserted starting from root as shown in Figure 3(b). Figures 3(c) and (d) show
the PLWAP-tree after sequences babac and abacc have been inserted into the
tree. Now, that the PLWAP tree is constructed, the algorithm traverses the tree
to construct a pre-order linkage of frequent header nodes, a, b and c. Starting
from the root using pre-order traversal mechanism to create the queue, we first

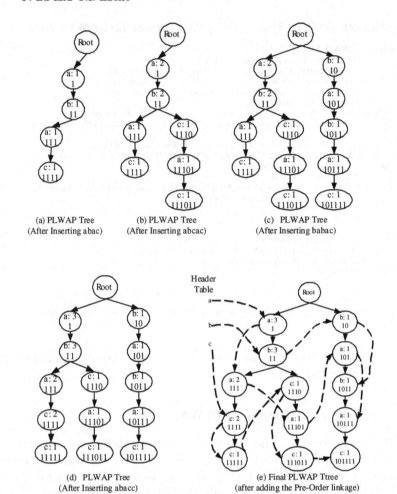

Fig. 3. Construction of PLWAP-Tree Using Pre-Order Traversal

add (a:3:1) into the a-queue. Then, visiting its left child, (b:3:11) will be added to the b-queue. After checking (b:3:11)'s left child, (a:2:111) is found and labeled as a, and it is added to the queue. Then, (c:2:1111) and (c:1:11111) are inserted into the c-queue. Since there is no more left child after (c:1:11111), algorithm goes backward, until it finds the sibling of (a:2:111). Thus, (c:1:1110) is inserted into c-queue. Figure 3(e) shows the completely constructed PLWAP tree with the pre-order linkages.

3.2 Example Mining of Constructed PLWAP Tree

The algorithm starts to find the frequent sequence with the frequent 1-sequence in the set of frequent events(FE) {a, b, c}. For example, the algorithm starts by

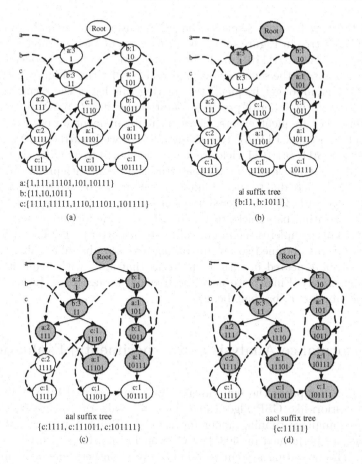

Fig. 4. Mining PLWAP-Tree to Find Frequent Sequence Starting with aa

mining the tree in Figure 4(a) for the first element in the header linkage list, *a* following the *a* link to find the first occurrencies of *a* nodes in a:3:1 and a:1:101 of the suffix trees of the root since this is the first time the whole tree is passed for mining a frequent 1-sequence. Now the list of mined frequent patterns F is {a} since the count of event *a* in this current suffix trees is 4 (sum of a:3:1 and a:1:101 counts). The mining of frequent 2-sequences that start with event *a* would continue with the next suffix trees of *a* rooted at {b:3:11, b:1:1011} shown in Figure 4(b) as unshadowed nodes. The objective here is to find if 2-sequences aa, ab and ac are frequent using these suffix trees. In order to confirm aa frequent, we need to confirm event *a* frequent in the current suffix tree set, and so on. Using the position codes, we continue to mine all frequent events in the suffix trees of a:3:1 and a:1:101, which are rooted at b:3:11 and b:1:1011 respectively. From Figure 4(b), we find the first occurrence of 'a' on each suffix tree, as a:2:111, a:1:11101 and a:1:10111 giving a total count of 4 to make *a* the next frequent event in sequence. Thus, a is appended to the last list of frequent sequence 'a', to form the new frequent sequence 'aa'. We continue to mine the conditional suffix PLWAP-tree in Figure 4(c). The suffix trees of these 'a' nodes which are rooted at c:2:1111, c:1:111011 and c:1:101111, give another c frequent event in sequence, to obtain the sequence 'aac'. The last suffix tree (Figure 4(d)) rooted at c:1:11111 is no longer frequent, terminating this leg of the recursive search. Backtracking in the order of previous conditional suffix PLWAP-tree mined, we search for other frequent events. Since no more frequent events are found in the conditional suffix PLWAP-tree in Figure 4(c), we backtrack to Figure 4(b), to find that b:3:11, b:1:1011 yield frequent event for b to give the next frequent sequence as ab. This leg of mining of patterns that begin with prefix subsequence ab is not shown here due to space limitation. So far, discovered frequent patterns are {a, aa, aac, ab}. The discovered frequent patterns in this ab leg are: aba, $abac$. Next the algorithm backtracks to Figure 4(b) to mine for the pattern ac using the c link. This completes the mining of frequent patterns starting with event a and the patterns obtained so far are {a, aa, aac, ab, aba, abac, abc, ac}. This process will be repeated in turn for patterns that start with frequent events b and c respectively. Finally, we have the frequent sequence set {a, aa, aac, ab, aba, abac, abc, ac, b, ba, bac, bc, c}.

4   Performance Analysis and Experimental Evaluation

This section compares the experimental performance of PLWAP, WAP-tree, and the Apriori-like GSP algorithms. The three algorithms are implemented with C++ language running under Inprise C++ Builder environment. All experiments are performed on 400MHz Celeron PC machine with 64 megabytes memory. The operating system is Windows 98. Synthetic datasets are generated using the publicly available synthetic data generation program of the IBM Quest data mining project at http://www.almaden.ibm.com/cs/quest/, which has been used in most sequential pattern mining studies [9,6]. The parameters shown below are used to generate the data sets.

$|D|$ Number of sequences in the database
$|C|$: Average length of the sequences
$|S|$: Average length of maximal potentially frequent sequence
$|N|$: number of events

For example, C10.S5.N2000.D60k means that $|C| = 10$, $|S| = 5$, $|N| = 2000$, and $|D| = 60k$. It represents a group of data with average length of the sequences as 10, the average length of maximal potentially frequent sequence is 5, the number of individual events in the database are 2000, and the total number of sequences in database is 60 thousand.

Experiment 1: Execution Time for Different Support

This experiment uses fixed size database and different minimum support to compare the performance of PLWAP, WAP and GSP algorithms. The datasets are described as C10.S5.N2000.D60k, and algorithms are tested with minimum supports between 0.8% and 10% on the 60 thousand sequences database.

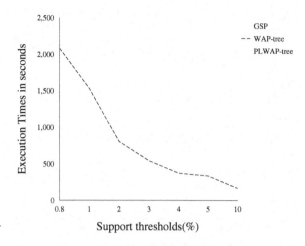

Fig. 5. Execution Times Trend with Different Minimum Supports

Table 2. Execution Times for Dataset at Different Minimum Supports

Algorithms	Runtime (in secs) at Different Supports						
	0.8	1	2	3	4	5	10
GSP	-				-	-	1678	852
WAP-tree	2080	1532	805	541	372	335	166
PLWAP-tree	492	395	266	179	127	104	69

Table 2 and Figure 5 show that the execution time of GSP increases sharply, when the minimum support decreases and that the PLWAP algorithm always uses less runtime than the WAP algorithm.

Experiment 2: Execution Times for Databases with Different Sizes
In this experiment, databases with different sizes from 20k to 100k with fixed minimum support of 5% were used. The five datasets are C10.S5.N2000.D20k, C10.S5.N2000.D40k, C10.S5.N2000.D60k, C10.S5.N2000.D80k and C10.S5.N2000.D100k. The execution times of the WAP and PLWAP algorithms are presented in Table 3 and Figure 6.

Table 3. Execution Times at Different Data Sizes on Support 0.5%

Algorithms	Different Changed Transaction Size				
(times in secs)	20K	40K	60K	80K	100K
WAP-tree	158	272	335	464	566
PLWAP-tree	54	81	104	146	185

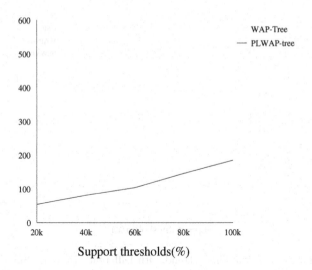

Fig. 6. Execution Times Trend with Different Data Sizes

Experiment 3: Different Length Sequences
In experiment 3 (Table 4), the performance of WAP and PLWAP algorithms for
the sequences with varying average lengths of 10, 20 and 30 was observed. Three
datasets used for the comparison are: C10.S5.N1500.D10k, C20.S8.N1500.D10k
and C30.S10.N1500.D10k. The minimum support is set at 1%.

Table 4. Execution Times with Different Sequence Lengths at Support 1%

Algorithms	Different Changed Transaction Size		
(times in secs)	10(5)	20(8)	30(10)
WAP-tree	402	1978	5516
PLWAP-tree	217	743	1751

5   Conclusions and Future Work

The algorithm PLWAP proposed in this paper, improves on mining efficiency, by
finding common prefix patterns instead of suffix patterns as done by WAP-tree
mining. To avoid recursively re-constructing intermediate WAP-trees, pre-order
frequent header node linkages and position codes are proposed. While the pre-
order linkage provides a way to traverse the event queue without going back-
wards, position codes are used to identify the position of nodes in the PLWAP
tree. With these two methods, the next frequent event in each suffix tree is
found without traversing the whole WAP-tree. Experiments show that mining

web log using PLWAP algorithm is much more efficient than with WAP-tree and GSP algorithms, especially when the average frequent sequence becomes longer and the original database becomes larger. Future work should consider applying PLWAP-tree mining techniques to distributed mining as well as to incremental mining of web logs and sequential patterns.

References

1. Agrawal, R., Srikant, R.: Mining Sequential Patterns. Proceedings of the 11th International Conference on Data Engineering, Taiwan, 1995.
2. Berendt, B., Spiliopoulou, M.: Analyzing Navigation Behaviour in Web Sites Integrating Multiple Information Systems. VLDB Journal, Special Issue on Databases and the Web, volume 9, number 1, 2000, pages 56–75.
3. Han, J., Kamber, M.: Data Mining-Concepts and Techniques, Morgan Kaufmann Publisher, 2001.
4. Masseglia, F., Poncelet, P., Cicchetti, R.: An Efficient Algorithm for Web Usage Mining. Networking and Information Systems Journal (NIS), volume 2, number 5–6, 1999, pages 571–603.
5. Nanopoulos, A., Manolopoulos, Y.: Mining Patterns from Graph Traversals. Data and Knowledge Engineering, volume 37, number 3, 2001, pages 243–266.
6. Pei, J., Han, J., Mortazavi-Asl, B., Zhu, H.: Mining Access Patterns Efficiently from Web Logs. Proceedings of the 2000 Pacific-Asia Conference on Knowledge Discovery and Data Mining (PAKDD'00), Kyoto, Japan, 2000.
7. Shaffer, C.A.: A Practical Introduction to Data Structures and Algorithm Analysis. Prentice Hall Inc., September 2000.
8. Spiliopoulou, M.: The Laborious Way from Data Mining to Web Mining. Journal of Computer Systems Science and Engineering, Special Issue on Semantics of the Web, volume 14, 1999, pages 113–126.
9. Srikant, R., Agrawal, R.: Mining Sequential Patterns: Generalizations and Performance Improvements. Proceedings of the Proceedings of the Fifth International Conference On Extending Database Technology (EDBT), Avignon, France, 1996.

An Integrated System of Mining HTML Texts and Filtering Structured Documents

Bo-Hyun Yun[1], Myung-Eun Lim[1], and Soo-Hyun Park[2]

[1]Dept. of Human Information Processing
Electronics and Telecommunications Research Institute
161, Kajong-Dong, Yusong-Gu, Daejon, 305-350, Korea
{ ybh, melim }@etri.re.kr
[2]School of Business IT, Kookmin University,
861-1, Cheongrung-dong, Sungbuk-ku, Seoul, 136-702, Korea
shpark21@kookmin.ac.kr

Abstract. This paper presents a method of mining HTML documents into structured documents and of filtering structured documents by using both slot weighting and token weighting. The goal of a mining algorithm is to find slot-token patterns in HTML documents. In order to express user interests in structured document filtering, slot and token are considered. Our preference computation algorithm applies vector similarity and Bayesian probability to filter structured documents. The experimental results show that it is important to consider hyperlinking and unlablelling in mining HTML texts; slot and token weighting can enhance the performance of structured document filtering.

1 Introduction

The Web has presented users with huge amounts of information, and some may feel they will miss something if they do not review all available data before making a decision. These needs results in HTML text mining and structured document filtering. The goal of mining HTML documents is to transform HTML texts into a structural format and thereby reducing the information in texts to slot-token patterns. Structured document filtering is to provide users with only information which satisfies user interests.

Conventional approaches of mining HTML texts can be divided into 3 types of systems such as FST(Finite-State Transducers) method[1, 5, 7], relational rule learning method [2, 6, 10], and knowledge-based method [4]. Knowledge-based method is easy to understand and implement but needs prior knowledge construction. Thus, we choose knowledge-based method for mining HTML texts with considering hyperlinking and unlabelling. The existing methods[8, 11] of filtering structured documents use a naïve Bayesian classifier and decision tree and Bayesian classification algorithm. However, these methods use only token weighting in slot-token patterns for structured documents.

In this paper, we propose an integrated system of mining HTML texts by converting them to slot-token patterns and filtering structured document. At first, Information extraction (IE) system extracts structured information such as slot-token

K.-Y. Whang, J. Jeon, K. Shim, J. Srivatava (Eds.): PAKDD 2003, LNAI 2637, pp. 350–355, 2003.
© Springer-Verlag Berlin Heidelberg 2003

patterns from Web pages. And then, our information filtering(IF) system predicts user preference of that information and filters only information which predicted user maybe interested in. We analyze user feedback from user logs and construct a profile using slot preference and token preference.

2 Mining HTML Texts and Filtering Structured Document

Fig. 1 shows our integrated system of mining HTML texts and filtering structured documents. IE system mines Web documents into structured document and IF system filters the extracted information to provides the filtered information to users.

Fig. 1. An Integrated System Configuration

Given HTML documents, our IE system interprets the document and mines the meaningful information fragment. The system finds the repeated patterns from the objects of the interpreted document. Using the repeated patterns, we induce the wrapper of the HTML document. In order to extract the information, wrapper interpreter loads the generated wrapper and maps the objects into the wrapper. Finally, the wrapper extracts the information and stores the extracted information as the XML based document.

Because of unlabeled data, mining HTML texts may often fail to extract structured information. In order to assign labels to unlabeled data, we assume that target documents consist of many labeled information and a little unlabeled information. To recognize the proper label of information, we observe the previous extraction results. We calculate each token's probability within the slot. The equation (1) shows the probability that tokens within a slot belongs to a labeled slot.

$$p(s_i \mid T_j) = p(s_i) * p(T_j \mid s_i) \cong p(s_i) * \frac{1}{v} \sum_{k=1}^{v} p(t_j \mid s_i) \tag{1}$$

Here, s_i means slot i, T_j means token set of jth slot in all templates, p is probability, and v is total number of templates in document. The probability that jth slot label is s equals the probability of token set in the slot j.

To refer sub-linked pages, we have to integrate main pages and sub-linked pages and extract the slot-token patterns. The integration phase analyzes the structure of sites and integrates main pages and sub-linked pages if the linked pages have the valid information. On the other hand, the extraction phase reads wrapper and extract information from main pages and sub-linked pages.

Structured document filtering can be divided into the learning part and the filtering part. In the learning stage, when user logs are obtained by user feedback, the indexer loads structured documents such as slot-token patterns and computes token and slot preference within documents. Token preference is the frequency of the tokens and slot preference is assigned with token preference. Finally, user profile is updated in learning stage. In filtering stage, when new structured documents come from IE system, the indexer extracts the tokens and token preference within each slot is calculated. And then, document similarity is calculated by using slot preference. Finally, filtered documents are suggested to users.

Slot preference is the degree of importance which user regards to be important for choosing information. We assume that if slot preference is high, users judge information value by that slot. On the contrary, if slot preference is low, that cannot help user choosing information.

To evaluate slot preference, we premise hypothesis as follows:

- There is at least one slot which is used for user to estimate information in a structured document.

  ex) Some users decide information according to 'genre' and 'cast' in the 'movie' document.

- User refers only slot with high preference to evaluate information.

  ex) Some user is interested in 'genre' and 'director' of the movie, but not in 'title' and 'casting'.

Base slot is used as a baseline for choosing information. If slot preference is sufficiently high, that slot becomes one of the base slots. The filtering system predicts preference of the document by using only tokens of base slots.

In vector space model, user preference is composed of some tokens which are selected from documents rated by users. Token weights represent user's interests. If users have a long-term interest – it can be positive or negative – to a certain token, then the *variance* of the token weight will be low. On the other hand, if users don't have any interest to a token, the weight will be represented variously.

The importance of the slot is high if slots contain many important tokens in the user profile; otherwise, the importance is low. We determine the slot preference by the number of important tokens within a slot. Important tokens are selected by its variance in each slot. The equation (1) means the token variance and the equation (2) is the equation of slot weighting.

$$\sigma_a = \frac{1}{N}\sum_{i=1}^{n}(w_{ai} - \overline{w}_a)^2 freq_{ai}, \ N = \sum_{i=1}^{n} freq_{ai} \tag{2}$$

σ_a is the variance of token a and w_{ai} is preference of token a in slot i. $freq$ is token frequency, n is the number of preference w, and N is the total number of tokens in all slots.

$$SW_m = \frac{\sum_{i=1}^{k} x_i}{k}, \ \begin{Bmatrix} if \ \sigma_i < threshold \ , \ x_i = 1, \\ else \ \ x_i = 0 \end{Bmatrix} \tag{3}$$

In the equation (3), SW_m is preference of slot m and k is total number of tokens in slot. x is constant means count of tokens. x is 1 if the variance of token is larger then threshold; otherwise, 0.

The filtering is done by calculating the similarity between documents and profiles [9]. The vectors of documents and profiles are generated and only base slot is used in calculating the similarity. The equation (4) shows the similarity of slot. S_{pa} is slot a's token vector of profile and S_{da} is vector of document. t_w is token's weight of document slot and p_w is token's weight of profile slot.

$$SS_a = S_{Pa} \cdot S_{Da} = \frac{\sum tw(i)\,pw(i)}{\sqrt{\sum tw(i)^2}\,\sqrt{\sum pw(i)^2}} \qquad (4)$$

The equation (5) shows the final document preference. SS_i is the slot similarity of slot i and SW_i is the slot preference. s is the number of slot whose preference weight is larger than threshold.

$$Sim(P, D) = \frac{\sum_{i=1}^{s} SS_i SW_i}{s} \qquad (5)$$

3 Experimental Results

The evaluation data of mining Web pages are seven movie sites such as Core Cinema and Joy Cinema, etc. Because movie sites have the characteristics to be updated periodically, wrapper induction is performed by recent data to detect slot-token patterns. Because our knowledge for wrapper induction is composed of Korean language, we test our method of mining Korean movie sites. We determine 12 of target slots such as title, genre, director, actor, grade, music, production, running time, and so on.

We evaluate three kinds of mining methods such as "knowledge only", "link extraction", and "label detection". "knowledge only" is the method of mining movie sites by using only knowledge without considering hyperlinking and unlabelling. "link extraction" means the method of using hyperlinking. "label detection" is the method of considering both hyperlinking and unlabelling.

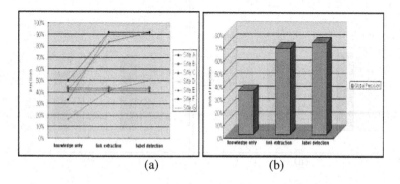

(a)   (b)

Fig. 2. Results of Mining HTML Texts

The precision of each site is in Fig. 2 (a) and global precision is in Fig. 2 (b). The performance of "link extraction" and "label detection" is better than that of "knowledge only" significantly. In other words, experimental results show that it is important to consider hyperlinking and unlablelling in mining HTML texts and our system can transform HTML texts into slot-token patterns.

The experimental data of filtering structured documents are *eachmovie* dataset[3]. This data includes a rating of users to the movie information. It consists of 2,811,983 ratings of 72,916 users to 1,628 movies. The range of rating is between 0 and 1. From that data, we choose only 6,044 user's data whose rating is more than 100. It can be thought that users with more than 100 ratings are reliable. Over 1,628 movies, 1,387 real data are gathered from IMDB site (http://www.imdb.com). This information contains *title, director, genre, plot outline, user comments, and cast overview* and so on.

To compare the relative performance, we test 3 different methods such as "unst, "st", and "sw". "unst" regards a structured document as an unstructured one. Slots are dealt with tokens without distinguishing slots and tokens. "st" deals with structured documents but doesn't use slot weighting. User profile is constructed by slots and tokens but has only weights of tokens. The document similarity is calculated by weights of tokens within the same slot in user profile and document. "st with sw" uses both slot weighting and token weighting. We select 1,000 of 6,044 users randomly and divide their rating data into training set and test set such as 1:1 and 3:1. This is reason that we see the change of performance according to the size of data. We test 50 times repeatedly at each method and get the precision and recall of each method.

The experimental results of filtering structured documents are shown in Fig. 3. Fig. 3 (a) and (b) shows the precision of "st" is higher than that of "unst" and the precision of "st with sw" is higher than that of "st". From this result, we can see that structured document filtering methods have to consider both slot and token weighting. Comparing Fig. 3 (a) and (b), we see that precision with 3:1 data ratio is a little higher than precision with 1:1 ratio. This means that increasing size of training data can improve the performance.

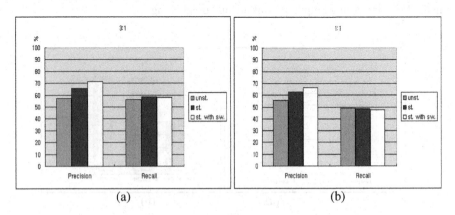

Fig. 2. Results of Filtering Structured Documents
(a) is the result of 3:1 ratio, and (b) is 1:1 ratio.

4 Conclusion

This paper has described a method of mining HTML documents into structured documents and of filtering structured documents by using both slot weighting and token weighting. Our integrated system mines HTML texts by IE system comprised of wrapper generation and interpretation. We used Bayes probability to solve the unlabelling problem and integrated main pages and sub-linked pages for hyperlinking. And then, structured documents are filtered by IF system using slot and token weighting.

The experimental results shows that it is important to consider hyperlinking and unlablelling in mining HTML texts and our system can transform HTML texts into slot-token patterns. In structured document filtering, slot weighting can enhance the performance of structured document filtering. For the future works, we try to evaluate how the total number of tokens affects slot weighting.

References

[1]   Chun-nan Hsu and Ming-tzung Dung, Generating Finite-State Transducers for semi-structured data extraction from the web. Information Systems vol.23, no.8, p 521–538, 1998.
[2]   Dayne Freitag, Toward General-Purpose Learning for Information Extraction, Proceedings of the 36th annual meeting of the Association for Computational Linguistics and 7th International Conference on Computational Linguistics, 1998.
[3]   eachmovie data download site , http://www.research.compaq.com/SRC/eachmovie/data/.
[4]   Heekyoung Seo, Jaeyoung Yang, and Joongmin Choi, Knowledge-based Wrapper Generation by Using XML, Workshop on Adaptive Text Extraction and Mining(ATEM 2001), pp. 1–8, Seattle, USA, 2001.
[5]   Ion Muslea, Steven Minton, Craig A. Knoblock, Hierarchical Wrapper Induction for Semistructured Information Soueces.
[6]   Mary Elaine Califf and Raymond J. Mooney, Relational Learning of Pattern-Match Rules for Information Extraction, Proceedings of the 16th National Conference on Artificial Intelligence, p. 328–334, Orlando, FL, July, 1999.
[7]   Naveen Ashish and Craig Knoblock, Semi-automatic Wrapper Generation for Internet Information Sources, Proceedings of the Second International Conference on Cooperative Information Systems, Charleston, SC, 1997.
[8]   Raymond J. Mooney. Content-Based Book Recommending Using Learning for Text Categorization, Proceedings of the 5th ACM conference on Digital Libraries, June 2000.
[9]   Robert. B. Allen, User models: theory, method, and practice, international journal on man-machine studies, vol.32, p. 511–543, 1990.
[10] Stephen Soderland, Learning Information Extraction Rules for Semi-structured and Free text. Machine Learning, 34(1-3):233–272, 1999.
[11] Yanlei Diao, Hongjun Lu, and Dekai Wu, A Comparative Study of Classification Based Personal E-mail Filtering, Proceedings of the 4th Pacific-Asia Conference on Knowledge Discovery and Data Mining, Kyoto, Japan, April 2000.

A New Sequential Mining Approach to XML Document Similarity Computation[1]

Ho-pong Leung, Fu-lai Chung, and Stephen Chi-fai Chan

Department of Computing, Hong Kong Polytechnic University,
Hunghom, Kowloon, Hong Kong
{csleung,cskchung,csschan}@comp.polyu.edu.hk

Abstract. There exist several methods to measuring the structural similarity among XML documents. The data mining approach seems to be a novel, interesting and promising one. In view of the deficiencies encountered by ignoring the hierarchical information in encoding the paths for mining, we propose a new sequential pattern mining scheme for XML document similarity computation. It makes use of the hierarchical information to computing the document structural similarity. In addition, it includes a post-processing step to reuse the mined patterns to estimate the similarity of unmatched elements so that another metric to qualify the similarity between XML documents can be introduced. Encouraging experimental results were obtained and reported.

1 Introduction

Measuring the structural similarity among XML documents has been an active area of research in the past few years and it is fundamental to many applications, such as integrating XML data source [1], XML dissemination [2] and repositories [3].

The concept of frequent tree pattern for determining XML similarity was introduced in [4,5]. These approaches make use of data mining techniques to find the repetitive document structure for determining the similarity between documents. In [4], Lee et al. defines the structural similarity as the number of paths that are common and similar between the hierarchical structures of the XML documents. This method provides accurate quantitative computation of similarity between XML documents and encouraging experimental results have been reported. However, the determination of similar paths does not take into consideration of the hierarchical information, i.e., the level of the hierarchy at which an element locates, several weaknesses of the method can be observed.

Firstly, the same common paths located at different levels of the hierarchies will have same similarity value. Secondly, Lee et al.'s method restricts measuring the element similarity in the pre-processing stage and prevents the use of the level information of common elements to discover the synonym elements in the mining stage, which can be used to further qualify the similarity between the XML

---

[1] This work was supported by the research project H-ZJ84 and the full version of this paper can be found in http://www.comp.polyu.edu.hk/~cskchung/Paper/PAKDD03.pdf.

K.-Y. Whang, J. Jeon, K. Shim, J. Srivatava (Eds.): PAKDD 2003, LNAI 2637, pp. 356–362, 2003.

documents. Lastly, ignoring the hierarchical information is too restrictive and incompatible to measuring the overall hierarchical structure of XML documents.

In this paper, we further pursue the sequential pattern mining approach to compute the structural similarity of XML documents. Here, the structural similarity means the number of paths and elements that are common and similar among the hierarchical structure of the XML documents. A simple method composed of a pre-processing step and a post-processing step (for the mining engine) is proposed and through which the XML semantics can be determined and hence the similarity between the XML documents can be computed appropriately. To overcome the aforementioned weaknesses of Lee et al.'s method, the proposed method takes into accounts of the change of level information in the tree hierarchy and elements along the path. It is beneficial to the similarity computation and also facilitates the determination of synonyms (common unmatched elements) between the paths. The remainder of this paper is structured as follows. Section 2 presents the problem of sequential mining of frequent XML document tree pattern and the proposed pre-processing method. The similarity computation and the way to identify the synonyms between XML documents are described in Section 3. Section 4 reports the experimental results. The final section concludes the paper and outlines the future work.

2 Mining Frequent XML Document Tree Patterns

In this section, we first introduce a preorder tree representation (PTR) [6] (pre-processing step) for the incorporation of hierarchical information in encoding the XML tree's paths. We then describe the sequential pattern mining approach to compute the similarity between two sets of encoded paths, i.e., two XML documents. In the rest of the paper, each XML document is represented as a labeled tree and the values of the elements or attributes in the tree will not be considered, i.e., only the structure of the XML document is considered.

As the XML's document is considered as an XML tree [7], the path can be represented by the elements from the root to the leaf. In this paper, we have adopted the PTR to encode the XML tree's path where the information of each node is listed before the nodes of its sub-tree and a path is simply represented by a list of nodes and their parentheses. In the PTR, each element followed by a left parenthesis denotes a change of level in the tree hierarchy, i.e., going down the tree. The left parenthesis is a start symbol of traversing from a non-leaf root of the tree. The right parentheses are meaningless and hence can be removed from the path. More importantly, it reduces the mining time significantly. Under this representation, the level and the structure of the tree can be presented by a set of paths. It contains not only the path's element information, but also the level of the hierarchy for every element in the path. This level of the hierarchy information indicates the amount and resolution of information. Besides, the level of the tree also facilitates the determination of common unmatched elements between the paths and it will be described in the next section. Therefore, if any such paths are found frequent by the mining algorithm, the structural similarity between XML documents can be computed by determining the number of paths and their levels of hierarchy that are common and similar.

The problem being considered here is based on the idea that each PTR encoded path can be viewed as a sequence and based upon which the sequential pattern mining

[8] can be applied to find the frequent tree patterns. We denote the PTR encoded path as $< x_1 x_2 \cdots x_n >$ and call such a sequence as "XML-sequence". An XML document contains a number of XML-sequences. Let the set of XML-sequences of an XML document be $\{s_1, s_2, \cdots, s_n\}$. Using the terminology of sequential pattern mining [8], a sequence is contained by another if it is a subsequence of that sequence. In a set of sequences, a sequence s_j is maximal if it is not contained by any other sequence.

Here, we are given a database D of XML documents, each of which consisting of a set of PTR encoded paths. The problem of mining XML sequence patterns is to find the maximal frequent sequences among all sequences satisfying the user-specified minimum support. Each such maximal frequent sequence represents a common structure or pattern of the XML documents. Unlike other data mining applications, the minimum support for finding the maximal sequence between the two XML documents must be 100% (for the concern of similarity computation here). Another difference is that the support count should be incremented only one per document even if the document contains the same set of items in two different paths.

3   Measuring Similarity between XML Documents

As described in the previous section, the sequential pattern mining algorithm [8] was adopted to find the maximal common paths (i.e., maximal frequent sequences) from the encoded XML paths of the two documents for comparison. Based upon the mining results, the similarity between them can be measured by the "ratio" between the maximal common paths and the extracted paths of the larger document, i.e. the one with more elements. In this way, the similarity will change more significantly with the XML document size and it also provides a more reasonable similarity value. In order to reduce the potentially extreme effect of the root element in the similarity computation, we separate the root from each path and the root is treated as a single path of the tree. Here, we assume that a pair of more similar XML documents should have a higher number of maximal common paths and more left parentheses in the paths. The similarity between two XML documents D_1 and D_2 is defined as follows:

$$Sim(D_1, D_2) = \frac{1}{N+1} \left[\left(\sum_{t=1}^{N} \frac{1}{N_t} \sum_{p=1}^{N_t} \frac{m_{t,p}}{M_{t,p}} \right) + MR \right] \tag{1}$$

where N is the total number of level 1 sub-trees in the larger document; N_t is the total number of paths in the t-th sub-tree; $M_{t,p}$ is the number of elements in the (t,p)-th path; $m_{t,p}$ is the number of common elements (obtained from the maximal frequent sequences) in the (t,p)-th path; and MR is either 0 or 1 to denote whether the two documents have the same root element.

Based upon the maximal frequent sequences being mined, we further propose to determine the possible positions of synonyms, i.e., common unmatched elements, and use them to calculate the similarity of these elements. We call this similarity as *element similarity* (ES) that can be used as an additional measure for document similarity computation. Here, a common unmatched element refers to the element located at a relatively similar position in the common path but denoted by another wording of the same meaning (synonyms). The ES is determined as follows (cf Fig. 1).

First, the maximal common path obtained from the mining stage is used to determine the most similar paths from each document. The most similar path means the longest path that contains the maximal frequent sequence. For notational convenience, let S_1 and S_2 be two sets of PTR encoded paths from two XML documents D_1 and D_2, between which we aim at determining the common paths and identifying the synonyms. Let $p_i \in S_1$ and $p_j \in S_2$ be two most similar paths and p_c be their maximal common path, i.e., $p_c \in p_i$ and $p_c \in p_j$.

Second, p_c is aligned with p_i and p_j and the candidates of common unmatched element from each path are determined. Let $e_{c,1} \dots e_{c,k}$ be an ordered elements set of the path p_c, where k is the number of elements in p_c and $e_{c,1}$ ($e_{c,k}$) is the first (last) element of path p_c. Algorithmically, when the elements of the common path $e_{c,r}=$"(" and $e_{c,r+1}=$"(" where $1 \leq r < k$ are located, a common unmatched element is assumed "exist" between the positions r and r+1 of the common path. With respect to p_i or p_j, the element(s) in between the corresponding positions (i.e. $e_{c,r}=$"(" and $e_{c,r+1}=$"(") is (are) the candidate(s) of common unmatched element or possible synonyms.

Third, a reference table is constructed for the candidates identified in the previous step and the element similarity is computed. The reference table stores information about leaf nodes (elements) of the sub-tree for each candidate common unmatched element. Since each element is assumed unique in each XML document and the XML structure is acyclic, the leaf nodes indicate the semantic or concept of their ancestors. Based on this concept, each XML document is considered as a concept hierarchy [9], containing a number of concepts organized into multiple levels. The concepts at higher levels have broader meanings than those at lower levels. A child concept contains more specific meaning than its parent concept. In general, the leaf elements represent the semantic of their parent element. We make use of this property to measure the semantic association between the candidate common unmatched elements from p_i and p_j and thus to compute the element similarity. The element similarity of candidate common unmatched elements e_a and e_b from p_i and p_j respectively is defined as:

$$Similarity(e_a, e_b) = \frac{N_a \cap N_b}{N_a \cup N_b} \qquad (2)$$

where N_a and N_b are the sets of leaf nodes from the reference table for e_a and e_b respectively. Obviously, $Similarity(e_a, e_b)$ falls in the real interval [0, 1].

The steps above determine the most relevant pair of common unmatched elements for documents D_1 and D_2 and the corresponding element similarity can be computed according to eq.(2). Such a similarity measure can be used to further qualify the XML document similarity. By replacing the zero contribution of the unmatched element(s) in the path in eq.(1) with the ES in eq.(2), a combined method called PTR&ES is proposed and the new similarity computation is defined as

$$Sim(D_1, D_2) = \frac{1}{N+1}\left[\left(\sum_{t=1}^{N}\frac{1}{N_t}\sum_{p=1}^{N_t}\frac{m_{i,p}+c_{i,p}}{M_{i,p}}\right) + MR\right] \qquad (3)$$

where $c_{t,p}$ is the sum of ES of the common unmatched elements in the (t,p)-th path:

$$c_{t,p} = \sum_{\forall e_i, e_j \ pairs} Similarity(e_i, e_j).$$ (4)

Readers may refer to the full version of this paper for an example and the details of the PTR&ES method.

4 Experimental Results

In this experiment, three DTDs were downloaded from ACM's SIGMOD Record homepage [10]: Ord(inaryIssuePage).dtd , Sig(modRecord).dtd and Rec(ord).dtd where Rec.dtd is a modified version of Sig.dtd. Specifically, the hierarchical structure between Rec.dtd and Ord.dtd is more similar to that between Sig.dtd and Ord.dtd. We used IBM's XML document generator [11] to accept the above DTDs as input and create the sets of XML documents for simulations. Based upon the three sets of XML documents with similar characteristics, their similarities were computed, reported and analyzed as follows. Lee et al.'s method [4] was chosen for comparisons.

Table 1. Results of XML document similarity in heterogeneous document experiment.

DocumentSet(x,y)		Method			DocumentSet(x,z)		Method		
x: Ordinary Issued-Page	y: Record	Lee et al.	PTR	PTR&ES	x: Ordinary Issued-Page	z: SigmodRecord	Lee et al.	PTR	PTR&ES
1	1	0.15	0.2787	0.3063	1	1	0.2109	0.3398	0.3729
	2	0.3793	0.4277	0.5052		2	0.3793	0.4138	0.4774
	3	0.3746	0.4253	0.4776		3	0.2073	0.3389	0.3617
	4	0.1606	0.3037	0.3283		4	0.3793	0.4138	0.4814
	5	0.1608	0.3050	0.3288		5	0.3793	0.4138	0.4732
2	1	0.1648	0.3	0.3405	2	1	0.3725	0.3980	0.4430
	2	0.3726	0.4085	0.4777		2	0.3769	0.4044	0.4780
	3	0.3700	0.4072	0.4583		3	0.3700	0.3967	0.4312
	4	0.1727	0.3212	0.3573		4	0.3726	0.3980	0.4602
	5	0.1724	0.3217	0.3565		5	0.3726	0.3980	0.4541
3	1	0.1444	0.2759	0.2968	3	1	0.2031	0.3359	0.3613
	2	0.3787	0.4231	0.4821		2	0.3787	0.4106	0.4597
	3	0.3787	0.4231	0.4779		3	0.2073	0.3389	0.3627
	4	0.1560	0.3015	0.3202		4	0.3787	0.4106	0.4628
	5	0.1565	0.3028	0.3210		5	0.2045	0.3295	0.3638
4	1	0.3770	0.4255	0.4570	4	1	0.3770	0.4126	0.4444
	2	0.3770	0.4255	0.4708		2	0.3770	0.4126	0.4523
	3	0.3770	0.4255	0.4689		3	0.3770	0.4126	0.4434
	4	0.1598	0.3034	0.3217		4	0.3770	0.4126	0.4539
	5	0.1601	0.3047	0.3225		5	0.3770	0.4126	0.4507
5	1	0.375	0.4218	0.4611	5	1	0.3715	0.4032	0.4347
	2	0.3715	0.4149	0.4596		2	0.375	0.4085	0.4581
	3	0.3694	0.4138	0.4460		3	0.3694	0.4022	0.4255
	4	0.1681	0.3189	0.3431		4	0.3715	0.4032	0.4442
	5	0.1681	0.3195	0.3430		5	0.3715	0.4032	0.4409

In this experiment, the similarities between documents of different DTDs were analyzed. The XML documents from Ord.dtd were adopted as the base documents while those from Rec.dtd and Sig.dtd were used as query documents. The experimental results are shown in Table 1. As the XML documents come from different DTDs, this is called *heterogeneous XML document similarity*. In the following, DocumentSet(x,y,z) is used to denote a comparative analysis of the similarity of document *x* from Ord.dtd with document *y* from Rec.dtd, i.e.,

DocumentSet(x,y) shown in Table 1, and similarity of document x with document z from Sig.dtd, i.e., DocumentSet(x,z) shown in Table 1. This analysis is used to determine which document (y or z) is more similar to document x. For DocumentSet(1,2,2), DocumentSet(2,3,3), DocumentSet(3,2,2), DocumentSet(4,1,1), DocumentSet(4,2,2), DocumentSet(4,3,3) and DocumentSet(5,3,3), the proposed PTR method was found superior to Lee et al.'s method. Lee et al.'s method shows that the similarity values of Rec-Ord and Sig-Ord are the same. It is because these document pairs contain the same number of common elements in the paths for the Ord.dtd's document. When comparing these two sets of documents, our PTR method can determine the most similar document from them. This is due to the considerations of the hierarchical information in the proposed method. As these elements do not locate at the same level of the two hierarchies, they are treated differently by our method and hence the similarity values are different.

For DocumentSet(2,2,2) and DocumentSet(5,2,2), the PTR method obtained a different result from Lee et al.'s method. The reason is that their method has found more common elements in the paths for the Sig-Ord case and hence generating higher similarity values. As mentioned in the beginning of this section, the hierarchical structure between Rec.dtd and Ord.dtd is more similar to that between Sig.dtd and Ord.dtd. Hence, the proposed PTR method has obtained more reasonable results. It has also been validated that documents with dissimilar hierarchical structures but many common paths (located at different levels of the hierarchies) will not have unreasonably high similarity values generated by the proposed PTR method. The aforementioned distinctive features of the PTR method are also possessed by the PTR&ES method because it takes use of the PTR method to encode the path.

The following comparative study is used to compare the similarities of a set of documents (z) from the Sig.dtd with a document x from Ord.dtd and determine which document (z) is more similar to document x. DocumentSet(x,z) is used to denote the similarity between document x from Ord.dtd with document z from Sig.dtd. For DocumentSet(4,1), DocumentSet(4,2), DocumentSet(4,3), DocumentSet(4,4) and DocumentSet(4,5), both PTR and Lee et al.'s methods show that the similarity values among them are the same. The reason is that all Sig.dtd's documents contain the same number of common elements and unmatched elements in the paths for the Ord.dtd's document. When comparing these two sets of documents, our PTR&ES method can determine the most similar document from them, i.e., Sig.dtd's document 4. It found common unmatched elements in the paths and hence generated non-zero unmatched element similarity values. This unmatched element similarity is used as additional information to calculate the distance among the XML documents for further distinguishing the most similar XML document among the XML documents with the same number of common elements and unmatched common elements.

5 Conclusions and Future Work

In this paper, a new pre-processing step for preparing XML documents for similarity computation using sequential pattern mining is proposed. It takes use of a preorder tree representation (PTR) to encode the XML tree's paths. It has the ability to include the element's semantic and hierarchical structure of document in the computation of similarity between XML documents. A novel post-processing step is also proposed to calculate the common unmatched element similarity obtained from the hierarchical information based mining stage. It estimates the element's semantic by their sub-tree or leaf node. The experimental results show that the proposed methods can overcome several weaknesses of Lee et al.'s method [4]. The combined PTR and element similarity (PTR&ES) method provides a further improvement in computing the XML documents' structural similarity and compensates the shortcoming of the PTR method and Lee et al.'s method. We believe that the proposed methods can provide valuable help for many applications.

Our future works include the investigation of a way to classify the XML documents by their hierarchical structure. The structural semantic can be used as the abstract of the document and filter on the document content. It would be interesting to investigate the impact both on performance as well as quality of the results.

References

1.   Galhardas H., Florescu D., Shasha D., Simon E., Saita C.A.: Declarative data cleaning Language, model, and algorithms. In Proc. of 28th Int. Conf. on Very Large Data Bases (VLDB), Hong Kong, China, pp. 371–380, August, 2002.
2.   Pereira J., Fabret F., Jacobsen H.A., Llirbat F., Shasha D.: WebFilter A High-throughput XML-based publish and subscribe system. In Proc. of 27th Int. Conf. on Very Large Data Bases, Roma, Italy, pp. 723–724, September, 2001.
3.   Hartmut Liefke, Dan Suciu: XMill An efficient compressor for XML data. In Proc. of the ACM SIGMOD Int. Conf. on Management of Data, Dallas, Texas, pp. 153–164, May, 2000.
4.   Lee J.W., Lee K. and Kim W.: Preparations for semantics-based XML mining. In Proc. of the 2001 IEEE Int. Conf. on Data Mining, San Jose, California, pp. 345–352, Dec., 2001.
5.   Chang C.H., Lui S.C. and Wu Y.C.: Applying pattern mining to Web information extraction. In Proc. of the Fifth Pacific-Asia Conf. on Knowledge Discovery and Data Mining, Hong Kong, China, pp. 4–16, April, 2001.
6.   Sedgewick R.: An introduction to the analysis of algorithms. Addison-Wesley Press, 1996.
7.   W3C's Document Object Model (DOM) home page [http:// www.w3.org/DOM/].
8.   Agrawal R. and Srikant R.: Mining sequential patterns. In Proc. of the Eleventh Int. Conf. on Data Engineering, Taipei, Taiwan, pp. 3–14, March, 1995.
9.   Meng W., Wang W., Sun H. and Yu C.: Concept Hierarchy Based Text Database Categorization. In J. of Knowledge and Information Systems 4(2): 132–150, 2002.
10. ACM SIGMOD Record home page [http://www.acm.org/ sigmod/record/xml].
11. IBM's XML Generator homepage [http://www. alphaworks.ibm.com].

Optimization of Fuzzy Rules for Classification Using Genetic Algorithm

Myung Won Kim[1], Joung Woo Ryu[1] ,Samkeun Kim[2], and Joong Geun Lee[3]

[1] School of Computing, Soongsil University, 1-1, Sangdo 5-Dong, Dongjak-Gu,
Seoul, Korea
mkim@comp.ssu.ac.kr,ryu0914@orgio.net
[2] Dept. of Computer Engineering, Hankyoung National University, 67, Seokjeong-dong, An-
sung, Kyonggi-do, Korea
skim@ce.hankyong.ac.kr
[3] FINANCE/SERVICE BUSSINESS UNIT, LG CNS Co.,Ltd., Prime Tower 10-1,
Hoehyun-Dong, 2-Ga, Jung-Gu, Seoul, Korea
leejg@lgcns.com

Abstract. In this paper, we propose an efficient fuzzy rule generation algorithm based on fuzzy decision tree for high accuracy and better comprehensibility. We combine the comprehensibility of rules generated based on decision tree such as ID3 and C4.5 and the expressive power of fuzzy sets for dealing with quantita-tive data. Particularly, fuzzy rules allow us to effectively classify patterns of non-axis-parallel decision boundaries, which are difficult to do using attribute-based classification methods. We also investigate the use of genetic algorithm to optimize fuzzy decision trees in accuracy and comprehensibility by determining an appropriate set of membership functions for quantitative data. We have ex-perimented our algorithm with several benchmark test data including manually generated two-class patterns, the iris data, the Wisconsin breast cancer data, and the credit screening data. The experiment results show that our method is more efficient in performance and comprehensibility of rules compared with methods including C4.5 and FID (Fuzzy ID3).

1 Introduction

With an extended use of computers, we can easily generate and collect data. As their volume and dimensionality increase, it becomes hard to analyze the data manually. Therefore there is a need to automatically analyze the data to discover knowledge from databases [1].
Data mining is a new technique which discovers useful knowledge from data. In data mining important evaluation criteria are efficiency and comprehensibility of knowl-edge. The discovered knowledge should well describe the characteristics of the data and it should be easy to understand in order to facilitate better understanding of the data and use it effectively.

K.-Y. Whang, J. Jeon, K. Shim, J. Srivatava (Eds.): PAKDD 2003, LNAI 2637, pp. 363–375, 2003.

A decision tree such as ID3 and C4.5 is one of the most widely used classification methods in data mining. One of the difficult problems in classification is to handle quantitative data appropriately. Conventionally, a quantitative attribute domain is divided into a set of crisp regions and by doing so the whole data space is partitioned into a set of subspaces(hyper-rectangles), each of which corresponds to a classification rule describing that a sample belonging to the subspace is classified into the representative class of the subspace. However, such a crisp partitioning is not natural to human and inefficient in performance because of the sharp boundary problem. Recently, fuzzy decision trees have been proposed to overcome this problem [2],[3]. It is well known that the fuzzy theory not only provides natural tool for describing quantitative data but also generally produces good performance in many applications. However, one of the difficulties with fuzzy decision trees is determining an appropriate set of membership functions corresponding fuzzy linguistic terms. Usually membership functions are given manually by human, however, it is difficult for even an expert to determine an appropriate set of membership functions when the volume and dimensionality of data increase.

In this paper we investigate combining the fuzzy theory and the conventional decision tree algorithm for accurate and comprehensible classification. We propose an efficient fuzzy rule generation algorithm called the fuzzy decision tree(FDT) algorithm for data mining, which integrates the comprehensibility of a decision tree and the expressive power of fuzzy sets. We also propose the use of genetic algorithm for optimal decision tree construction by determining an appropriate set of fuzzy sets for quantitative data.

The paper is organized as follows: in the next section we describe fuzzy rules and fuzzy inference we use for classification, followed by fuzzy decision tree construction. In section 3, we describe the use of genetic algorithm for determining an appropriate set of membership functions. In section 4, we describe experiments of our algorithm with a variety of the benchmark data sets including the iris data and the Wisconsin breast cancer data, the credit screening data, and others.

2   Fuzzy Inference

2.1   Fuzzy Classification Rules

We use a simple form of fuzzy rules and inference for better human's comprehensibility. Each fuzzy rule is composed of an antecedent part and a consequent part. Each fuzzy rule also is associated with a CF(Certainty Factor) to represent the degree of belief that the consequent is drawn from the antecedent satisfied. Equation (1) is a typical form of fuzzy classification rules used in our approach.

$$R_i \; : \; if \; A_{i1} \, is V_{i1} \; and \; A_{i2} \, is V_{i2} ... and \; A_{im} \, is V_{im} \; then \; 'Class' is k \; (CF_i) \qquad (1)$$

In equation (1) A_{ik} represents an attribute and V_{ik} denotes a fuzzy linguistic term representing a fuzzy set associated with attribute A_{ik}. Application of such a rule to a sample $X = (x_1, x_2, ..., x_n)$ results the confidence with which X is classified into class k from the antecedent satisfied. In this paper among a variety of fuzzy inference methods we adopt the standard method as described in the following:

1) *min* is used to combine the degrees of satisfaction of individual conditions of an antecedent;

2) *product* is used to propagate the degree of satisfaction of an antecedent to a consequent;

3) *max* is applied for aggregating the results of individual rule applications.

For a given datum $X = (x_1, x_2, ..., x_n)$, according to our method the confidence of class k is obtained as

$$Conf_k(X) = \max_{R_i \in R(k)} \left\{ \left(\min_j \mu_{V_{ij}}(x_{ij}) \right) \cdot CF_i \right\} \tag{2}$$

In equation (2) $R(k)$ is the set of all rules that classify patterns into class k (their consequent parts are *'Class* is k') . The class of the maximum $Conf_k(X)$ is the final classification of X.

2.2 Membership Functions for Fuzzy Sets

Membership functions are very important in fuzzy rules. They affect not only the performance of fuzzy rule based systems but also the comprehensibility of rules. Triangular, trapezoidal, and Gaussian membership functions are widely used, however, in this paper we adopt triangular membership functions. A triangular membership function can be represented by a triple of numbers (l, c, r), where l, c, and r represent the left, the center, and the right points of the triangular membership function, respectively (Fig. 1).

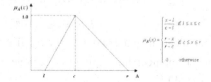

Fig. 1. Triangular membership function

In this paper we investigate the use of genetic algorithm to automatically generate an appropriate set of membership functions for a given set of data. Membership functions are generated in such a way that they are optimized in the sense that the generated rules are efficient and comprehensible and it is described in section 3.

2.3 Fuzzy Decision Tree

A fuzzy decision tree is similar to a (crisp) decision tree. It is composed of nodes and arcs representing attributes and attribute values or value sets, respectively. The major difference is that in a fuzzy decision tree each arc corresponds to a fuzzy set, which is usually represented by a fuzzy linguistic term. Also in a fuzzy decision tree a terminal node represents a class and it is associated with a certainty factor representing the confidence of the decision corresponding to the terminal node. In the fuzzy decision tree a decision is made by aggregating the conclusions of multiple paths(rules) fired as equation (2) describes while in the crisp decision tree only a single path is fired for a decision.

Let us represent a datum $X = (x_1, x_2, , , x_n)$ where $x_i \in S_i$ and S_i represents the domain of attribute i. Now the whole data space can be represented by $W = S_1 \times S_2 \times ... \times S_n$. Suppose we have a fuzzy decision tree T. We can consider that each node n in T is associated with a fuzzy subspace W_n of W (we call it a node space) and it can be defined as following:

$$W_n = \begin{cases} W & \text{if } n \text{ is the root node;} \\ W_{p(n)} \cap W_{(p(n),n)} & \text{otherwise.} \end{cases}$$

where $p(n)$ is the parent node of n and $W_{(m,n)}$ is a fuzzy subspace(we call it an *arc space*) of W associated with an arc (m,n) defined as following:

$$W_{(m,n)} = U_1 \times U_2 \times , , , \times U_n$$

where $U_i = \begin{cases} F_{(m,n)} & \text{if } i = att(m); \\ S_i & \text{otherwise.} \end{cases}$

Here, $F_{(m,n)}$ is a fuzzy set corresponding to arc (m,n) and $att(m)$ is the attribute represented by node m. Let $v_n(X)$ represent the membership that X belongs to the node space of n, then we have the following according to the above definitions:

$$v_n(X) = \mu_{W_n}(X) = \min(v_{p(n)}(X), \mu_{W_{(p(n),n)}}(X))$$
$$= \min(v_{p(n)}(X), \mu_{F_{(p(n),n)}}(x_i)) \text{ where } i = att(p(n)).$$

Now let $P = (n_0, n_1, , , n_k)$ be a path in T where n_0 is the root node.

$$v_{n_k}(X) = \min_i(\mu_{F_{(n_i, n_{i+1})}}(x_j)) \text{ where } j = att(n_i).$$

Each node n in P represents an attribute and arc (n_i, n_{i+1}) corresponds to a fuzzy set equation (3) calculates the membership degree that sample X belongs to fuzzy subspace S_p corresponding to path P.

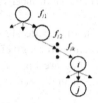

Fig. 2. Fuzzy decision tree construction

$$E_m(i) = \sum (q_{mn} I_n) \tag{3}$$

$$I_n = -\sum_{k \in C} \left(p_n(k) \log_2 p_n(k) \right) \tag{4}$$

$$p_n(k) = \frac{\sum_{Y \in D_k} v_n(Y)}{\sum_{X \in D} v_n(X)} \tag{5}$$

$$q_{mn} = \frac{\sum_{Y \in D} v_n(Y)}{\sum_{X \in D} v_m(X)} \tag{6}$$

The entropy of attribute i for node m, $E_m(i)$ is a measure indicating how good attribute i is to the node. It is defined as in equation (3), where C is the set of classes, D is the set of training data. D_k denotes the set of training samples that are classified into class k.

In the above equations, n is a child node of node m, corresponding to a fuzzy membership function for an attribute i. Equation (4) calculates the information gain for node n, equation (5) represents the probability that node m represents class k, and equation (6) represents the possibility that for a given attribute F a training data fits to node n compared to other sibling nodes.

Our fuzzy decision tree construction algorithm is as follows.

<Step 1>

(1) $\dfrac{1}{|D|} \sum_{X \in D} v_m(X) \le \theta_s$

(2) $\dfrac{\sum_{X \in D_{k^*}} v_m(Y)}{\sum_{X \in D} v_m(X)} \ge \theta_d$ where $k^* = \arg \max_{k \in C} (\sum_{Y \in D_k} v_m(Y))$

(3) no more attributes are available.

If one of the following conditions is satisfied, then make node m a terminal node. In condition (2) class k^* represents the majority class of the corresponding fuzzy subspace. In this case it is the terminal node whose class is k^* and the associated

$$CF = \frac{\sum_{Y \in D_{k^*}} v_m(Y)}{\sum_{X \in D} v_m(X)}$$

CF is determined by

<Step 2>

Otherwise,

(1) Let $E_m(i^*) = \min_i (E_m(i))$

(2) For each fuzzy membership function of attribute i^*,
    make a child node from the node associated with i^*.

(3) Go to Step.1 and apply the algorithm to all newly generated nodes, recursivly.

In the above the threshold parameters θ_s and θ_d determine when we terminate the pattern space partitioning. Condition (1) prohibits generating rules for sufficiently sparse fuzzy subspaces while condition (2) prohibits generating rules for the minor classes having insufficient number of patterns in a fuzzy subspace. In general the fuzzy partitioning method divides the whole pattern space into L^{N_F} subspaces, where N_F is the number of attributes and L is the average number of membership functions associated with an attribute. Each subspace also can contain $|C|$ classes of samples. In this case we may have an excessive number of fuzzy rules to be generated. The threshold parameters θ_s and θ_d are used to control overfitting by prohibiting too much detail rules to be generated.

For a fuzzy decision tree constructed each path from the root node to a terminal node of the tree corresponds to a fuzzy rule. After constructing a fuzzy decision tree, we determine an appropriate linguistic expression for each fuzzy membership function. Our fuzzy decision tree construction method described in this section is similar to FID proposed [4],[5].

3 Membership Function Generation Using Genetic Algorithm

In fuzzy rules membership functions are important since they affect both of accuracy and comprehensibility of rules. However, membership functions are usually given by the human and it is difficult even for an expert to determine an appropriate set of membership functions when the volume and dimensionality of data increase. In this paper, we propose the use of genetic algorithm to determine an optimal set of membership functions for a given classification problem.

3.1 Genetic Algorithm

Genetic algorithm is an efficient search method simulating natural evolution, which is characterized by survival of the fittest.

Our fuzzy decision tree construction using genetic algorithm is following.

(1) Generate an initial population of chromosomes of membership functions;
(2) Construct fuzzy decision trees using the membership functions;
(3) Evaluate each individual membership function set by testing its corresponding fuzzy decision tree;
(4) Test it the termination condition is satisfied;
(5) If the termination condition is satisfied, then exit;
(6) Otherwise, generate a new population of chromosomes of membership functions by applying genetic operators, and go to (2).

We use genetic algorithm to generate an appropriate set of membership functions for quantitative data values. Membership functions should be appropriate in the sense that they result in a good performance and they result in as simple a decision tree as possible.

In our genetic algorithm a chromosome is of the form $(\phi_1, \phi_2, ..., \phi_h)$ where ϕ_i represents a set of membership functions associated with attribute i, which is, in turn, of the form $\{\mu_1^i, \mu_2^i, ..., \mu_k^i\}$ where μ_j^i represents a triplet of membership function for attribute i. The number of membership functions for an attribute may not necessarily be fixed.

3.2 Genetic Operations

We have genetic operations such as crossover, mutation, splitting and deletion as described in the following.

1) *crossover*: it generates new chromosomes by exchanging the whole sets of membership functions for a randomly selected attribute of the parent chromosomes.

2) *mutation*: randomly selected membership functions for a randomly selected attribute in a chromosome are mutated by adding random (Gaussian) noise to the left, the center, and the right points of an individual membership function;

3) *addition*: for any attribute in a chromosome, its associated set of membership functions are analyzed and new appropriate membership functions are added if necessary.

4) *merging*: any two close membership functions of any attribute of a chromosome are merged into one.

3.3 Fitness Evaluation

If membership functions are determined for each of attributes, we generate an fuzzy decision tree according to the algorithm described in section 2.3. We use the performance of the generated fuzzy decision tree in fitness evaluation of a chromosome. Suppose a fuzzy decision tree $\tau(i)$ is generated from the membership functions represented by chromosome i. The fitness score of chromosome i is given by

$$Fit(i) = \alpha P(\tau(i)) - \beta C(\tau(i)) \tag{7}$$

where $P(\tau(i))$ represents the performance of decision tree $\tau(i)$ and $C(\tau(i))$ represents the complexity of $\tau(i)$, measured in terms of the number of nodes of $\tau(i)$. We also apply the roulette wheel method for selecting candidate chromosomes for the crossover operation. For more efficient search we first generate a set of fuzzy membership functions by histogram analysis of training data and introduce it to the initial population for genetic algorithm.

We also can use genetic algorithm to generate fuzzy rules by evolving the form of rules(selection of attributes and their associated values) and their associated membership functions simultaneously. Genetic algorithm often generates efficient fuzzy rules, however, it is time-consuming and our method trades off between classification accuracy and computational time. In our approach we only apply genetic algorithm to generate membership functions for continuous value attributes and by doing so we trade off classification accuracy and computational time.

4 Experiments

We have conducted several experiments with various data sets including manually generated generalized XOR, and several benchmark data sets in the UCI machine learning databases[6] such as Iris, Wisconsin breast cancer, Credit screening, Heart disease, Sonar, Bupa, and Pima. We compare different decision tree algorithms and classification methods such as ID3 and C4.5. We also compare our algorithm with FID3.1, which is a fuzzy decision tree algorithm with automatic membership function generation proposed by [5]. We describe our experiments in the following.

4.1 Generalized XOR

In this experiment we use manually generated data sets including the slashed two-class pattern, the generalized XOR data and the 90-degree rotated generalized XOR data. For the slashed pattern and the rotated generalized XOR data we notice that our algorithm can well classify patterns with non-axis-parallel decision boundaries. Such patterns are difficulty to classify efficiently for the conventional ID3 and C4.5 type decision tree based classification methods. For the generalized XOR data, the histogram analysis fails to generate an appropriate set of membership functions, however, the genetic algorithm well succeeds in doing it(Fig. 4(c)). For all cases our algorithm generates simple fuzzy rules which classify patterns 100% correctly. Fig. 5 illustrates four fuzzy rules generated for the rotated generalized XOR. Each rule corresponds to one of four pattern blocks as shown in Fig. 3(c). We also notice that fuzzy rules have a kind of abstraction capability in that fuzzy rules in linguistic terms describe rough pattern classification, while all the detail classification is taken care of by membership functions.

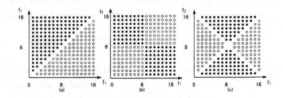

Fig. 3. (a) Slashed two-class pattern; (b) Generalized XOR; (c) 90-degree rotated generalized XOR

4.2 Characteristics of Pruning Parameters

The Wisconsin Breast Cancer data obtained from the University of Wisconsin Hospitals at Madison classifies 2 different kinds of breast cancer: the benign breast cancer(c_1), and the malignant breast cancer(c_2), based on 10 attributes: sample code number(f_1), clump thickness(f_2), uniformity of cell size(f_3), uniformity of cell shape(f_3),

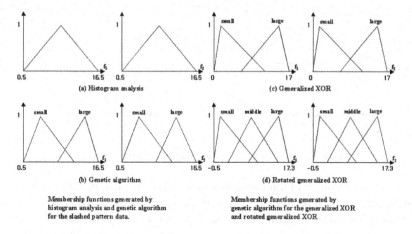

Fig. 4. Membership functions

> if f_1 is middle and f_2 is small then class is C_1 (0.75).
> if f_1 is middle and f_2 is large then class is C_1 (0.76).
> if f_1 is small and f_2 is middle then class is C_2 (0.75).
> if f_1 is large and f_2 is middle then class is C_2 (0.78).

Fig. 5. Fuzzy rules generated for the rotated generalized XOR. Associated membership functions are shown in Fig. 7(b). C1: black circle, C2: white circle.

marginal adhesion(f_5), single epithelial cell size(f_6), bare nuclei(f_7), bland chromatin(f_8), normal nucleoli(f_9), mitoses(f_{10}). The total number of data is 367 and 14 data have missing values. In this experiment we discard those data with missing values so we use only 353 data for our experiment. To construct the fuzzy decision tree we use 106 data for class 1 and 94 data for class 2. We use all the 353 data to test the constructed fuzzy decision tree. In this experiment we have θ_s and θ_d set to 0.35, and 0.7, respectively. The number of generated fuzzy rules is 5 and the classification rate is 91.8%. Fig.6 shows classification rates and the number of generated fuzzy rules as varying θ_s and θ_d, respectively, with the other parameter fixed. We can obtain the desired performance and an appropriate number of rules by controlling the threshold parameters properly.

4.3 Comparison of FDT and Other Fuzzy Rule Algorithms

We used the Iris data sets to compare the performance of FDT and fuzzy rule generation algorithms. The iris data classifies 3 different kinds of Iris, setosa(c_1), versicolor(c_2), and virginica(c_3), based on 4 attributes: length of sepal(f_1), width of sepal(f_2), length of petal(f_3), and width of petal(f_4). The number of training data for each class is 50 and the total number of data is 150. To construct a fuzzy decision tree we sample

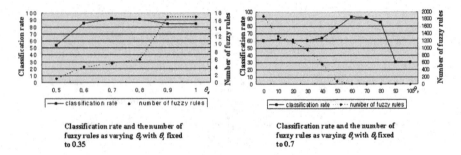

Fig. 6. Classification rate and the number of fuzzy rules

randomly the same number of training data for each class. We use all 150 data for test. To analyze classification performance we vary the size of training data to 21, 30, 60, 90 for training. For all the experiments we have θ_s and θ_d fixed to 0.1 and 0.7, respectively. Table 1 shows the classification rate and the number of generated fuzzy rules(in parenthesis) of the FDT algorithm and those described in [7],[8] as varying the size of training data.

Table 1. Comparison of classifications rate and the number of generated funzzy rules for various algoritms

The size of training data	Simple fuzzy grid[7]	Distributed fuzzy if-then rules[7]	CF criterion[7]	NM criterion [7]	RM criterion [7]	Jang and Choi [8]	FDT
21	91.3% (455)	92.4% (8328)	91.1% (253)	89.8% (71)	89.6% (72)	88.3% (32)	91.3% (3)
30	92.7% (1727)	93.8% (20512)	91.8% (307)	93.0% (83)	93.3% (87)	91.6% (36)	95.3% (3)
60	93.5% (2452)	95.4% (63069)	93.8% (307)	93.9% (105)	94.1% (107)	93.3% (46)	96.0% (3)
90	94.5% (3440)	95.8% (140498)	95.1% (528)	94.8% (150)	94.6% (150)	95.0% (46)	96.0% (3)

We notice that the proposed algorithm generates a smaller number of fuzzy rules and its classification rate is higher compared to other methods. The generated fuzzy rules are as follow:

if petal(f_4) is small(-0.18,0.3,0.65) then setosa(c_1) (1);
if petal(f_4) is middle(0.80,1.4,1.95) then vesicolor(c_2) (0.84);
if petal(f_4) is large(1.3,2.01,2.7) then virginica(c_3) (0.81).

In the above rules, the fuzzy linguistic terms such as 'small', 'middle', and 'large' are given by the human.

Rules generated by the methods described in [7] have unnecessarily complicated and often an excessive number of rules, even exceeding the training data size, are generated. The simple fuzzy grid method generates fuzzy rules corresponding to fuzzy subspaces determined by partitioning each attribute space into fuzzy sets. The distributed fuzzy if-then rule method generates fuzzy rules of different fuzzy partitions of the pattern space and employs all those rules simultaneously for classifying a pattern. [7] proposes a method of successive partitioning subspaces selected by different criteria such as the certainty factor(CF), the number of misclassified patterns(NM), and the rate of misclassified patterns(RM) by the rule corresponding to the fuzzy subspace. The fuzzy rules generated by [8] is represented in connection weights which causes difficulties in understanding. However it is easy to understand the fuzzy rules generated by the FDT algorithm because their antecedent parts are simple and they are small in number.

In addition, we have experimented our algorithm with several other set of data including the credit screening data(approval of credit cards), the heart disease data(diagnosis of heart disease), and the sonar data(target object classification by analyzing the echoed sonar signals). In these experiments, we use genetic algorithm to generate membership functions for the FDT algorithm. Table 2 compares the performance of FDT with that of C4.5 [9],[10] (release 7 and 8) in classification accuracy and tree size. Accuracies and tree sizes are averaged using 10-fold cross-validation for each data set. It is clearly shown that our algorithm is more efficient than C4.5 both in classification accuracy and tree size.

Table 2. Comparison of FDT and C4.5(release 7 and 8) ($\pm x$ indicates the standard deviation.)

Data set	C4.5(Rel. 7)		C4.5(Rel. 8)		FDT	
	accuracy(%)	tree size	acuracy(%)	tree size	accuracy(%)	tree size
Breast cancer	94.71±0.09	20.3±0.5	94.74±0.19	25.0±0.5	96.66±0.02	5.9±2.8
Credit screening	84.20±0.30	57.3±1.2	85.30±0.20	33.2±1.1	86.18±0.03	4±0.0
Heart disease	75.10±0.40	45.3±0.3	77.00±0.50	39.9±0.4	77.8±0.05	22±6.0
Iris	95.13±0.20	9.3±0.1	95.20±0.17	8.5±0.0	97.99±0.03	4.2±0.4
Sonar	71.60±0.60	33.1±0.5	74.40±0.70	28.4±0.2	77.52±0.06	14.1±3.1

However, it should be noticed that for FDT the tree sizes are more sensitive to training data compared with C4.5. It can be expected by considering the nature of genetic algorithm. It is because that in genetic algorithm membership functions can evolve quite differently from others depending the training data and the initial population of chromosomes. For each data set in only one or two cases out of 10 relatively big trees are constructed.

In next experiment, we compare our algorithm with C4.5 and FID3.1 using data sets Iris, Bupa and Pima. All experiments were performed using 10-fold cross-validation. Table 3 compares the performance in accuracy and tree size of FDT and FID3.1 [5]. It is clearly shown that FDT is consistently more efficient than both C4.5 and FID3.1, in terms of both accuracy and tree size.

Table 3. Comparison of C4.5, FID3.1 and FDT

	C4.5		FID3.1		FDT	
	accuracy (%)	tree size	accuracy(%)	tree size	accuracy (%)	tree size
Iris	94.0	5.0	96.0	5.0	97.99	4.2
Bupa	67.9	34.4	70.2	28.9	70.69	12.1
Pima	74.7	45.2	71.56	23.1	76.90	4.2

5 Conclusions

In this paper, we propose an efficient fuzzy rule generation algorithm based on fuzzy decision tree for data mining. Our method provides the efficiency and the comprehensibility of the generated fuzzy rules, which are important to data mining. Particularly, fuzzy rules allow us to effectively classify patterns of non-axis-parallel decision boundaries using membership functions properly, which is difficult to do using attribute-based classification methods. In fuzzy rule generation it is important to determine an appropriate set of membership functions associated with attributes, however, it is difficult to do manually. We also propose the use of genetic algorithm for automatic generation of an optimal set of membership functions. We have experimented our algorithm with several benchmark data sets including the iris data, the Wisconsin breast cancer data, the credit screening data, and others. The experiment results show that our method is more efficient in classification accuracy and compactness of rules compared with other methods including C4.5 and FID3.1.

We are now applying our algorithm for generating fuzzy rules describing the blast furnace operation in steel making. Our preliminary experiments have shown that the algorithm is powerful for many real world applications. We also plan to work for improving the algorithm to be able to handle a large volume of data efficiently, which is often an important requirement for data mining. Particularly, coevolution can replace genetic algorithm and some performance improvement is expected.

Acknowledgement. This paper was supported by Brain Science and Engineering Research Program sponsored by Korean Ministry of Science and Technology.

References

1. Fayyad, U., Mannila, H., and Piatetsky-Shapiro, G., Data Mining and Knowledge Discovery. Kluwer Academic Publishers (1997)
2. Suárez, A. and Lutsko, J.F., Globally Optimal Fuzzy Decision Trees for Classification and Regression, IEEE Trans. on Pattern Analysis and Machine Intelligence, 21(12), (1999) 1297–1311
3. Zeidler, J. and Schlosser M., Continuous-Valued Attributes in Fuzzy Decision Trees., Proc. of the 6th Int. Conf. on Information Processing and Management of Uncertainty in Knowledge-Based Systems, (1996) 395–400

4. Janikow, C.Z., Fuzzy Decision Tree : Issues and Methods, IEEE Trans. on Systems, Man and Cybernetics – Part B: Cybernetics, 28(1),(1998) 1–14
5. Janikow, C.Z. and Faifer, M., Fuzzy Partitioning with FID3.1, Proc. of the 18th International Confernece of the North American Fuzzy Information Processing Society, IEEE 1999. (1999) 467–471
6. Blake, C.L. and Merz, C.J., UCI Repository of machine learning databases. Irvine, CA: University of California, Department of Information and Computer Science (1998).
7. Ishibuchi, H., Nozaki, K., and Tanaka, H., Effective fuzzy partition of pattern space for classification problems, Fuzzy Sets and Systems, Vol.59. (1993) 295–304
8. Jang, D.S. and Choi, H.I., Automatic Generation of Fuzzy Rules with Fuzzy Associative Memory, Proc. of the ISCA 5th Int. Conf. (1996) 182–186
9. Quinlan, J.R.,Improved use of continuous attributes in C4.5, Journal of Artificial Intelligence Research, 4, (1996) 77–90
10. Quinlan, J.R.,C4.5: Programs for Machine Learning, Morgan Kaufmann Publishers (1993)

Fast Pattern Selection for Support Vector Classifiers

Hyunjung Shin and Sungzoon Cho

Department of Industrial Engineering, Seoul National University,
San 56-1, Shillim-Dong, Kwanak-Gu, 151-742, Seoul, Korea
{hjshin72, zoon}@snu.ac.kr

Abstract. Training SVM requires large memory and long cpu time
when the pattern set is large. To alleviate the computational burden in
SVM training, we propose a fast preprocessing algorithm which selects
only the patterns near the decision boundary. Preliminary simulation
results were promising: Up to two orders of magnitude, training time
reduction was achieved including the preprocessing, without any loss in
classification accuracies.

1 Introduction

One of the strongest points of Support Vector Machine (SVM) theory is the fact
that SVM quadratic programming (QP) problem is quite simple [14]. In SVM
QP formulation, the dimension of kernel matrix ($M \times M$) is equal to the number
of training patterns(M). Unfortunately, however, this innate characteristic yields
a difficulty when a large scale real-world problem is given. This enormous matrix
cannot easily fit into the memory of a standard computer. Moreover, training
cannot be finished in a reasonable time even if we manage to load the matrix on
the memory.

A standard QP solver has time complexity of order $O(M^3)$ [10]. Here, for the
sake of simplicity, we do not consider the dimension of input pattern. A small
training set can be solved with such standard QP solvers: MINOS, CPLEX,
LOQO and MATLAB QP routines. In order to attack the large scale SVM QP
problem, the decomposition methods or iterative methods have been suggested
which break down the large QP problem into a series of smaller QP problems.
They include the Chunking algorithm, Sequential Minimal Optimization (SMO)
algorithm, SVM Light algorithm and Successive Over Relaxation (SOR) algorithm [5]. The general time complexity of those decomposition methods is approximately (*the number of iterations*)$\cdot$ $O(Mq+q^3)$: if q is the size of the working
set, each iteration costs $O(Mq)$ to calculate q columns of the kernel matrix, about
$O(q^3)$ for solving the subproblem, and $O(Mq)$ for finding the next working set
and updating the gradient. "The number of iterations" is supposed to increase
as the number of training patterns increases.

K.-Y. Whang, J. Jeon, K. Shim, J. Srivatava (Eds.): PAKDD 2003, LNAI 2637, pp. 376–387, 2003.
© Springer-Verlag Berlin Heidelberg 2003

One way to circumvent this computational burden is to select only the training patterns, in advance, that are more likely to be support vectors. In a classification problem, the support vectors are distributed near the decision boundary and participate in the constitution of the margins thereabout. Therefore, selecting those patterns (potential support vectors) prior to SVM training is quite desirable (see Fig. 1).

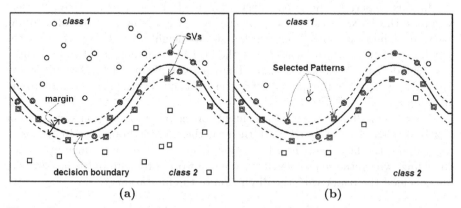

Fig. 1. Pattern selection: a large training set shown in (a) is condensed to a small training set (b) which is composed of only potential support vectors near the decision boundary.

Up to now, there have been considerable researches about pattern selection near the decision boundary for classification problem. Shin and Cho selected the clean patterns near the decision boundary based on the bias and variance of outputs of a network ensemble [12]. Lyhyaoui et al. implemented RBF classifiers, somewhat like SVMs, by selecting patterns near the decision boundary. They proposed 1-nearest neighbor method in opposite class after class-wise clustering [7]. To lessen the burden of the MLP training, Choi and Rockett used kNN classifier to estimate the posterior probabilities for pattern selection. But one major drawback is that it takes approximately $O(M^2)$ to estimate the posterior probabilities [3]. A pattern selection approach, more focused on SVMs, was proposed by Almeida et al. They conducted k-means clustering on the entire training set. All patterns were selected for heterogeneous clusters in their class membership, while only centroids were selected for homogeneous clusters [1].

Recently, we proposed to select the patterns near the decision boundary based on the neighborhood properties [11]. We utilized k nearest neighbors to look around the pattern's periphery. The first idea stems from the fact that a pattern located near the decision boundary tends to have more heterogeneous neighbors with respect to their class membership. The degree of proximity to the decision boundary is estimated by Neighbors_Entropy. A larger Neighbors_Entropy value indicates that the pattern is close to the decision boundary. The second idea arises from the fact that a pattern on the right side of the decision boundary tends to belong to the same class as its neighbors. Among the patterns with positive Neighbors_Entropy value, potential noisy patterns are eliminated by the ratio of the neighbors whose label matches that of the pattern. A smaller ratio indicates that the pattern is potentially incorrectly labeled. The approach reduced the number of patterns significantly, thus reduced the training time. However, the pattern selection process evaluating the kNNs for all patterns, was time consuming. A naive algorithm took $O(M^2)$, so the pattern selection process became a near bottleneck.

In this paper, we propose a fast version of the algorithm. Here, we just compute the kNNs of the patterns near the decision boundary, not all training patterns. The idea comes from another simple neighborhood property that the neighbors of the pattern located near the decision boundary tend to be located near the decision boundary as well. The time complexity of the fast algorithm is $O(bM)$, where b is the number of patterns in the "overlapped" region around decision boundary.

Now, we briefly summarize the contents of the paper. In section 2, we present our basic idea of selecting the patterns near the decision boundary. In section 3, we propose the fast algorithm. In section 4, we show the experimental results, involving two synthetic problems and one real world problem. In the last section, we conclude the paper with the discussion of the limitations.

2   Selection Criteria Based on Neighborhood Properties

The first neighborhood property is that a pattern located near the decision boundary tends to have heterogeneous neighbors with respect to their class labels. Thus, the degree of pattern x's proximity to the decision boundary can be estimated by "$Neighbors\_Entropy(x)$", which is defined as the entropy of pattern x's k-nearest neighbors' class labels (see Fig. 2). A pattern with a positive Neighbors_Entropy(x) value is assumed to be located near the decision boundary. Note that the measure considers the pattern's neighbors only, not the pattern itself.

The second neighborhood property is that a pattern on the right side of the decision boundary tends to belong to the same class as its neighbors. If a pattern's own label does not match the majority label of its neighbors', it is likely to be incorrectly labeled. The measure "$Neighbors\_Match(x)$" is defined as the ratio of x's neighbors whose label matches that of x. Only those pattern xs are selected whose Neighbors_Match$(x) > \beta \cdot \frac{1}{J}$ (J is the number of classes and $0 < \beta \leq 1$). In other words, potential noisy patterns are eliminated. The parameter β controls the selectivity. We have empirically found a value of 0.5. Based on those neighborhood properties we propose following selection criteria.

[**Selecting Criteria**] Neighbors_Entropy$(x) > 0$ and Neighbors_Match$(x) > \beta \cdot \frac{1}{J}$.

3   Fast Implementation of the Proposed Method

A naive algorithm was represented in [11] where the kNNs of all patterns were evaluated. This algorithm is easy to implement and also performs well as long as the size of training set, M, is relatively small.

LabelProbability(x) {

/* For x, calculate the label probabilities of kNN(x) over J classes, $\{C_1, C_2, \ldots, C_J\}$, where kNN(x) is defined as the set of k nearest neighbors of x. */

$k_j = |\{x' \in C_j | x' \in k\text{NN}(x)\}|, \quad j = 1, \ldots, J.$

return $\left(P_j = \frac{k_j}{k}, \forall j\right).$

}

Neighbors_Entropy(x) {

/* Calculate the neighbors-entropy of x with its nearest neighbors' labels. In all calculations, $0\,log_J\frac{1}{0}$ is defined to be 0. */

Do **LabelProbability(x)**.

return $\left(\sum_{j=1}^{J} P_j \cdot log_J\frac{1}{P_j}\right).$

}

Neighbors_Match(x) {

/* Calculate the neighbors-match of x. j^* is defined as the label of x itself.*/

$j^* = \underset{j}{\arg}\{C_j \mid x \in C_j, j = 1, \ldots, J\}.$

Do **LabelProbability(x)**.

return (P_{j^*}).

}

Fig. 2. Neighbors_Entropy and Neighbors_Match functions

However, when the size of training set is large, the computational cost increases in proportion to the size. Let's assume that the distance between any two points in d-dimensional space can be computed in $O(d)$. Then to find the nearest neighbors for each pattern takes sum of "distance computation time (DT)" and "search time (ST)". The total time complexity of the naive algorithm, therefore, is $O(\,M \cdot (DT + ST)\,)$. Roughly speaking, it is $O(M^2)$.

DT : Distance Computation Time	$O(\,d \cdot (M-1)\,)$
ST : Search(query) Time	$\min\{O(\,(M-1) \cdot \log(M-1)\,), O(\,k \cdot (M-1)\,)\}$
Total Time Complexity	$O(\,M \cdot (DT + ST)\,) \approx O(M^2)$

There is a considerable literature on nearest neighbor searching for the case of the huge data size and the high dimensional. Most approaches focus on reducing DT or ST. See [4] and [13] for reducing DT, and [2] and [8] for reducing ST.

Our approach, on the other hand, focuses on reducing M of $O(M \cdot (DT + ST))$. The idea comes from yet neighborhood property that the neighbors of the pattern located near the decision boundary tend to be the patterns near the decision boundary themselves. We start with a set of randomly selected patterns. Once we find some of the patterns near the decision boundary, we examine the its neighbors. This successive evaluation of the neighbors of a current set of patterns will stop when all the patterns near the decision boundary are chosen and evaluated. By following [Spanning Criteria], it is determined whether the next evaluation will be spanned to the pattern's neighbors or not. This "selective kNN spanning" procedures is shown in Fig. 3.

[**Spanning Criteria**] Neighbors_Entropy(x) > 0

4   Experiments

The proposed algorithm was tested on two artificial binary-class problems and one well-known real world problem. All simulations were carried on a Pentium 800 MHz PC with 256 MB memory.

4.1   Synthetic Problems

The two problems were devised for the performance proof according to the different decision boundary characteristics. We used Gunn's SVM Matlab Toolbox [15].

Continuous XOR Problem with Four Gaussian Densities: The first problem is a continuous XOR problem. From the four Gaussian densities, a total of 600 training patterns were generated. There are about 10% patterns in the overlapped region between the densities. The characteristic of this problem is that the density of patterns gets sparser when it goes closer to the decision boundary.

$$C_1 = \left\{ x \mid x \in C_{1A} \cup C_{1B}, \begin{bmatrix} -3 \\ -3 \end{bmatrix} \leq x \leq \begin{bmatrix} 3 \\ 3 \end{bmatrix} \right\},$$

$$C_2 = \left\{ x \mid x \in C_{2A} \cup C_{2B}, \begin{bmatrix} -3 \\ -3 \end{bmatrix} \leq x \leq \begin{bmatrix} 3 \\ 3 \end{bmatrix} \right\}$$

where C_{1A}, C_{1B}, C_{2A} and C_{2B} were

$$C_{1A} = \left\{ x \mid N \left(\begin{bmatrix} 1 \\ 1 \end{bmatrix}, \begin{bmatrix} 0.5^2 & 0 \\ 0 & 0.5^2 \end{bmatrix} \right) \right\}, \; C_{1B} = \left\{ x \mid N \left(\begin{bmatrix} -1 \\ -1 \end{bmatrix}, \begin{bmatrix} 0.5^2 & 0 \\ 0 & 0.5^2 \end{bmatrix} \right) \right\},$$

$$C_{2A} = \left\{ x \mid N \left(\begin{bmatrix} -1 \\ 1 \end{bmatrix}, \begin{bmatrix} 0.5^2 & 0 \\ 0 & 0.5^2 \end{bmatrix} \right) \right\}, \; C_{2B} = \left\{ x \mid N \left(\begin{bmatrix} 1 \\ -1 \end{bmatrix}, \begin{bmatrix} 0.5^2 & 0 \\ 0 & 0.5^2 \end{bmatrix} \right) \right\}.$$

$\mathbf{D}$	The original training set.
$\mathbf{D}_e^i$	The set to be evaluated (spanned or non-spanned) at i th step.
$\mathbf{D}_o^i$	The set of the patterns to be spanned at i th step from $\mathbf{D}_e^i$. Each element will find its k nearest neighbors to constitute the next evaluation set, $\mathbf{D}_e^{i+1}$.
$\mathbf{D}_x^i$	The set of the patterns not to be spanned at i th step from $\mathbf{D}_e^i$.
$\mathbf{D}_s^i$	The set of the selected patterns at i th step from $\mathbf{D}_o^i$.
$\mathbf{S}_o^i$	The accumulated set of the spanned patterns after subsequent evaluations till $(i\text{-}1)$ th step.
$\mathbf{S}_x^i$	The accumulated set of the non-spanned patterns after subsequent evaluations till $(i\text{-}1)$ th step.
$\mathbf{SS}^i$	The accumulated set of the selected patterns till $(i\text{-}1)$ th step, the final set will be returned as a aimed set.
$k\mathbf{NN}(\boldsymbol{x})$	The set of k-nearest neighbors of $\boldsymbol{x}$, whose cardinality is k.

SelectiveKnnSpanning() {

[0] Initialize $\mathbf{D}_e^0$ with randomly chosen patterns from $\mathbf{D}$.
K and J are global constant variables, and $P_1, \cdots, P_J$ are global variables.
$i \leftarrow 0, \quad \mathbf{S}_o^0 \leftarrow \emptyset, \quad \mathbf{S}_x^0 \leftarrow \emptyset, \quad \mathbf{SS}^0 \leftarrow \emptyset.$

while $\mathbf{D}_e^i \neq \emptyset$ do
[1] Calculate **Neighbors_Entropy**$(\boldsymbol{x})$ and **Neighbors_Match**$(\boldsymbol{x})$ of $\boldsymbol{x} \in \mathbf{D}_e^i$.

[2] Choose $\boldsymbol{x}$ satisfying [Spanning Criteria].
$\mathbf{D}_o^i \leftarrow \{\boldsymbol{x} \mid Neighbors\_Entropy\,(\boldsymbol{x}) > 0, \ \boldsymbol{x} \in \mathbf{D}_e^i\}.$
$\mathbf{D}_x^i \leftarrow \mathbf{D}_e^i - \mathbf{D}_o^i.$

[3] Select $\boldsymbol{x}$ satisfying [Selecting Criteria].
$\mathbf{D}_s^i \leftarrow \{\boldsymbol{x} \mid Neighbors\_Match\,(\boldsymbol{x}) > \beta \cdot \frac{1}{J}, \ \boldsymbol{x} \in \mathbf{D}_o^i\}.$

[4] Update the pattern sets: the spanned, the non-spanned and the selected.
$\mathbf{S}_o^{i+1} \leftarrow \mathbf{S}_o^i \cup \mathbf{D}_o^i, \ \mathbf{S}_x^{i+1} \leftarrow \mathbf{S}_x^i \cup \mathbf{D}_x^i, \ \mathbf{SS}^{i+1} \leftarrow \mathbf{SS}^i \cup \mathbf{D}_s^i.$

[5] Constitute the next evaluation set $\mathbf{D}_e^{i+1}$.
$\mathbf{D}_e^{i+1} \leftarrow \bigcup_{\boldsymbol{x} \in \mathbf{D}_o^i} k\mathbf{NN}(\boldsymbol{x}) - (\mathbf{S}_o^{i+1} \cup \mathbf{S}_x^{i+1}).$
[6] $i \leftarrow i + 1.$
end
return $\mathbf{SS}^i$

}

Fig. 3. Selective kNN spanning algorithm

Sine Function Problem: In the second problem, the input patterns were generated from a two-dimensional uniform distribution, and then the class labels were determined by whether the input is located above or below a sine decision function. To make the density near the decision boundary thicker, four different Gaussian noises were added along the decision boundary, i.e., $N(\mu, s^2)$ where μ is an arbitrary point on the decision boundary and s is a Gaussian width parameter(0.1, 0.3, 0.8, 1.0). Among 500 training patterns, 20% are located in the overlapped region. Therefore, the characteristic of this problem is that the density of patterns increases near the decision boundary.

$$C_1 = \left\{ x \mid x_2 > \sin(3x_1 + 0.8)^2, \begin{bmatrix} 0 \\ -2.5 \end{bmatrix} \leq \begin{bmatrix} x_1 \\ x_2 \end{bmatrix} \leq \begin{bmatrix} 1 \\ 2.5 \end{bmatrix} \right\},$$

$$C_2 = \left\{ x \mid x_2 \leq \sin(3x_1 + 0.8)^2, \begin{bmatrix} 0 \\ -2.5 \end{bmatrix} \leq \begin{bmatrix} x_1 \\ x_2 \end{bmatrix} \leq \begin{bmatrix} 1 \\ 2.5 \end{bmatrix} \right\}.$$

Fig. 4(a) and Fig. 4(b) shows the selected patterns against the original ones after normalization ranging from -1 to 1. The value of k was set as 5.

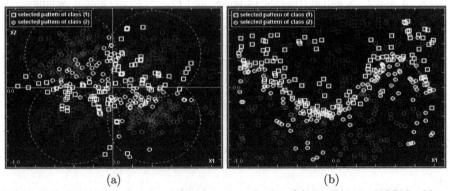

(a)   (b)

Fig. 4. The selected patterns against the original ones: (a) Continuous XOR Problem and (b) Sine Function Problem. The selected patterns, shown as outlined circles or squares, are scattered against the original ones.

Only 180(30%) and 264 (52.8%) of the original patterns were selected, respectively. The number of the selected patterns by fast algorithm was identical to that by the naive algorithm. The reduction ratio in Sine Function Problem is rather smaller when it is compared with that of Continuous XOR problem. That is due to the difference in densities near the decision boundary.

A total of 300 test patterns were generated from the statistically identical distribution as its training set for both problems. We built 50 SVMs with all patterns and 50 SVMs with the selected patterns according to every combination of hyper-parameters. Five RBF kernels with different width parameters(σ=0.25, 0.5, 1, 2, 3) and five polynomial kernels with different degree parameters(p=1, 2, 3, 4, 5) were adopted. As for the tolerance parameter, five levels (C= 0.1,

1,10, 100, 1000) were used. This optimal parameter searching is laborious and time-consuming if individual training takes a long time.

SVM performances "with all patterns" (A) and "with the selected patterns" (S) were compared in terms of the test error rate, the execution time, and the number of support vectors. Fig. 5 shows SVM test error rates in all parameter combinations we adopted. In each cell, upper number indicates test error rate for A and lower for S. Gray cell means that S performed better than or equal to A, while white cell means that A performed better. An interesting fact is that SVM performance with S is more sensitive to parameter variation. This phenomenon is shown substantially with ill-posed parameter combination (white cells). For the parameters where SVM with A performs well, SVM with S also works well.

(a) Continuous XOR Problem

Kernel / C	Rbf (Width Parameter) 0.25	0.5	1	2	3	Polynomial (Degree Parameter) 1	2	3	4	5
0.1	11.33 / 11.33	11.67 / 29.33	9.67 / 50.00	25.33 / 50.00	34.33 / 50.00	51.67 / 50.00	10.67 / 50.00	11.67 / 32.67	10.00 / 11.67	10.33 / 10.67
1.0	10.00 / 10.33	12.00 / 12.00	11.00 / 31.33	10.67 / 50.00	34.33 / 50.00	34.00 / 50.00	10.00 / 15.67	10.67 / 12.33	11.33 / 11.33	11.33 / 11.00
10	10.00 / 10.00	11.33 / 10.33	10.00 / 10.00	12.00 / 35.33	10.00 / 50.00	31.67 / 50.00	9.67 / 10.00	10.33 / 9.67	11.67 / 10.33	12.00 / 10.00
100	10.00 / 9.00	11.00 / 10.33	9.67 / 9.67 *	10.67 / 10.33	10.33 / 22.67	31.67 / 50.00	10.00 / 9.00	10.00 / 10.00	11.67 / 10.33	12.33 / 10.67
1000	10.07 / 11.00	10.33 / 10.67	11.33 / 10.33	10.33 / 10.33	10.00 / 10.00	31.07 / 50.00	10.33 / 9.00	9.07 / 10.00	12.33 / 11.33	12.07 / 11.67

(b) Sine Function Problem

Kernel / C	Rbf (Width Parameter) 0.25	0.5	1	2	3	Polynomial (Degree Parameter) 1	2	3	4	5
0.1	16.33 / 16.67	18.33 / 18.33	20.00 / 20.00	20.67 / 33.33	20.33 / 55.33	20.67 / 22.67	21.00 / 19.33	20.67 / 19.67	16.67 / 16.33	16.00 / 15.67
1	13.67 / 13.33	16.33 / 15.67	19.33 / 19.33	20.00 / 20.33	20.67 / 22.33	20.33 / 22.00	19.67 / 19.33	17.33 / 16.67	14.00 / 14.00	13.67 / 13.67
10	14.67 / 14.67	13.33 / 12.67	16.67 / 14.67	19.00 / 19.67	20.67 / 20.67	20.33 / 21.67	20.00 / 20.00	15.67 / 15.33	14.00 / 13.67	14.33 / 14.33
100	16.67 / 15.33	14.33 / 13.33	13.33 / 13.67	18.83 / 18.67	19.67 / 19.33	20.33 / 21.67	20.00 / 18.67	13.67 / 13.07 *	13.33 / 13.33	14.33 / 14.33
1000	18.00 / 18.00	13.33 / 13.33	14.00 / 14.33	15.00 / 14.67	19.00 / 17.67	20.33 / 20.33	20.00 / 18.67	15.00 / 12.33	14.33 / 14.33	14.67 / 14.00

Fig. 5. Test error comparison for SVM experimental results: (a) Continuous XOR Problem and (b) Sine Function Problem. In each cell, upper number indicates test error rate for A and lower for S.

Table 1 compares the best performance of each SVM, A and S. The corresponding parameter combination is marked '*' in Fig. 5. The SVM with the selected patterns was trained much faster thanks to fewer training patterns. Less important support vectors were eliminated in building the margins. Employing fewer SVs results in a smaller recall time. All this happened while the generalization ability was not degraded at al.

Table 1. Best Result Comparison

	Continuous XOR Problem A(All)	S(Selected)	Sine Function Problem A(All)	S(Selected)
Execution Time (sec.)	454.83	3.85	267.76	8.79
Margin	0.0612	0.0442	0.1904	0.0874
No. of Training Patterns	600	180	500	264
No. of Support Vectors	167	84	250	136
Training Error(%)	10.50	13.89	19.00	19.32
Test Error(%)	9.67	9.67	13.33	13.33

Table 2 shows the execution time from Fig. 5. The average SVM training time of A was 472.20(sec) for Continuous XOR problem and 263.84(sec) for Sine

Function Problem. It is quite comparable 4.45(sec) and 9.13(sec) of S. Pattern selection time of S took 0.88(sec) and 0.66(sec) by the naive algorithm. The proposed fast algorithm reduced the selection time to 0.21(sec) and 0.17(sec), respectively. Therefore, by conducting pattern selection procedure just once, we could shrink SVM training time by one or two orders of magnitude.

Table 2. Execution Time Comparison

		No. of patterns	No. of SVs(avg.)	Pattern Selection CPU time	SVM Avg. CPU time
Continuous	A (All)	600	292.20	-	472.20
XOR	S (Selected)	180	120.72	0.88 (Naive) 0.21 (Fast)	4.45
Sine	A (All)	500	293.60	-	263.84
Function	S (Selected)	264	181.76	0.66 (Naive) 0.17 (Fast)	9.13

Fig. 6 shows the decision boundaries and margins of the SVMs both with all patterns and with the selected patterns of Continuous XOR problem. (Because of page limitation, we omit the figures for Sine Function problem.) Decision boundary is depicted as a solid line and the margins are defined by the dotted lines in both sides of it. Support vectors are outlined. Note that in the aspects of generalization performance, the decision boundaries are all but similar in both plots.

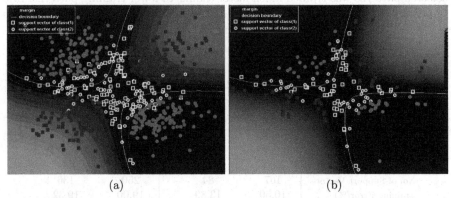

(a)   (b)

Fig. 6. Patterns and decision boundaries between (a) SVM with all patterns and (b) SVM with the selected patterns for continuous XOR Problem.

4.2   Real Problem: MNIST Dataset

For real world data benchmarking test, we used MNIST dataset of handwritten digits from [15]. It consists of 60,000 patterns for training and 10,000 patterns for testing. All binary images are size-normalized and coded by gray-valued 28x28 pixels in the range between -1 and 1, therefore input dimension is 784.

Nine binary classifiers of MNIST were trained: class 8 is paired with each of the rest. We used OSU SVM Classifier Matlab Toolbox since the kernel matrix was too large for the standard QP rountine to fit into memory [15]. A Fifth-order polynomial kernel ($p=5$), C value of 10, and KKT(Karush-Kuhn-Tucker) tolerance of 0.02 were used same as in [10], and the value of K was set to 50.

The results for the training runs both with all patterns and with the selected patterns are shown in Table 3. First, we compare the results with regard to the number of training patterns and also the number of support vectors. The pattern selection chose an average 16.75% of patterns from the original training sets.

Table 3. MNIST Result Comparison

	Paired Classes								
	0-8	**1-8**	**2-8**	**3-8**	**4-8**	**5-8**	**6-8**	**7-8**	**9-8**
A(All Patterns)									
No. Of Patterns	11774	12593	11809	11982	11693	11272	11769	12116	11800
No. Of SVs	538	455	988	1253	576	1039	594	578	823
SVM Exe. Time	201.74	190.76	368.82	477.25	212.18	379.59	222.84	222.40	308.73
Test Error(%)	0.51	0.09	0.34	0.50	0.10	0.16	0.31	0.10	0.41
S(Selected Patterns)									
No. Of Patterns	1006	812	2589	4089	1138	3959	1135	999	1997
No. Of SVs	364	253	828	1024	383	882	421	398	631
SVM Exe. Time	11.11	5.92	33.03	49.84	12.11	49.74	14.69	13.55	26.61
Pattern Sel. Time	46.11	43.33	78.81	97.89	48.13	93.42	44.27	40.14	59.62
Test Error(%)	0.41	0.95	0.34	0.45	0.15	0.21	0.31	0.18	0.43

When all patterns were used, only 6.44% of the training patterns were used as support vectors. With selected patterns, 32.09% of its training patterns were support vectors. In terms of utilization, the pattern selection did a good job (see Fig. 7(a)).

Second, SVM execution time was significantly reduced from 287.15(sec) to 24.07(sec) on average after the pattern selection procedure. Including the pattern selection time, the total time was, on average, 85.37(sec). Consider that the SVM training is performed several times to find the optimal parameters. On the other hand, the pattern selection procedure is performed only once (see Fig. 7(b)).

Now we finally compare the SVM test error rates between A (with all patterns) and S (with the selected patterns). First, average test error rate over nine classifiers were 0.28% for the former and 0.38% for the latter. But note that most SVM classifiers of S performed all but same as the ones of A except 1-8 classifier. This exception is mainly due to the unique characteristic of digit '1' images. Individual pattern belongs to '1' class is a sparse vector. Only a few fields of a

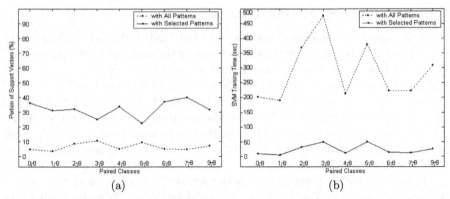

Fig. 7. MNIST Result Comparison: (a) The portion of support vectors in training set, and (b) SVM execution time.

vector have significant gray-scaled value, and the rest of the fields have value of '0'. Hence, it might affect the finding the nearest neighbors in opposite class(8 class) during the pattern selection procedure. This conjecture is accompanied by the number of selected patterns of '1' digit class. Only 95 patterns were selected from 6742 patterns in '1' digit class. In the meantime, average test error rate for the rest was 0.30% for A and 0.31% for S.

5   Conclusion

To alleviate the computational burden in SVM training, we proposed a fast pattern selection algorithm which selects the patterns near the decision boundary based on the neighborhood properties. Through several simulations, we obtained promising results: SVM with the selected patterns was trained much faster with fewer redundant support vectors, without any loss in test performance.
The time complexity analysis of the proposed algorithm will be carried out in the near future.

Related to the proposed algorithm, we would like to address two limitations here. First, it is not easy to find the right value for β in [Selecting Criteria]. β controls the selectivity by kNN classifier. In small dimensional space, β can be set to a large value(i.e. $\beta = 1$), since kNN classifier performs well. But in large dimensional space, a large value of β is somewhat risky. Because kNN classifier's accuracy is degraded. Empirically, up to about 20 dimensional space, $\beta = 1$ was not problematic. This coincides with the recommendation in [9].

Second, the proposed algorithm works efficiently under the following circumstances: when the classes are overlapped and when the pattern is stored as a non-sparse vector. Neither condition is met for '1' digit class of MNIST data set. It is known as a linearly separable class from all other digit classes. It is also the most sparse matrix among 10 digit classes [6] [10].

References

[1]   Almeida, M.B., Braga, A. and Braga J.P.(2000). SVM-KM: speeding SVMs learning with a priori cluster selection and k-means, *Proc. of the 6th Brazilian Symposium on Neural Networks*, pp. 162–167.

[2]   Arya, S., Mount, D.M., Netanyahu, N.S. and Silverman, R., (1998). An Optimal Algorithm for Approximate Nearest Neighbor Searching in Fixed Dimensions, *Journal of the ACM*, vol. 45, no. 6, pp. 891–923.

[3]   Choi, S.H. and Rockett, P., (2002). The Training of Neural Classifiers with Condensed Dataset, *IEEE Transactions on Systems, Man, and Cybernetics – PART B: Cybernetics*, vol. 32, no. 2, pp.202–207.

[4]   Grother, P.J.,Candela, G.T. and Blue, J.L, (1997). Fast Implementations of Nearest Neighbor Classifiers, *Pattern Recognition*, vol. 30, no. 3, pp. 459–465.

[5]   Hearst, M.A., Scholkopf, B., Dumais, S., Osuna, E., and Platt, J., (1998). Trends and Controversies - Support Vector Machines, *IEEE Intelligent Systems*, vol. 13, pp. 18–28.

[6]   Liu C.L., and Nakagawa M., (2001). Evaluation of Prototype Learning Algorithms for Nearest-Neighbor Classifier in Application to Handwritten Character Recognition, *Pattern Recognition*, vol.34, pp. 601–615.

[7]   Lyhyaoui, A., Martinez, M., Mora, I., Vazquez, M., Sancho, J. and Figueiras-Vaidal, A.R., (1999). Sample Selection Via Clustering to Construct Support Vector-Like Classifiers, *IEEE Transactions on Neural Networks*, vol. 10, no. 6, pp. 1474–1481.

[8]   Masuyama, N., Kudo, M., Toyama, J. and Shimbo, M., (1999). Termination Conditions for a Fast k-Nearest Neighbor Method, *Proc. of 3rd International Conference on Knoweldge-Based Intelligent Information Engineering Systems*, Adelaide, Australia, pp. 443–446.

[9]   Mitchell, T.M., (1997). *Machine Learning*, McGraw Hill, See also Lecture Slides on Chapter8 for the Book at http://www-2.cs.cmu.edu/ tom/mlbook-chapter-slides.html.

[10] Platt, J.C. (1999). Fast Training of Support Vector Machines Using Sequential Minimal Optimization, *Advances in Kernel Methods: Support Vector Machines*, MIT press, Cambridge, MA, pp. 185–208.

[11] Shin, H.J. and Cho, S.Z., (2002). Pattern Selection For Support Vector Classifiers, *Proc. of the 3rd International Conference on Intelligent Data Engineering and Automated Learning (IDEAL)*, Manchester, UK, pp. 469–474.

[12] Shin, H.J. and Cho, S.Z., (2002). Pattern Selection Using the Bias and Variance of Ensemble, *Journal of the Korean Institute of Industrial Engineers*, vol. 28, no. 1, pp. 112–127, 2002.

[13] Short, R., and Fukunaga, (1981). The Optimal Distance Measure for Nearest Neighbor Classification, *IEEE Transactions on Information and Theory*, vol. IT-27, no. 5, pp. 622–627.

[14] Vapnik, V., (2000). *The Nature of Statistical Learning Theory*, Springer-Verlag New York, Inc. 2nd eds.

[15] http://www.kernel-machines.org/

Averaged Boosting: A Noise-Robust Ensemble Method*

Yongdai Kim

Department of Statistics, Ewha Womans University
Seoul, Korea, 120-750
ydkim@mm.ewha.ac.kr
http://home.ewha.ac.kr/ ydkim/

Abstract. A new noise robust ensemble method called "Averaged Boosting (A-Boosting)" is proposed. Using the hypothetical ensemble algorithm in Hilbert space, we explain that A-Boosting can be understood as a method of constructing a sequence of hypotheses and coefficients such that the average of the product of the base hypotheses and coefficients converges to the desirable function. Empirical studies showed that A-Boosting outperforms Bagging for low noise cases and is more robust than AdaBoost to label noise.

1   Introduction

Bagging (Breiman [2]) and AdaBoost (Freund and Schapire [7]) are two most commonly used ensemble algorithms. A number of empirical studies including those by Quinlan [10], Bauer and Kohavi [1], Optiz and Maclin [9] and Ditterich [4] indicated that although AdaBoost is more accurate than Bagging in most cases, AdaBoost may overfit high noise cases, thus decreasing its performance.

In this paper, we propose a new noise-robust ensemble method, named "Averaged Boosting (A-Boosting)," which combines the advantages of both the Bagging and AdaBoost method; that is, A-Boosting improves Bagging for low-noise cases and outperforms AdaBoost for high-noise cases.

In section 2, the A-Boosting algorithm is presented. Using a hypothetical ensemble algorithm, we show that A-Boosting can be understood as an algorithm that generates a sequence of ensemble models converging to the optimal function. In section 3, numerical results on several data sets are presented. The paper concludes with a brief discussion in section 4.

2   A-Boosting

2.1   The Algorithm

The A-Boosting algorithm is given as follows.

---

* This research is supported in part by US Air Force Research Grant F62562-02-P-0547 and in part by KOSEF through Statistical Research Center for Complex Systems at Seoul National University.

K.-Y. Whang, J. Jeon, K. Shim, J. Srivatava (Eds.): PAKDD 2003, LNAI 2637, pp. 388–393, 2003.
© Springer-Verlag Berlin Heidelberg 2003

Algorithm 2. A-Boosting

1. Start with weights $w_i = 1/n, \; i = 1, \ldots, n$.
2. Repeat for $m = 1, \ldots, M$;
   a) Fit a hypothesis f_m on $\mathcal{F}$ minimizing the weighted misclassification error rate with weights $\{w_i\}$ on the training examples.
   b) Calculate the misclassification error ϵ_m of f_m on the original training examples by
   $$\epsilon_m = \sum_{i=1}^{n} I(y_i \neq f_m(\mathbf{x}_i))/n.$$
   c) Let $\beta_m = \log((1 - \epsilon_m)/\epsilon_m)/2$
   d) Update $w_i = \exp(-y_i \sum_{t=1}^{m} \beta_t f_t(\mathbf{x}_i)/m), \; i = 1, \ldots, n$ and renormalize so that $\sum_i w_i = 1$.
3. Output the classifier $\mathrm{sign}(\sum_{m=1}^{M} \beta_m f_m(\mathbf{x}))$.

There are two key differences between the A-Boosting and AdaBoost algorithms. The first is that the A-Boosting algorithm uses the average of the product of the base hypotheses and coefficients while AdaBoost uses the sum of it. The second difference is that the A-boost algorithm calculates the coefficient based on the error rate of the current hypothesis on the original training example while the AdaBoost algorithm uses the updated weights. The A-Boosting algorithm in Hilbert space given in the next subsection explains why such modifications are necessary.

2.2   A-Boosting in Hilbert Space

Consider a probability measure Q on R^p defined by

$$Q(d\mathbf{x}) = \frac{\sqrt{p(\mathbf{x})(1 - p(\mathbf{x}))}P(d\mathbf{x})}{\int_{R^p} \sqrt{p(\mathbf{x})(1 - p(\mathbf{x}))}P(d\mathbf{x})}$$

where $p(\mathbf{x}) = \Pr(Y = 1 | \mathbf{X} = \mathbf{x})$ and $P(d\mathbf{x})$ is the marginal distribution of $\mathbf{X}$. Let $\mathcal{F}$ be the set of functions in $L_2(Q)$ with norm 1 and let f^* be the target function to be estimated. Assume that $\|f^*\| < \infty$. The A-Boosting algorithm in Hilbert space is given as follows.

Algorithm 3. A-Boosting in Hibert space

1. Set $f_0(\mathbf{x}) \equiv 0$ and $\beta_0 = 0$.
2. Repeat $m = 1, \ldots, M$
   a) Set f_m by
   $$f_m = \mathrm{argmin}_{f \in \mathcal{F}} \left\| f^* - \sum_{t=0}^{m-1} \beta_t f_t/(m - 1) - f \right\|.$$

b) Find β_m;

$$\beta_m = \mathrm{argmin}_{a \in R} \|f^* - af_m\|.$$

3. Estimate $f^*(\mathbf{x})$ by $\sum_{t=1}^M \beta_t f_t(\mathbf{x})/M$.

The following theorem proves the convergence of the above algorithm.

Theorem 1. *In Algorithm 3,*

$$\left\| \sum_{t=1}^M \beta_t f_t(\mathbf{x})/M - f^*(\mathbf{x}) \right\| \to 0$$

as $M \to \infty$.

Now, we turn to the classfication problem. Consider the population exponential criteria $C(f) = E(\exp(-Yf(\mathbf{X})))$ for given function f. Let $p(\mathbf{x}) = \Pr(Y = 1|\mathbf{X} = \mathbf{x})$ and let $f^*(\mathbf{x}) = \log(p(\mathbf{x})/(1-p(\mathbf{x}))/2$. Lemma 1 in Friedman et al. [5] proves that $C(f)$ attains its minimum at $f = f^*$.

Taylor expansion yields

$$C(f) = \int_{R^p} \left(\exp(f(\mathbf{x}) - f^*(\mathbf{x})) + \exp(f^*(\mathbf{x}) - f(\mathbf{x})) \right) Q(dx)$$

$$= \int_{R^p} 2 + (f^*(\mathbf{x}) - f(\mathbf{x}))^2 + o((f^*(\mathbf{x}) - f(\mathbf{x}))^2) Q(dx) \qquad (1)$$

$$\approx 2Q(R^p) + C\|F - f\|^2$$

if we ignore the $o((f^*(\mathbf{x}) - f(\mathbf{x}))^2)$ term. Hence, the population exponential criterion is approximately the same as the square of $L_2(Q)$ norm, and hence the minimizer of $\|f^* - f\|$ on $f \in \mathcal{F}$ is approximately the same as the minimizer of the population exponential criterion $C(f)$. Using this property, the modification of Algorithm 3 with the population exponential criterion becomes

Algorithm 4. A-Boosting with the population exponential criterion

---

1. Set $f_0(\mathbf{x}) \equiv 0$ and $\beta_0 = 0$.
2. Repeat $m = 1, \ldots, M$
   a) Set f_m by

   $$f_m = \mathrm{argmin}_{f \in \mathcal{F}} C\left(\sum_{t=0}^{m-1} \beta_t f_t + f \right).$$

   b) Find β_m;

   $$\beta_m = \mathrm{argmin}_{a \in R} C(af_m).$$

3. Estimate $f^*(\mathbf{x})$ by $\sum_{t=1}^M \beta_t f_t(\mathbf{x})/M$.

Now, the A-Boosting algorithm is obtained by replacing the population exponential criteria $C(f)$ by the exponential criteria $C_n(f)$ and letting $\mathcal{F}$ be the set of base hypotheses. Note that the norm of any base hypothesis is 1.

3   Empirical Studies

For base hypotheses, unpruned trees are used in A-Boosting and Bagging. For AdaBoost, the optimal tree size is selected using the validation set a priori. For the final decision, 300 base hypotheses are combined. We analyzed 14 benchmark datasets from the UC-Irvine machine learning archive. For datasets without test samples, the generalization errors were calculated by 10 repetitions of 10-fold cross validation.

Table 1 presents the generalization errors. A-Boosting improved Bagging in most cases though AdaBoost had the lowest level of generalization errors in general. However, note that AdaBoost used optimally selected tree sizes while nothing was tuned a priori for A-Boosting. Hence, the degree of superiority of AdaBoost over A-Boosting should be discounted.

Table 2 presents the generalization errors when 10% label noise was added. A-Boosting outperformed AdaBoost in many cases (9 out of 14 cases) and was comparable to Bagging. Also, the increases in error rates due to the label noise indcate that A-Boosting is much more robust to label noise that AdaBoost. Bagging is the most robust.

In summary, our empirical results confirm that (i) A-Boosting improves Bagging in most cases, (ii) A-Boosting is much more robust to label noise than AdaBoost, and (iii) A-Boosting is comparable to Bagging for high noise cases.

In practice, non-experts in ensemble methods can use A-Boosting, as almost nothing needs to be tuned a priori. In contrast, AdaBoost requires tuning to find both the optimal number of iterations in addition to the optimal complexity of base hypotheses a priori.

Table 1. Generalization errors

ID	Data Set	A-Boosting	AdaBoost	Bagging
1	Breast Cancer	0.0348	0.0322	0.0332
2	Pima-Indian-Diabetes	0.2400	0.2598	0.2415
3	German	0.2328	0.2481	0.2351
4	Glass	0.2261	0.2204	0.2333
5	House-vote-84	0.0489	0.0439	0.0467
6	Image	0.0566	0.0485	0.0638
7	Ionosphere	0.0698	0.0597	0.0802
8	kr-vs-kp	0.0031	0.0034	0.0075
9	Letter	0.0465	0.0252	0.0640
10	Satimage	0.0880	0.0775	0.0975
11	Sonar	0.1615	0.1180	0.1825
12	Vehicle	0.2411	0.2107	0.2516
13	Vowel	0.4588	0.5844	0.4740
14	Waveform	0.1842	0.1826	0.1842

Table 2. Generalization errors with 10% label noise (Increases of error rates due to label noise, %)

ID	Data Set	A-Boosting	AdaBoost	Bagging
1	Breast Cancer	0.1300 (273.56)	0.1483 (360.55)	0.1279 (285.24)
2	Pima-Indian-Diabetes	0.3024 (26.00)	0.3103 (19.43)	0.3010 (24.63)
3	German	0.3026 (29.98)	0.3291 (32.64)	0.2995 (27.39)
4	Glass	0.3147 (39.18)	0.3047 (38.24)	0.3109 (33.26)
5	House-vote-84	0.1433 (193.04)	0.1818 (314.12)	0.1402 (200.21)
6	Image	0.1402 (147.70)	0.1761 (263.03)	0.1404 (120.06)
7	Ionosphere	0.1685 (141.40)	0.1837 (207.70)	0.1657 (106.60)
8	kr-vs-kp	0.1159 (3638.70)	0.1065 (3032.35)	0.1228 (1537.33)
9	Letter	0.1525 (227.95)	0.1580 (526.98)	0.1505 (135.15)
10	Satimage	0.1575 (78.97)	0.1520 (96.12)	0.1645 (68.71)
11	Sonar	0.2575 (59.44)	0.2305 (95.33)	0.2660 (45.75)
12	Vehicle	0.2905 (20.48)	0.2786 (32.22)	0.2965 (17.84)
13	Vowel	0.5303 (15.58)	0.5497 (-5.93)	0.5368 (13.24)
14	Waveform	0.2398 (30.18)	0.2500 (36.91)	0.2464 (33.76)

4   Discussions

In this paper, we proposed a noise-robust ensemble method called A-Boosting. Using the hypothetical ensemble algorithm in Hilbert space, we explained that A-Boosting can be understood as a method of constructing a sequence of ensemble models converging to the optimal model. Empirical results showed than A-Boosting outperforms Bagging for low noise cases and is much more robust to label noise than AdaBoost.

Most explanations of AdaBoost are related to the numerical optimization algorithms of a certain cost functions (Schapire and Singer [13], Friedman et al. [5], Mason et al. [8], Friedman [6]). One exception is Breiman [3], who conjectures that the success of AdaBoost is its ability to generate an ergodic sequence of base hypotheses whose average converges to the optimal function. A-Boosting can be considered as a deterministic ensemble algorithm of implementing Breiman's idea.

References

1. Bauer, E., Kohavi, R.: An empirical comparison of voting classification algorithms: Bagging, Boosting and Variants. Machine Learning. **36** (1999) 105–139.
2. Breiman, L.: Bagging predictors. Machine Learning. **24** (1996) 123–140.
3. Breiman, L.: Random forests. Machine Learning. **45** (2001) 5–32.
4. Dieterich, T.G.: An experimental comparison of three methods for constructing ensembles of decision trees: Bagging, Boosting and Randomization. Machine Learning, **40** (2000) 139–157.
5. Friedman, J.H., Hastie, T., Tibshirani, R.: Additive logistic regression: a statistical view of boosting. Annals of Statistics. **38** (2000) 337–374.

6. Friedman, J.H.: Greedy function approximation: a gradient boosting machine. Annals of Statistics. **39** (2001) 1189–1232.
7. Freund, Y., Schapire, R.: A decision-theoretic generalization of on-line learning and an application to boosting. Journal of Computer and System Sciences. **55** (1997) 119–139.
8. Mason, L., Baxter, J., Bartlett, P.L. and Frean, M.: Functional gradient techniques for combining hypotheses, In: A.J. Smola, P. Bartlett, B. Schölkopf and C. Shuurmans (Eds.): Advances in Large Margin Classifiers, Cambridge, MA: MIT Press (2000).
9. Opitz, D., Maclin, R.: Popular ensemble methods: An empirical study. Journal of Artificial Intelligence Research. **11** (1999) 169–198.
10. Quinlan, J.: Boosting first-order learning. In: Arikawa S., Sharma (Eds.): LNAI, Vol. 1160: Proceedings of the 7th International Workshop on Algorithmic Learning Theory, Berlin Springer (1996) 143–155.
11. Räsch, G., Onoda, T., Müller, K.R.: Soft margins for AdaBoost. Machine Learning. **42** (2001) 287–320.
12. Schapire, R., Freund, Y., Bartlett, P., Lee, W.: Boosting the margin: a new explanation for the effectiveness of voting methods. Annals of Statistics. **26** (1998) 1651–1686.
13. Schapire, R., Singer, Y.: Improved boosting algorithms using confidence-rated predictions. Machine Learning. **37** (1999) 297–336.

A   Appendix: The Proof of Theorem 1

We will prove that

$$\|f^* - H_m\| \le \frac{2\|f^*\|}{\sqrt{m}} \tag{2}$$

by use of the mathematical induction. For $m = 1$, (2) is trivially true since $\|f^* - \beta_m f_m\| \le \|f^*\|$. Suppose (2) holds for $m > 1$. Then, we have

$$\|f^* - H_{m+1}\|^2 = \left\| \frac{m}{m+1}(f^* - H_m) + \frac{1}{m+1}(f^* - \beta_{m+1}f_{m+1}) \right\|^2$$

$$= \frac{m^2}{(m+1)^2}\|f^* - H_m\|^2 + \frac{1}{(m+1)^2}\|f^* - \beta_{m+1}f_{m+1}\|^2$$

$$\le \frac{(2\|f^*\|)^2}{m+1},$$

since $< f^* - H_m, f^* - \beta_{m+1}f_{m+1} >= 0$. $\square$

Improving Performance of Decision Tree Algorithms with Multi-edited Nearest Neighbor Rule

Chen-zhou Ye, Jie Yang, Li-xiu Yao, and Nian-yi Chen

Institute of Image Processing & Pattern Recognition, Shanghai Jiaotong University,
Shanghai 200030, China
jieyang@online.sh.cn

Abstract. The paper proposed a new method based on the multi-edited nearest neighbor rule to prevent decision tree algorithms from growing a tree of unnecessary large size and hence partially alleviate the problem of "over-training". For this purpose, two useful prosperities of the multi-edited nearest neighbor rule are investigated. Experiments show that the method proposed could drastically reduce the size of resulting trees, significantly enhance their understandability, and meanwhile improve the test accuracy when the control parameter takes an appropriate value.

1 Introduction

Although constitute a large family in the field of data mining and exhibit advantages of flexibility, efficiency, and understandability, decision tree (DT) algorithms such as CART, ID3, and C4.5 [1,2] seem to be more vulnerable to the so-called "over-training" problem. At the final stage of training, they tend to generate a large number of non-leaf nodes to split relatively small subsets of the training dataset, most of which mainly contain mislabeling noises (samples that should be assigned to some class but was mistakenly assigned to another class) and/or samples located in mixed regions (in which area there is no clear boundary between samples of different classes). Since the statistical information derived from those small subsets is usually unreliable, serious mismatching between the newly generated nodes and the truth of data tends to occur. If we can prevent DT algorithms from entering this stage, a tree of unnecessary large size can be avoided and the "over-training" problem can be partially alleviated.

2 Our Method

Before introducing our method, we briefly introduce the multi-edited nearest neighbor rule (MENN) [2]:

1. According to the ratio of r ($0 < r < 1$), randomly extract samples from the original dataset X_{all} (containing n_{all} samples) to form the reference dataset X_{ref} (containing $n_{ref} = n_{all} \cdot r$ samples), let the test dataset $X_{test} = X_{all}$;

K.-Y. Whang, J. Jeon, K. Shim, J. Srivatava (Eds.): PAKDD 2003, LNAI 2637, pp. 394–398, 2003.

2.   Delete from X_{test} those samples that are misclassified by the standard nearest neighbor algorithm taking X_{ref} as the reference dataset. The samples left in X_{test} consist the edited reference dataset X_{edited} (containing n_{edited} samples), let $X_{all} = X_{edited}$;

3.   When there is sample being deleted, set $i=0$, and turn to step 1; if not, and i is less than a predetermined value i_{max}, set $i=i+1$ and turn to step 1; Otherwise, terminate the process.

In the following, we will show that when $n_{all} \rightarrow +\infty$ (and so do the n_{ref} and n_{edited}) and the editing times $q \rightarrow +\infty$ (although the MENN algorithm is sure to be terminated within finite steps), the MENN algorithm will purify mixed regions according to the *Bayes* rule and eliminate mislabeling noises when some prerequisites are satisfied. For the purpose of simplicity, only two classes ω_1 and ω_2 are involved here.

Property 1: For the sample x located in the mixed region of two classes, if $p(\omega_i \mid x) > p(\omega_j \mid x)$, $i, j \in \{1,2\}$ and $i \neq j$, the MENN algorithm will sweep off the samples of ω_j, which obeys the *Bayes* rule.

Proof: Suppose after the kth (k≥1) editing, posterior probabilities of two classes are $p_k(\omega_l \mid x)$ $l \in \{1,2\}$. After q th (q≥1) editing, we get:

$$p_q(\omega_l \mid x) = \frac{p^{2^q}(\omega_l \mid x)}{p^{2^q}(\omega_i \mid x) + p^{2^q}(\omega_j \mid x)} \tag{1}$$

Since $p(\omega_i \mid x) > p(\omega_j \mid x)$, it is easy to prove that as $q \rightarrow +\infty$, $p_q(\omega_i \mid x) \rightarrow 1$ and $p_q(\omega_j \mid x) \rightarrow 0$.

□

Suppose $p(t_l \mid x)$ $l \in \{1,2\}$ is the probability of x that came form ω_l and was correctly labeled as a sample of ω_l, and suppose $p(m_l \mid x)$ is the probability of x that did not come form ω_l but was mistakenly recorded as a sample of ω_l (a mislabeling noise). The following property holds:

Property 2: If $p(\omega_i \mid x) > p(\omega_j \mid x)$ and $\dfrac{p(m_i \mid x)}{p(\omega_i \mid x)} < \dfrac{p(m_j \mid x)}{p(\omega_j \mid x)}$, $i, j \in \{1,2\}$, $i \neq j$, the MENN algorithm will eliminate the noise incurred by mislabeling.

Proof: After the k th editing, posterior probabilities of two classes are:

$$p_k(\omega_l \mid x) = p_k(t_l \mid x) + p_k(m_l \mid x) \quad l \in \{1,2\} \tag{2}$$

Obviously, $\sum_{l=1}^{2} [p_k(t_l \mid x) + p_k(m_l \mid x)]$ 1 holds. The conditional probability of mislabeling noise is:

$$p_k(n \mid x) = \frac{p_k(m_1 \mid x) + p_k(m_2 \mid x)}{[p_k(t_1 \mid x) + p_k(m_1 \mid x)] + [p_k(t_2 \mid x) + p_k(m_2 \mid x)]} \tag{3}$$

After the $k+1th$ editing, we get:

$$p_{k+1}(n \mid x) = \frac{p_k(m_1 \mid x)[p_k(t_1 \mid x) + p_k(m_1 \mid x)] + p_k(m_2 \mid x)[p_k(t_2 \mid x) + p_k(m_2 \mid x)]}{[p_k(t_1 \mid x) + p_k(m_1 \mid x)]^2 + [p_k(t_2 \mid x) + p_k(m_2 \mid x)]^2} \tag{4}$$

In order to establish:

$$p_{k+1}(n \mid x) < p_k(n \mid x) \tag{5}$$

the following inequality, which is the equivalent of (5), should be satisfied:

$$[p_k(\omega_1 \mid x) - p_k(\omega_2 \mid x)][p_k(\omega_1 \mid x)p_k(m_2 \mid x) - p_k(\omega_2 \mid x)p_k(m_1 \mid x)] > 0 \tag{6}$$

Due to the arbitrariness of k and the same effect of editing for each possible k, it is easy to find that if $p(\omega_1 \mid x) > p(\omega_2 \mid x)$ and $\dfrac{p(m_2 \mid x)}{p(\omega_2 \mid x)} > \dfrac{p(m_1 \mid x)}{p(\omega_1 \mid x)}$ hold or

$p(\omega_1 \mid x) < p(\omega_2 \mid x)$ and $\dfrac{p(m_1 \mid x)}{p(\omega_1 \mid x)} > \dfrac{p(m_2 \mid x)}{p(\omega_2 \mid x)}$ hold, the inequality (6) and in turn the

inequality (5) will be established. Furthermore, after q ($q \cdot 1)th$ editing, the conditional probability of mislabeling noise is:

$$p_q(n \mid x) = \frac{\{p(m_1 \mid x)[p(t_1 \mid x) + p(m_1 \mid x)]\}^{2^{q-1}} + \{p(m_2 \mid x)[p(t_2 \mid x) + p(m_2 \mid x)]\}^{2^{q-1}}}{[p(t_1 \mid x) + p(m_1 \mid x)]^{2^q} + [p(t_2 \mid x) + p(m_2 \mid x)]^{2^q}} \tag{7}$$

Since inequality (6) is satisfied, the fact of $p(\omega_1 \mid x) > p(\omega_2 \mid x)$ or $p(\omega_2 \mid x) > p(\omega_1 \mid x)$ can lead to $p(m_1 \mid x) < p(\omega_1 \mid x)$ or $p(m_2 \mid x) < p(\omega_2 \mid x)$ respectively. Thus, as $q \to +\infty$, $p_q(n \mid x) \to 0$. $\qquad\square$

In our method, we take the original training dataset as X_{all}, multi-edit it with MENN algorithm, and take X_{edited} as the final training dataset of DT algorithm. According to the property 1 and the property 2, the multi-edited dataset, under ideal conditions, can prevent DT algorithms from wasting inefficient and possibly detrimental efforts in dealing with mixed regions and mislabeling noises. In order to resemble the ideal conditions in practice, the original dataset should satisfy the following requirements: 1) Sample number of each class is large enough; 2) Distribution of each class should not be distorted; 3) Accuracy in labeling samples, especially for those belonging to the majority class is acceptable.

3 Experiments

We applied our method to three DT algorithms: CART, ID3, and C4.5 (all in binary separation mode and without any pruning step). 10-fold cross validation test was carried out on five different datasets (see Table 1, the datasets are available at the UCI repository of machine learning databases). During each fold, a multi-edited training dataset was generated after the standard training dataset was multi-edited by the MENN algorithm (The drop in sample number is illustrated in Table 1). Table 2 illustrates the performance of DT algorithms trained respectively by the standard training dataset and the multi-edited one. In the table, the size of resulting tree equals to the number of non-leaf nodes, and the training accuracy was got from the standard training dataset. Influence of the control parameter r was investigated by comparing three different settings. Euclidean distance calculated on normalized attributes was used in the nearest neighbor algorithm embedded in MENN.

Table 1. Average number of training samples used in each fold of test

No. of Dataset	Dataset	Before editing	After editing		
			$r=0.5$	$r=0.33$	$r=0.25$
1	Australian	621	485.8	467.5	450.9
2	Breast cancer	615	585.6	579.8	574.3
3	Diabetes	692	376.8	338.6	325.1
4	Heart disease	243	165.7	164.1	139.8
5	Iris Plants	135	128.0	124.0	121.5

Table 2. Performance of DT algorithms before and after the training samples being multi-edited

Data Set		Avg size of tree (r=)				Avg training accuracy(r=)				Avg test accuracy(r=)			
		-	0.5	0.33	0.25	-	0.5	0.33	0.25	-	0.5	0.33	0.25
C	1	528	166	109	57.8	100	87.6	86.3	85.9	77.4	84.8	84.6	84.4
A	2	74.2	41.2	31.3	30.6	100	97.3	96.8	96.7	94.0	94.6	95.4	94.9
R	3	424	249	132	70.3	100	76.9	74.5	71.7	63.7	69.3	71.7	70.0
T	4	159	103	99.1	62.5	100	85.8	83.6	81.0	61.1	72.6	77.8	73.0
	5	48.1	6.0	2.1	2.1	100	96.7	96.0	95.9	89.3	93.3	93.3	92.7
I	1	72.8	6.9	3.3	2.2	100	87.5	86.1	85.5	82.8	85.5	84.4	83.6
D	2	26.3	7.7	6	6.1	100	97.1	96.7	96.7	94.1	95.7	96.0	95.3
3	3	112	12.7	6.5	3.4	100	76.9	73.3	70.0	69.1	74.1	70.8	67.1
	4	39.0	7.6	7.2	4.7	100	82.7	81.6	79.8	75.2	75.2	78.5	73.7
	5	7.4	3.0	2.1	2.1	100	96.8	96.0	95.9	92.7	96.0	93.3	92.7
C	1	87.2	9.4	3.9	2.2	99.1	87.8	86.0	85.6	80.3	85.1	84.1	83.8
4.	2	28.9	7.4	6.5	6.4	99.8	97.0	96.5	96.4	95.4	95.7	95.7	95.2
5	3	153	13.8	8.5	5.2	99.0	76.0	74.4	70.8	72.6	73.0	72.9	67.5
	4	45.9	11.0	8.6	6.1	98.8	83.6	81.7	79.7	73.0	76.3	78.9	74.8
	5	7.2	3.1	2.1	2.1	99.5	96.7	96.0	95.9	88.0	94.0	93.3	92.7

After the standard training datasets were multi-edited, a drastic drop in the average size of trees can be observed in all of the cases. The maximum reduction even surpasses 95%, which is much more significant than the maximum reduction of the training sample number. Trained by multi-edited datasets, both ID3 and C4.5 can generate trees of very small size (mostly, less than 10), which make their solutions truly understandable. Although, the editing process brings a loss in training accuracy, test accuracy of DT algorithms—the item we truly care about—increases in most of the cases. Actually, only when r decreases to 0.25, can ID3 and C4.5 contribute several exceptions. The biggest gain of test accuracy caused by sample editing surpasses 20% (in the case of CART on *Heart disease* when $r=0.33$), and obvious improvement can be observed in many non-exceptions.

There is no obvious link between the value of r and the resulting performance of DT algorithms. Although, in all of the cases, there is a choice of r that can lead a specific type of DT algorithm to generate a better solution, different DT algorithms prefer different values of r on different datasets. Roughly speaking, a moderate value of r can work well in most of the cases.

4 Conclusions

Compared with the early stop criterion method [2] and the pruning strategy [1,2] that, for most of the time, only focus on small subsets of training samples and treat them in a uniform way, our method tends to treat the training dataset as a whole and can treat different parts of it in a way obeying the *Bayes* rule. Besides, our method requires no modification to the DT algorithm involved. As for the strategy that constructs ensemble of trees [3], our method displays its prominent advantage in maintaining and enhancing the understandability of the final solution. A strategy that is similar to ours but is based on neural network and re-sampling technique can be found in [4]. However, our method displays its simplicity in implementation and avoids introducing new uncertainties to the training process. A further comparison between the performances of above techniques is our future work.

References

1. Apté Chidanand, Weiss Sholom: Data Mining with Decision Trees and Decision Rules. Future Generation Computer Systems, 13 (1997) 197–210
2. Richard O. Duda, Peter E. Hart, David G. Stork: Pattern Classification (2nd edition). John Wiley&Sons, New York (2001)
3. Thomas G. Dietterich: An Experimental Comparison of Three Methods for Constructing Ensembles of Decision Trees: Bagging, Boosting, and Randomization. Machine Learning, 40 (2000) 139–157
4. Krishnan R., Sivakumar G., Bhattacharya P.: Extracting Decision Trees from Trained Neural Networks. Pattern Recognition, 32 (1999) 1999–2009

HOT: Hypergraph-Based Outlier Test for Categorical Data*

Li Wei[1], Weining Qian[**1], Aoying Zhou[1], Wen Jin[2], and Jeffrey X. Yu[3]

[1] Department of Computer Science and Engineering, Fudan University
{lwei, wnqian, ayzhou}@fudan.edu.cn,
[2] Department of Computer Science, Simon Fraser University
wjin@cs.sfu.ca,
[3] Department of Systems Engineering and Engineering Management,
The Chinese University of Hong Kong
yu@se.cuhk.edu.hk

Abstract. As a widely used data mining technique, outlier detection is a process which aims at finding anomalies with good explanations. Most existing methods are designed for numeric data. They will have problems with real-life applications that contain categorical data. In this paper, we introduce a novel outlier mining method based on a hypergraph model. Since hypergraphs precisely capture the distribution characteristics in data subspaces, this method is effective in identifying anomalies in dense subspaces and presents good interpretations for the local outlierness. By selecting the most relevant subspaces, the problem of "curse of dimensionality" in very large databases can also be ameliorated. Furthermore, the connectivity property is used to replace the distance metrics, so that the distance-based computation is not needed anymore, which enhances the robustness for handling missing-value data. The fact, that connectivity computation facilitates the aggregation operations supported by most SQL-compatible database systems, makes the mining process much efficient. Finally, experiments and analysis show that our method can find outliers in categorical data with good performance and quality.

1 Introduction

Outlier detection is one of the major technologies in data mining, whose task is to find small groups of data objects that are exceptional when compared with rest large amount of data. Outlier mining has strong application background in telecommunication, financial fraud detection, and data cleaning, since the patterns lying behind the outliers are usually interesting for helping the decision makers to make profit or improve the service quality.

* The work was partially supported by the "973" National Fundamental Research Programme of China (Grant No. G1998030414), National "863" Hi-Tech Programme of China (Grant No. 2002AA413310), grants from Ministry of Education of China, and Fok Ying Tunk Education Foundation.
** The author is partially supported by Microsoft Research Fellowship.

K.-Y. Whang, J. Jeon, K. Shim, J. Srivatava (Eds.): PAKDD 2003, LNAI 2637, pp. 399–410, 2003.

A descriptive definition of outliers is firstly given by Hawkins [8]. Although some different definitions have been adopted by researchers, they may have problems when being applied to real-life data. In real applications, data are usually mix-typed, which means they contain both numeric and categorical data. Since most current definitions are based on distance, such as distance-based outlier [11,12], local outlier [5], or density in cells, such as high-dimensional outlier [1], they cannot handle categorical data effectively. The following example shows the difficulties of processing categorical data.

Example 1. Consider a ten-record, five-dimensional customer dataset shown in Table 1. We are interested in dimensions *Age-range*, *Car*, and *Salary-level*, which may be useful for analyzing the latent behavior of the customers. The occurrences of the combinations of the attribute values are close to each other - there are two instances of (*'Middle'*, *'Sedan'*, *'Low'*), three of (*'Young'*, *'Sports'*, *'Low'*), one of (*'Young'*, *'Sedan'*, *'High'*), one of (*'Young'*, *'Sports'*, *'High'*), and three of (*'Middle'*, *'Sedan'*, *'High'*). So it is hard to figure out the outliers based on the occurrence.

With the dimensionality and the number of possible values of each attribute increase, we cannot even calculate out the occurrence of all combinations, due to *curse of dimensionality* and *combinational explosion*. Furthermore, in some applications, data may contain missing values, which means that for some objects, some attribute values are unknown. And distance- or density-in-cell- based definitions cannot process this kind of data. *How to define the local outliers in mix-typed, high-dimensional data? Is there an approach to find the outliers efficiently?* This paper introduces a possible solution to these two questions.

Table 1. Customer Data

RID	Age-range	Car	Salary-level
1	Middle	Sedan	Low
2	Middle	Sedan	High
3	Young	Sedan	High
4	Middle	Sedan	Low
5	Young	Sports	High
6	Young	Sports	Low
7	Middle	Sedan	High
8	Young	Sports	Low
9	Middle	Sedan	High
10	Young	Sports	Low

Table 2. Hypergraph modeling

Hyper-edgeID	Frequent itemsets	Vertices
1	('Middle', *, *)	1, 2, 4, 7, 9
2	('Young', *, *)	3, 5, 6, 8, 10
3	(*, 'Sedan', *)	1, 2, 3, 4, 7, 9
4	(*, *, 'Low')	1, 4, 6, 8, 10
5	(*, *, 'High')	2, 3, 5, 7, 9
6	('Middle', 'Sedan', *)	1, 2, 4, 7, 9

The major contribution of this paper is as follows:

– We propose a definition for outliers in high-dimensional categorical data, which not only considers the locality of the whole data space, but also of the dimensions, so that the outliers can be easily explained;

- We provide an efficient algorithm for finding this kind of outliers, which is robust for missing values;
- The techniques for handling real life data are discussed, which include preprocessing for numeric data, handling missing values, postprocessing for pruning banal outliers, and explaining and managing of outliers and the corresponding knowledge;
- We introduce a quantified method for measuring the outliers, which can be used to analyze the quality of outliers being found.

The rest of the paper is organized as follows. Section 2 provides the hypergraph model and the definitions of outliers formally. In section 3, the algorithm for mining outliers is presented with the enhancement for handling real-life data. The empirical study of the proposed method is given in section 4. After a brief introduction to related work in section 5, section 6 is for concluding remarks.

2   Problem Statement

To describe the hypergraph model, based-on which the outlier is defined, the notions to be used are listed in Table 3 first.

Table 3. Notations

Notion	Meaning
N	The number of objects in database DB.
$\|DS\|$	The number of elements in set DS.
A, A_i	Each denotes an attribute.
B, C	Each denotes a set of attributes.
v_o^i	The value of attribute A_i in object o.
A^{DS}	The set of values of A appearing in dataset DS.
$S_A^{DS}(x)$	Given $x \in A$ and dataset DS, it is $\|\{o\|v_o^A = x, o \in DS\}\|$

Table 4. Selected deviation values

RID	Age-range	Car	Salary-level	$Dev_{he}(o, Car)$
3	Young	Sedan	High	-0.708
5	Young	Sports	High	0.708
6	Young	Sports	Low	0.708
8	Young	Sports	Low	0.708
10	Young	Sports	Low	0.708

Definition 1. *A hypergraph $H = (V, HE)$ is a generalized graph, where V is a set of vertices and HE is a set of hyperedges. Each hyperedge is a set of vertices that contains more than two vertices. In our model, each vertex $v \in V$ corresponds to a data object in the dataset, and each hyperedge $he \in E$ denotes a group of objects that all contain a frequent itemset.*

Definition 2. *For data object o in a hyperedge he and attribute A, the **deviation of the data object o on attribute A w.r.t.** he is defined as*

$$Dev^{he}(o, A) = \frac{S_A^{he}(x_o) - \mu_{S_A^{he}}}{\sigma_{S_A^{he}}}, \text{ where } \mu_{S_A^{he}} = \frac{1}{\|A^{he}\|} \sum_{x \in A} S_A^{he}(x) \text{ is the average}$$

value of $S_A^{he}(x)$ for all $x \in A^{he}$, and $\sigma_{S_A^{he}} = \sqrt{\frac{1}{\|A^{he}\|} \sum (S_A^{he}(x) - \mu_{S_A^{he}})^2}$ is the standard deviation of $S_A^{he}(x)$ for all $x \in A^{he}$.

Definition 3. *Given a hyperedge he, a data object o in it is defined as an **outlier with common attributes** C **and outlying attribute** A, in which C is the set of attributes that have values appear in the frequent itemset corresponding to he, if $Dev^{he}(o, A) < \theta$. The threshold of deviation θ determines how abnormal the outliers will be. Usually, θ is set to a negative value.*

Example 2. Consider Example 1. Assume that the minimum support count is set to five, we can get the hyperedges shown in Table 2. The frequent itemsets in the table are presented in trinaries, whose elements denote the values in *Age-range*, *Car*, and *Salary-level* respectively. Furthermore, the '*' denotes that any value of the corresponding attribute does not appear in the itemset. The numbers in the vertices column represent *RIDs* of the objects that appear in each hyperedge. We calculate the deviation of objects in the second hyperedge on attribute *Car*. The result is shown in Table 4. According to the outlier definition, object 3 is discerned as an outlier. The explanation is that, data records with *Age-range*='*Young*' usually have that *Car*='*Sports*', but the third object is different since it has *Car*='*Sedan*'. Therefore, it is an outlier.

This example shows that although objects are always sparse in the whole space, some of them are anomalies when observed from certain viewpoint. And hyperedges are the appropriate viewpoints. Firstly, the large frequency of the hypergraph guarantees that the objects gathered in each hyperedge are *common* from certain view. Secondly, a hyperedge determines not only the locality of objects but also the dimensions. And, since only part of the dimensions are considered in each hyperedge, the objects with missing-values can also be examined in the hyperedges which are not related to the attribute the missing values belong to. Furthermore, the example shows that discriminating the common attributes and outlying attributes when finding outliers is important for searching anomalous objects. Therefore, in this paper, we study the problem of, *given the minimum support threshold min_sup and the deviation threshold θ, finding all outliers according to definition 3.*

3   Algorithm

The main algorithm for mining outliers is shown in Algorithm 1. The process for mining outliers can be roughly divided into three steps.

Step 1 *Building the hierarchy of the hyperedges*

Line 1 and 2 of main algorithm find the frequent itemsets and build the hierarchy of them. For one k-frequent-itemset I_k and one $(k+1)$-frequent-itemset I_{k+1}, if

---

Algorithm 1 HypergraphBasedOutlierTest

Input: DB, min_sup, θ
Output: Ouliers: (o, C, A), C is the set of common attributes,
A is the outlying attribute
 1: Mine_frequent_itemsets(DB,min_sup);
 2: Build hypergraph and construct the hierarchy;
 3: **for** each node i in level 1 **do**
 4:    Construct multi-dim. array M_i;
 5: **end for**
 6: **for** level $l = 2$ to n **do**
 7:    **for** each node i in level l **do**
 8:       Choose one of i's ancestor j;
 9:       Construct multi-dim. array M_i from M_j;
10:       FindOutlier(he_i,M_i,θ);
11:    **end for**
12: **end for**

$I_k \subset I_{k+1}$, then I_k is I_{k+1}'s ancestor in the hierarchy. And, i-frequent-itemset is in ith level. We employ Apriori [2] for finding the frequent itemsets. Note that Apriori tests all subsets of I_{k+1}, including I_k, when finding I_{k+1} [2]. Our algorithm just records the subset relationships, so that the two steps are integrated together.

Step 2 *Constructing multidimensional array*

For each frequent itemset $I = \{A_1 = a_1, A_2 = a_2, ..., A_p = a_p\}$, we construct a multi-dimensional array M, whose dimensions are attributes other than $A_1, A_2, ..., A_p$, and coordinates are the identities of values of corresponding attributes. Each entry in the array is the count of objects fall in the hyperedge whose attribute values are equal to the coordinates respectively. More formally, the entry of the array, named as *amount* in the following paper, according to frequent itemset I above with coordinates $(a_{p+1}, a_{p+2}, ..., a_k)$ is $\|\{o|o.A_i = a_i, i = 1, 2, ..., k\}\|$, in which $A_i, i = p+1, ..., k$ are the attributes that have no value appear in I.

Assume that i and j are two nodes in the hierarchy, and j is one of i's ancestor, which means $j \subset i$, and $\|i\| = \|j\| + 1$. M_i and M_j are their multi-dimensional arrays respectively. M_j is stored in a table $(A'_1, A'_2, ..., A'_{k-p}, amount)$, in which $i - j = \{A'_{k-p} = a\}$, and M_i will be stored in table $(A'_1, A'_2, ..., A'_{k-p-1}, amount)$. Then, we get M_i from M_j like this:
SELECT $A'_1, A'_2, ..., A'_{k-p-1}, sum(amount)$ INTO M_i FROM M_j WHERE $A'_{k-p} = a$

Step 3 *Finding outliers in the array*

Given a multi-dimensional array, the process to traverse the array to find outliers is shown in Algorithm 2. For each dimension, it calculates the occurrence of each value. Then, the deviation of each value is tested, so that outliers are found.

Algorithm 2 FindOutlier

Input: he, M, θ
Output: (o, C, A)

1: Set C as the set of attributes forming he;
2: **for** each dimension A_i in M **do**
3:    Calculate $S_{A_i}^{he}(v)$ for each value v in A_i^{he};
4:    Calculate $\mu_{S_{A_i}^{he}}$;
5:    **for** each value v in A_i^{he} **do**
6:      **if** $Dev^{he}(v, A_i) < \theta$ **then**
7:        **for** each object $o \in he$ with v in A_i **do**
8:          Output (o, C, A_i);
9:        **end for**
10:     **end if**
11:   **end for**
12: **end for**

Heuristic 1 *When choosing ancestors to generate multidimensional array (line 8 of Algorithm 1), we choose the ancestor with minimum records.*

Although any ancestor of a node can be chosen for computing the multi-dimensional array, using the smallest one is the most efficient choice, since choosing the smallest one can reduce the number of entries to be examined, and minimize the I/O cost.

Heuristic 2 *If both i and j are nodes in the hierarchy, and j is an ancestor of i (so, i has one more item $A = a$ than j), then, when finding outliers in the hyperedge corresponding to j, we do not execute the test of line 6 in Algorithm 2.*

Since any object o in i has quite a lot of similar objects on attribute A in j, for i is frequent, they may not be outliers with outlying attribute A. In all of our experiments, whether using this heuristic doesn't affect the outliers found.

3.1   Analysis

The performance analysis for frequent itemset mining using Apriori is given in [2]. To construct a multi-dimensional array for node i, the complexity is $O(n_j)$, in which n_j is the number of entries in multi-dimensional array of j.

In the outer iteration of Algorithm 2, the time complexity to calculate $S_{A_i}^{he}(v)$, $\mu_{S_{A_i}^{he}}$ and outliers are $O(n_i)$, $O(\|A_i^{he}\|)$, and $O(\|A_i^{he}\|)$ respectively. n_i denotes the number of entries in multi-dimensional array of i. The outer iteration will be executed at most $(k - p)$ times, where k is the dimensionality of the database, and p is the number of itemsets corresponding to the hyperedge. Note that in each hyperedge, n_i is always larger than any $\|A^{he}\|$, and smaller than $\|he\|$, the total time complexity is $O(\|HE\| \cdot k \cdot \max\{\|he\|\})$, where $\|HE\|$ is the number of hyperedges found by Apriori, after the hyperedges are found. The algorithm needs to store the multi-dimensional array for each hyperedge. Therefore, the space complexity is $O(\|HE\| \cdot k \cdot \max\{\|he\|\})$.

3.2 Enhancement for Real Applications

The data sets in real-life applications are usually complex. They have not only categorical data but also numeric data. Sometimes, they are *incomplete*. And in most cases, the data sets are very huge. Explanation and postprocessing are important. Furthermore, in real applications, expert knowledge is usually a valuable resource. In this section, we discuss the techniques for handling data with these characteristics in HOT.

Handling numeric data. To process numeric data, we apply the widely used binning technique [9] and choose equal-width method for its insensitiveness to outliers. The number of bins is set to the maximum cardinality of all categorical attributes, to make sure the bins are enough for discriminating the different characteristics of different objects.

Handling missing-value data. HOT is robust to incomplete data, since the relationships between two objects are tested on attributes one by one, instead of on distance. An incomplete object will not be considered when the attribute containing the missing-value is tested. However, it may still be found as outlier with other outlying attributes. Meanwhile, this object is also considered in hyperedges it falls in, so that it still contributes in finding other outliers that have common attribute values with it.

Existing knowledge integration. HOT can take advantage of two kinds of expert knowledge: *horizontal* and *vertical*. Horizontal knowledge means the information of grouping of objects. The groups of objects can be added into the hypergraph as new hyperedges or even just use these groups as hyperedges. Vertical knowledge means the information of interested attributes. The attributes interested by experts or users can be viewed as *class labels*, and their attributes as *class attributes*. Then, the HOT changes from an *unsupervised* algorithm to a *supervised* one. Only the class attributes values are tested in the algorithm.

Pruning banal outliers. Outliers found by HOT may be of unequal interest. Some outliers are consistent with the "already-known" exceptions. Others are out of the expectation and are more valuable. To distinguish the two kinds of outliers, we give out following definitions.

Definition 4. *Given a hyperedge he and the outlier o in it with common attributes C and outlying attribute A, $\frac{\|he\| \cdot S_A^{he}(x_o)}{N \cdot S_A^{DB}(x_o)}$ is called the **degree of interest of outlier o with common attributes C and outlying attribute A**, denoted as $Doi_{he}(o, A)$.*

Definition 5. *Given an outlier o in he with outlying attribute A, if $Doi_{he}(o, A) \geq \delta$, where δ is the threshold of interest, in he o is an **interesting outlier according to A**, otherwise it is a **banal outlier according to A**.*

The degree of interest can be regarded as a measurement of the difference between reality and estimation. The larger the value is, the more surprising the outlier is. A pruning process can be integrated easily into HOT algorithm to filter *banal outliers*, as postprocessing.

Mining result explanation. HOT provides sufficient information for explaining an outlier: common attributes, outlying attribute and deviation. Meanwhile, the values in outlying attributes can be provided to users. This kind of knowledge is useful in applications such as data cleaning. Furthermore, the deviation values can help users to find possible correct values.

4   Experimental Result

4.1   Experiments on Real-Life Data

The experimental environment is a Pentium 4 PC workstation with 256MB RAM running Microsoft Windows 2000. We test the effectiveness of HOT on the Mushroom and Flag datasets obtained from UCI repository [13].

Mushroom Dataset. Mushroom dataset is an 8124-record dataset, which includes descriptions of hypothetical samples of gilled mushrooms in the Agaricus and Lepiota Family. Each mushroom is identified as definitely edible, definitely poisonous, or of unknown edibility and not recommended. Besides the classification propriety, there are other 22 attributes for each mushroom, some of which contain missing values.

When finding frequent itemsets, the minimal support is set to 40 percent and 565 frequent itemsets are generated. When the threshold of deviation is set to -0.849 to discover outliers deviating greatly from other data, we find that among 3968 records satisfying the frequent itemset $\{Veil = p, Ring = p\}$, only 4 records have the attribute value $Cap - surface = g$, as shown in Table 5.

Table 5. The deviation of data records in hyperedge of $\{Veil = p, Ring = p\}$

Cap-surface	Number of occurrence	$Dev^{he}(o, Cap - surface)$
f	1248	0.22
g	4	-0.85
s	1212	0.19
y	1504	0.44

Table 6. The deviation of data records in hyperedge $\{Bruises = f, Gill - attachment = f, Gill - spacing = c, Veil = p\}$

Edibility	Number of occurrence	$Dev^{he}(o, Edibility)$
p	3170	0.71
e	160	-0.71

When the threshold is set to -0.7, another interesting kind of outliers is as follows. Totally, there are 3330 records comply the frequent itemset $\{Bruises = f, Gill - attachment = f, Gill - spacing = c, Veil = p\}$, most of which are poisonous. But there are 160 mushrooms of this kind are edible and so are regarded as outliers. Table 6 illustrates the condition clearly. These outliers are not only interesting but also useful. When the knowledge is applied to practice, it will gain much benefit.

Flag Dataset. Flag dataset contains details of 194 nations and their flags. There are overall 30 attributes, 10 of which are numeric-valued, others are categorical. Numeric-valued attributes are ignored in our experiments.

We set the minimal support to 60 percent and 75 frequent itemsets are found. When the threshold of deviation is set to -0.71, 13 data objects are detected as outliers, which belong to two types. One kind is that most countries (111 countries) whose flags have white color but have neither crescent moon symbol nor letters, are not located in South America. However, we find 11 countries having these attributes are in South American. The other kind is that gold color appears in two countries' flags which have no black or orange colors, while it is not present in the flags of other 120 countries without black or orange colors.

4.2 Quantitative Empirical Study

Definition 6. *In a database DB, for an object $o(v_o^1, v_o^2, ..., v_o^k)$, its **density in whole space** is defined as $density_o^{all} = \|\{p|p \in DB, v_p^i = v_o^i, i = 1, ..., k\}\|/\|DB\|$. Given an attribute set $B = \{A_{i_1}, A_{i_2}, ...A_{i_l}\} \subseteq A$, o's **subspace density over** B is defined as $density_o^B = \|\{p|p \in DB, v_p^{i_j} = v_o^{i_j}, j = 1, 2, ...l\}\|/\|DB\|$.*

We study an outlier o in hyperedge he with common attributes C and outlying attribute A_o from following four aspects:

- o's density in whole space vs. minimum, maximum, and average density in whole space for all $p \in DB$;
- o's density in whole space vs. minimum, maximum, and average density in whole space for all $p \in he$;
- o's subspace density over $C \cup \{A_o\}$ vs. minimum, maximum, and average subspace density over $A \cup \{A_o\}$ for all $p \in DB$;
- o's subspace density over $C \cup \{A_o\}$ vs. minimum, maximum, and average subspace density over $A \cup \{A_o\}$ for all $p \in he$;

The experiments still run on mushroom dataset, with the threshold of deviation set to -0.84. 32 *outlier class* (we call outliers in the same hyperedge belong to the same *outlier class*) are found as well as 76 outliers. Figure 1 shows the result for comparison of subspace density between outliers and all data. The x-axis denotes the 32 outlier classes found by HOT, the y-axis denotes the logarithmic subspace density. Although outliers found by HOT always have very low subspace density in the database, they are not always the most *sparse* ones. This situation happens for two reasons: 1) Some objects with very low subspace density over some attributes do not appear in any hyperedge. These data can be treated as noises. 2) Some objects appear in certain hyperedges, but they do not have outlying attributes. It means that in those hyperedges, most distinct values in attributes other than common ones occur few times. Therefore, no object is special when compared to other objects similar to it.

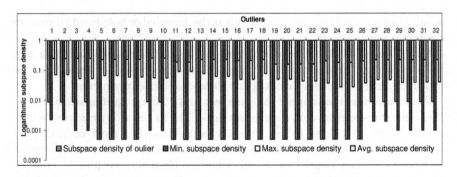

Fig. 1. o's subspace density over $A \cup \{A_o\}$ vs. minimum, maximum, average subspace density over $A \cup \{A_o\}$ for all $p \in DB$

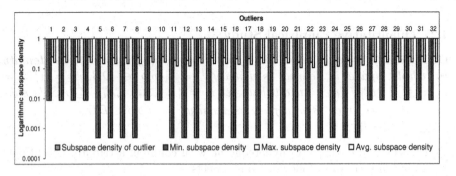

Fig. 2. o's subspace density over $A \cup \{A_o\}$ vs. minimum, maximum, average subspace density over $A \cup \{A_o\}$ for all $p \in he$

Figure 2 shows the result of comparison of subspace density between outliers and all data in the hyperedge the outlier falls in. Different from 1, outliers' subspace densities are always the lowest ones in the hyperedge. This property is guaranteed by our definition and algorithm. It ensures that the most anomalous data can be found in each local dense area.

5   Related Work

Outlier detection is firstly studied in the field of statistics [3,8]. Many techniques have been developed to find outliers, which can be categorized into distribution-based [3,8] and depth-based [14,16]. Recent research proved that these methods are not suitable for data mining applications, since that they are either ineffective and inefficient for multidimensional data, or need a priori knowledge about the distribution underlying the data [11]. Traditional clustering algorithms focus on minimizing the affect of outliers on cluster-finding, such as DBSCAN [6] and CURE [7]. Outliers are eliminated without further analyzing in these methods, so that they are only by-products which are not cared about.

In recent years, outliers themselves draw much attention, and outlier detection is studied intensively by the data mining community [11,12,4,15,5,10,17, 1]. **Distance-based outlier** detection is to find data that are sparse within the hyper-sphere having given radius [11]. Researchers also developed efficient algorithms for mining distance-based outliers [11,15]. However, since these algorithms are based on distance computation, they may fall down when processing categorical data or data sets with missing values. **Graph-based spatial outlier** detection is to find outliers in spatial graphs based on statistical test [17]. However, both the attributes for locating a spatial object and the attribute value along with each spatial object are assumed to be numeric.

In [4] and [5], the authors argue that the outlying characteristics should be studied in local area, where data points usually share similar distribution property, namely density. This kind of methods is called **local outlier** detection. Both efficient algorithms [4,10] and theoretical background [5] have been studied. However, it should be noted that in high-dimensional space, data are almost always sparse, so that density-based methods may suffer the problems that all or none of data points are outliers. Similar condition holds in categorical datasets. Furthermore, the density definition employed also bases on distance computation. As the result, it is inapplicable for incomplete real-life data sets.

Multi- and high-dimensional data make the outlier mining problem more complex because of the impact of *curse of dimensionality* on algorithms' both performance and effectiveness. In [12], Knorr and Ng tried to find the smallest attributes to explain why an object is exceptional, and is it dominated by other outliers. However, Aggarwal and Yu argued that this approach may be expensive for high-dimensional data [1]. Therefore, they proposed a definition for outliers in low-dimensional projections and developed an evolutionary algorithm for finding outliers in projected subspaces. Both of these two methods consider the outlier in global. The sparsity or deviation property is studied in the whole data set, so that they cannot find outliers relatively exceptional according to the objects near it. Moreover, as other existing outlier detection methods, they are both designed for numeric data, and cannot handling data set with missing values.

6   Conclusion

In this paper, we present a novel definition for outliers that captures the local property of objects in partial dimensions. The definition has the advantages that 1) it captured outlying characteristics of categorical data effectively; 2) it is suitable for incomplete data, for its independent to traditional distance definition; 3) the knowledge, which includes common attribute, outlying attribute, and deviation, are defined along with the outlier, so that the mining result is easy for explanation. Therefore, it is suitable for modelling anomalies in real applications, such as fraud detection or data cleaning for commercial data. Both the algorithm for mining such kind of outliers and the techniques for applying it in real-life dataset are introduced. Furthermore, a method for analyzing outlier mining results in subspaces is developed. By using this analyzing method, our experimental result shows that HOT can find interesting, although may not be most sparse, objects in subspaces. Both qualitative and quantitative empirical studies support the conclusion that our definition of outliers can capture the anomalous properties in categorical and high-dimensional data finely. To the best of our knowledge, this is the first trial to finding outliers in categorical data.

References

1. C. Aggarwal and P. Yu. Outlier detection for high dimensional data. In *Proc. of SIGMOD'2001*, pages 37–47, 2001.
2. R. Agrawal and R. Srikant. Fast algorithms for mining association rules in large databases. In *Proc. of VLDB'94*, pages 487–499, 1994.
3. V. Barnett and T. Lewis. *Outliers In Statistical Data*. John Wiley, Reading, New York, 1994.
4. M. Breunig, H.-P. Kriegel, R. Ng, and J. Sander. Optics-of: Identifying local outliers. In *Proc. of PKDD'99*, pages 262–270, 1999.
5. M. Breunig, H.-P. Kriegel, R. Ng, and J. Sander. Lof: Identifying density-based local outliers. In *Proc. of SIGMOD'2000*, pages 93–104, 2000.
6. M. Ester, H.-P. Kriegel, J. Sander, and X. Xu. A density-based algorithm for discovering clusters in large spatial databases with noise. In *Proc. of KDD'96*, pages 226–231, 1996.
7. S. Guha, R. Rastogi, and K. Shim. Cure: An efficient clustering algorithm for large databases. In *Proc. of SIGMOD'98*, pages 73–84, 1998.
8. D. Hawkins. *Identification of Outliers*. Chapman and Hall, Reading, London, 1980.
9. F. Hussain, H. Liu, C. L. Tan, and M. Dash. Discretization: An enabling technique. *Technical Report TRC6/99, National University of Singapore, School of Computing*, 1999.
10. W. Jin, A. K. Tung, and J. Han. Mining top-n local outliers in large databases. In *Proc. of KDD'2001*, pages 293–298, 2001.
11. E. Knorr and R. Ng. Algorithms for mining distance-based outliers in large datasets. In *Proc. of VLDB'98*, pages 392–403, 1998.
12. E. Knorr and R. Ng. Finding intensional knowledge of distance-based outliers. In *Proc. of VLDB'99*, pages 211–222, 1999.
13. G. Merz and P. Murphy. Uci repository of machine learning databases. *Technical Report, University of California, Department of Information and Computer Science: http://www.ics.uci.edu/mlearn/MLRepository.html*, 1996.
14. F. Preparata and M. Shamos. *Computational Geometry: an Introduction*. Springer-Verlag, Reading, New York, 1988.
15. S. Ramaswamy, R. Rastogi, and K. Shim. Efficient algorithms for mining outliers from large data sets. In *Proc. of SIGMOD'2000*, pages 427–438, 2000.
16. I. Ruts and P. Rousseeuw. Computing depth contours of bivariate point clouds. *Journal of Computational Statistics and data Analysis*, 23:153–168, 1996.
17. S. Shekhar, C.-T. Lu, and P. Zhang. Detecting graph-based spatial outliers: Algorithms and applications (a summary of results). In *Proc. of KDD'2001*, 2001.

A Method for Aggregating Partitions, Applications in K.D.D.

Pierre-Emmanuel Jouve and Nicolas Nicoloyannis

LABORATOIRE ERIC (Equipe de Recherche en Ingénierie des Connaissances)
Université Lumière – Lyon2
Bâtiment L, 5 avenue Pierre Mendès-France
69 676 BRON cedex FRANCE
{Pierre.Jouve, Nicolas.Nicoloyannis}@eric.univ-lyon2.fr
http://eric.univ-lyon2.fr

Abstract. K.D.D. (Knowledge Discovery in Databases) methods and methodologies nearly all imply the retrieval of one or several structures of a data set. In practice, using those methods may give rise to a bunch of problems (excessive computing time, parameters settings, ...). We show in this paper that some of these problems can be solved via the construction of a global structure starting from a set of sub-structures. We thus propose a method for aggregating a set of partial structures into a global one and then present how this method can be used for solving several traditional practical problems of K.D.D. ...

1 Introduction

K.D.D. (Knowledge Discovery in Databases) methods and methodologies nearly all imply the retrieval of one or several structures of a data set. Using those methods may give rise to a bunch of problems which we only evoke later. However, we advance and show later on that some of these problems can be solved via the construction of a global structure starting from a set of sub-structures. Thus, we propose in this article a method for aggregating a set of partial structures into a global one, then we present how this method can be used for solving several traditional practical problems of K.D.D. (such as excessive computing time for clustering processes).

2 Considered Structures, Introductory Concepts

Many more or less complex mathematical formulations may be used to represent the structure of a data set (lattices, graphs, covers, partitions). We will consider here one of the simplest, but certainly one of the most used: partitions. (Thereafter, the term structure refers to a partition of a set of objects) We now introduce notations and formalisms used in this paper.

Notation 1 $O = \{o_i, i = 1..n\}$ *a set of objects,*
$C_k^O = \{o_i^{C_k^O}, i = 1..n_{C_k^O}\}$ *a set of objects of O,*

K.-Y. Whang, J. Jeon, K. Shim, J. Srivatava (Eds.): PAKDD 2003, LNAI 2637, pp. 411–422, 2003.

$P_w = \{C_1^O, ..., C_h^O\}$ a partition of O in h groups. $(\forall i = 1..h, \forall j = 1..h, j \neq i, \forall o \in C_i^O, o \notin C_j^O)$

Notation 2 We represent $P_w = \{C_1^O, ..., C_h^O\}$, a partition of the set O in h groups by the mean of its $n \times n$ connexity matrix $(MC_{P_w})_{i,j}$ defined as follows:
$(MC_{P_w})_{i,j} = 1$ if $\exists k \in \{1,..,h\}$ such that $o_i \in C_k^O$ and $o_j \in C_k^O$
$(MC_{P_w})_{i,j} = 0$ otherwise

Definition 1 Let (P_1, P_2) be a couple of structures (partitions), with $P_1 = \{C_1^{O_1}, ..., C_l^{O_1}\}$ and $P_2 = \{C_1^{O_2}, ..., C_m^{O_2}\}$. The couple (P_1, P_2) is of:

- A-type if and only if $O_1 = O_2$.
- B-type if and only if $O_1 \subseteq O_2$. (a A-Type couple of structures is therefore a B-type couple of structures)
- C-type if and only if $\exists o \in O_1$ and $o \in O_2$. (a A or B-Type couple of structures is therefore a C-type couple of structures)
- D-type if and only if $\forall o \in O_1$, $o \notin O2$.

We now extend these definitions to the case of a couple constituted by a set of structures and a unique structure.

Definition 2 Let $E = \{P_1, ..., P_z\}$ be a set of structures and P_y be a structure. The couple (E, P_y) is of:

- A-type if and only if : $\forall P_i \in E$, the couple (P_i, P_y) is of A-type.
- B-type if and only if : $\forall P_i \in E$, the couple (P_i, P_y) is of B-type.
- C-type if and only if : $\forall P_i \in E$, the couple (P_i, P_y) is of C-type.
- D-type if and only if : $\forall P_i \in E$, the couple (P_i, P_y) is of D-type.

3   Adequacy between Structures

3.1   Adequacy for a Couple of Structures

In order to represent the adequacy between 2 structures P_1 and P_2 (with $P_1 = \{C_1^{O_1}, ..., C_l^{O_1}\}$ and $P_2 = \{C_1^{O_2}, ..., C_m^{O_2}\}$), we use a couple of values $(val1, val2)$ from which we build a measure of adequacy between structures. $val1 = Agg(P_1, P_2)$ represents the number of agreements between the two structures while $val2 = DisAgg(P_1, P_2)$ represents the number of disagreements between the 2 structures; more formally they are defined as follows:

$$Agg(P_1, P_2) = val1 = \sum_{o_i \in O_1, o_j \in O_2, o_i \neq o_j} \delta_1(o_i, o_j) \qquad (1)$$

$$DisAgg(P_1, P_2) = val2 = \sum_{o_i \in O_1, o_j \in O_2, o_i \neq o_j} \delta_2(o_i, o_j) \qquad (2)$$

$$\delta_1(o_i, o_j) = \begin{cases} 1 \text{ if}: \exists C_f^{O_1} \ (f \in \{1, ..., l\}) \text{ such that } o_i \in C_f^{O_1} \text{ and } o_j \in C_f^{O_1} \\ \quad \text{and if } \exists C_g^{O_2} \ (g \in \{1, ..., m\}) \text{ such that } o_i \in C_g^{O_2} \text{ and } o_j \in C_g^{O_2} \\ 1 \text{ if}: o_i \in O_1 \text{ and } o_j \in O_1 \text{ and } \nexists C_f^{O_1} \text{ such that } o_i \in C_f^{O_1} \text{ and } o_j \in C_f^{O_1} \\ \quad \text{and if } o_i \in O_2 \text{ and } o_j \in O_2 \text{ and } \nexists C_g^{O_2} \text{ such that } o_i \in C_g^{O_2} \text{ and } o_j \in C_g^{O_2} \\ 0 \text{ otherwise} \end{cases} \quad (3)$$

$$\delta_2(o_i, o_j) = \begin{cases} 1 - \delta_1(o_i, o_j) \text{ if}: o_i \in O_1 \text{ and } o_j \in O_1 \text{ and if } o_i \in O_2 \text{ and } o_j \in O_2 \\ 0 \text{ otherwise} \end{cases} \quad (4)$$

Definition 3 *We can now define a measure of adequacy between two structures P_1, P_2 : $Adq(P_1, P_2)$*

- *$Adq(P_1, P_2) = val2/(val1 + val2)$ if (P_1, P_2) is of A, B or C-Type (val1 + val2 is never null). $Adq(P_1, P_2) \in [0, 1]$.*
- *$Adq(P_1, P_2) = 0$ if (P_1, P_2) is of D-type (val1 + val2 is always null).*

Consequently, the closer to zero is $Adq(P_1, P_2)$ the more the two structures can be considered to be in adequacy, however, if (P_1, P_2) is a couple of D-type a null value does not signify a good adequacy between the two structures, we clarify those two points in the following remarks.

Remarks:

- if (P_1, P_2) is of A-type, $Adq(P_1, P_2) = 0$ means that the two structures are the same.
- if (P_1, P_2) is of B or C-type, $Adq(P_1, P_2) = 0$ means that the two structures are the same over the objects they share. (In the case of B-type we may say that P_1 is a sub-structure of P_2.)
- if (P_1, P_2) is of D-type then $Adq(P_1, P_2)$ is mandatory null, that does not mean that the two structures are the same over the objects they share but simply means that they do not have any object in common.

3.2 Adequacy between a Structure and a Set of Structures

In order to represent the adequacy between a set of structures $E = \{P_1, \ldots, P_z\}$ and a single particular structure P_y, we use a couple of values $(val3, val4)$ which we use to build a measure of adequacy between E and P_y. $val3 = Agg(E, P_y)$ stands for the number of agreements between P_y and the structures of E whereas $val4 = DisAgg(E, P_y)$ represents the number of disagreements between P_y and the structures of E; more formally we define those values as follows:

$$Agg(E, P_y) = val3 = \sum_{P_i \in E} Agg(P_i, P_y) \quad (5)$$

$$DisAgg(E, P_y) = val4 = \sum_{P_i \in E} DisAgg(P_i, P_y) \quad (6)$$

Definition 4 *We define $Adq(E, P_y)$ the measure of adequacy between E and P_y :*

- $Adq(E, P_y) = val4/(val3 + val4)$ if (E, P_y) is of A, B or C-type
  (val3 + val4 is always non null). $Adq(E, P_y) \in [0, 1]$.
- $Adq(E, P_y) = 0$ if (E, P_y) is of D-type (val3 + val4 is always null).

Consequently, the closer to zero is $Adq(E, P_y)$ the more E and P_y can be considered to be in adequacy, however, if (E, P_y) is a couple of D-type a null value does not signify a good adequacy between E and P_y, we clarify those two points in the following remarks.

Remarks:

- if (E, P_y) is of A-type, $Adq(E, P_y) = 0$ means that P_y and each structure of E are the same.
- if (E, P_y) is of B or C-type, $Adq(E, P_y) = 0$ means that P_y and each structure of E are the same over the objects they share. (In the case of B-type we may say that each structure of E is a sub-structure of P_y.)
- if (E, P_y) is of D-type then $Adq(E, P_y)$ is mandatory null, that does not mean that each structure of E and P_y are the same over the objects they share but simply means that they do not have any object in common.
- The use of the measure $Adq(E, P_y)$ for a B-type couple (E, P_y) should be correlated with works dealing with validation of clusterings [4].

4 Building an Optimal Global Structure According to a Set of Structures and to the Adequacy Criterion Adq

Assume now that we have a set of structures $E\{P_1, ..., P_z\}$, it may be interesting for solving some K.D.D. problems (we present some of these problems in section 6 page 419) to determine the structure P_y such as (E, Py) is of B-type (or of A-type, if each couple $(P_i \in E, P_y)$ is of A-type) such that (E, Py) minimizes $Adq(E, Py)$ that is to say such that the adequacy between E and P_y is maximized.

Since this problem is combinatorial, the computational cost associated to the search for an optimal solution is extremely high if the structures of E possess a great number of objects. We thus propose a greedy algorithm that discovers a solution to this problem, which, if it is not always optimal, constitutes a solution of good quality.

4.1 Determining Objects of P_y

Theorem 1 $E = \{P_1^{O_1}, ..., P_z^{O_z}\}$ a set of structures,
if P_y is a structure such that $(E, P_y^{O_u})$ is of B-type then all objects belonging to the structures of E should also belong to $P_y : \forall i(i = 1..z), O_i \subseteq O_y$

Remark:

- Due to the practical problems we want to solve, we consider that $O_y = \bigcup_{i(i=1..z)} O_i$ and so P_y verifies the condition imposed by theorem 1.

4.2   An Algorithm for Finding P_y

The proposed algorithm proceeds according to the following principle:

- for each couple (o_i, o_j) of objects belonging to P_y we first evaluate how many times (s) the two objects have been joined together within a same class into structures of E and then we evaluate how many times (d) the two objets have been separated into two different classes of structures of E. For each couple, values s and d give an idea of the majority treatment of the couple of objects into structures of E (that is to say it shows if o_i and o_j were more often joined into a same class or separated in two different classes).
- By means of these informations, we determine objects that should be separated into two different classes of the final global structure P_y or joined into a same class of P_y in priority in order to maximize the adequacy between E and P_y. Actually, the higher is $|s - d|$ for a couple of objects the more P_y should respect the majority treatment of structures of E for the two objects (separation or union) if we want $adq(E, P_y)$ to be minimized (that is to say if we want to maximize the adequacy between E and P_y).
- So, we use a greedy method that build P_y connexity matrix element by element according to previously joined and separated objects and by considering couple of objects according to the descending order onto their values for $|s - d|$.

Remark:

- This algorithm may not lead to a unique structure but to a set of structures that are equivalent as far as the adequacy criterion with the set E is concern.

The algorithm is presented in a more formal way by means of pseudo-code see Algorithm1 page 416. We should finally note that its complexity is quadratic according to the number of objects of the searched structure. (The sorting stage is only quadratic according to the number of objects since the set of different possible values of elements of $M1$ is known: $\forall(i, j)M1_{i,j} \in [0, z]$ with z the number of structures of E.)

4.3   An Example

We illustrate the way our algorithm proceeds on one example:
Let us consider the set of structures $E = \{P_1^{O_1}, P_2^{O_2}, P_3^{O_3}, P_4^{O_4}, P_5^{O_5}\}$ with:
$O_1 = \{o_1, o_2, o_4, o_5\}$, $P_1^{O_1} = \{\{o_1, o_2\}, \{o_4\}, \{o_5\}\}$
$O_2 = \{o_2, o_3, o_5, o_6\}$, $P_2^{O_2} = \{\{o_2, o_3\}, \{o_5, o_6\}\}$
$O_3 = \{o_2, o_5, o_6\}$, $P_3^{O_3} = \{\{o_2\}, \{o_5\}, \{o_6\}\}$
$O_4 = \{o_3, o_4, o_5\}$, $P_4^{O_4} = \{\{o_3\}, \{o_4\}, \{o_5\}\}$
$O_5 = \{o_5, o_6\}$, $P_5^{O_5} = \{\{o_5, o_6\}\}$

We want to build $P_y^{O_v}$ such that $(E, P_y^{O_v})$ is a couple of B-type and $adq(E, P_y^{O_v})$ is minimized. According to definition 1 and the remark made on

Algorithm 1 Heuristic

Parameters and Settings

$E = \{P_1^{O_1}, ..., P_z^{O_z}\}$, $O_y = \bigcup_{i(i=1..z)} O_i$, $O_y = \{o_i, i = 1..n\}$,

$M1$ a $n \times n$ matrix, $M2$ a $(n(n-1)/2) \times 4$ matrix

BEGIN

1. Create $M1$

for all (i,j) such that $i < j$ **do**

$\quad (M1)_{i,j} = \sum_{P_k^{O_k} \in E} phi_{P_k^{O_k}}(o_i, o_j)$

$\quad (M1)_{j,i} = \sum_{P_k^{O_k} \in E} |phi_{P_k^{O_k}}(o_i, o_j)|$

$\quad phi_{P_k^{O_k}}(o_i, o_j) = \begin{cases} 1 \text{ if: } \exists C_p^{O_k} \text{ such that } o_i \in C_p^{O_k} \text{ and } o_j \in C_p^{O_k} \\ -1 \text{ if : } o_i \in O_k \text{ and } o_j \in O_k \text{ and } \nexists C_q^{O_k} \text{ such that } o_i \in C_q^{O_k} \text{ and } o_j \in C_q^{O_k} \\ 0 \text{ otherwise} \end{cases}$

end for

2. Create $M2$ with $M1$ (sorting elements of the lower diagonal part of $M1$ according to the descending order):

$(M2)_{k,1} = (M1)_{j,i}$ $(j > i)$, $(M2)_{k,2} = i$, $(M2)_{k,3} = j$, $(M2)_{k,4} = (M1)_{i,j}$, with $\forall w \geq 0$, $(M2)_{v,1} \geq (M2)_{v+w,1}$

3. Build $P_y^{O_y}$ connexity matrix and stock it in M1:

Clear M1 and set each of its elements to -1: $\forall i, \forall j, M1_{i,j} = -1$

for $k = 1$ to $(n(n-1)/2)$ **do**

$\quad$ **if** $(M2_{k,1} > 0)$ and $(M1_{M2_{k,2}, M2_{k,3}} = -1)$ **then**

$\quad\quad$ **if** $(M2_{k,4} > 0)$ **then**

$\quad\quad\quad M1_{M2_{k,2}, M2_{k,3}} = 1$

$\quad\quad\quad$ Possible Completion of M1 by transitivity:

$\quad\quad\quad$ *each object that was formerly joined to $o_{M2_{k,2}}$ or to $o_{M2_{k,3}}$ as well as $o_{M2_{k,2}}$ and*

$\quad\quad\quad$ *$o_{M2_{k,3}}$ themselves are joined together*

$\quad\quad\quad$ *(setting each corresponding elements of M1 to 1)*

$\quad\quad$ **else**

$\quad\quad\quad M1_{M2_{k,2}, M2_{k,3}} = 0$

$\quad\quad\quad$ Possible Completion of M1 by transitivity:

$\quad\quad\quad$ *each object that was formerly joined to $o_{M2_{k,2}}$ and $o_{M2_{k,2}}$ itself are separated from each*

$\quad\quad\quad$ *object that was formerly joined to $o_{M2_{k,3}}$ and $o_{M2_{k,3}}$ itself*

$\quad\quad\quad$ *(setting each corresponding elements of M1 to 0)*

$\quad\quad$ **end if**

$\quad$ **end if**

end for

END.

4.1, we define O_y as follows: $O_y = \bigcup_{i(i=1..5)} O_i = \{o_1, o_2, o_3, o_4, o_5, o_6\}$. Then the algorithm previously described can create $M1$, then it creates $M2$, and finally builds $P_y^{O_y}$ (see figure 1). The final result is the connexity matrix of $P_y^{O_y}$.

5   Some Experimental Results

We present here some experimental results obtained by treating "1984 United States Congressional Voting Records Database" from the UCI dataset repository [5]. This data set includes votes for each of the 435 U.S. House of Representatives Congressmen on 16 key votes. The results we present correspond to the following experimentation:

- An initial clustering is made by considering the whole data set (all objects (the 435 congressmen) and all attributes (the 16 key votes) are considered). The clustering method used (a non hierarchical method) needs to set one

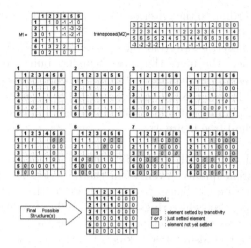

Fig. 1. An Example

parameter, and the final clustering it processes does not depend on any initialization process or number of iteration (that is to say that several runs of this method with the same parameter setting on the same set of objects would lead to the same final clustering), furthermore the resulting clustering is a partition of the objects of the data set. We call this clustering reference clustering.

- We then practice several series of clusterings on data sets composed by a part of the objects and a part of the attributes of the original data set, each series correspond to 200 different clusterings on data sets composed by the same number of objects (a random sample of the set of objects of the original dataset) and the same number of attributes (a random sample of the set of attributes of the original dataset). The different sampling levels are 5%, 10%, 20%, 30%, 40%, 50%, 60%, 70%, 80%, 90%, 100% as for the objects and 30%, 40%, 50%, 60%, 70%, 80%, 90%, 100% as far as the attributes are concerned. So, for each couple of values (x,y) (x=sampling level for the objects, y=sampling level for the attributes) 200 different subsets of the original data set (including $round(x \times 435)$ objects and $round(y \times 16)$ attributes) have been randomly composed. Then for each series of 200 data sets, we practiced clustering on each data set (using the same method and the same parameter setting used for the reference clustering), and we finally built the clustering corresponding to the optimal global structure according to the set of 200 clusterings and to the adequacy criterion Adq. (In our terminology, we build an optimal global structure P_y according to the set E of 200 clusterings, and the couple (E, P_y) is of B-type.)
- We have then evaluated the adequacy of each final clustering with the reference clustering (see figure 2) as well as the quality of each of these final

clusterings according to the criterion to be minimized underlying to the clustering method (see figure 3).

(Remark: the more the criterion value is weak, the better is the clustering, as a consequence, the more the ratio "Quality built clustering / Quality reference clustering" is weak the better is the built clustering).

– Finally, we evaluated the adequacy of each final clustering with the best final clustering according to the criterion to be minimized underlying to the clustering method (see figure 4).

These tests exhibit a really high efficiency of the method in building a final clustering of great quality[1] even when the data sets are composed of very few objects and attributes. Thus, we can notice that the best clustering is the clustering build upon 200 clusterings for which only 10% of the object of the original data set are present in each partial clustering and 90% of the attributes are, furthermore we can also notice a kind of stability of the quality of results since the quality is only degrading for extremely small sample of objects and attributes. We consequently conclude that our heuristic seems really efficient.

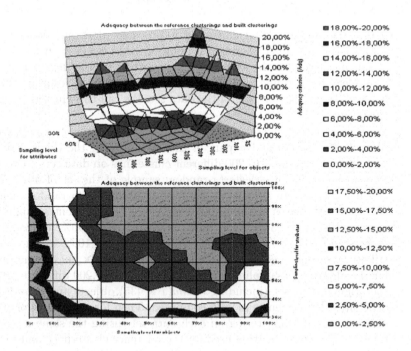

Fig. 2. Adequacy between Built Clusterings and the Reference Clustering

---

[1] We assume that the quality of a clustering may be evaluated with two points: "Are the considered clustering and the reference clustering in adequacy ?" and "Does the considered clustering present a good value for the criterion to be optimized underlying to the clustering method?".

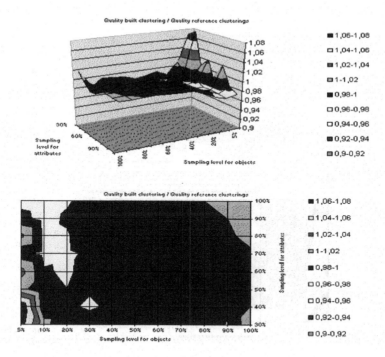

Fig. 3. Experimental Results: Quality of Built Clusterings / Quality of Reference Clusterings

6   Applications in K.D.D.: Applications to Clustering

Our method may constitue a good tool for solving different KDD problems, we focus here on 4 problems that may arise when one is looking for clustering a data set:

1. Some highly used clustering methods such as K-Means are characterized by an iterative heuristic search for a solution to an optimization problem, but the obtained solution is not always optimal and highly depends on the initialization process and on the number of iterations. A common solution for solving this problem is to run several clustering processes and to select the best result of those processes, we propose here to use our method for aggregating all those clusterings (see figure 5) instead of choosing the best. As the tests of the previous section have shown it may lead to better clusterings. (In our terminology, the problem is here to build an optimal global structure P_y according to a set of structures E, and the couple (E, P_y) is of A-type.)

2. Excessive computing time is also one of the more classical problem that faces a Data Miner. We think that our method associated to a parallelization of several clustering processes of sub-sets of a data set may be used to reduce

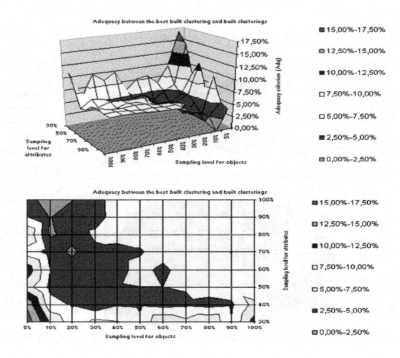

Fig. 4. Adequacy Between Built Clusterings and the Best Built Clustering according to the criterion to be optimized underlying to the clustering method.

computing time since its complexity is not high. As shown in figure 6, instead of processing clustering on a whole data set, drawing samples of this data set and processing simultaneously clusterings over these data sets on different computers and finally aggregating all those clusterings into a single one may lead to a reduction of the computing time by decreasing heavily the computing time associated to clustering. (In our methodology this problem may be of 2 different types: if the subsets are composed of samples of objects and all the attributes or of samples of objects and samples of attributes then the problem is to build an optimal global structure P_y according to a set of structures E, and the couple (E, P_y) is of B-type; if the subsets are composed of samples attributes and all the objects then the problem is to build an optimal global structure P_y according to a set of structures E, and the couple (E, P_y) is of A-type).

3. An other traditional problem is parameters setting, the setting stage (for instance determining the final number of clusters of the clustering) is crucial in a clustering process but it is also something pretty hard. One common solution for solving this problem is to run several clustering processes and to post-treat them in order to select the one that is considered as the best. We propose a solution similar to the one given in the first point: to use our method for aggregating those clusterings and to get a new one instead of choosing the best.(same kind of problem as the one of the first point)

4. Choosing a Clustering method may also be something particularly hard. Our algorithm may avoid this choice by aggregating clusterings resulting from several clustering processes using different methods. Moreover, we can also create meta-models by proceeding in a similar way. (In our terminology, this kind of problems may be of two types such as point 2)

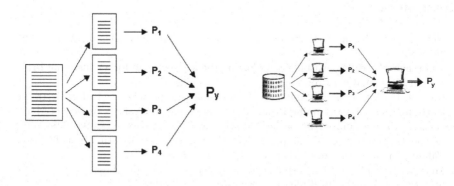

Fig. 5. Aggregating Different   **Fig. 6.** Reduction of Computing Time, A Parallel Architecture

7   Summary, Related and Future Works

We have presented a method for aggregating partitions and some of its potential utilization within K.D.D.. This method tries to find the partition which is the most in adequacy with a set of partitions. The measure that is used for representing the adequacy is not new since some nearly equivalent measure have been presented in [1], but the heuristic that we propose for solving the optimisation problem is completely original and its calculative cost is far less high than those of exact algorithms and heuristics presented in [1][3][2][6]. The experimental evaluations that were made have exhibited that its potential utilization within K.D.D. process associated to a parallelization of several clustering processes may be really interesting since it may allow reduction of computing time (clustering processes may only take into account a small part of objects and attributes of a data set), it may allow obtaining best results, it permits to reduce obscure human interaction with the computer (the user may only choose a set of possible parameters and not a unique parameter setting) and it permits to reduce results post treatment (when dealing with many solutions, the user has got (in most cases) to choose the final solution, our method will propose a unique solution).

Actually the experimental evaluations presented in this paper do not explicitly show the quality of this method as far as reduction of computing time and reduction of obscure human interaction are concerned, but they show the quality of the method for the other two points, however we would like to perform more experimentations. Consequently, we are currently working on a larger set of evaluations of our method.

References

1. Gordon, A.D., Vichi M.: Partitions of Partitions. in Journal of Classification. (1998) 265–285
2. Grötschel, M., Wakabayashi, Y.: A Cutting Plane Algorithm for a Clustering Problem. in Mathematical Programming, Vol. 45. (1989) 59–96
3. Hansen, P., Jaumard, B., Sanlaville, E.: in Diday, Lechevallier, Schader, Bertrand, Burtschy (Eds): Partitionning Problems in Clustering Analysis: A Review of Mathematical Programming Approaches. edn Springer-Verlag, Berlin. (1994) 228–240
4. Levine, E., Domany, E.: Resampling Method for Unsupervised Estimation of Cluster Validity. in Neural Computation, Vol. 13. (2001) 2573–2593
5. Merz, C., Murphy, P.: UCI repository of machine learning databases. http://www.ics.uci.edu/#mlearn/mlrepository.html. (1996)
6. Nicoloyannis, N., Terrenoire, M., Tounissoux, D.: An Optimisation Model for Aggregating Preferences: A Simulated Annealing Approach. in Health and System Science, Vol. 2. Hermes (1998) 33–44

Efficiently Computing Iceberg Cubes with Complex Constraints through Bounding

Pauline L.H. Chou and Xiuzhen Zhang

School of Computer Science and Information Technology
RMIT University, Melbourne, VIC 3001, Australia
{lchou, zhang}@cs.rmit.edu.au

Abstract. Data cubes facilitate fast On-Line Analytical Processing (OLAP). Iceberg cubes are special cubes comprise only the multi-dimensional groups satisfying some user-specified constraints. Previous algorithms have focused on iceberg cubes defined by relatively simple constraints such as "$COUNT(*) \geq \delta$" and "$COUNT(*) \geq \delta$ AND $AVG(Profit) \geq \alpha$". We propose an algorithm *I-Cubing* that computes iceberg cubes defined by complex constraints involving multiple predicates of aggregates such as "$COUNT(*) \geq \delta$ AND $(AVG(Profit) \geq \alpha$ OR $AVG(profit) \leq \beta)$". State-of-the-art iceberg cubing algorithms: *BUC* cannot handle such cases whereas *H-Cubing* has to incur extra cost. Our proposed bounding technique can prune for *all* the given constraints at once without extra cost. Experiments show that bounding has superior pruning power and *I-Cubing* is twice as fast as *H-Cubing*. Furthermore, *I-Cubing* performs equally well with more complex constraints.

1 Introduction

On-Line Analytical Processing (OLAP) systems [2] are used by managers, executives, and researchers to make decisions. A data cube is [3] the multi-dimensional generalisation of the group-by operator in SQL. Given a relation R defined by dimensional and measure attributes, CUBE(R) consists of aggregates of groups over all combinations of dimensions. Since the size of a data cube grows exponentially with the number of dimensions, it is impractical to completely compute a high-dimensional cube [1,5]. The *Iceberg Cube* was proposed [1] to selectively compute only those groups that satisfy some user-specified constraints. A constraint is monotonic [1] so that if a group fails the constraint, all its sub-groups also fail the constraint. For example, " COUNT(*) $\geq \delta$" is a monotonic constraint while " COUNT(*) $\leq \delta$" is a non-monotonic constraint. All constraints involving AVERAGE are non-monotonic.

Bottom-Up Cubing (BUC) [1] and *H-Cubing* [4] are the two current algorithms for computing iceberg cubes. BUC can only compute iceberg cubes defined by simple monotonic constraints such as "COUNT(*) $\geq \delta$ and SUM(Profit)

---

[1] In literature there are monotonic aggregate functions [1] and anti-monotonic constraints [4].

K.-Y. Whang, J. Jeon, K. Shim, J. Srivatava (Eds.): PAKDD 2003, LNAI 2637, pp. 423–429, 2003.
© Springer-Verlag Berlin Heidelberg 2003

$\geq \alpha$". H-Cubing was proposed to compute iceberg cubes with non-monotonic constraints such as "COUNT(*) $\geq k$ and AVG($Profit$) $\geq \delta$". As AVERAGE is not monotonic, a weaker but monotonic function *top-k average* was proposed for effective pruning.   However, top-k average can not be easily extended to pruning for multiple non-monotonic constraints. As is noted in [4], to compute an iceberg cube defined by "AVG(M) $\geq \delta$ and AVG(M) $\leq \delta$", the top-k average as well as the bottom-k average need to be computed, which incurs more space and computation.

In this study, we propose an approach to computing iceberg cubes defined by complex constraints involving multiple predicates on monotonic and non-monotonic constraints. We propose bounding for pruning, which do not incur extra cost under more complex constraints. The algorithm Iceberg-cubing (I-cubing) uses Iceberg-tree (I-tree) for implementing bounding effectively. Different from the H-tree of H-cubing, which keeps all dimension-values, I-tree only keeps the frequent dimension-values where the threshold is common for all iceberg queries and is specified from users' experience. An I-tree is usually smaller than a H-tree.

2   Pruning for Complex Constraints through Bounding

To compute iceberg cubes with complex constraints, our main idea is to find the upper and lower bounds of the aggregates of a group which are also the upper and lower bounds of all its sub-groups. These bounds are used to prune for various constraints, as summarised in the table below, where M is some aggregate function, $M_{upper}(g)$ is the upper bound of group g and its subgroups, and $M_{lower}(g)$ the lower bound.

Constraint	Pruning Condition
$M \geq \delta$	$M_{upper}(g) < \delta$
$M \leq \delta$	$M_{lower}(g) > \delta$
$\delta_1 \leq M$ and $M \leq \delta_2$ ($\delta_1 \leq \delta_2$)	$M_{lower}(g) > \delta_2$ or $M_{upper}(g) < \delta_1$
$M \leq \delta_1$ or $\delta_2 \leq M$ ($\delta_1 \leq \delta_2$)	$M_{lower}(g) > \delta_1$ and $M_{upper}(g) < \delta_2$

While pruning conditions can be easily observed, an interesting question is how to find such upper and lower bounds of groups and their sub-groups. We observe that bounds can be easily derived if groups are so formed: A group g is formed by multiple disjoint sets of tuples, let each such set be a **bundle**. Sub-groups of g are formed by subsets of bundles forming g. For most common aggregates the upper and lower bounds can be directly derived from the bundles, as shown in Table 1. Note that sum and count values of groups are required for AVERAGE while only one value is required for all other aggregates.

3   Bounding Groups in the Iceberg Tree

We call the groups of iceberg cubes *iceberg groups*. Given a set of constraints, only combinations of the dimension values that pass the monotonic constraints

Table 1. The upper and lower bounds of common aggregate functions

Aggregate		Bound
SUM of Positive Numbers	Upper:	The SUM of all bundles
	Lower:	The smallest SUM among the bundles
COUNT	Upper:	The COUNT of all bundles
	Lower:	The smallest COUNT among all bundles
MAX	Upper:	The largest MAX among the bundles
	Lower:	The smallest MAX among the bundles
MIN	Upper:	The largest MIN among the bundles
	Lower:	The smallest MIN among the bundles
AVERAGE	Upper:	The largest AVERAGE among the bundles
	Lower:	The smallest AVERAGE among the bundles
SUM of Positive and Negative Numbers	Upper:	The sum of all positive SUM among the bundles
	Lower:	The sum of all negative SUM among the bundles if there exist any, otherwise the smallest SUM among the bundles

can possibly produce iceberg groups. We call such a dimension value an *iceberg dimension value* or simply *idv*. An **Iceberg Tree** (**I-tree** is a prefix tree which keeps all *idv*'s: The *idv*'s are sorted into frequency descending order after the first scan of the dataset. The dataset is scanned the second time to insert the *idv*'s of each tuple to the tree in their sorted order. The same *idvs* are connected by the side-links to its header table entry. An example I-tree is shown in Figure 1.

Iceberg groups are formed by searching an I-tree. In Figure 1, the group (Computer) is collectively represented by the 4 branches having "Computer" which form a virtual tree of "Computer". The conditional header table constructed from the virtual tree includes "Melbourne", "Jan", and "May" as they also appear on the virtual tree of "Computer". Each of them together with "Computer" forms a 2-*d* sub-group of (Computer). The 3-*d* sub-groups are searched the same way on the virtual trees of their 2-*d* super-groups. Higher dimensional sub-groups are searched in the same way recursively. Note that (Computer) is formed by the 4 branches, among which 2 form its sub-group (Computer, Melbourne). Also, each branch represents a set of tuples. Hence, a branch is equivalent to the concept of a *bundle*. Note that the bounds are obtained easily as the by-products when traversing the side-links to construct the conditional header table.

Due to space limitation, we omit the formal proof but use AVERAGE as an example to illustrate that the upper and lower bounds of a group *g* and its sub-groups lie in the branches for a group. The same principle is applicable to all the aggregates listed in Table 1. In the example I-tree in Figure 1, among the branches forming (Computer) , the 2 bold branches that have both "Computer" and "Melbourne"nodes constitute a sub-group (Computer, Melbourne) of (Computer). According to Table 1, the upper and lower bound averages of (Computer) are the largest and smallest averages among the 4 branches, which

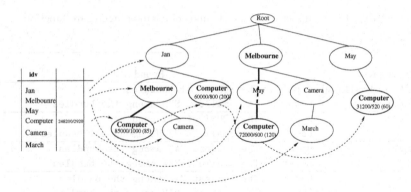

Fig. 1. Bounding Averages of groups in the I-tree

form the range [60, 200]. Note that they also bound the subgroup (Computer, Melbourne), whose smallest and largest averages are 85 and 120 respectively.

4   Iceberg Cubing

We propose an algorithm **Iceberg Cubing (I-Cubing)** which prunes through bounding on the I-tree for computing iceberg cubes. The algorithm is shown in Figure 2. We use one monotonic and one non-monotonic constraint as an example to demonstrate our algorithm. More constraints can be specified for the same algorithm. **I-Cubing** starts with the construction of the I-tree, which keeps only the *idv*'s satisfying the monotonic constraint in the tree and the header H. In line 4, a 1-d group is a candidate iceberg group only if its upper and lower bounds pass the non-monotonic constraint; in line 5, further computation decides a 1-d group is an iceberg group if its aggregate also passes the non-monotonic constraint. In lines 6–7, the sub-groups of candidate iceberg groups are searched by constructing the conditional header table.

Procedure **Iceberg-Miner** is called for every non-empty conditional header table generated. **Iceberg-Miner** searches the sub-groups of a group. In lines 1–3, since the conditional header is constructed using the monotonic constraint, they are checked only against the non-monotonic constraint. The way in which we search for the sub-groups of a group a_i with m aggregate dimensions simulates the way how 1-dimensional groups are derived from the header table (line 4). The dimension-values under the condition of a_i derives the sub-groups of a_i with $m+1$ dimensions. These dimension-values are collected via the sidelinks of a_i (line 4.a). Pruning of sub-groups is achieved here by including only those conditional dimension-values that pass all constraints (line 4.b). In line 4.c, ancestors of the same dimension-values are connected by side-links for further conditional header table construction. Each time a conditional header table is constructed, a set of potential iceberg groups are directly derived from the entries. The procedure is called recursively (line 5) for every non-empty conditional header table for the search of higher dimensional sub-groups.

Input: a) relational table T with dimensional attributes and measure e,
        b) a monotonic constraint:"$S(e) \geq \tau$", and non-monotonic constraints
            "$X(e) \geq \delta_1$ OR $X(e) \leq \delta_2$", where $S(e)$ and $X(e)$ are
                                            aggregate functions.

Output: Iceberg groups satisfying the input constraints.
1)   construct the I-tree for T using "$S(e) \geq \tau$";
2)   **foreach** $a_i \in H$ **do** $//H$ is the header of the I-tree;
3)   let g_{a_i} denote the 1-d group represented by a_i in H
           let $X_{upper}(g_{a_i})$ be the upper and $X_{lower}(g_{a_i})$ the lower bound of g_{a_i}
4)   **if** $X_{upper}(g_{a_i}) \geq \delta_1$ OR $X_{lower}(g_{a_i}) \leq \delta_2$ **then**
5)   **if** $g_{a_i}.X(e) \geq \delta_1$ OR $g_{a_i}.X(e) \leq \delta_2$ **then** Output g_{a_i};
6)   then construct H_{a_i}, the conditional header table based on a_i;
7)   **if** H_{a_i} is not empty **then** call Iceberg-Miner(H_{a_i});

Iceberg-Miner(H_C)
// H_C *is the Header conditioned on the set of dimension-values* C,
1)   **foreach** $a_i \in H_C$ **do**
2)   let g_{a_i} denote the group formed by a_i under C;
3)   **if** $g_{a_i}.X(e) \geq \delta_1$ OR $g_{a_i}.X(e) \leq \delta_2$ **then** Output g_{a_i};
4)   construct the **conditional** Header $H_{C_{a_i}}$ based on C and a_i;
        a) follow the side links of a_i to collect all a_i's ancestor dimension-values
           and their aggregates conditioned on a_i ;
           record the upper and lower bounds of each unique ancestors of a_i;
        // *pruning with bounding*
        b) select only those dimension-values having $S(e) \geq \tau$
           and $X_{upper}(g_{a_i}) \geq \delta_1$ OR $X_{lower}(g_{a_i}) \leq \delta_2$ to form $H_{C_{a_i}}$;
        c) connect the ancestors to the corresponding entries in $H_{C_{a_i}}$
                                        with side links;
    //search higher dimensional sub-groups.
5)   **if** $H_{C_{a_i}}$ is not empty **then** Iceberg-Miner($H_{C_{a_i}}$);

Fig. 2. Algorithm Iceberg Cubing

5   Experiments

5.1   Bounding Average and Top-k Average on the H-Tree

In order to have a fair comparison of the bounding and the top-k average for pruning, we apply both techniques on the H-tree for the average constraint. The computation time is shown in Figure 3. We fix the count threshold to 0.1% and vary the average thresholds. Experiments show that bounding is faster than the top-k average for pruning for various average thresholds. The difference is more significant at lower average thresholds (which is large iceberg cubes). Bounding is 1.4 times faster when the average threshold is below 200 (20% of the value range). Bounding achieves more severe pruning and boosts the performance of H-Cubing.

Experiments were conducted on the real-world weather data [1]. We compare I-Cubing, which prunes on the I-tree through bounding, with H-Cubing, which uses top-k average pruning on the H-tree. The constraints of the iceberg cube are "COUNT(*) $\geq k$" AND "AVG(e) $\geq \delta$". We fix the count threshold to 0.1%, vary the average threshold, and compare the computation time of the two algorithms. The results are shown in Figure 3. I-Cubing is always faster than H-Cubing for various average thresholds. The lower the average threshold, the more significant the improvement of I-Cubing over H-Cubing. I-Cubing is more than twice as fast when the average is below 60% of the value range. When the average threshold is low, the iceberg cube is large. The larger the iceberg cube is, the more time H-Cubing requires on maintaining top-k average values. However, the cost of bounding is not affected as much by the size of iceberg cubes. Our experiments have also confirmed that I-Cubing is very effective for computing iceberg cubes with complex constraints. As shown in Figure 4, as no notable extra cost has been incurred under an additional constraint on average, H-cubing uses almost the same computation time for the iceberg cube with 2 constraints as for the one with single constraint.

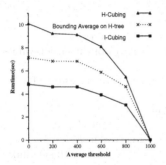

Fig. 3. I-Cubing and H-Cubing for average constraints

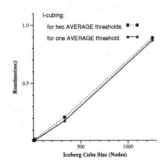

Fig. 4. I-Cubing for average constraints of various complexity

6   Conclusions and Future Work

We have examined the efficient computation of iceberg cubes defined by multiple predicates on common monotonic and non-monotonic constraints. Experiments show that our solution is flexible for constraints of various complexity and it is significantly faster than other approaches. The aggregate functions covered in this study are not exhaustive. It is still difficult to compute even more complex aggregate functions such as standard deviation and variance. Future work will study the efficient computation of iceberg cubes of these functions.

Acknowledgement. We thank Justin Zobel for his helpful comments.

References

1. K. Beyer and R. Ramakrishnan. Bottom-up computation of sparse and Iceberg CUBE. In *SIGMOD'99*, pages 359–370, 1999.
2. S. Chaudhuri and U. Dayal. An overview of data warehousing and olap technology. *SIGMOD Record*, Volume 26, Number 1, pages 65–74, 1997.
3. J. Gray, S. Chaudhuri, A. Bosworth, A. Layman, D. Reichart, M. Venkatrao, F. Pellow and H. Pirahesh. Data cube: A relational aggregation operator generalizing group-by, cross-tab, and sub-totals. *Journal of Data Mining and Knowledge Discovery*, Volume 1, Number 1, pages 29–53, 1997.
4. J. Han, J. Pei, G. Dong and K. Wang. Efficient computation of iceberg cubes with complex measures. In *SIGMOD'2001*, 2001.
5. V. Harinarayan, A. Rajaraman and J. D. Ullman. Implementing data cubes efficiently. In *SIGMOD'96*, pages 205–216, 1996.

Extraction of Tag Tree Patterns with Contractible Variables from Irregular Semistructured Data

Tetsuhiro Miyahara[1], Yusuke Suzuki[2], Takayoshi Shoudai[2],
Tomoyuki Uchida[1], Sachio Hirokawa[3], Kenichi Takahashi[1], and Hiroaki Ueda[1]

[1] Faculty of Information Sciences,
Hiroshima City University, Hiroshima 731-3194, Japan
{miyahara,takahasi,ueda}@its.hiroshima-cu.ac.jp.,
uchida@cs.hiroshima-cu.ac.jp
[2] Department of Informatics, Kyushu University, Kasuga 816-8580, Japan
{y-suzuki,shoudai}@i.kyushu-u.ac.jp
[3] Computing and Communications Center,
Kyushu University, Fukuoka 812-8581, Japan
hirokawa@cc.kyushu-u.ac.jp

Abstract. Information Extraction from semistructured data becomes more and more important. In order to extract meaningful or interesting contents from semistructured data, we need to extract common structured patterns from semistructured data. Many semistructured data have irregularities such as missing or erroneous data. A tag tree pattern is an edge labeled tree with ordered children which has tree structures of tags and structured variables. An edge label is a tag, a keyword or a wildcard, and a variable can be substituted by an arbitrary tree. Especially, a contractible variable matches any subtree including a singleton vertex. So a tag tree pattern is suited for representing common tree structured patterns in irregular semistructured data. We present a new method for extracting characteristic tag tree patterns from irregular semistructured data by using an algorithm for finding a least generalized tag tree pattern explaining given data. We report some experiments of applying this method to extracting characteristic tag tree patterns from irregular semistructured data.

1 Introduction

In this paper, we present a new method for extracting characteristic tag tree patterns from irregular semistructured data which are considered to be positive data. Web documents such as HTML files and XML files have no rigid structure and are called semistructured data. According to the Object Exchange Model [1], we treat semistructured data as tree structured data. To represent a tree structured pattern common to such tree structured data, we propose a *tag tree pattern*, which is a rooted tree consisting of ordered children, structured variables

K.-Y. Whang, J. Jeon, K. Shim, J. Srivatava (Eds.): PAKDD 2003, LNAI 2637, pp. 430–436, 2003.
© Springer-Verlag Berlin Heidelberg 2003

and edges labeled with tags, keywords or wildcards. A variable can be substituted by an arbitrary tree.

Many semistructured data have irregularities such as missing or erroneous data. In the Object Exchange Model, many essential data are represented as leaves or subtrees. So we introduce a new type of variable, called a *contractible variable*, which is regarded as an anonymous subtree in a tag tree pattern and matches any subtree including a singleton vertex. A usual variable, called an *uncontractible variable*, in a tag tree pattern does not match any singleton vertex.

Since a variable can be replaced by an arbitrary tree, overgeneralized patterns explaining given positive data are meaningless. Then, in order to extract meaningful information from irregular or incomplete tree structured data such as semistructured Web documents, we need to find one of the least generalized tag tree patterns. Consider the examples in Fig. 1. Let T_i' be the corresponding tree which is obtained by retaining the edge labels such as "Sec1" or "SubSec3.1" and ignoring the other edge labels in T_i. The tag tree pattern t_1 explains trees T_1', T_2' and T_3'. That is T_1', T_2' and T_3' are obtained from t_1 by substituting the variables of t_1 with trees. Further, t_1 is a least generalized tag tree pattern. The tag tree pattern t_2 also explains the three trees. But t_2 explains any tree with two or more vertices. So t_2 is overgeneralized and meaningless.

A tag tree pattern is different from other representations of tree structured patterns such as in [2,12] in that a tag tree pattern has structured variables which can be substituted by arbitrary trees. Recently, Information Extraction has been extensively studied [3,5]. But most studies are for free-text documents. Information Extraction or wrapper extraction from high-level data such as semistructured data or tables is a hot topic in the field of Web learning or Web mining [4,11]. Our extraction method is an application of the methods for extracting characteristic patterns from tree structured data, which are studied extensively [2,6,7,12].

2   Tag Tree Patterns and Extraction Method

2.1   Ordered Term Trees

Let $T = (V_T, E_T)$ be a rooted tree with ordered children (or simply a *tree*) which has a set V_T of vertices and a set E_T of edges. Let E_g and H_g be a partition of E_T, i.e., $E_g \cup H_g = E_T$ and $E_g \cap H_g = \emptyset$. And let $V_g = V_T$. A triplet $g = (V_g, E_g, H_g)$ is called a *term tree*, and elements in V_g, E_g and H_g are called a *vertex*, an *edge* and a *variable*, respectively. We assume that every edge and variable of a term tree is labeled with some words from specified languages. A label of a variable is called a *variable label*. Λ and X denote a set of edge labels and a set of variable labels, respectively, where $\Lambda \cap X = \phi$. For a term tree g and its vertices v_1 and v_i, a *path* from v_1 to v_i is a sequence $v_1, v_2, \ldots, v_i$ of distinct vertices of g such that for any j with $1 \leq j < i$, there exists an edge or a variable which consists of v_j and v_{j+1}. If there is an edge or a variable which consists of v and v' such that v lies on the path from the root to v', then v is said to be the *parent* of v' and

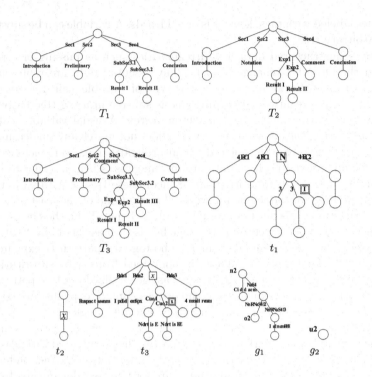

Fig. 1. Tag tree patterns t_1, t_2 and t_3 and trees T_1, T_2 and T_3. An uncontractible (resp. contractible) variable is represented by a single (resp. double) lined box with lines to its elements. The symbol "?" is a wildcard of an edge label. The label inside a box is the variable label of the variable.

v' is a *child* of v. We use a notation $[v, v']$ to represent a variable $\{v, v'\} \in H_g$ such that v is the parent of v'. Then we call v the *parent port* of $[v, v']$ and v' the *child port* of $[v, v']$.

Let X^c be a distinguished subset of X. We call variable labels in X^c *contractible variable labels*. A contractible variable label can be attached to a variable whose child port is a leaf. We call a variable with a contractible variable label a **contractible variable**, which is allowed to substitute a tree with a singleton vertex, as stated later. We call a variable which is not a contractible variable an uncontractible variable. For a variable $[v, v']$, when we pay attention to the kind of the variable, we denote by $[v, v']^c$ and $[v, v']^u$ a contractible variable and an uncontractible variable, respectively.

A term tree g is called *ordered* if every internal vertex u in g has a total ordering on all children of u. The ordering on the children of u is denoted by $<^g_u$. An ordered term tree g is called *regular* if all variables in H_g have mutually distinct variable labels in X.

An ordered term tree with no variable is called a **ground ordered term tree**, which is a standard ordered tree. OT_Λ denotes the set of all ground ordered term trees whose edge labels are in Λ. OTT^c_Λ denotes the set of all ordered term trees which have contractible or uncontractible variables, and edge labels in Λ. In this paper, we treat only regular ordered term trees with contractible or uncontractible variables. Therefore we call them **term trees** simply.

Let $f = (V_f, E_f, H_f)$ and $g = (V_g, E_g, H_g)$ be term trees. We say that f and g are *isomorphic*, if there is a bijection φ from V_f to V_g such that the following conditions (i)-(iv) hold: (i) The root of f is mapped to the root of g by φ. (ii) $\{u, v\} \in E_f$ if and only if $\{\varphi(u), \varphi(v)\} \in E_g$ and the two edges have the same edge label. (iii) $[u, v] \in H_f$ if and only if $[\varphi(u), \varphi(v)] \in H_g$. In particular, $[u, v]^c \in H_f$ if and only if $[\varphi(u), \varphi(v)]^c \in H_g$. (iv) For any internal vertex u in f which has more than one child, and for any two children u' and u'' of u, $u' <_u^f u''$ if and only if $\varphi(u') <_{\varphi(u)}^g \varphi(u'')$.

Let $\sigma = [u, u']$ be a list of two vertices in g where u is the root of g and u' is a leaf of g. The form $x := [g, \sigma]$ is called a *binding* for x. If x is a contractible variable label in X^c, g may be a tree with a singleton vertex u and thus $\sigma = [u, u]$. It is the only case that a tree with a singleton vertex is allowed for a binding. A new term tree $f\{x := [g, \sigma]\}$ is obtained by applying the binding $x := [g, \sigma]$ to f in the following way. Let $e = [v, v']$ be a variable in f with the variable label x. Let g' be one copy of g and w, w' the vertices of g' corresponding to u, u' of g, respectively. For the variable $e = [v, v']$, we attach g' to f by removing the variable e from H_f and by identifying the vertices v, v' with the vertices w, w' of g', respectively. If g is a tree with a singleton vertex, i.e., $u = u'$, then v becomes identical to v' after applying the binding. A *substitution* θ is a finite collection of bindings $\{x_1 := [g_1, \sigma_1], \ldots, x_n := [g_n, \sigma_n]\}$, where x_i's are mutually distinct variable labels in X. The term tree $f\theta$, called the *instance* of f by θ, is obtained by applying the all bindings $x_i := [g_i, \sigma_i]$ on f simultaneously.

For example, let t_3 be a term tree described in Fig. 1 and $\theta = \{x := [g_1, [u_1, v_1]], y := [g_2, [u_2, u_2]]\}$ be a substitution, where g_1 and g_2 are trees in Fig. 1. Then the instance $t_3\theta$ of the term tree t_3 by θ is isomorphic to the tree T_3 in Fig. 1.

2.2   Tag Tree Patterns

Let Λ_{Tag} and Λ_{KW} be two languages which consist of infinitely or finitely many words, where $\Lambda_{Tag} \cap \Lambda_{KW} = \emptyset$. We call words in Λ_{Tag} and Λ_{KW} a **tag** and a **keyword**, respectively. A **tag tree pattern** is a term tree such that each edge label on it is any of a tag, a keyword, and a special symbol "?", which is a wildcard of an edge label. A tag tree pattern with no variable is called a **ground tag tree pattern**.

For an edge $\{v, v'\}$ of a tag tree pattern and an edge $\{u, u'\}$ of a tree, we say that $\{v, v'\}$ *matches* $\{u, u'\}$ if the following conditions (i)-(iii) hold: (i) If the edge label of $\{v, v'\}$ is a tag, then the edge label of $\{u, u'\}$ is the same tag or a tag which is considered to be identical under an equality relation on tags. (ii) If the edge label of $\{v, v'\}$ is a keyword, then the edge label of $\{u, u'\}$ is the same keyword. (iii) If the edge label of $\{v, v'\}$ is "?", then we don't care the edge label of $\{u, u'\}$.

A ground tag tree pattern $\pi = (V_\pi, E_\pi, \emptyset)$ *matches* a tree $T = (V_T, E_T)$ if there exists a bijection φ from V_π to V_T such that the following conditions (i)-(iv) hold: (i) The root of π is mapped to the root of T by φ. (ii) $\{v, v'\} \in E_\pi$ if and only if $\{\varphi(v), \varphi(v')\} \in E_T$. (iii) For all $\{v, v'\} \in E_\pi$, $\{v, v'\}$ matches $\{\varphi(v), \varphi(v')\}$. (iv) For any internal vertex u in π which has more than one child, and for any two children u' and u'' of u, $u' <_u^\pi u''$ if and only if $\varphi(u') <_{\varphi(u)}^T \varphi(u'')$. A tag tree pattern π **matches** a tree T if there exists a substitution θ such that $\pi\theta$ is a ground tag tree pattern and $\pi\theta$ matches T. Then the *language* $L_\Lambda(\pi)$, which is the descriptive power of a tag tree pattern π, is defined as $L_\Lambda(\pi) = \{$a tree T in $\mathcal{OT}_\Lambda \mid \pi$ matches $T\}$ where $\Lambda = \Lambda_{Tag} \cup \Lambda_{KW}$.

2.3   Extraction of Least Generalized Tag Tree Patterns

We propose a new method for extracting characteristic tag tree patterns from irregular semistructured data which are considered to be positive tree structured data. A tag tree pattern π is a **least generalized tag tree pattern** explaining a given set S of trees which are considered to be positive data, if (i) $S \subseteq L_\Lambda(\pi)$ (π explains S) and (ii) there is no tag tree pattern π' satisfying that $S \subseteq L_\Lambda(\pi') \subsetneqq L_\Lambda(\pi)$. The problem for finding a least generalized tag tree pattern for a given set of trees is discussed as the minimal language problem (MINL for short) in the field of computational learning theory [8,10].

Our extraction method finds a least generalized tag tree pattern explaining a given set of trees S, by using a polynomial time algorithm for solving the MINL problem [10]. First, the algorithm finds a least generalized tag tree pattern t which consists of only uncontractible variables and explains S. Secondly, it finds a tag tree pattern t' which is obtained from t, by replacing a variable in t with an edge labeled with an edge label or a wildcard, or a contractible variable, if the obtained tag tree pattern t' explains S. The algorithm then repeatedly applies the replacing to the tag tree pattern until no more replacing is applicable. Finally it outputs the resulting tag tree pattern. The extraction method uses a polynomial time matching algorithm for deciding whether or not a given tag tree pattern matches a given tree for hypothesis checking [10]. The matching algorithm is an extension of the polynomial time matching algorithms [8,9].

3   Implementation and Experimental Results

We have implemented the extraction method in Section 2.3, which finds a least general tag tree pattern explaining the given semistructured data. The implementation is by GCL2.2 and on a Sun workstation Ultra-10 clock 333MHz. In Fig. 2 we report some experiments on sample files of semistructured data. In these experiments, an input tree represents a subtree of a parsed tree of an HTML/XML file. The tree structure and HTML/XML tags in a parsed tree are preserved in the corresponding input tree. Attributes and their values are ignored. No equality relation on tags is assumed. All string data in a parsed

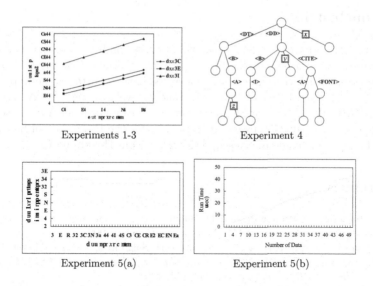

Experiments 1-3	Experiment 4
Experiment 5(a)	Experiment 5(b)

Fig. 2. Experimental results for extracting least generalized tag tree patterns from semistructured data.

tree are converted to the same dummy keyword, in order to pay attention to structures of tags in a parsed tree.

In Exp. 1 to 3 , we made samples of artificial HTML files in order to evaluate our method. The input file for Exp. 1 consists of trees with about 40 vertices. The input file for Exp. 2 consists of 90 % of trees with about 40 vertices and 10 % of trees with about 20 vertices. The input file for Exp. 3 consists of 90 % of trees with about 40 vertices and 10 % of trees with about 70 vertices. The graphs for Exp. 1 to 3 show the running time of the method with varying the number of data for the three experiments. The numbers of vertices of the obtained tag tree patterns of the method are almost same for the three experiments. This shows that the method has robustness for irregularities of sample semistructured data.

In Exp. 4, the sample HTML file is a result of a search engine of a web site with a local search function (http://www.ael.org). The sample file consists of 10 trees with about 18 vertices. A tree in the sample file is a record of bibliographic data. The obtained tag tree pattern in Fig. 2 (Exp. 4) is a least generalized tree pattern explaining the sample file. An edge with no edge label represents an edge with the dummy keyword. So the obtained tag tree pattern is considered to be a wrapper for such tree structured data. In Exp. 5, the sample XML file is from the DBLP bibliographic database (http://dblp.uni-trier.de/xml/dblp.xml). The sample file consists of 50 trees with about 18 vertices. The graphs in Exp. 5(a) and 5(b) show the numbers of vertices of the obtained tag tree pattern and the running time of the method with varying the number of data. The obtained tag tree patterns from the files with 42 or more number of data are the same. This shows that small number of data are sufficient for the method to extract a characteristic tag tree pattern from such bibliographic data.

4   Conclusions

In this paper, we have studied Information Extraction from semistructured data. We have proposed a new method for extracting characteristic tag tree patterns from irregular semistructured data. Also, we have reported some experiments of applying this method to extracting least generalized tag tree patterns from HTML/XML files. This work is partly supported by Grant-in-Aid for Scientific Research (C) No.13680459 from Japan Society for the Promotion of Science and Grant for Special Academic Research No.2101 from Hiroshima City University.

References

1. S. Abiteboul, P. Buneman, and D. Suciu. *Data on the Web: From Relations to Semistructured Data and XML.* Morgan Kaufmann, 2000.
2. T. Asai, K. Abe, S. Kawasoe, H. Arimura, H. Sakamoto, and S. Arikawa. Efficient substructure discovery from large semi-structured data. *Proc. 2nd SIAM Int. Conf. Data Mining (SDM-2002)*, pages 158–174, 2002.
3. C.-H. Chang, S.-C. Lui, and Y.-C. Wu. Applying pattern mining to web information extraction. *Proc. PAKDD-2001, Springer-Verlag, LNAI 2035*, pages 4–15, 2001.
4. W.W. Cohen, H. Mathew, and S.J. Lee. A flexible learning system for wrapping tables and lists in HTML documents. *Proc. WWW 2002*, pages 1–21, 2002.
5. N. Kushmerick. Wrapper induction: efficiency and expressiveness. *Artificial Intelligence*, 118:15–68, 2000.
6. T. Miyahara, T. Shoudai, T. Uchida, K. Takahashi, and H. Ueda. Discovery of frequent tree structured patterns in semistructured web documents. *Proc. PAKDD-2001, Springer-Verlag, LNAI 2035*, pages 47–52, 2001.
7. T. Miyahara, Y. Suzuki, T. Shoudai, T. Uchida, K. Takahashi, and H. Ueda. Discovery of frequent tag tree patterns in semistructured web documents. *Proc. PAKDD-2002, Springer-Verlag, LNAI 2336*, pages 341–355, 2002.
8. Y. Suzuki, R. Akanuma, T. Shoudai, T. Miyahara, and T. Uchida. Polynomial time inductive inference of ordered tree patterns with internal structured variables from positive data. *Proc. COLT-2002, Springer-Verlag, LNAI 2375*, pages 169–184, 2002.
9. Y. Suzuki, T. Shoudai, T. Miyahara, and T. Uchida. A polynomial time matching algorithm of structured ordered tree patterns for data mining from semistructured data. *Proc. ILP-2002, Springer-Verlag, LNAI (to appear)*, 2003.
10. Y. Suzuki, T. Shoudai, T. Miyahara, T. Uchida, and S. Hirokawa. Polynomial time inductive inference of ordered term trees with contractible variables from positive data. *Proc. LA Winter Symposium, Kyoto, Japan*, pages 13–1 – 13–11, 2003.
11. T. Taguchi, K. Koga, and S. Hirokawa. Integration of search sites of the World Wide Web. *Proc. of CUM, Vol.2*, pages 25–32, 2000.
12. K. Wang and H. Liu. Discovering structural association of semistructured data. *IEEE Trans. Knowledge and Data Engineering*, 12:353–371, 2000.

Step-by-Step Regression: A More Efficient Alternative for Polynomial Multiple Linear Regression in Stream Cube

Chao Liu[1], Ming Zhang[1], Minrui Zheng[1], and Yixin Chen[2]

[1] School of Electronics Engineering and Computer Science
Peking University, Beijing, 100871, China
{Chaoliu, mzhang, zhengmr}@db.pku.edu.cn
[2] Department of Computer Science, University of Illinois at Urbana-Champaign,
Urbana, IL, 61801, U.S.A.
chen21@students.uiuc.edu

Abstract. Facing tremendous and potentially infinite stream data, it is impossible to record them entirely. Thus synopses are required to be generated timely to capture the underlying model for stream management systems. Traditionally, curve fitting through *Multiple Linear Regression* (MLR) is a powerful and efficient modeling tool. In order to further accelerate its processing efficiency, we propose *Step-by-step Regression* (SR) as a more efficient alternative. As revealed in experiments, it speeds up for more than 40 times. In addition, inspired by previous work, we integrated SR into cube environment through similar compression technique to perform online analytical processing and mining over data stream. Finally, experiments show that SR not only significantly alleviates the computation pressure on the front ends of data stream management systems, but also results in a much smaller stream cube for on line analysis and real-time surveillance.

1 Introduction

Recent years have witnessed an exploration of stream data in many emerging application domains, such as the network installations of large Telecom and Internet service providers, where the data arrive fast in large volume. For example, AT&T collects 100GBs of NetFlow data each day, more than 1MB per second [1]! Such large volume and rapid incoming speed set an exciting challenge for stream data management systems. Aiming at avoiding recording the entire data stream (which is potentially impossible), certain synopsis is in urgent need to summarize data stream. Fortunately, decades of development in time series analysis has distinguished some useful ones, such as Discrete Fourier Transformation (DFT) [2, 3] and Discrete Wavelet Transformation (DWT) [4, 5]. Although they work well for classic time series data, they seem somewhat incompetent when facing large volume of stream data because short per-item processing time is one of the key criteria imposed by the nature of stream data [1]. In response, tradeoff between accuracy and time is often adopted to provide approximate query answers [6], such as trend analysis. In addition, this tradeoff just suits the exploratory nature of data stream. Guided by this idea, we

K.-Y. Whang, J. Jeon, K. Shim, J. Srivatava (Eds.): PAKDD 2003, LNAI 2637, pp. 437–448, 2003.

adopt a division strategy, and propose SR as an alternative for MLR in the case of polynomial regression, expecting it to achieve the equivalent or similar effect in practice but with lighter computation consumption. Besides time improvement, this division strategy also lowers the need for memory usage, which is another critical requirement for stream data processing system [1]. For clarification, we use the following simplest example to examine the notions of SR in comparison with MLR.

Example 1: Given those sample points in Fig. 1, we are interested in how sharply the changing speed changes quantitatively.

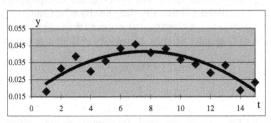

Fig. 1. Sample time series data

Traditionally, we can use MLR, say quadratic regression for example, getting the fitting curve $y = at^2 + bt + c$.

Because mathematically how sharply the changing speed changes is measured by the second order derivative, we can take a as the desired measurement because the second order derivative of y is $y'' = 2a$, and it has no difference with a but a constant coefficient.

In another place, if we adopt SR, we observe the problem from a different angle. As what we want is how sharply the changing speed changes, we can first get the measure for changing speed, and after that we compute how sharply this measure changes. Because we usually use the slope to represent the changing speed, we carry out our method in the following manner.

First, we sequentially divide the entire time series into n segments, on each of which we perform one order regression and get the fitting curves

$$y = k_i t + b_i \quad (i = 1, 2 \cdots n).$$

Clearly, k_i quantitatively measures how sharply the time series data change in the *ith* segment. Then we perform another one order regression on these k_i, getting the fitting curve $s = pk + q$.

Finally, we claim that p is the desired quantitative measure. For the two methods used here, we call the former *quadratic* and the latter *2linear*. Intuitively *2linear* adopts a division strategy, which can be expected to boost efficiency. The in-depth examination will be presented in the rest of the paper.

Moreover, inspired by previous works [7, 8, 9], in this paper, we also proved the feasibility of integrating SR with the OLAP environment. In previous work, *Linearly Compression Representation* (LCR) is gained through MLR, and is registered as the regression value in every cell. In this paper, we get through SR the *Step-by-step Regression Representation* (SRR), and substituted it for LCR in the cube. As shown in theory and experiments, this substitution results in a much smaller cube than that registering LCR.

Although SR outperforms MLR on both time and space, a key question must be answered. That is "Can SR gain the equivalent or similar effects as MLR?" We will deal with this problem in the latter part of this paper, and reach the final conclusion that they are significantly correlated.

In summary, driven by the challenges and criteria for stream data mining applications [1], we explored stream data with the following contributions.

1.   We proposed SR as a more efficient alternative for polynomial MLR by adopting a division strategy. It resulted in a more concise synopsis (SRR) than LCR, which is generated through MLR.
2.   This division strategy leads to a significant improvement on time and memory usage for online computation, perfectly suiting the two key criteria imposed by the nature of stream data. Experiments show that this improvement on processing time is at least 40 times and meanwhile with less memory usage.
3.   Inspired by new idea of "OLAPing data stream", we proved the same feasibility of SRR to substitute LCR in stream cube [8], which means integrating SR into OLAP technology.
4.   Because SRR is more concise than LCR, it reduces the cube size to as small as 66.7% of the original. In addition, concise storage strategy we proposed further exploits this reduction ratio to less than 22.2%.

The rest of the paper is organized as follows. Section 2 introduces the general theories about SR. Feasibility for Integrating SR into OLAP environment is demonstrated in Section 3. In Section 4, we report our empirical study on both time and space. The correlation analysis between SR and MLR, together with other issues is discussed in Section 5. Finally, Section 6 comes to our conclusion.

2   General Theories of SR

Because we have proposed and extended SR no more than polynomial regression hitherto, we confine our discussion to polynomial regression in this paper. Therefore, we first give a quick review about the theory about MLR in the background of polynomial regression in section 2.1. Then in section 2.2, we introduce the shift invariant as a key concept for SR. Finally, we examine the general algorithm for SR computation in section 2.3.

2.1 Quick Review of MLR

In real world practice, we are usually most concerned with the main trend over tremendous data, such as increases, decreases or how sharply it varies. From another point of view, main trends can also be taken as a measure for abnormal extent because it assesses how different this set of data is in comparison with the common cases. For example, if the regression slope over a certain segment of data is bigger than others, we can reasonably deduce that this segment is more abnormal than the others. For convenience, we give the following definition.

Definition 1: *Abnormal Extent Indicator* (AEI): A numeric value indicating the abnormal extent of certain segment of data.

However, standard SQL functions, such as *avg*, *max*, *min* etc, are not competent for this aim. Previous works [10, 11] demonstrated that MLR is a frequently used and powerful tool in this aspect. Due to its fundamental role in SR, we first give a quick review through a simple example (for in-depth knowledge, please consult [11, 12]).

Example 2: Without loss of generality, we here model n sample points through k-1 order polynomial regression. Each sample point is in the form of (t_i, y_i), where t is the index of the sample value y_i.

Solution: For k-1 order polynomial regression, the fitting curve is in the form of

$$y = \eta_0 + \eta_1 t + \eta_2 t^2 + \ldots\ldots + \eta_{k-2} t^{k-2} + \eta_{k-1} t^{k-1}. \tag{1}$$

If we denote $\eta^T = (\eta_0 \quad \eta_1 \quad \cdots \quad \eta_{k-1})$, the following theorem gives the proper estimate of η.

Theorem 1: If the inverse of $(U^T U)$ exists, Ordinary Least Square (OLS) estimate of η is unique and given by

$$\eta' = (U^T U)^{-1} U^T Y, \text{ where } U = \begin{pmatrix} 1 & t_1 & t_1^2 & \cdots & t_1^{k-1} \\ 1 & t_2 & t_2^2 & \cdots & t_2^{k-1} \\ \vdots & \vdots & \vdots & \cdots & \vdots \\ 1 & t_n & t_n^2 & \cdots & t_n^{k-1} \end{pmatrix}. \tag{2}$$

2.2 Shift Invariant in Polynomials

Now we have determined the fitting curve (1) through (2), but not all the elements in η are equally useful. So we try to locate which part represents our interests.

When considering a fitting curve, the last thing we are concerned is commonly the curve position. Therefore, in order to capture the main trend from the regression function, we should eliminate those tokens that are relevant to curve position. Naturally, this task is equivalent to finding the shift invariant. The following definition and theorem deal with this problem.

Definition 2: *Shift invariant in polynomials*:

Given some notations $u = (1 \quad t \quad t^- \quad \cdots \quad t^-)$,

$$\eta_p^T = (\eta_{p,0} \quad \eta_{p,1} \quad \cdots \quad \eta_{p,n}), \eta_q^T = (\eta_{q,0} \quad \eta_{q,1} \quad \cdots \quad \eta_{q,n}),$$

and two polynomial fitting curves

$$y_p = P^{(n)}(t) = u\eta_p, \ y_q = Q^{(n)}(t) = u\eta_q,$$

where y_q is the resulting curve of y_p after the follow shift transformation

$$T: \begin{Bmatrix} t' = t - t_0 \\ y' = y - y_0 \end{Bmatrix},$$

Then *if* $\exists i \in \{1, 2 \cdots n\}$ *st* $\forall t_0, y_0 \in R, \eta_{p,i} = \eta_{q,i}$ holds, $\eta_{p,i}$ is the shift invariant in the given polynomial y_p.

With the definition above, we now probe the very token that we are interested in.

Theorem 2: In polynomial fitting curve

$$y = \eta_0 + \eta_1 t + \eta_2 t^2 + \ldots\ldots + \eta_{k-2} t^{k-2} + \eta_{k-1} t^{k-1},$$

only η_{k-1} is the shift invariant.

Proof: Here we eliminate the proof, which is rather direct once that shift transformation is performed based on the definition of shift invariant.

2.3   Algorithms for SR

With the preparation of MLR and shift invariant, the two key concepts concerned with SR, we are ready to examine how SR works in this subsection. Its main idea is performing regression in a step-by-step manner, taking advantage of the shift invariants in every step regardless of all the other regression coefficients. To illustrate it more clearly, we propose a general example as follows.

General Example: Now we have s sample points, each of which has the same form as in Example 2. Our objective is to obtain AEI over these s sample points. In the case of n order polynomial regression, the desired AEI is the last element in η'. We take it as η_n, and η'is gained through (2). In the following, we will examine how could SR accomplish the equivalent task.

Without loss of generality, we decided to solve the proposed problem through m steps, and on the ith step, we perform n_i order regression. Thus we have

$$\sum_{i=1}^{m} n_i = n. \tag{3}$$

In correspondence to the above division of n, we also divide those s sample points.

$$s = \prod_{i=1}^{m} a_i \text{ where } a_i \geq n_i. \tag{4}$$

Then in the ith step, we perform n_i order regression on every data group, which is composed of a_i values. For convenience, we call (3) and (4) the *division scheme* for SR, which is uniform for a specific applications. With the two division strategies (3) and (4), we begin to examine SR in a step-by-step manner.

First, we denote $Y_1^T = \begin{pmatrix} y_1 & y_2 & \cdots & y_s \end{pmatrix}$, which represents the initial sample points, then we divide Y_1 for every a_1 elements sequentially, getting $\prod_{i=2}^{m} a_i$ sub vectors. Then we get

$$\eta_i = (U_1^T U_1)^{-1} U_1^T Y_{1,i} = \begin{pmatrix} \eta_{i,0} & \eta_{i,1} & \cdots & \eta_{i,n_1} \end{pmatrix}^T (i = 1,2 \cdots \prod_{j=2}^{m} a_j),$$

where U_1 is determined by n_1 and a_1 as in (2). Picking out the shift invariant η_{i,n_1} in every η_i ($i = 1,2 \ldots \prod_{j=2}^{m} a_j$), we arrange them in the ascending order of i, constructing a new vector Y_2. Then we repeat the same operations on Y_2 as we have done on Y_1, building up another vector Y_3, which becomes the base for further such repetitions. In the end, after repeating such operation m times, we finally get the ultimate regression vector

$$\eta_m = (U_m^T U_m)^{-1} U_m^T Y_m = \begin{pmatrix} \eta_0 & \eta_1 & \cdots & \eta_{n_m} \end{pmatrix}^T, \tag{5}$$

and its shift invariant η_{n_m} is the just desired AEI through SR.

3   Integrating SR into OLAP Environment

Based on the success of OLAP technology [13, 14] and regression analysis, "multi-dimensional analysis over data stream" has been newly proposed. Now we use a simple example to illustrate this interesting idea.

Example 3: Suppose a powerful sever is used to collect all the ATM transactions throughout the country. Each record is in the form of *(city, transaction amount, minute)*. Our aim is to analyze how sharply the transaction amount varies in certain regions (e.g. city, province, etc) during a certain period of time (e.g. hour, day, etc). Usually, one order regression slope is suitable for this aim.

In the notion of multi-dimensional analysis, we consider "time" as a common attribute as "location", and construct a cube with these two dimensions. Different from traditional data cubes, we do not store the exact "transaction amount" in each cell, but the regression slope instead. Specifically, we collect 60 tuples within an hour, and perform regression on them to get the slope, which we will register in the corresponding cell in the cube. In this way, we could finally construct the base cuboid *(city, slope, hour)*.

Based on this base cuboid, we can aggregate certain cities to a specific province, and get the slope value of this province directly from that of its component cities. Moreover, if we have the slope values of a specific city for the past 24 hours, we can gain the slope of this city for the past day from them. In this case, we call "location" as a standard dimension and "time" as a regression dimension because "time" is our regression variable. Thus we have obtained a powerful environment for multi-dimensional analysis over data stream, and denote it as *stream cube* [8]. For consistency, we use the same definitions of regression and standard dimension as in [8]. For space concern, we do not elaborate on these definitions.

Definition 3: *Step-by-step Regression Representation* (SRR) is the set $\eta_m \cup \theta_m$, where η_m is computed through (5), and $\theta_m = U_m^T U_m$.

Now that SRR is equivalent to LCR in the stream cube, we can materialize this cube in the same manner as LCR and have the following theorem.

Theorem 3: *If we materialize the SRR of the base cells of a data cube, we can compute the SRR of all other cells in the data cube.*

This theorem can be proved with proofs of the feasibility of aggregation over standard and regression dimensions, which we will discuss in the following two sections. In addition, we are sorry to eliminate all the proofs in this section because we cannot fill those matrix formulae into the limited space.

3.1   Aggregation on Standard Dimensions

Theorem 4: *The algorithm for SR can be present as* $\eta_m = AY_1$,
where A is a matrix representing the whole computation procedure.
Proof: We still use the General Example in section 2.3 and denote
$$U_i^+ = (U_i^T U_i)^{-1} U_i^T \ (i=1,2...m).$$

If we denote the last line vector of U_m^+ as v_i^+ and follow the algorithm for SR computation as in Section 2.3. We get Y_{i+1} from Y_i through

$$Y_{i+1} = T_i Y_i \text{ where } T_i = \begin{pmatrix} v_i^+ & 0 & \cdots & 0 \\ 0 & v_i^+ & \ddots & 0 \\ \vdots & \ddots & \ddots & \vdots \\ 0 & 0 & \cdots & v_i^+ \end{pmatrix}.$$

Therefore $\eta_m = U_m^+ Y_m = U_m^+ T_{m-1} Y_{m-1} = U_m^+ T_{m-1} T_{m-2} Y_{m-2} = \cdots = U_m^+ T_{m-1} T_{m-2} \cdots T_2 T_1 Y_1 = AY$ where A represents the whole computation procedure.

Theorem 5 [*lossless aggregation on standard dimensions*]: For aggregation on a standard dimension, the SRR for the aggregated cell c_a can be derived from that of the component cells c_i using the following formulae.

(a) $\eta_m^a = \sum_{i=1}^{n} \eta_m^i$ (6)

(b) $\theta_m^a = \theta_m^i$ (i=1, 2, ... n)   (7)

Proof: (a)   If we compute the SRR of c_a from scratch, we first aggregated those sample points. It means $Y_a = \sum_{i=1}^{n} Y_i$. Therefore, according to Theorem 4, we have

$$\eta_m^a = AY_a = A\sum_{i=1}^{n} Y_i = \sum_{i=1}^{n} AY_i = \sum_{i=1}^{n} \eta_m^i,$$

which means we can get the SRR of aggregated cell from its component cells directly. Moreover, the above equation holds because A is determined by the division scheme, which stays the same for all the cells in a specific stream cube.

(b)   Because of the uniform division scheme for all the cells c_i, U_m stays the same. Thus $\theta_m^a = U_m^T U_m = \theta_m^i$ $(i = 1, 2 \ldots n)$.

3.2   Aggregation on Regression Dimensions

Theorem 6 [*Lossless aggregation on regression dimensions*] For aggregation on a regression dimension, the SRR of the aggregated cell can be derived from that of the component cells using the following equations:

(a) $\eta_m^a = (\sum_{i=1}^{n} \theta^i)^{-1} (\sum_{i=1}^{n} \theta^i \eta_m^i)$ (8)

(b) $\theta^a = \sum_{i=1}^{n} \theta^i$ (9)

In fact, we have hitherto proved Theorem 3 because the proofs for Theorem 5 and 6 together constitute the proof for Theorem 3. In other words, we have proved the feasibility of integrating SR into the OLAP environment as that of MLR.

3.3   Concise Storage Strategy

With careful examination of (7) and (9), we find that the auxiliary parameter θ_m in SRR changes only when we aggregate over regression dimensions. And this observation result also holds for LCR. Therefore, we can deprive those θ_m from every cell and only register them for every unit on the regression dimension, saving a lot of space. As this improvement represents little novelty, we do not elaborate it, but just show its effects through experiments later. Finally we call this improvement *concise storage strategy*.

4   Performance Study

In this section, we report our experimental comparison on both time and space consumption to show SR's superiorities to MLR.

All the experiments are carried out on a 433MHz Celeron PC machine with 152 megabytes main memory, running Microsoft Windows 2000 Professional. All the implementations are coded using Microsoft Visual C++ 6.0.

We designed two experiments for performance evaluation: Experiment 1 is for the processing time and the other is for the comparison of stream cube size. Our test beds are based on two randomly selected time series datasets: *dailyibm* and *dailysap* [15]. By repeating *dailyibm* and *dailysap* many times, we constructed two huge and potentially infinite datasets, *dibm* and *dsap*, which simulated data stream in Experiment 1. For Experiment 2, we first coded a data generator to get uniform distributed datasets with any number of dimensions of any cardinality. Then, we combined these huge random datasets with *dibm* and *dsap* to obtain the utmost simulated stream datasets: *streamsap* and *streamibm*, in Experiment 2.

Finally, we note that although we have carried out experiments on all the datasets generated from *dailysap* and *dailyibm*, we just report the results for *dailysap* because the results on *dailyibm* are similar.

4.1   Time Performance

Experiment 1: In practice, buffer is usually used to temporarily store those arrived data for batch processing (e.g. in data stream management systems). Therefore, we vary the buffer size to observe the needed processing time for *2linear* and *quadratic*.

Fig. 2 clearly exhibits the distinct comparison on time efficiency between *2linear* and *quadratic* although the figure looks somewhat strange. Exactly, regression analysis on the time costs of *2linear* and *quadratic* gave the regression function

$$t_{quadratic} = 44.771 t_{2linear} + 0.0351,$$

which tells us *2linear* can get the AEI more than 40 times faster than *quadratic*.

Now we turn to the strange aspect in Fig. 2. Due to the significant comparison, Fig. 2 does not show *2linear*'s variance in the same scale for *quadratic*. In Fig. 3, *2linear* also exhibits the similar abnormity. Because the time superiorities enjoyed by *2linear* is obvious, we surmise that the variances come from the same factors other than the algorithms (SR and MLR) themselves. Therefore, we try to resolve abnormality through correlation analysis because if their time costs are correlated, we can reasonably believe that there exists some mechanism determining such variance.

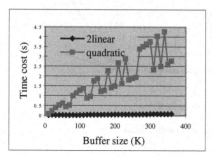

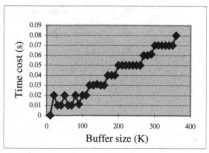

Fig. 2. Processing time comparison   **Fig. 3.** Time cost of *2linear*

Through calculation, we found that the correlation coefficient between the two series of time cost is as high as 0.871. Furthermore, F-test tells us that

$$F = (n-2)\frac{R^2}{1-R^2} = (36-2)\frac{0.871^2}{1-0.871^2} = 113.393 >> 9.18 = F_{0.005}(1,30) > F_{0.005}(1,34)$$

which means this correlation is highly significant. Therefore, we can claim that there should be certain mechanism that results in such variances for both methods. However, due to our limited researches in operating system, we regret that we can not exactly give the utmost answer but just suspect that memory management may be the final reason. Finally, we sincerely hope and appreciate other researchers' help to thoroughly work it out.

4.2   Space Performance

Experiment 2: In this experiment, we examine three stream cubes, which are generated through different regression methods and storage strategies. For clarification, we here describe their characters.

Cube1:   concise form of SRR through *2linear*
Cube2:   concise form of LCR through *quadratic*
Cube3:   initial form of LCR through *quadratic*

Then comparison between Cube1 and Cube2 tells us how much *2linear* reduces the cube size in comparison with *quadratic*, and comparison between Cube2 and Cube3 shows the effect of concise storage strategy. In this experiment, we set the cardinality for each dimension to 5, and vary the number of dimension for observation. Fig. 4 shows the comparison graphically. Furthermore, we could prove:

$$Size_{cube1} : Size_{cube2} : Size_{cube3} = 2 : 3 : 9 .$$

Thus the superiorities of SR and the concise storage strategy are very apparent.

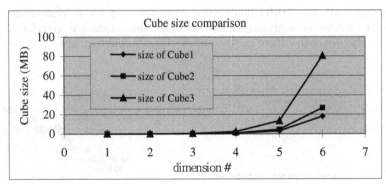

Fig. 4. Cube size comparison

4.3   Further Superiorities of SR in Practice

When it comes to practice, real world computation environment further distinguishes SR for the following two main reasons.

First, SR needs smaller size of memory for online computation. SR breaks the whole huge computation apart into small ones, thus smaller buffer and matrices are expected. For example, in order to compute the AEI over 3600 sample points, *2linear* needs a buffer of 60 units and a matrix in the size of 60×2. In comparison, *quadratic* requires a buffer size of 3600 and a matrix of 3600×3, which is as large as 80 times as that of *2linear*. We believe this reduction on memory usage is one of the contributors for the overall time improvement (Fig. 2) because the operating system does not have to swap data between memory and disk so frequently.

Secondly, SR can generate the synopsis AEI in a more simultaneous manner. For example, if we aim at computing the AEI for every hour, *quadratic* runs at the end of each hour, manipulating 3600 tuples at one time. *2linear,* on the other hand, runs every minute, getting the AEI from 60 tuples. Finally, when one hour finishes, 60 AEIs for the past 60 minutes are ready, and then another one order regression can be performed on these AEIs, giving out the final desired AEI. Therefore, SR not only breaks the huge computation burden apart but also distributes them evenly.

5   Discussion

Although we have made significant improvements on both time and space, there is still one critical question we must face: "Can SR achieve the equivalent or similar effects as MLR?" Clearly, if the AEI obtained through these two approaches have no relevance, the results mean nothing no matter how faster SR runs than MLR and how smaller the final cube is. Therefore, in this section, we will first probe this correlation through statistics methods. Then in section 5.2, we discuss possible extensions for our future work.

5.1   Effectiveness Correlation

First, we define the measure for effectiveness. In our cases, we define it as the proper estimate of AEI over a specific segment of series data. In other words, given several segments of data, SR and MLR each gives a series of AEI. If the two series of AEI are relevant, we consider the two methods relevant. For the same reason as in Section 4, we also use Example 1 as a specific scenario.

Experiment 3: For every dataset (*dailysap* and *dailyibm*), we evenly divided it sequentially into 15 segments, each of which contains 200 sample points. Then we apply *2linear* and *quadratic* to these segments of data, getting two series of AEIs. Finally, we analyze the relevance between the two methods through the correlation between the two series of AEIs.

On *dailysap*, the calculated correlation coefficient between the two series of AEIs is 0.823. Furthermore, F-test proved that such correlation is highly significant. Finally, we declare that the experiment on *dailyibm* gave out the identical conclusion. Due to the limited space, we here eliminate the detailed procedures.

5.2   Possible Extensions

Because our work is somewhat interesting, we have the following extensions for our future works.
(1)   In this paper, we explored regression efficiency in the background of polynomial regression. Therefore, we are interested in the cases for other function family.
(2)   Based on the correlation analysis between *2linear* and *quadratic* in this paper, we would also explore the correlation between SR and MLR in general cases.
(3)   As denoted by (3) and (4), SR possesses freedom to choose division schemes. So "which is the best one?" is a key question for better performance in practice.
(4)   In this paper, we have demonstrated the feasibility to register SR into OLAP technology; however, "can there be other attributes or synopses that can be integrated into OLAP environment?" also deserves attention to fully OLAPing data stream.
(5)   For the effectiveness concern, in this paper we compared SR with MLR and got the conclusion that they are significantly correlated. However, what is the effect of SR in comparison with those classic methods (e.g. ARIMA model) deserves more attention in the future.

6   Conclusion

In this paper, we proposed Step-by-step Regression (SR) as a more efficient alternative for polynomial Multiple Linear Regression (MLR) by adopting a division strategy. In theory and experiments, SR outperforms MLR on both processing time and memory usage, which alleviates the heavy computation burden on the front ends of data stream management systems. Moreover, we successfully integrated SR into OLAP environment to perform multi-dimensional analysis over data stream. As revealed through experiments, we significantly reduced the size of data cube in comparison with previous works that registers LCR. This reduction will further boost the efficiency for OLAP operations over data streams. Finally, we proved the correlation between MLR and SR statistically as the solid base for our contributions.

Acknowledgement. Our work was funded in part by NSFC grant 60221120144 as well as Development Plan of State Key Fundamental Research (Projects 973) grant G199903270.

References

1. M. Garofalakis, J. Gehrke, and R. Rastogi. Querying and Mining Data Streams: You Only Get One Look. SIGMOD'02.
2. D. Rafiei and A. Mendelzon. Efficient retrieval of similar time sequences using DFT. In Proc. of the 5th Intl. Conf. on Found. of Data Org. and Alg. FODO '98
3. T. Kahveci, A. K. Singh, and A. Gurel. An Efficient Index Structure for Shift and Scale Invariant Search of Multi-Attribute Time Sequences. ICDE'02
4. K. P. Chan and A.W. Fu. Efficient time series matching by wavelets. ICDE'99.
5. I. Popivanov and R. J. Miller. Similarity search over time series data using wavelets ICDE'02
6. Y. Zhu, D. Shasha. StatStream: Statistical Monitoring of Thousands of Data Streams in Real Time. VLDB'02.
7. Y. Chen, G. Dong, J. Han, B. W. Wah, and J. Wang. Multi-Dimensional Regression Analysis of Time-Series Data Streams. VLDB'02
8. Y. Chen, G. Dong, J. Han, J. Pei, B. W. Wah, and J. Wang. Stream Cube: An Architecture for Multi-Dimensional Analysis of Data Streams. Submitted for publication.
9. Y. Chen, G. Dong, J. Han, J. Pei, B. W. Wah and J. Wang. Online Analytical Processing Stream Data: Is It Feasible?" DMKD'02.
10. DataDay, Kathleen & R. A. Devlin. The payoff to work without pay: volunteer work as an investment in human capital. Canadian Journal of Economics, vol. 31(5) Nov. 1998 pp 1179–91.
11. R. D. Cook and Sanford Weisberg. Applied Regression Including Computing and Graphics. 1999, John Wiley and Sons.
12. Thomas H. Cormen, Charles E. Leiserson, and Ronald L.Rivest. Introduction to Algorithms. MIT Press, ambridge, Massachusetts, 1997.
13. J. Gray, S. Chaudhuri, A. Bosworth, A. Layman, D. Reichart, M. Venkatrao, F. Pellow, and H. Pirahesh. Data cube: A relational aggregation operator generalizing group-by, cross-tab and sub-totals. Data Mining and Knowledge Discovery, 1:29–54, 1997.
14. S. Chaudhuri and U. Dayal. An overview of data warehousing and OLAP technology. ACM SIGMOD Record, 26:65–74, 1997.
15. http://www-personal.buseco.monash.edu.au/~hyndman/TSDL/

Progressive Weighted Miner: An Efficient Method for Time-Constraint Mining

Chang-Hung Lee[1], Jian Chih Ou[2], and Ming-Syan Chen[2]

[1] BenQ Corporation 18, Jihu Road, Neihu,Taipei 114, Taiwan, R.O.C.,
MichaelCHLee@BenQ.com
[2] Department of Electrical Engineering, National Taiwan University, Taipei, Taiwan,
ROC
mschen@cc.ee.ntu.edu.tw, alex@arbor.ee.ntu.edu.tw

Abstract. The discovery of association relationship among the data in
a huge database has been known to be useful in selective marketing,
decision analysis, and business management. A significant amount of
research effort has been elaborated upon the development of efficient
algorithms for data mining. However, without fully considering the time-
variant characteristics of items and transactions, it is noted that some
discovered rules may be expired from users' interest. In other words, some
discovered knowledge may be obsolete and of little use, especially when
we perform the mining schemes on a transaction database of short life
cycle products. This aspect is, however, rarely addressed in prior studies.
To remedy this, we broaden in this paper the horizon of frequent
pattern mining by introducing a weighted model of *transaction-weighted
association rules* in a time-variant database. Specifically, we propose
an efficient *Progressive Weighted Miner* (abbreviatedly as *PWM*)
algorithm to perform the mining for this problem as well as conduct the
corresponding performance studies. In algorithm *PWM*, the importance
of each transaction period is first reflected by a proper weight assigned
by the user. Then, *PWM* partitions the time-variant database in light
of weighted periods of transactions and performs weighted mining.
Algorithm *PWM* is designed to progressively accumulate the itemset
counts based on the intrinsic partitioning characteristics and employ a
filtering threshold in each partition to early prune out those cumulatively infrequent 2-itemsets. With this design, algorithm *PWM* is able
to efficiently produce weighted association rules for applications where
different time periods are assigned with different weights and lead to
results of more interest.

Keywords. Data mining, time-constraint, time-variant, weighted association rules

1 Introduction

The discovery of association relationship among the data in a huge database has
been known to be useful in selective marketing, decision analysis, and business

K.-Y. Whang, J. Jeon, K. Shim, J. Srivatava (Eds.): PAKDD 2003, LNAI 2637, pp. 449–460, 2003.

management [6,11]. A popular area of applications is the market basket analysis, which studies the buying behaviors of customers by searching for sets of items that are frequently purchased *either* together *or* in sequence. For a given pair of confidence and support thresholds, the problem of mining association rules is to identify all association rules that have confidence and support greater than the corresponding minimum support threshold (denoted as min_supp) and minimum confidence threshold (denoted as min_conf). Association rule mining algorithms [1] work in two steps: (1) generate all frequent itemsets that satisfy min_supp; (2) generate all association rules that satisfy min_conf using the frequent itemsets.

Note that the data mining process may produce thousands of rules, many of which are uninteresting to users. Hence, several applications have called for the use of constrained rule mining [4,12,13,23]. Specifically, in constraint-based mining, which is performed under the guidance of various of constraints provided by the user. The constraints addressed in the prior works include the following: (1) Knowledge type-constraints [20]; (2) Data constraints [4]; (3) Interestingness constraints [13]; and (4) Rule constraints [19,23]. Such constraints may be expressed as meta-rules (rule templates), as the maximum or minimum number of predicates that can occur in the rule antecedent or consequent, or as relationships among attributes, attribute values, and/or aggregates. Recently, many constraint-based mining works have focused on the use of rule constraints. This form of constraint-based mining allows users of specifying the rules to be mined according to their need, thereby leading to much more useful mining results. According to our observation, these kinds of rule-constraint problems are based on the concept of embedding a variety of *item-constraints* in the mining process.

On the other hand, a time-variant database, as shown in Figure 1, consists of values or events varying with time. Time-variant databases are popular in many applications, such as daily fluctuations of a stock market, traces of a dynamic production process, scientific experiments, medical treatments, weather records, to name a few. In our opinion, the existing model of the constraint-based association rule mining is not able to efficiently handle the time-variant database due to two fundamental problems, i.e., (1) lack of consideration of the *exhibition period* of each individual transaction; (2) lack of an intelligent support counting basis for each item. Note that the traditional mining process treats transactions in different time periods indifferently and handles them along the same procedure. However, since different transactions have different exhibition periods in a time-variant database, only considering the occurrence count of each item might not lead to interesting mining results. This problem can be further explained by the example below.

Example 1.1: In a transaction database as shown in Figure 2, the minimum transaction support and confidence are assumed to be $min\_supp = 30\%$ and $min\_conf = 75\%$, respectively. A set of time-variant database indicates the transaction records from January 2002 to March 2002. The starting date of each transaction item is also given. Based on the traditional mining techniques, the *support threshold* is denoted as $min\_S^T = \lceil 12 \times 0.3 \rceil = 4$ where 12 is the size of transaction set $\mathcal{D}$. It can be seen that only $\{B,\ C,\ D,\ E,\ BC\}$ can be termed

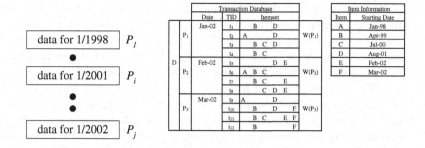

Fig. 1. A time-variant transaction database

Fig. 2. An illustrative transaction database

as frequent itemsets since their occurrences in this transaction database are all larger than the value of support threshold min_S^T. Thus, rule $C \Rightarrow B$ is termed as a frequent association rule with support $supp(C \cup B) = 41.67\%$ and confidence $conf(C \Rightarrow B) = 83.33\%$. However, it can be observed from Figure 2 that an early product intrinsically possesses a higher likelihood to be determined as a frequent itemset. In addition, different transactions are usually of different importance to the user. This aspect is not well explored by prior works.

Since the early work in [1], several efficient algorithms to mine association rules have been developed. These studies cover a broad spectrum of topics including: (1) fast algorithms based on the level-wise Apriori framework [2,18], partitioning [16], and FP-growth methods [10]; (2) incremental updating [9,15]; (3) mining of generalized multi-dimensional [24] and multi-level rules [21]; (4) constraint-based rule mining [13,23]; and (5) temporal association rule discovery [3,5,8,14]; While these are important results toward enabling the integration of association mining and fast searching algorithms, e.g., BFS and DFS which are classified in [11], we note that these mining methods cannot effectively be applied to the problem of *time-constraint mining* on a *time-variant database* which is of increasing importance. Note that one straightforward approach to addressing the above issues is to employ the item-constraints [13,23] and/or multiple supports strategies [17,22], i.e., new coming items have higher weights for their item occurrences. However, as noted in [17,22] these approaches will encounter another problem, i.e., there is no proper confidence threshold in such cases for the corresponding rule generation.

Consequently, we broaden in this paper the horizon of frequent pattern mining by introducing a weighted model of *transaction-weighted association rules* (abbreviatedly as *weighted association rules*) in a time-variant database. Specifically, we propose an efficient *Progressive Weighted Miner* (abbreviatedly as *PWM*) algorithm to perform the mining for this problem. In algorithm *PWM*, the importance of each transaction period is first reflected by a proper weight assigned by the user. Then, *PWM* partitions the time-variant database in light of weighted periods of transactions and performs weighted mining. Explicitly, algorithm PWM explores the mining of *weighted associa-tion rules*, denoted by $(X \Rightarrow Y)^W$, which is produced by two newly defined concepts of *weighted − support* and *weighted − confidence* in light of the corresponding weights in individual transactions. Basically, an association rule $X \Rightarrow Y$ is termed to be a frequent weighted association rule $(X \Rightarrow Y)^W$ if and only if its weighted support is larger than minimum support required, i.e., $supp^W(X \cup Y) > min\_supp$, and the weighted confidence $conf^W(X \Rightarrow Y)$ is larger than minimum confidence needed, i.e., $conf^W(X \Rightarrow Y) > min\_conf$. In-stead of using the *traditional support threshold* $min\_S^T = \lceil |\mathcal{D}| \times min\_supp \rceil$ as a minimum support threshold for each item, a weighted minimum support, de-noted by $min\_S^W = \{\Sigma |P_i| \times W(P_i)\} \times min\_supp$, is employed for the mining of weighted association rules, where $|P_i|$ and $W(P_i)$ represent the amount of par-tial transactions and their corresponding weight values by a *weighting function* $W(\cdot)$ in the weighted period P_i of the database $\mathcal{D}$. Let $N_{P_i}(X)$ be the number of transactions in partition P_i that contain itemset X. The support value of an itemset X can then be formulated as $S^W(X) = \Sigma N_{P_i}(X) \times W(P_i)$. As a result, the weighted support ratio of an itemset X is $supp^W(X) = \frac{S^W(X)}{\Sigma |P_i| \times W(P_i)}$.

Example 1.2: Let us follow Example 1.1 with the given $min\_supp = 30\%$ and $min\_conf = 75\%$. Consider $W(P_1) = 0.5$, $W(P_2) = 1$, and $W(P_3) = 2$, we have this newly defined support threshold as $min\_S^W = \{4 \times 0.5 + 4 \times 1 + 4 \times 2\} \times 0.3 = 4.2$, we have weighted association rules, i.e., $(C \Rightarrow B)^W$ with relative weighted support $supp^W(C \cup B) = 35.7\%$ and confidence $conf^W(C \Rightarrow B) = \frac{supp^W(C \cup B)}{supp^W(C)} = 83.3\%$ and $(F \Rightarrow B)^W$ with relative weighted support $supp^W$ $(F \cup B) = 42.8\%$ and confidence $conf^W(F \Rightarrow B) = 100\%$. More details can be found in a complete example in Section 3.2.

Explicitly, PWM first partitions the transaction database in light of weighted periods of transactions and then progressively accumulates the occurrence count of each candidate 2-itemset based on the intrinsic partitioning characteristics. With this design, algorithm PWM is able to efficiently produce weighted as-sociation rules for applications where different time periods are assigned with different weights. Algorithm PWM is also designed to employ a filtering thresh-old in each partition to early prune out those cumulatively infrequent 2-itemsets. The feature that the number of candidate 2-itemsets generated by PWM is very close to the actual number of frequent 2-itemsets allows us of employing the scan reduction technique by generating C_ks from C_2 directly to effectively reduce the number of database scans. In fact, the number of the candidate itemsets C_ks generated by PWM approaches to its theoretical minimum, i.e., the number of actual frequent k-itemsets, as the value of the minimal support increases. Specif-ically, the execution time of PWM is, in orders of magnitude, smaller than those required by $Apriori^W$.

Note that those constraint-based rule mining methods that allow users of deriving rules of interest by providing meta-rules and item-constraints [13,23] are not applicable to solving the weighted mining problem addressed in this paper since the constraints we consider are on individual *transactions* rather than on *items*. Indeed, the problem of mining weighted association rules will be degenerated to the traditional one of mining association rules if the weighting function is assigned to be $W(\cdot) = 1$, meaning that the model we consider can be viewed as a general framework of prior studies. In this paper, we not only explore the new model of weighted association rules in a time-variant database, but also propose an efficient *Progressive Weighted Miner* methodology to perform the mining for this problem. These features distinguish this paper from others.

The rest of this paper is organized as follows. Problem description is given in Section 2. Algorithm *PWM* is described in Section 3. This paper concludes with Section 4.

2   Problem Description

Let n be the number of partitions with a time granularity, e.g., *business-week, month, quarter, year*, etc., in database $\mathcal{D}$. In the model considered, P_i denotes the part of the transaction database where $P_i \subseteq \mathcal{D}$. Explicitly, we explore in this paper the mining of *transaction-weighted association rules* (abbreviatedly as *weighted association rules*), i.e., $(X \Rightarrow Y)^W$, where $X \Rightarrow Y$ is produced by the concepts of *weighted − support* and *weighted − confidence*. Further, instead of using the *traditional support threshold* $min\_S^T = \lceil |\mathcal{D}| \times min\_supp \rceil$ as a minimum support threshold for each item in Figure 2, a weighted minimum support for mining an association rules is determined by $min\_S^W = \{\Sigma |P_i| \times W(P_i)\} \times min\_supp$ where $|P_i|$ and $W(P_i)$ represent the amount of partial transactions and their corresponding weight values by a *weighting function* $W(\cdot)$ in the weighted period P_i of the database $\mathcal{D}$. Formally, we have the following definitions.

Definition 1: Let $N_{P_i}(X)$ be the number of transactions in partition P_i that contain itemset X. Consequently, the weighted support value of an itemset X can be formulated as $S^W(X) = \Sigma N_{P_i}(X) \times W(P_i)$. As a result, the weighted support ratio of an itemset X is $supp^W(X) = \frac{S^W(X)}{\Sigma |P_i| \times W(P_i)}$.

In accordance with Definition 1, an itemset X is termed to be frequent when the weighted occurrence frequency of X is larger than the value of min_supp required, i.e., $supp^W(X) > min\_supp$, in transaction set $\mathcal{D}$. The weighted confidence of a weighted association rule $(X \Rightarrow Y)^W$ is then defined below.

Definition 2: $conf^W(X \Rightarrow Y) = \frac{supp^W(X \bigcup Y)}{supp^W(X)}$.

Definition 3: An association rule $X \Rightarrow Y$ is termed a frequent weighted association rule $(X \Rightarrow Y)^W$ if and only if its weighted support is larger than minimum support required, i.e., $supp^W(X \cup Y) > min\_supp$, and the weighted confidence $conf^W(X \Rightarrow Y)$ is larger than minimum confidence needed, i.e., $conf^W(X \Rightarrow Y) > min\_conf$.

Example 2.1: Recall the illustrative weighted association rules, e.g., $(F \Rightarrow B)^W$ with relative weighted support $supp^W(F \cup B) = 42.8\%$ and confidence

$conf^W(F \Rightarrow B) = 100\%$, in Example 1.2. In accordance with Definition 3, the implication $(F \Rightarrow B)^W$ is termed as a frequent weighted association rule if and only if $supp^W(F \cup B) > min\_supp$ and $conf^W(F \Rightarrow B) > min\_conf$. Consequently, we have to determine if $supp^W(F) > min\_supp$ and $supp^W(FB) > min\_supp$ for discovering the newly identified association rule $(F \Rightarrow B)^W$.

Note that with this weighted association rule $(F \Rightarrow B)^W$, we are able to discover that a new coming product, e.g., a high resolution digital camera, may promote the selling of an existing product, e.g., a color printer. This important information, as pointed out earlier, will not be discovered by the traditional mining schemes, showing the usefulness of this novel model of weighted association rule mining. Once, $\mathcal{F}^W = \{ X \subseteq \mathcal{I} \mid X \text{ is } frequent\}$, the set of all frequent itemsets together with their support values is known, deriving the the desired weighted association rules is straightforward. For every $X \in \mathcal{F}^W$, one can simply check the confidence of all rules $(X \Rightarrow Y)^W$ and drop those whose $\frac{supp^W(X \cup Y)}{supp^W(X)} < min\_conf$ for rule generation. Therefore, in the rest of this paper we concentrate our discussion on the algorithms for mining frequent itemsets.

It is worth mentioning that there is no restriction imposed on the weighting functions assigned by users. In fact, in addition to the time periods of transactions, other attributes of transactions, such as ownership, transaction length, etc., can also be incorporated into the determination of weights for individual transactions.

3   Progressive Weighted Mining

It is noted that most of the previous studies, including those in [1,9,18], belong to Apriori-like approaches. Basically, an Apriori-like approach is based on an anti-monotone Apriori heuristic [1], i.e., if any itemset of length k is not frequent in the database, its length $(k + 1)$ super-itemset will never be frequent. The essential idea is to iteratively generate the set of candidate itemsets of length $(k + 1)$ from the set of frequent itemsets of length k (for $k \geq 1$), and to check their corresponding occurrence frequencies in the database. As a result, if the largest *frequent* itemset is a j-itemset, then an Apriori-like algorithm may need to scan the database up to $(j+1)$ times. This is the basic concept of an extended version of Apriori-based algorithm, referred to as $Apriori^W$.

In [7], the technique of scan-reduction was proposed and shown to result in prominent performance improvement. By scan reduction, C_k is generated from $C_{k-1} \star C_{k-1}$ instead of from $L_{k-1} \star L_{k-1}$. Clearly, a C_3' generated from $C_2 \star C_2$, instead of from $L_2 \star L_2$, will have a size greater than $|C_3|$ where C_3 is generated from $L_2 \star L_2$. However, if $|C_3'|$ is not much larger than $|C_3|$, and both C_2 and C_3 can be stored in main memory, we can find L_2 and L_3 together when the next scan of the database is performed, thereby saving one round of database scan. It can be seen that using this concept, one can determine all L_ks by as few as two scans of the database (i.e., one initial scan to determine L_1 and a final scan to determine all other frequent itemsets), assuming that C_k' for $k \geq 3$ is generated from C_{k-1}' and all C_k' for $k > 2$ can be kept in the memory.

3.1   Algorithm of PWM

In general, databases are too large to be held in main memory. Thus, the data mining techniques applied to very large databases have to be highly scalable for efficient execution. As mentioned above, by partitioning a transaction database into several partitions, algorithm *PWM* is devised to employ a progressive filtering scheme in each partition to deal with the candidate itemset generation and process one partition at a time. For ease of exposition, the processing of a partition is termed a *phase* of processing. Under *PWM*, the cumulative information in the prior phases is selectively carried over toward the generation of candidate itemsets in the subsequent phases. After the processing of a phase, algorithm *PWM* outputs a progressive candidate set of itemsets, their occurrence counts and the corresponding partial supports required.

The procedure of algorithm *PWM* is outlined below, where algorithm *PWM* is decomposed into four sub-procedures for ease of description. C_2 is the set of progressive candidate 2-itemsets generated by database $\mathcal{D}$. Recall that $N_{P_i}(X)$ is the number of transactions in partition P_i that contain itemset X and $W(P_i)$ is the corresponding weight of partition P_i.

Algorithm *PWM (n, min_supp)*

Procedure I:InitialPartition
1. $|D| = \sum_{i=1,n} |P_i|$;

Procedure II: Candidate 2-Itemset Generation
1. **begin for** $i = 1$ **to** n // 1st scan of D
2.   **begin for** each 2-itemset $X_2 \in P_i$
3.   **if** ($X_2 \notin C_2$)
4. $X_2.count = N_{P_i}(X_2) \times W(P_i)$;
5. $X_2.start = i$;
6.   **if** ($X_2.count \geq min\_supp \times |P_i| \times W(P_i)$)
7. $C_2 = C_2 \cup X_2$;
8.   **if** ($X_2 \in C_2$)
9. $X_2.count = X_2.count + N_{P_i}(X_2) \times W(P_i)$;
10.   **if** ($X_2.count_i min\_supp \times \sum_{m=X_2.start,i} (|P_m| \times W(P_m))$)
11. $C_2 = C_2 - X_2$;
12.   **end**
13. **end**

Procedure III: Candidate k-Itemset Generation
1. **begin while** ($C_k \neq \emptyset$ & $k \geq 2$)
2. $C_{k+1} = C_k \star C_k$;
3. $k = k + 1$;
4. **end**

Procedure IV: Frequent Itemset Generation
1. **begin for** $i = 1$ **to** n
2.   **begin for** each itemset $X_k \in C_k$
3. $X_k.count = X_k.count + N_{P_i}(X_k) \times W(P_i)$;
4. **end**
5. **begin for** each itemset $X_k \in C_k$

6.   **if** $(X_k.count \geq min\_supp \times \sum_{m=1,n} (|P_m| \times W(P_m)))$

7. $L_k = L_k \cup X_k$;

8. **end**

9. **return** L_k;

Procedure II (Candidate 2-Itemset Generation) first scans partition P_i, for $i = 1$ to n, to find the set of all local frequent 2-itemsets in P_i. Note that C_2 is a superset of the set of all frequent 2-itemsets in $\mathcal{D}$. Algorithm PWM constructs C_2 incrementally by adding candidate 2-itemsets to C_2 as well as counting the number of occurrences for each candidate 2-itemset X_2 in C_2. If the cumulative occurrences of a candidate 2-itemset X_2 does not meet the partial minimum support, X_2 is removed from the progressive C_2. In *Procedure II (Candidate 2-Itemset Generation)*, algorithm PWM processes one partition at a time for all partitions. The number of occurrences of an itemset X_2 and its starting partition which keeps its first occurrence in C_2 are recorded in $X_2.count$ and $X_2.start$, respectively. As such, in the end of processing P_i, an itemset X_2 will be kept in C_2 only if $X_2.count > min\_supp \times \sum_{m=X_2.start,i}(|P_m| \times W(P_m))$. Next, in *Procedure III (Candidate k-Itemset Generation)*, with the scan reduction scheme [18], C_2 produced by the first scan of database is employed to generate C_ks in main memory.

Then, from *Procedure IV (Frequent Itemset Generation)* we begin the second database scan to calculate the support of each itemset in C_k and to find out which candidate itemsets are really frequent itemsets in database $\mathcal{D}$. As a result, those itemsets whose $X_k.count \geq min\_supp \times \sum_{m=1,n}(|P_m| \times W(P_m))$ are the frequent itemsets L_ks. Finally, according to these output L_ks in Step 9, all kinds of weighted association rules implied in database $\mathcal{D}$ can be generated in a straightforward manner.

Note that PWM is able to filter out false candidate itemsets in P_i with a hash table. Same as in [18], using a hash table to prune candidate 2-itemsets, i.e., C_2, in each accumulative ongoing partition set P_i of transaction database, the CPU and memory overhead of PWM can be further reduced. Owing to the small number of candidate sets generated, the scan reduction technique can be applied efficiently. As a result, only two scans of the database are required.

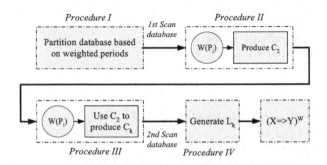

Fig. 3. The flowchart of PWM

3.2 An Illustrative Example of PWM

Recall the transaction database shown in Figure 2 where the transaction database $\mathcal{D}$ is assumed to be segmented into three partitions P_1, P_2 and P_3, which correspond to the three time granularities from January 2001 to March 2001. Suppose that $min\_supp = 30\%$ and $min\_conf = 75\%$. In addition, the weight value of each partition is given as follows: $W(P_1) = 0.5$, $W(P_2) = 1$, and $W(P_3) = 2$. The operation of algorithm PWM can be best understood by an illustrative example described below. The flowchart of PWM is given in Figure 3.

Specifically, each partition is scanned sequentially for the generation of candidate 2-itemsets in the first scan of the database $\mathcal{D}$. After scanning the first segment of 4 transactions, i.e., partition P_1, 2-itemsets $\{BD, BC, CD, AD\}$ are sequentially generated as shown in Figure 4. In addition, each potential candidate itemset $c \in C_2$ has two attributes: (1) $c.start$ which contains the partition number of the corresponding starting partition when c was added to C_2, and (2) $c.count$ which contains the number of *weighted occurrences* of c. Since there are four transactions in P_1, the partial weighted minimal support is $min\_S^W(P_1) = 4 \times 0.3 \times 0.5 = 0.6$. Such a partial weighted minimal support is called the *filtering threshold*. Itemsets whose occurrence counts are below the filtering threshold are removed. Then, as shown in Figure 4, only $\{BD, BC\}$, marked by "$\bigcirc$", remain as candidate itemsets whose information is then carried over to the next phase P_2 of processing.

Similarly, after scanning partition P_2, the occurrence counts of potential candidate 2-itemsets are recorded. From Figure 4, it is noted that since there are also 4 transactions in P_2, the filtering threshold of those itemsets carried out from the previous phase is $min\_S^W(P_1 + P_2) = 4 \times 0.3 \times 0.5 + 4 \times 0.3 \times 1 = 1.8$ and that of newly identified candidate itemsets is $min\_S^W(P_2) = 4 \times 0.3 \times 1 = 1.2$. It can be seen that we have 3 candidate itemsets in C_2 after the processing of partition P_2.

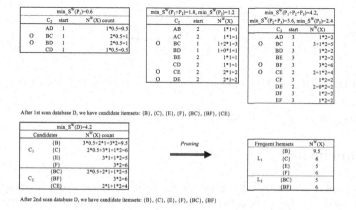

Fig. 4. Frequent itemsets generation for mining weighted association rules by PWM

Finally, partition P_3 is processed by algorithm PWM. The resulting candidate 2-itemsets are $C_2 = \{BC, CE, BF\}$ as shown in Figure 4. Note that though appearing in the previous phase P_2, itemset $\{DE\}$ is removed from C_2 once P_3 is taken into account since its occurrence count does not meet the filtering threshold then, i.e., $2 < 3.6$. Consequently, we have 3 candidate 2-itemsets generated by PWM. Note that only 3 candidate 2-itemsets are generated by PWM.

After generating C_2 from the first scan of database $\mathcal{D}$, we employ the scan reduction technique [18] and use C_2 to generate C_k. As discussed earlier, since the $|C_2|$ generated by PWM is very close to the theoretical minimum, i.e., $|L_2|$, the $|C_3'|$ is not much larger than $|C_3|$. Similarly, the $|C_k'|$ is close to $|C_k|$. Since $C_2 = \{BC, CE, BF\}$, no candidate k-itemset is generated in this example for $k \geq 3$. Thus, $C_1' = \{B, C, E, F\}$ and $C_2' = \{BC, CE, BF\}$, where all C_k's can be stored in main memory. Then, we can find L_ks ($k = 1, 2, ..., m$) together when the second scan of the database $\mathcal{D}$ is performed. Finally we get $L_2 = \{BC, BF\}$ in Figure 4.

It is important to note that if we adopt single $min\_supp = 30\%$ by Apriori, then the itemset $\{BF\}$ will not be large since its occurrence in this transaction database is 3 which is smaller than $min\_S^T = \lceil 12 \times 0.3 \rceil = 4$. However, itemset $\{BF\}$ appears very frequently in the most recent partition of the database of which the weight is relatively large, thus discovering more desirable information. It can be seen that the algorithm Apriori is not able to discover the information behind the new coming data in the transaction database.

Note that since there is no candidate k-itemset ($k \geq 2$) containing A or D in this example, A and D are not necessary taken as potential itemsets for generating weighted association rules. In other words, we can skip them from the set of candidate itemsets C_k's. Finally, all occurrence counts of C_k's can be calculated by the second database scan. As shown in Figure 5, after all frequent k-itemsets are identified, the corresponding weighted association rules can be derived in a straightforward manner. Explicitly, the weighted association rule of $(X \Rightarrow Y)^W$ holds if $conf^W(X \Rightarrow Y) \geq min\_conf$.

Rules	Support	Confidence
B ⇒ C	5/(4*0.5+4*1+4*2)=35.7%	5/9.5=52.6%
B ⇒ F	6/(4*0.5+4*1+4*2)=42.8%	6/9.5=63.1%
C ⇒ B	5/(4*0.5+4*1+4*2)=35.7%	5/6=83.3%
F ⇒ B	6/(4*0.5+4*1+4*2)=42.8%	6/6=100%

Pruning ⇒

Rules	Support	Confidence
C ⇒ B	35.7%	83.3%
F ⇒ B	42.8%	100.0%

Fig. 5. The weighted association rule generation from frequent itemsets

4   Conclusion

In this paper, we explored a new model of mining weighted association rules, i.e., $(X \Rightarrow Y)^W$, in a transaction database and developed algorithm PWM to generate the weighted association rules as well as conducted related performance studies. In algorithm PWM, the importance of each transaction period was first reflected by a proper weight assigned by the user. Then, *PWM* partitioned the time-variant database in light of weighted periods of transactions and performed weighted mining. Algorithm *PWM* was designed to progressively accumulate the itemset counts based on the intrinsic partitioning characteristics and employed a filtering threshold in each partition to early prune out those cumulatively infrequent 2-itemsets. With this design, algorithm *PWM* was able to efficiently produce weighted association rules for applications where different time periods were assigned with different weights, leading to more interesting results.

Acknowledgements. The authors are supported in part by the Ministry of Education Project No. 89E-FA06-2-4-7 and the National Science Council, Project No. NSC 91-2213-E-002-034 and NSC 91-2213-E-002-045, Taiwan, Republic of China.

References

1. R. Agrawal, T. Imielinski, and A. Swami. Mining Association Rules between Sets of Items in Large Databases. *Proc. of ACM SIGMOD*, pages 207–216, May 1993.
2. R. Agrawal and R. Srikant. Fast Algorithms for Mining Association Rules in Large Databases. *Proc. of the 20th International Conference on Very Large Data Bases*, pages 478–499, September 1994.
3. J. M. Ale and G. Rossi. An Approach to Discovering Temporal Association Rules. *ACM Symposium on Applied Computing*, 2000.
4. A. M. Ayad, N. M. El-Makky, and Y. Taha. Incremental mining of constrained association rules. *Proc. of the First SIAM Conference on Data Mining*, 2001.
5. C.-Y. Chang, M.-S. Chen, and C.-H. Lee. Mining General Temporal Association Rules for Items with Different Exhibition Periods. *Proc. of the IEEE 2nd Intern'l Conf. on Data Mining (ICDM-2002)*, December.
6. M.-S. Chen, J. Han, and P. S. Yu. Data Mining: An Overview from Database Perspective. *IEEE Transactions on Knowledge and Data Engineering*, 8(6):866–883, December 1996.
7. M.-S. Chen, J.-S. Park, and P. S. Yu. Efficient Data Mining for Path Traversal Patterns. *IEEE Transactions on Knowledge and Data Engineering*, 10(2):209–221, April 1998.
8. X. Chen and I. Petr. Discovering Temporal Association Rules: Algorithms, Language and System. *Proc. of 2000 Int. Conf. on Data Engineering*, 2000.
9. D. Cheung, J. Han, V. Ng, and C. Y. Wong. Maintenance of Discovered Association Rules in Large Databases: An Incremental Updating Technique. *Proc. of 1996 Int'l Conf. on Data Engineering*, pages 106–114, February 1996.
10. J. Han and J. Pei. Mining Frequent Patterns by Pattern-Growth: Methodology and Implications. *ACM SIGKDD Explorations (Special Issue on Scaleble Data Mining Algorithms)*, December 2000.
11. J. Hipp, U. Güntzer, and G. Nakhaeizadeh. Algorithms for association rule mining – a general survey and comparison. *SIGKDD Explorations*, 2(1):58–64, July 2000.
12. D. Kifer, C. Bucila, J. Gehrke, and W. White. Dualminer: A dual-pruning algorithm for itemsets with constraints. *Proc. of the 8th ACM SIGKDD International Conference on Knowledge Discovery and Data Mining*, 2002.

13. L. V. S. Lakshmanan, R. Ng, J. Han, and A. Pang. Optimization of Constrained Frequent Set Queries with 2-Variable Constraints. *Proc. of 1999 ACM-SIGMOD Conf. on Management of Data*, pages 157–168, June 1999.

14. C.-H. Lee, C.-R. Lin, and M.-S. Chen. On Mining General Temporal Association Rules in a Publication Database. *Proc. of 2001 IEEE International Conference on Data Mining*, November 2001.

15. C.-H. Lee, C.-R. Lin, and M.-S. Chen. Sliding-Window Filtering: An Efficient Algorithm for Incremental Mining. *Proc. of the Tenth ACM Intern'l Conf. on Information and Knowledge Management*, November 2001.

16. J.-L. Lin and M. H. Dunham. Mining Association Rules: Anti-Skew Algorithms. *Proc. of 1998 Int'l Conf. on Data Engineering*, pages 486–493, 1998.

17. B. Liu, W. Hsu, and Y. Ma. Mining Association Rules with Multiple Minimum Supports. *Proc. of 1999 Int. Conf. on Knowledge Discovery and Data Mining*, August 1999.

18. J.-S. Park, M.-S. Chen, and P. S. Yu. Using a Hash-Based Method with Transaction Trimming for Mining Association Rules. *IEEE Transactions on Knowledge and Data Engineering*, 9(5):813–825, October 1997.

19. J. Pei and J. Han. Can We Push More Constraints into Frequent Pattern Mining? *Proc. of 2000 Int. Conf. on Knowledge Discovery and Data Mining*, August 2000.

20. J. Pei, J. Han, and L.V.S. Lakshmanan. Mining Frequent Itemsets with Convertible Constraints. *Proc. of 2001 Int. Conf. on Data Engineering*, 2001.

21. R. Srikant and R. Agrawal. Mining Generalized Association Rules. *Proc. of the 21th International Conference on Very Large Data Bases*, pages 407–419, September 1995.

22. K. Wang, Y. He, and J. Han. Mining Frequent Itemsets Using Support Constraints. *Proc. of 2000 Int. Conf. on Very Large Data Bases*, September 2000.

23. W. Wang, J. Yang, and P. S. Yu. Efficient mining of weighted association rules (WAR). *Proc. of the Seventh ACM SIGKDD International Conference on Knowledge Discovery and Data Mining*, 2000.

24. C. Yang, U. Fayyad, and P. Bradley. Efficient discovery of error-tolerant frequent itemsets in high dimensions. *Proc. of the Seventh ACM SIGKDD International Conference on Knowledge Discovery and Data Mining*, 2001.

Mining Open Source Software (OSS) Data Using Association Rules Network

Sanjay Chawla[1], Bavani Arunasalam[1], and Joseph Davis[1]

Knowledge Management Research Group
School of Information Technologies
University of Sydney, NSW, Australia
{chawla,bavani,jdavis}@it.usyd.edu.au

Abstract. The Open Source Software(OSS) movement has attracted considerable attention in the last few years. In this paper we report our results of *mining* data acquired from SourceForge.net, the largest open source software hosting website. In the process we introduce Association Rules Network(ARN), a (hyper)graphical model to represent a special class of association rules. Using ARNs we discover important relationships between the attributes of successful OSS projects. We verify and validate these relationships using Factor Analysis, a classical statistical technique related to Singular Value Decomposition(SVD).

Keywords: Open Source Software, Association Rule, Networks, Hypergraph clustering, Factor Analysis.

1 Introduction and Motivation

In this paper we detail our exploration through Open Source Software(OSS) data in search for *patterns* with the goal of modeling the nature of OSS development. In the process we will use and modify several standard data mining techniques in conjunction with classical statisical techniques. In particular we will introduce the concept of an Association Rules Network(ARN)[1] which provide insight into the interelationships between relational data attributes in general and OSS data in particular. The relationships discovered using ARNs will be validated using Factor Analysis, a classical statistical method. Our work will be a step towards validating an often cited maxim in the data mining community: *Data mining is for hypothesis generation and statistics for hypothesis verification* [8,7].

The OSS movement has attracted considerable attention in recent years, primarily because it is perceived to offer a non-propietary and socially beneficial model of software development. The source code for this software is freely available and not subject to limitations on possible modifications or distribution. The main idea behind open source is that good software evolves when a dedicated community of programmers and developers can read, redistribute, and modify the source code for a piece of software. Members of

---

[1] This term has been used before on http://www.statsoftinc.com, where it used for a graphical representation for generic association rules. Our terminology is, as we will explain, for representing a subclass of association rules as directed hypergraphs.

K.-Y. Whang, J. Jeon, K. Shim, J. Srivatava (Eds.): PAKDD 2003, LNAI 2637, pp. 461–466, 2003.

this community improve and adapt the software and fix bugs. It is claimed that this rapid, evolutionary process produces better quality software than the traditional hierarchical and closed model [5,10,11].

SourceForge.net is currenlty the world's largest open source development website with the largest repository of Open Source code and applications available on the Internet. SourceForge.net provides free services to OS developers, including project hosting, version control, bug and issue tracking, project management, backups and archives and communication and collaboration resources. As of March 2002 SourceForge.net had 40,002 registered project and 424,862 registered users.

The main contributions of this paper are as follows. We introduce Association Rules Network(ARN) as a paradigm for modeling association rules whose consequents are singletons. The items, which form the nodes of the network are then clustered using min-cut graph partitioning and the clusters are validated using an independent technique, namely Factor Analysis.

The rest of the paper is as follows. In Section 2 we motivate ARNs with the help of an example and provide a formal definition. Section 3 is devoted to related work. In Section 4 we generate an ARN from OSS data and discuss our results. We conclude in Section 5 with a summary and directions for future work.

2   Association Rules Network

In this section we introduce Association Rules Network(ARN). We assume that the reader is familiar with basic association rules terminology [1,4]. An ARN is a (hyper)graphical model to represent certain classes of rules, namely rules whose consequents (right-hand sides) are singletons. We found that that in the case of OSS data, ARNs provide good insights into the inter-relationships between the attributes of the software projects. But ARNs can be viewed as more abstract concept applicable to diverse data sets.

Example: Consider a relation $R(A, B, C, D, E, F)$ where the attributes are binary-valued. Assume the following association rules were derived from the relation R using an association rule algorithm

$$\{B = 1, C = 1\} \rightarrow \{A = 1\}$$
$$\{F = 0\} \quad \rightarrow \{B = 1\}$$
$$\{D = 1\} \quad \rightarrow \{A = 1\}$$
$$\{F = 0, E = 1\} \rightarrow \{C = 1\}$$
$$\{E = 1, G = 0\} \rightarrow \{D = 1\}$$
$$\{A = 1, G = 1\} \rightarrow \{E = 1\}$$

Suppose we fix the consequent $\{A = 1\}$. Then we can recursively build a network of association rules which flow into $\{A = 1\}$. This is shown in Figure 1(a). Notice the edges of this network are hyperedges. A hyperedge is a generalization of an edge which can span more than two nodes of a graph. As has been noted before [6] a natural way to model itemset transactions is to use hypergraphs. This is because a transaction can consist of more than two items. Also notice that the last rule is not part of the ARN.

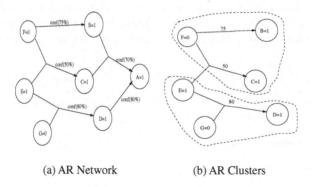

(a) AR Network   (b) AR Clusters

Fig. 1. (a) Associaton Rules Network(ARN)(b) Items clustered using mincut hypergraph partitioning.

This is because once we fix the first consequent $\{A = 1\}$ we want to exclude rules in which this consequent appears as an antecedent.

Definition 1. *A Hypergraph $G = (V, E)$ is a pair where V is the set of nodes and $E \subset 2^V$. Each element of E is called a hyperedge. A directed hyperedge $e = \{v_1, \ldots v_{k-1}; v_k\}$ is a hyperedge with a distinguished node v_k. A directed Hypergraph is a hypergraph with directed hyperedges.*

Definition 2. *Given a set of association rules R which satisfy the minimum support and confidence threshold. An association rule network, $ARN(R, z)$, is a weighted directed hypergraph $G = (V \cup z, E)$, where z is a distinguished sink item(node) such that either G is an $\emptyset$ OR*

1. *Each hyperedge E corresponds to a rule in R whose consequent is a singleton.*
2. *There is a hyperedge which corresponds to a rule r_o whose consequent is the singleton item z.*
3. *The distinguished vertex z is reachable from any other vertex in G*
4. *Any vertex $p \neq z$ is not reachable from z.*
5. *The weight on the edges correspond to the confidence of the rule that they encapsulate.*

Lemma 1. *Given a rule set R and a distinguished item z, Algorithm 1 will build an $ARN(R, z)$.*

Proof: The algorithm builds the $ARN(R, z)$ using a standard breadth-first strategy. The algorithm and the proof of the lemma are detailed in the long version of this paper [2].

2.1 Clustering

An Association Rule Network(ARN) provide information about the hierarchical relationships between the variables(attributes) of the system under investigation. A natural thing to follow-up is to cluster the items that are elements of the ARN. Hypergraph clustering based on min-cut hypergraph partitioning is a well researched topic and has been used for VLSI design. Han et. al. [6] have used it for itemset clustering.

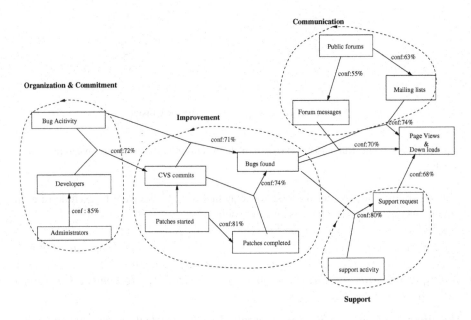

Fig. 2. Association Rule Network(ARN) generated from OSS data. The ARN is a directed hypergraph. Each node corresponds to the item(attribute=high). The distinguished sink node is *Page Views and Downloads*. Each edge corresponds to an association rule. The minimum support level was 6% and the minimum confidence was 50%. The four clusters derived using min-cut hypergraph partitioning on the ARN. The four clusters have a natural meaning as indicated in the figure.

The result of clustering items in Figure 1(a) are shown in Figure 1(b). We looked for two clusters and dropped the sink consequent $\{A = 1\}$ before clustering. In the next section we will show that the attributes whose instances(items) end up in the same cluster are also grouped together using Factor Analysis. We argue that our approach is a testament to the inductive power of data mining techniques.

3   Related Work

Association rules and fast algorithms to mine them were introduced by [1]. Our work is more closely aligned with [9] and [6]. Liu, Hsu and Ma [9] have integrated classification with association rules into *class association rules*(CARs). The key operation in their approach finds all ruleitems of the $< condset, y >$ where *condset* is the set of items and y is a class-label. We extend this by recursively constructing a network of ruletimes where the antecedents in one stage play the role of consequents in the next stage. Han, Karypis and Kumar [6] have used hypergraphs to represent itemsets. They have also used

Table 1. C1 to C4 represent the factors(latent variables) into which the original 12 variables are mapped. For example the first line should be read as variable "number of patches completed" was mapped to factor C1 using factor analysis and cluster 1 using min-cut hypergraph partitioning. The actual number, 0.968 represents the loading of the variable on the factor. The table shows that all but two variables are mapped identically by two very different techniques.

Variable	C1	C2	C3	C4
Number of patches completed	.968(1)			
Number of patches started	.967(1)			
Number of bugs found	.791(1)			
Number of developers		.800(2)		
Number of Adiministrators		.756(2)		
Number of CVS commits		.557(1)		
The percentage of bugs fixed		.449(2)		
Number of forum messages			.711(4)	
Number of support requests			.686(3)	
Pct. of support req. completed			.521(3)	
Number of public forums				.903(4)
Number of mailing lists	.517(4)			.561(4)

min-cut hypergraph partitioning to cluster itemsets. We apply their techniques on ARNs which are a special class of directed hypergraphs. Factor Analysis is a standard technique in statistics and is closely related to Singular Value Decomposition. Our treatment follows [8].

4   Results

We have carried out extensive experimentation on OSS data using techniques described in the previous section. Details are available in the full paper [2]. Here we briefly outline some important results.

We generated an ARN from the OSS data using a support level of 6% and a confidence level of 50%. Given a set of rules R and an item (attribute-value pair) z we find all rules in R in which z appears as a consequent. We recursively extract more rules where the antecedents in the previous step now play the role of consequents. The resulting ARN, shown in Figure 2, clearly shows that ARN naturally discover attributes which work in concert. For example, the three variables: high activity in public forums, forum messages and mailing list encapsulate communication activity related to the project and have a first-order effect on the success of the software project. This validates the importance of effective communication which has often been citied as a predictor for successful software engineering projects [3].

As described in Section 2.2 we partitioned the ARN into four components using mincut hypergraph partitioning. We also applied Factor Analysis, a technique similar to Singular Value Decomposition(SVD) except that it is applied on the correlation matrix as opposed to the covariance matrix on the original numerical valued dataset. The result of clustering are shown in Figure 2. The items naturally fall in clusters which we have labeled Communication, Support, Improvement and Organization/Commitment. The result of factor analysis and an explanation is shown in Table 1. It seems remarkable that two completely distinct techniques result in very similar results.

5   Summary, Conclusion, and Future Work

We have introduced the concept of an Association Rules Network(ARN) to model association rules whose consequents are singletons. We have generated ARNs from OSS data acquired from SourceForge.net and have discovered important relationships which characterize successful OSS projects. These relationships have been validated using Factor Analysis, a classical statistical technique. An interesting area of future work, which we are currently investigating, is the relationship between ARNs and Bayesian Networks.

References

1. Rakesh Agrawal and Ramakrishnan Srikant. Fast algorithms for mining association rules. In Jorge B. Bocca, Matthias Jarke, and Carlo Zaniolo, editors, *Proc. 20th Int. Conf. Very Large Data Bases, VLDB*, pages 487–499. Morgan Kaufmann, 12–15 1994.
2. Sanjay Chawla, Bavani Arunasalam, and Joseph Davis. Mining open source software(oss) data using association rules network. Technical Report TR 535, School of IT, University of Sydney, Sydney, NSW, Australia, 2003.
3. A. Dutoit and B. Bruegge. Communication metrics for software development. *IEEE Transactions On Software Engineering*, 24(8):615–628, 1998.
4. L. Feng, J. Yu, H. Lu, and J. Han. A template model for multi-dimensional, inter-transactional association rules. *VLDB Journal*, 11(2):153–175, 2002.
5. R.L Glass. The sociology of open source: of cults and cultures. *IEEE Software*, 17(3):104–105, 2000.
6. Eui-Hong Han, George Karypis, Vipin Kumar, and Bamshad Mobasher. Clustering based on association rule hypergraphs. In *Proceedings SIGMOD Workshop Research Issues on Data Mining and Knowledge Discovery(DMKD '97)*, 1997.
7. Han, J., Kamber, M., 2001. *Data Mining, Concepts and Trends*. Morgan Kaufmann.
8. Hand, D., Mannila, H.,Smyth, P., 2001. *Principles of Data Mining*. M.I.T Press.
9. Bing Liu, Wynne Hsu, and Yiming Ma. Integrating classification and association rule mining. In *Knowledge Discovery and Data Mining*, pages 80–86, 1998.
10. E.S. Raymond. *The Cathedral and Bazaar:Musings on Open Source and Linux by an Accidental Revolutionary*. O'Reilly, 2001.
11. L. Torwalds. The linux edge. *Communications of the ACM*, 42(4):38–39, 1999.

Parallel FP-Growth on PC Cluster

Iko Pramudiono and Masaru Kitsuregawa

Institute of Industrial Science, The University of Tokyo
4-6-1 Komaba, Meguro-ku, Tokyo 153-8505, Japan
{iko,kitsure}@tkl.iis.u-tokyo.ac.jp

Abstract. FP-growth has become a popular algorithm to mine frequent patterns. Its metadata FP-tree has allowed significant performance improvement over previously reported algorithms. However that special data structure also restrict the ability for further extensions. There is also potential problem when FP-tree can not fit into the memory. In this paper, we report parallel execution of FP-growth. We examine the bottlenecks of the parallelization and also method to balance the execution efficiently on shared-nothing environment.

1 Introduction

Frequent pattern mining has become one popular data mining technique. It also becomes the fundamental technique for other important data mining tasks such as association rule, correlation and sequential pattern.

FP-growth has set new standard for frequent pattern mining [4]. The compression of transaction database into on-memory data structure called FP-tree benefits FP-growth with performance better than previously reported algorithms such as Apriori [2].

Further performance improvement can be expected from parallel execution. Parallel engine is essential for large scale data warehouse. Particularly, development of parallel algorithms on large scale shared nothing environment such as PC cluster has attracted a lot of attention since it is a promising platform for high performance data mining. However parallel algorithm for complex data structure like FP-tree is much harder to implement compared to sequential program or shared-memory parallel system.

Section 2 lists related works on frequent pattern mining and its parallel executions. Section 3 briefly describes the underlying sequential FP-growth algorithm. In section 4 we explain our approaches for parallel execution of FP-growth on shared-nothing environment and we give the evaluation in section 5. Section 6 concludes the paper.

2 Related Works

Apriori is the first algorithm which addresses mining frequent pattern in 1995, particularly to generate association rules [2]. Many variants of Apriori based algorithms are developed since then.

K.-Y. Whang, J. Jeon, K. Shim, J. Srivatava (Eds.): PAKDD 2003, LNAI 2637, pp. 467–473, 2003.

Pioneering works on parallel algorithm for frequent pattern mining were done in [3,5]. A better memory utilization schema called Hash Partitioned Apriori (HPA) was proposed in [6].

Some alternatives to Apriori-like "generate-and-test" paradigm were proposed such as TreeProjection[1]. However it was FP-growth that brought the momentum for the new generation of frequent pattern mining algorithms[4].

3   FP-Growth

The FP-growth algorithm can be divided into two phases : the construction of FP-tree and mining frequent patterns from FP-tree [4].

3.1   Construction of FP-Tree

The construction of FP-tree requires two scans on transaction database. The first scan accumulates the support of each item and then selects items that satisfy minimum support, i.e. frequent 1-itemsets. Those items are sorted in frequency descending order to form F-list. The second scan constructs FP-tree.

First, the transactions are reordered according to F-list, while non-frequent items are stripped off. Then reordered transactions are inserted into FP-tree. The order of items is important since in FP-tree itemset with same prefix shares same nodes. If the node corresponds to the items in transaction exists the count of the node is increased, otherwise a new node is generated and the count is set to 1.

FP-tree also has a frequent-item header table that holds head of node-links, that connect nodes of same item in FP-tree. The node-links facilitate item traversal during mining of frequent pattern.

3.2   FP-Growth

Input of FP-growth algorithm is FP-tree and the minimum support. To find all frequent patterns whose support are higher than minimum support, FP-growth traverses nodes in the FP-tree starting from the least frequent item in F-list. The node-link originating from each item in the frequent-item header table connects the same item in FP-tree.

While visiting each node, FP-growth also collects the prefix-path of the node, that is the set of items on the path from the node to the root of the tree. FPGrowth also stores the count on the node as the count of the prefix path. The prefix paths form the so called *conditional pattern base* of that item.

The conditional pattern base is a small database of patterns which co-occur with the item. Then FP-growth create small FP-tree from the conditional pattern base called *conditional FP-tree*. The process is recursively iterated until no conditional pattern base can be generated and all frequent patterns that consist the item are discovered.

The same iterative process are repeated for other frequent items in the F-list.

4   Parallel Execution of FP-Growth

Since the processing of a conditional pattern base is independent of the processing of other conditional pattern base, it is natural to consider it as the execution unit for the parallel processing.

Here we describe a simple parallel version of FP-growth. We assume that the transaction database is distributed evenly among nodes.

4.1   Trivial Parallelization

The basic idea is each node accumulates a complete conditional pattern base and processes it independently until the completion before receiving other conditional pattern base.

Pseudocode for this algorithm is depicted in Fig. 1 (right) and the illustration is given in Fig. 1 (left).

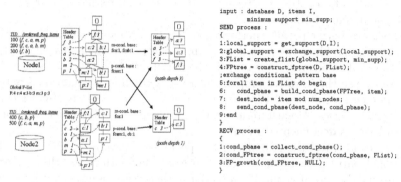

```
input : database D, items I,
         minimum support min_supp;
SEND process :
{
1:local_support = get_support(D,I);
2:global_support = exchange_support(local_support);
3:FList = create_flist(global_support, min_supp);
4:FPtree = construct_fptree(D, FList);
;exchange conditional pattern base
5:forall item in FList do begin
6:    cond_pbase = build_cond_pbase(FPTree, item);
7:    dest_node = item mod num_nodes;
8:    send_cond_pbase(dest_node, cond_pbase);
9:end
}
RECV process :
{
1:cond_pbase = collect_cond_pbase();
2:cond_FPtree = construct_fptree(cond_pbase, FList);
3:FP-growth(cond_FPtree, NULL);
}
```

Fig.1. Trivial parallel execution : illustration (left) pseudo code (right)

After the first scan of transaction database the support count of all items are exchanged to determine globally frequent items. Then each node builds F-list since it also have global support count. Notice that each node will have the identical F-list. At the second database scan, each node builds local FP-tree from local transaction database with respect to the global F-list.

To find all frequent patterns, the From the local FP-tree, local conditional pattern bases are generated. But instead processing conditional pattern base locally, we use hash function to determine which node should process it. We can do this because of the accumulation of local conditional pattern base results in the global conditional FP-tree.

4.2   Path Depth

It is obvious to achieve good parallelization, we have to consider the granularity of the execution unit or parallel task. Granularity is the amount of computation done in parallel relative to the size of the whole program.

When the execution unit is the processing of a conditional pattern base, the granularity is determined by number of iterations to generate subsequent conditional pattern bases. The number of iteration is exponentially proportional with the depth of the longest frequent path in the conditional pattern base. Thus here we define *path depth* as the measure of the granularity.

Definition 1. *Path depth is the longest path in the conditional pattern base whose count satisfies minimum support count.*

Notice that path depth is similar with the term "longest pass" in Apriori based algorithms. The path depth also can be calculated during the creation of FP-tree.

Since the granularity differs greatly, many nodes with smaller granularity will have to wait busy nodes with large granularity. This wastes CPU time and reduces scalability. It is confirmed by Fig. 2 (left) that shows the execution of the trivial parallel scheme given in the previous subsection. The line represents the CPU utilization ratio in percentage. Here other nodes have to wait node 1 (pc031) completes its task.

To achieve better parallel performance, we have to split parallel tasks with large granularity. Since the path depth can be calculated when creating FP-tree, we can predict in advance how to split the parallel tasks.

Here we use the iterative property of FP-growth that a conditional pattern base can create conditional FP-tree, which in turn can generate smaller conditional pattern bases. At each iteration, the path depth of subsequent conditional pattern bases is decremented by one.

So we can control the granularity by specifying a *minimum path depth*. Any conditional pattern base whose path depth is smaller than the threshold will be immediately executed until completion, otherwise it is executed only until the generation of subsequent conditional patterns bases. Then the generated conditional pattern bases are stored, some of them might be executed at the same node or sent to other idle nodes. After employing the path depth adjustment we get a more balanced execution as shown in Fig. 2 (right).

5   Implementation and Performance Evaluation

5.1   Implementation

As the shared nothing environment for this experiment we use PC cluster of 32 nodes that interconnected by 100Base-TX Ethernet Switch. Each PC node runs the Solaris 8 operating system on Pentium III 800Mhz with 128 MB of main memory.

Three processes are running on each node :

1. SEND process create FP-tree, send conditional pattern base
2. RECV process receive conditional pattern base, process conditional pattern base after exchanging finish

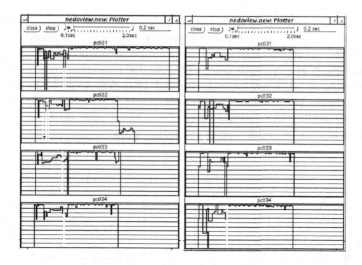

Fig. 2. Trivial execution (left) with path depth (right) (T25.I10.D100K 0.11%)

3. EXEC process process conditional pattern base in background when exchanging

There are also small COORD processes that receive requests for conditional pattern base from idle nodes and coordinate how to distribute them.

5.2   Performance Evaluation

For the performance evaluation, we use synthetically generated dataset as described in Apriori paper [2]. In this dataset, the average transaction size and average maximal potentially frequent itemset size are set to 25 and 20 respectively. While the number of transactions in the dataset is set to 100K with 10K items.

We have varied the minimum path depth to see how it affects performance. The experiments are conducted on 1, 2, 4, 8, 16 and 32 nodes. The execution time for minimum support of 0.1% is shown in Fig. 3 (left) The best time of 40 seconds is achieved when minimum path depth is set to 12 using 32 nodes. On single node, the experiment requires 904 seconds in average.

Fig. 3 (right) shows that path depth greatly affects the speedup achieved by the parallel execution. The trivial parallelization, denoted by "Simple", performs worst since almost no speedup achieved after 4 nodes. This is obvious since the execution time is bounded by the busiest node, that is node that has to process conditional pattern base with highest path depth.

When the minimum path depth is too low such as as "pdepth min = 5", the speedup ratio is not improved because there are too many small conditional pattern bases that have to be stored thus the overhead is too large. On the other hand, when the minimum path depht is too high, as represented by "pdepth min = 20", the granularity is too large so that the load is not balanced sufficiently.

When the minimum path depth is optimum, sufficiently good speedup ratio can be achieved. For "pdepth min = 12", parallel execution on 8 nodes can gain speedup ratio of 7.3. Even on 16 nodes and 32 nodes, we still can get 13.4 and 22.6 times faster performance respectively.

However finding the optimal value of minimum path depth is not a trivial task yet, and it is becoming one of our future work.

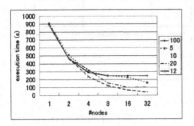 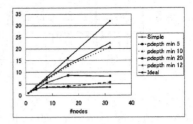

Fig.3. Execution time(left) Speedup ratio(right) for T25.I20.D100K 0.1%

6   Conclusion

We have reported the development of parallel algorithm or FP-growth that designed to run on shared-nothing environment. The algorithm has been implemented on top of PC cluster system with 32 nodes. We have also introduced a novel notion of *path depth* to break down the granularity of parallel processing of conditional pattern bases.

Although the data structure of FP-tree is complex and naturally not suitable for parallel processing on shared-nothing environment, the experiments show our algorithm can achieve reasonably good speedup ratio.

We are going to make direct comparison with other parallel algorithms using various datasets to explore the suitability of our algorithm.

References

1. R. Agarwal, C. Aggarwal and V.V.V. Prasad "A Tree Projection Algorithm for Generation of Frequent Itemsets". In *J. Parallel and Distributed Computing*, 2000
2. R. Agrawal and R. Srikant. "Fast Algorithms for Mining Association Rules". In *Proceedings of the 20th International Conference on VLDB*, pp. 487–499, September 1994.
3. R. Agrawal and J. C. Shafer. "Parallel Mining of Associaton Rules". In *IEEE Transaction on Knowledge and Data Engineering*, Vol. 8, No. 6, pp. 962–969, December, 1996.

4. J. Han, J. Pei and Y. Yin "Mining Frequent Pattern without Candidate Generation" In *Proc. of the ACM SIGMOD Conference on Management of Data*, 2000
5. J.S.Park, M.-S.Chen, P.S.Yu "Efficient Parallel Algorithms for Mining Association Rules" In *Proc. of 4th International Conference on Information and Knowledge Management (CIKM'95)*, pp. 31–36, November, 1995
6. T. Shintani and M. Kitsuregawa "Hash Based Parallel Algorithms for Mining Association Rules". In *IEEE Fourth International Conference on Parallel and Distributed Information Systems*, pp. 19–30, December 1996.

Active Feature Selection Using Classes

Huan Liu[1], Lei Yu[1], Manoranjan Dash[2], and Hiroshi Motoda[3]

[1] Department of Computer Science & Engineering
Arizona State University, Tempe, AZ 85287-5406
{hliu,leiyu}@asu.edu
[2] Department of Elec. & Computer Engineering
Northwestern University, Evanston, IL 60201-3118
manoranj@ece.northwestern.edu
[3] Institute of Scientific & Industrial Research
Osaka University, Ibaraki, Osaka 567-0047, Japan
motoda@sanken.osaka-u.ac.jp

Abstract. Feature selection is frequently used in data pre-processing for data mining. When the training data set is too large, sampling is commonly used to overcome the difficulty. This work investigates the applicability of active sampling in feature selection in a filter model setting. Our objective is to partition data by taking advantage of class information so as to achieve the same or better performance for feature selection with fewer but more relevant instances than random sampling. Two versions of active feature selection that employ class information are proposed and empirically evaluated. In comparison with random sampling, we conduct extensive experiments with benchmark data sets, and analyze reasons why class-based active feature selection works in the way it does. The results will help us deal with large data sets and provide ideas to scale up other feature selection algorithms.

1 Introduction

Feature selection is a frequently used technique in data pre-processing for data mining. It is the process of choosing a subset of original features by removing irrelevant and/or redundant ones. Feature selection has shown its significant impact in dealing with large dimensionality with many irrelevant features [1, 2] in data mining. By extracting as much information as possible from a given data set while keeping the smallest number of features, feature selection can remove irrelevant features, reduce potential hypothesis space, increase efficiency of the mining task, improve predictive accuracy, and enhance comprehensibility of mining results [3,4]. Different feature selection methods broadly fall into the *filter model* [5] and the *wrapper model* [6]. The filter model relies on general characteristics of the training data to select some features without involving any learning algorithm. The wrapper model requires one predetermined learning algorithm and uses the performance of the learning algorithm to evaluate and determine which features should be selected. The wrapper model tends to give superior performance as it finds features better suited to the predetermined

K.-Y. Whang, J. Jeon, K. Shim, J. Srivatava (Eds.): PAKDD 2003, LNAI 2637, pp. 474–485, 2003.

learning algorithm, but it also tends to be computationally more expensive [7]. When the training data becomes very large, the filter model is usually a good choice due to its computational efficiency and neutral bias toward any learning algorithm. Many feature selection algorithms [8,9] are being developed to answer challenging research issues: from handling a huge number of instances, large dimensionality, to dealing with data without class labels.

This work is concerned about a huge number of instances with class labels. Traditional feature selection methods perform dimensionality reduction using whatever training data is given to them. When the number of instances becomes very large, sampling is a common approach to overcome the difficulty [10,11]. However, random sampling is blind. It selects instances at random without considering the characteristics of the training data. In this work, we explore the possibility of *active feature selection* that can influence which instances are used for feature selection by exploiting some characteristics of the data. Our objective is to actively select instances with higher probabilities to be informative in determining feature relevance so as to improve the performance of feature selection without increasing the number of sampled instances. Active sampling used in active feature selection chooses instances in two steps: first, it partitions the data according to some *homogeneity* criterion; and second, it randomly selects instances from these partitions. Therefore, the problem of active feature selection boils down to how we can partition the data to actively choose useful instances for feature selection.

2   Feature Selection and Data Partitioning

In this work, we attempt to apply active sampling to feature selection which exploits data characteristics by sampling from subpopulations[1]. Each subpopulation is formed according to a homogeneity criterion. Since this work is dealing with a large number of instances in feature selection, when choosing a feature selection method to demonstrate the concept of active sampling, *efficiency* is a critical factor. We adopt a well received, efficient filter algorithm *Relief* [12,13] which can select statistically relevant features in linear time in the number of features and the number of instances. We first describe *Relief*, and then examine two ways of partitioning data into subpopulations using class information.

2.1   Relief for Feature Selection

The key idea of *Relief* (shown in Fig. 1) is to estimate the quality of features according to how well their values distinguish between the instances of the same and different classes that are near each other. *Relief* chooses features with n largest weights as relevant ones. For this purpose, given a randomly selected instance X from a data set S with k features, *Relief* searches the data set for

---

[1] We follow the practice in data mining that a given data set is treated as a population although it is only a sample of the true population, and subsets are subpopulations.

its two nearest neighbors: one from the same class, called nearest hit H, and the other from a different class, called nearest miss M. It updates the quality estimation $W[A_i]$ for all the features A_i depending on the difference $diff()$ on their values for X, M, and H. The process is repeated for m times, where m is a user-defined parameter [12,13]. Normalization with m in calculation of $W[A_i]$ guarantees that all weights are in the interval of [-1,1].

Given m - number of sampled instances, and k - number of features,

1. set all weights $W[A_i] = 0.0$;
2. for $j = 1$ to m do begin
3.   randomly select an instance X;
4.   find nearest hit H and nearest miss M;
5.   for $i = 1$ to k do begin
6. $W[A_i] = W[A_i] - diff(A_i, X, H)/m$
            $+ diff(A_i, X, M)/m$;
7.   end;
8. end;

Fig. 1. Original Relief algorithm.

Time complexity of *Relief* for a data set with N instances is $O(mkN)$. Clearly, efficiency is one of the major advantages of the *Relief* family over other algorithms. With m being a constant, the time complexity becomes $O(kN)$. However, since m is the number of instances for approximating probabilities, a larger m implies more reliable approximations. When N is very large, it often requires that $m \ll N$. The m instances are chosen randomly in *Relief*. Given a small constant m, we ask if by active sampling, we can improve approximations to close to those using N instances.

2.2   Data Partitioning Based on Class Information

Intuitively, since *Relief* searches for nearest hits and nearest misses, class information is important and can be used to form subpopulations for active sampling. We investigate two ways of using class information below, and indicate why this scheme of active feature selection should work in helping reduce data size without performance deterioration.

Stratified sampling. In a typical stratified sampling [10], the population of N instances is first divided into L subpopulations of $N_1, N_2, ..., N_L$ instances, respectively. These subpopulations are non-overlapping, and together they comprise the whole population, i.e., $N_1 + N_2 + ... + N_L = N$. If a simple random sampling is taken in each stratum, the whole procedure is called *stratified random sampling*. One of the reasons that stratification is a common technique is that stratification may produce a gain in precision in the estimates of characteristics of the whole population. It may be possible to divide a heterogeneous population into subpopulations, each of which is internally homogeneous. If each stratum is homogeneous, a precise estimate of any stratum statistics can be obtained from a small sample in that stratum. These estimates can then be combined into a precise estimate for the whole population.

In dealing with a data set with class information, a straightforward way to form strata is to divide the data according to their class labels. If there are j classes $(c_1, c_2, ..., c_j)$, we can stratify the data into j strata of sizes $N_1, N_2, ..., N_j$. That is, the number of strata is determined by the number of classes. Each stratum contains instances with the same class. There are only two strata if $j = 2$. Time complexity of stratifying the data into j classes is $O(jN)$.

Entropy-based partitioning. Having only j classes, if one wishes to create more than j strata, different approaches should be explored. The key to stratification is to form homogeneous subpopulations. In order to divide the data into more than j subpopulations, finer strata need to be formed. We can group instances that are similar to each other into subpopulations of pure classes. This can be achieved by (1) dividing the data using feature values, (2) then measuring each subpopulation's *entropy*, and (3) continuing the first two steps until each subpopulation is pure or it runs out of features to divide subpopulations. This idea of applying entropy to measure purity of a partition is frequently used in classification tasks [14]:

$$entropy(p_1, p_2, ..., p_j) = -\sum_{i=1}^{j} p_i \log p_i \ ,$$

where p_i is a fraction estimating the prior probability of class c_i. One can now form subpopulations (or partitions) based on feature values[2]. After using q values of feature A_i to divide the data into q partitions, the expected entropy is the sum of the weighted entropy values of the partitions:

$$entropy_{A_i} = \sum_{o=1}^{q} w_o * entropy_o(p_1, p_2, ..., p_j), \quad \sum_{o=1}^{q} w_o = 1 \ ,$$

where w_o is the percentage of the data that fall into partition o. It can be shown that for a pure partition (or all data points in the partition belong to one class), its entropy value $entropy_o()$ is 0. Hence, achieving pure partitions amounts to minimizing $entropy_{A_i}$. This partitioning process can continue until all partitions are pure or no further partitioning can be done. The formed partitions are in theory better than simple stratification described earlier because instances in the same partitions share the same feature values determined by partitioning. In other words, entropy-based partitioning uses feature values to partition data taking into account of class information. So, instances in a partition are close to each other besides having the same class value in an ideal case. It is also

---

[2] For continuous values, we find an optimal point that binarizes the values and minimizes the resulting entropy.

reasonable to anticipate that this entropy-based partitioning will result in more partitions than simple stratification for non-trivial data sets. Time complexity of entropy-based partitioning is $O(kN \log N)$ which is more expensive than that of stratified sampling $(O(jN))$.

With the above two methods for data partitioning, we are now able to partition a given data set into different partitions and then randomly sample from these partitions for *Relief* in choosing m instances as in Fig. 1. Since the number of instances in each stratum is different, to select a total of m instances, the number of instances sampled from a stratum (say, ith one) is proportional to the size (N_i) of the stratum, i.e., its percentage is $p_i\% = N_i/N$. Given m, the number of instances (m_i) sampled from each stratum is determined by $m * p_i\%$ for the ith stratum, and then random sampling is performed for each stratum to find m_i instances. The reason why stratified sampling works should apply to both methods and we expect gains of applying active sampling for feature selection. We present the details below.

3   Class-Based Active Feature Selection

The two partitioning methods described in Section 2 constitute two versions of active sampling for *Relief* in Fig. 2: (1) *ReliefC* - strata are first formed based on class values, then m instances are randomly sampled from the strata, and the rest remains the same as in *Relief*; and (2) *ReliefE* - data is first partitioned using entropy minimization, then m instances are randomly sampled from the partitions, and the rest remains the same as in *Relief*. *ReliefC* in Line 3a of Fig. 2 uses $p\%$ instances randomly sampled from the strata of different classes, and *ReliefE* in Line 3b selects $p\%$ instances randomly sampled from the partitions determined by entropy minimization when splitting data along feature values, where $p\% \approx m/N$. Each partition has a distinct signature which is a combination of different feature values. The length of a signature is in the order of $\log N$.

In the following, we will conduct an empirical study to verify our hypothesis that active sampling should allow us to achieve feature selection with fewer instances (i.e., smaller m). If *ReliefC* and *ReliefE* can achieve what they are designed for, we want to establish which one is more effective. As a reference for comparison, we use *ReliefF* [13,15] which extends *Relief* in many ways: it searches for several nearest neighbors to be robust to noise, and handles multiple classes.

4   Empirical Study

We wish to compare *ReliefC* and *ReliefE* with *ReliefF* and evaluate their gains in selecting m instances. Since m is the number of instances used to approximating probabilities, a larger m implies more reliable approximations. In [13], m is set to N to circumvent the issue of optimal m when many extensions of *Relief* are evaluated. In this work, however, we cannot assume that it is always possible to let $m = N$ as this would make the time complexity of *Relief* become $O(kN^2)$.

Given $p\%$ - percentage of N data for *ReliefC* and *ReliefE*, and k - number of features,

1. set all weights $W[A_i] = 0.0$;
2. do one of the following:
   a. stratifying data using classes; // active sampling for *ReliefC*
   b. entropy-based partitioning; // active sampling for *ReliefE*
3. corresponding to 2a and 2b,
   a. $m = \sum$ (sample $p\%$ data from each stratum);
   b. $m = \sum$ (sample $p\%$ data from each partition);
4. for $j = 1$ to m do begin
5.   pick instance X_j;
6.   find nearest hit H and nearest miss M;
7.   for $i = 1$ to k do begin
8. $W[A_i] = W[A_i] - diff(A_i, X_j, H)/m$
                   $+ diff(A_i, X_j, M)/m$;
9.   end;
10. end;

Fig. 2. Algorithms *ReliefC* and *ReliefE* with active sampling.

It is obvious that an optimal ranking of features can be obtained according to the weights $W[A_i]$ by running *ReliefF* with $m = N$. This optimal ranked list (or set) of features is named $S_{F,N}$. Let Y be one of $\{C, E\}$ and *ReliefY* denote either *ReliefC* or *ReliefE*. Given various sizes of m and the subsets $S_{F,m}$ and $S_{Y,m}$ selected by *ReliefF* and *ReliefY* respectively, the performance of *ReliefY* w.r.t. *ReliefF* can be measured in two aspects: (1) compare which subset ($S_{F,m}$ or $S_{Y,m}$) is more similar to $S_{F,N}$; and (2) compare whether features of the subsets $S_{F,m}$ and $S_{Y,m}$ are in the same order of the features in $S_{F,N}$. Aspect (1) is designed for feature subset selection to see if two subsets contain the same features; aspect (2) is for feature ranking and is more stringent than the first aspect. It compares the two ordered lists produced by *ReliefY* and by *ReliefF* of $m < N$ with reference to *ReliefF* of $m = N$. We discuss issues of performance measures below.

4.1 Measuring Goodness of Selected Features

A goodness measure of selected features should satisfy (1) its value improves as m increases; (2) its value reaches the best when $m = N$; and (3) it is a function of the features of the data. Given an optimal ranked list $S_{F,N}$, a target set T is defined as the optimal subset of features which contains the top n weighted features in $S_{F,N}$. For a data set without knowledge of the number of relevant features, T is chosen as the top n features whose weights $\geq \gamma$, where γ is a threshold equal to $W[i]$ (the i-th largest weight in $S_{F,N}$ and the gap defined by $W[i]$ and $W[i+1]$ is sufficiently large (e.g., greater than the average gap among $k - 1$ gaps). To compare the performance of both *ReliefF* and *ReliefY* with different sizes of m, we can define a performance measure $\mathcal{P}(S_{F,N}, R)$ where

R can be either $S_{F,m}$ or $S_{Y,m}$ with varying size m. We examine below three measures for $\mathcal{P}()$.

Precision. Precision is computed as the number of features in T that are also in R_n (the subset of top n features in R), normalized by dividing the number of features in T:

$$\frac{|\{x : x \in T \wedge x \in R_n\}|}{|T|} .$$

Precision ranges from 0 to 1, where the value of 1 is achieved when subsets T and R_n are equal.

Distance Measure. Precision treats all features in T equally without considering the ordering of features. One way of considering the ordering of features in the two subsets is named Distance Measure (DM) which is the sum of distances of the same features in R and T. The distance of a feature between two sets is the difference of their positions in the ranking. Let $S'_{F,N}$ be $S_{F,N}$ in reverse order. The maximum possible ranking distance between the two sets $S_{F,N}$ and $S'_{F,N}$ that share the same features is:

$$D_{max} = \sum_{\forall A_i \in S_{F,N}} |position(A_i \in S_{F,N}) - position(A_i \in S'_{F,N})| , \; and$$

$$DM = \frac{\sum\limits_{\forall A_i \in T} |position(A_i \in T) - position(A_i \in R)|}{D_{max}} .$$

Since the subset R_n may not contain all the features in T, we use the full set R in the definition of DM. D_{max} is used to normalize DM so that it ranges from 0 to 1, where the value of 0 is achieved if the two sets T and R_n have identical ranking, otherwise, DM is larger than 0.

Raw Distance. A straightforward performance measure is to directly calculate the sum of the differences of weights for the same features in the optimal ranking list $S_{F,N}$ and R. We name it Raw Distance (RD):

$$\sum_{i=1}^{k} |W_S[A_i] - W_R[A_i]| ,$$

where $W_S[A_i]$ and $W_R[A_i]$ are associated with $S_{F,N}$ and R, respectively. RD considers all the k features in the two sets. Thus, this measure avoids choosing a threshold for γ. When it is used for comparing the results of *ReliefF* with those of *ReliefE* and *ReliefC*, it serves the purpose well, although it cannot be used for measuring the performance of subset selection as it uses all features.

4.2   Data and Experiments

The experiments are conducted using Weka's implementation of *ReliefF* [15]. *ReliefC* and *ReliefE* are also implemented in Weka. All together 12 data sets from the UC Irvine machine learning data repository [16] and the UCI KDD

Table 1. Summary of bench-mark data sets.

Title	# Total Instances	# Total Features	# Total Classes
WDBC	569	30 plus class	2
Balance	625	4 plus class	3
Pima-Indian	768	8 plus class	2
Vehicle	846	18 plus class	4
German	1000	24 plus class	2
Segment	2310	19 plus class	7
Abalone	4177	8 plus class	3
Satimage	4435	36 plus class	6
Waveform	5000	40 plus class	3
Page-Blocks	5473	10 plus class	5
CoIL2000	5822	85 plus class	2
Shuttle	14500	8 plus class	7

Archive [17] are used in experiments. All have numeric features with varied numbers of instances (from 569 to 14500), number of features (from 4 to 85), and number of classes (from 2 to 7). The data sets are summarized in Table 1.

We use increasing percentages of data with three versions of *Relief* and have *ReliefF* with $m = N$ as their reference point. Each experiment is conducted as follows: For each data set,

1. Run *ReliefF* using all the instances, and obtain the ranked list of features according to their weights, i.e., $S_{F,N}$. The parameter for k-nearest neighbor search in *ReliefF* is set to 5 (neighbors). This parameter remains the same for all the experiments.
2. Specify five increasing percentage values P_i where $1 \le i \le 5$.
3. Run *ReliefE*, *ReliefC*, and *ReliefF* with each P_i determined in Step 2. For each P_i, run each algorithm 30 times and calculate Precision, Distance, and Raw Distance each time, and obtain their average values after 30 runs. Curves are plotted with average results.

4.3   Results and Discussions

Intuitively, if active sampling works and the data is divided sensibly, we should observe that the finer the data is divided, the more gain there should be in performance improvement. For each data set, we obtain three curves for the three versions of *Relief* for each performance measure. Fig. 3 demonstrates two illustrative sets of average results for Precision, Distance, and Raw Distance. Recall that for Precision, 1 is the best possible value, while for Distance and Raw Distance, 0 is the best possible value. As shown in Fig. 3, for the Segment Data, we notice that all three versions of *Relief* perform equally well in Precision, but differently in Distance and Raw Distance. Precision is 1 indicates that all features selected are the same as if we use the whole N instances for selection. When Distance and Raw Distance have values greater than 0, it indicates that

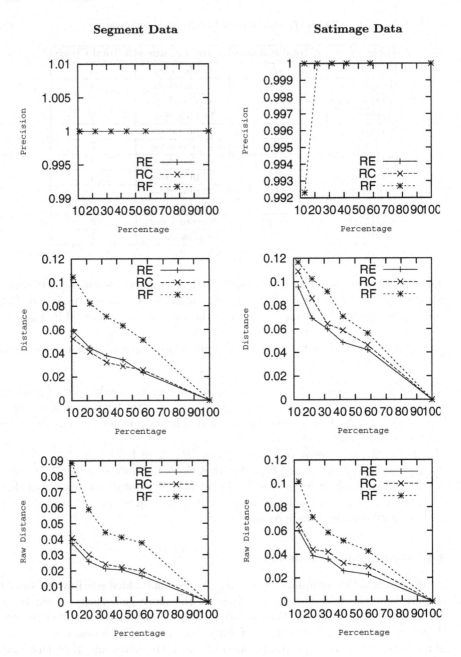

Fig. 3. Two illustrative sets of performance results on Segment and Satimage data sets.

the selected results are not exactly the same as that of using N instances. For the two sets of results (Segment and Satimage), we note one interesting difference between the two that *ReliefE* is, in general, worse than *ReliefC* for Segment in Distance Measure. It can be observed in all three versions of *Relief* that the more instances used, the better the performance of feature selection. A similar trend can also be observed for the Satimage data. It is clear that the two versions of active sampling generate different effects in feature selection.

Table 2. Precision, Distance, Raw Distance results: applying *ReliefE* (RE), *ReliefC* (RC) and ReliefF (RF) to feature selection on bench-mark data sets. We underline those figures that do not obey the general trend $\mathcal{P}(RE) \geq \mathcal{P}(RC) \geq \mathcal{P}(RF)$ and boldface those that are worse than $\mathcal{P}(RF)$.

	Precision			Distance			Raw Distance		
	RE	RC	RF	RE	RC	RF	RE	RC	RF
WDBC	0.992	**0.990**	0.991	<u>0.137</u>	0.136	0.139	0.105	0.111	0.111
Balance	0.891	0.884	0.864	0.317	0.398	0.468	0.030	0.032	0.037
Pima-Indian	0.919	0.914	0.906	**0.249**	**0.249**	0.248	0.019	**0.020**	0.019
Vehicle	0.999	0.999	0.996	0.150	0.183	0.206	0.041	0.048	0.052
German	0.907	0.907	0.898	0.345	0.352	0.360	0.146	0.149	0.154
Segment	1.0	1.0	1.0	<u>0.040</u>	0.036	0.074	0.025	0.027	0.054
Abalone	0.956	0.953	0.947	0.251	**0.277**	0.257	0.002	0.003	0.003
Satimage	1.0	1.0	0.998	0.064	0.073	0.088	0.039	0.042	0.065
Waveform	1.0	1.0	1.0	0.070	0.072	0.080	0.043	0.045	0.047
Page-Blocks	1.0	1.0	1.0	**0.214**	**0.220**	0.202	0.006	0.006	0.006
CoIL2000	1.0	1.0	1.0	0.054	0.056	0.060	0.102	0.106	0.110
Shuttle	1.0	1.0	1.0	0.0	0.0	0.0	0.002	0.002	0.003

Table 2 presents a summary of performance measures $\mathcal{P}_i$ averaged over five percentage values P_i and $1 \leq i \leq 5$, i.e.,

$$val_{Avg} = (\sum_{i=1}^{5} \mathcal{P}_i)/5 \ .$$

It is different from the results demonstrated in Fig. 3 that shows the progressive trends. Table 2 only provides one number (val_{Avg}) for each version of *Relief* (RE, RC, RF) and for each performance measure. In general, we observe that $\mathcal{P}(RE) \geq \mathcal{P}(RC) \geq \mathcal{P}(RF)$ where $\geq$ means "better than or as good as". In Precision, *ReliefC* has one case that is worse than *ReliefF*. In Distance, *ReliefE* has two cases that are worse than *ReliefC*, and two cases that are worse than *ReliefF*; *ReliefC* has 3 cases that are worse than *ReliefF*. In Raw Distance, *ReliefC* has one case that is worse than *ReliefF*. It is clear that for the 12 data sets, *ReliefE* is better than or as good as *ReliefC* with two exceptions in all three measures; and *ReliefC* is better than or as good as *ReliefF* with four exceptions in all three measures. The unexpected are (1) *ReliefE* does not show significant superiority over *ReliefC*; and (2) in a few cases, *ReliefE* performs worse than *ReliefC* as shown in Table 2 and Fig. 3, although *ReliefE* uses both feature values and class information to partition data. *ReliefE* and *ReliefC* can usually gain more for data sets with more classes.

The two versions of active sampling incur different overheads. As discussed earlier, between the two, the extra cost incurred by *ReliefC* is smaller - its time complexity is $O(jN)$. Time complexity of entropy-based partitioning is $O(kN \log N)$ which is more expensive than $O(jN)$ and usually $k \gg j$, where k is number of features. However, the additional costs for both versions incur only once. Because for 8 out of 12 cases *ReliefC* is better than *ReliefF*, with its low overhead, *ReliefC* can be chosen over *ReliefF* for feature selection. In *ReliefF*, it takes $O(N)$ to find the nearest neighbors; using signatures, it costs $O(\log N)$ for *ReliefE* to do the same. Therefore, the one-time loss of time in obtaining signatures can normally be compensated by the savings of at least m times of searching for neighbors. In addition, as an example seen in Fig. 3 (Segment Data), *ReliefE* and *ReliefC* using 10% of the data can achieve a performance similar to *ReliefF* using 50% of the data in terms of Distance and Raw Distance.

The three performance measures are defined with different purposes. For feature selection, in effect, Precision is sufficient as we do not care about the ordering of selected features. Many cells have value 1 in Table 2 for Precision measure. This means that for these data sets, the feature selection algorithms work very well - the usual smallest percentage ($P_1 = 10\%$) can accomplish the task: in order to have the average value to be 1, each $\mathcal{P}_i$ in $(\sum_{i=1}^{5} \mathcal{P}_i)/5$ should be 1. Similarly, we can infer that when the average Precision value is very close to 1, it indicates that active sampling can usually work with a smaller percentage of data. Since *Relief* is a ranking algorithm, when it selects features, it provides additional ordering information about selected features. When the value of Precision is 1, it does not mean that the orders of the two feature subsets are the same. We can further employ Distance (with a threshold γ for selecting features) and Raw Distance to examine whether their orders are also the same.

5   Concluding Remarks

In order to maintain the performance while reducing the number of required instances used for feature selection, active feature selection is proposed, implemented, and experimentally evaluated using a widely used algorithm *Relief*. Active sampling exploits data characteristics to first partition data, and then randomly sample data from the partitions. Two versions of active sampling for feature selection are investigated: (a) *ReliefC* - stratification using class labels and (b) *ReliefE* - entropy-based partitioning. Version (a) uses only class information, and version (b) splits data according to feature values while minimizing each split's entropy. The empirical study suggests that (1) active sampling can be realized by sampling from partitions, and the theory of stratified sampling in *Statistics* suggests some reasons why it works; (2) active sampling helps improve feature selection performance - the same performance can be achieved with fewer instances; and (3) between the two versions, in general, *ReliefE* performs better than *ReliefC* in all three measures (Precision, Distance, and Raw Distance).

Acknowledgments. We gratefully thank Bret Ehlert and Feifang Hu for their contributions to this work. This work is in part based on the project supported by National Science Foundation under Grant No. IIS-0127815 for H. Liu.

References

1. R. Kohavi and G.H. John. Wrappers for feature subset selection. *Artificial Intelligence*, 97(1–2):273–324, 1997.
2. H. Liu and H. Motoda. *Feature Selection for Knowledge Discovery & Data Mining.* Boston: Kluwer Academic Publishers, 1998.
3. M. Dash and H. Liu. Feature selection methods for classifications. *Intelligent Data Analysis: An International Journal*, 1(3), 1997.
4. L. Talavera. Feature selection as a preprocessing step for hierarchical clustering. In *Proceedings of Internationl Conference on Machine Learning (ICML'99)*, 1999.
5. U.M. Fayyad and K.B. Irani. The attribute selection Problem in decision tree generation. In AAAI-92, *Proceedings of the Ninth National Conference on Artificial Intelligence*, pages 104–110. AAAI Press/The MIT Press, 1992.
6. G.H. John, R. Kohavi, and K. Pfleger. Irrelevant feature and the subset selection Problem. In W.W. Cohen and Hirsh H., editors, *Machine Learning: Proceedings of the Eleventh International Conference*, pages 121–129, New Brunswick, N.J., 1994. Rutgers University.
7. P. Langley. Selection of relevant features in machine learning. In *Proceedings of the AAAI Fall Symposium on Relevance*. AAAI Press, 1994.
8. P.S. Bradley and O. L. Mangasarian. Feature selection via concave minimization and support vector machines. In *Proceedings of Fifteenth International Conference on Machine Learning*, pages 82–90, 1998.
9. M.A. Hall. Correlation-based feature selection for discrete and numeric class machine learning. In *Proceedings of Seventeenth International Conference on Machine Learning (ICML-00)*. Morgan Kaufmann Publishers, 2000.
10. W.G. Cochran. *Sampling Techniques.* John Wiley & Sons, 1977.
11. B. Gu, F. Hu, and H. Liu. *Sampling: Knowing Whole from Its Part*, pages 21–38. Boston: Kluwer Academic Publishers, 2001.
12. K. Kira and L.A. Rendell. The feature selection Problem: Traditional methods and a new algorithm. In *Proceedings of the Tenth National Conference on Artificial Intelligence*, pages 129–134. Menlo Park: AAAI Press/The MIT Press, 1992.
13. I. Kononenko. Estimating attributes: Analysis and extension of RELIEF. In F. Bergadano and L. De Raedt, editors, *Proceedings of the European Conference on Machine Learning*, April 6–8, pages 171–182, Catania, Italy, 1994. Berlin: Springer.
14. J.R. Quinlan. *C4.5: Programs for Machine Learning.* Morgan Kaufmann, 1993.
15. I.H. Witten and E. Frank. *Data Mining – Practical Machine Learning Tools and Techniques with JAVA Implementations.* Morgan Kaufmann Publishers, 2000.
16. C.L. Blake and C.J. Merz. UCI repository of machine learning databases, 1998. http://www.ics.uci.edu/~mlearn/MLRepository.html.
17. S.D. Bay. The UCI KDD archive, 1999. http://kdd.ics.uci.edu.

Electricity Based External Similarity of Categorical Attributes*

Christopher R. Palmer[1] and Christos Faloutsos[2]

[1] Vivisimo, Inc., 2435 Beechwood Blvd, Pittsburgh, PA, palmer@vivisimo.com
[2] Carnegie Mellon University, 5000 Forbes Ave, Pittsburgh, PA,
christos@cs.cmu.edu

Abstract. Similarity or distance measures are fundamental and critical properties for data mining tools. Categorical attributes abound in databases. The *Car Make, Gender, Occupation*, etc. fields in a automobile insurance database are very informative. Sadly, categorical data is not easily amenable to similarity computations. A domain expert might manually specify some or all of the similarity relationships, but this is error-prone and not feasible for attributes with large domains, nor is it useful for cross-attribute similarities, such as between *Gender* and *Occupation. External similarity* functions define a similarity between, say, *Car Makes* by looking at how they co-occur with the other categorical attributes. We exploit a rich duality between random walks on graphs and electrical circuits to develop *REP*, an external similarity function. *REP* is theoretically grounded while the only prior work was ad-hoc. The usefulness of *REP* is shown in two experiments. First, we cluster categorical attribute values showing improved inferred relationships. Second, we use *REP* effectively as a nearest neighbour classifier.

1 Introduction

Inter-object similarity is a fundamental property required for data mining applications. A great deal of categorical data is under-utilized because it is difficult to define categorical similarity functions. For example, an automobile insurance database would likely contain a categorical field for the manufacturers, such as *Toyota, Hyundai* or *Porsche*. To address the lack of similarity information, a domain expert might define a 3x3 similarity matrix to facilitate data mining tasks. Such an approach is error prone and nearly impossible with a realistic number of manufacturers. If the manufacturers' names appear in isolation, there is little else that can be done. However, the insurance data base is itself a valuable

* This material is based upon work supported by the National Science Foundation under Grants No. IIS-9817496, IIS-9988876, IIS-0083148, IIS-0113089, IIS-0209107 IIS-0205224 and by the Defense Advanced Research Projects Agency under Contract No. N66001-00-1-8936. Additional funding was provided by donations from Intel. Any opinions, findings, and conclusions or recommendations expressed in this material are those of the author(s) and do not necessarily reflect the views of the National Science Foundation, DARPA, or other funding parties.

K.-Y. Whang, J. Jeon, K. Shim, J. Srivatava (Eds.): PAKDD 2003, LNAI 2637, pp. 486–500, 2003.
© Springer-Verlag Berlin Heidelberg 2003

resource for determining the similarity between the manufacturers with such information as *Age*, *Gender*, *Occupation*, and *Number of Accidents*. The goal of this research is to effectively use these additional attributes to infer a superior similarity function. This is known as an *External Similarity* function [3].

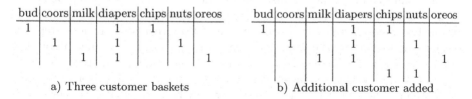

bud	coors	milk	diapers	chips	nuts	oreos
1			1	1		
	1		1		1	
		1	1			1

a) Three customer baskets

bud	coors	milk	diapers	chips	nuts	oreos
1			1	1		
	1		1		1	
		1	1			1
				1	1	

b) Additional customer added

Fig. 1. Toy-example shopping basket data

Our proposed similarity function is based on random walks in graphs (equivalently, current in circuits). By viewing a categorical table as a graph, we find natural "recursive" similarities. To motivate the need for this recursiveness, consider the toy shopping baskets in Fig. 1a) and b). The problem is to determine if *Bud* is more similar to *Coors* or to *Milk*. Using the first table, we must consider them equally similar because there is no other information (each appears with *Diapers* but have no other common data). However, in the second table, a fourth customer purchased both *Nuts* and *Chips* which provides an indirect similarity between *Bud* and *Coors* which could be used to determine that they are indeed more similar to each other than to *Milk*. To actually use this indirect similarity, we could run the following random process. Initially begin with *Bud*. Randomly select an item that co-occurs with *Bud* in the shopping basket table. Repeat with the newly selected item. Then *Bud* and *Coors* are similar if the expected number of steps between *Bud* and *Coors* is small. Section 4 explores random walk approaches such as this one (which are flawed). Section 5 uses these flawed attempts to motivate our proposed similarity function, called *REP*.

REP has some excellent properties. It is theoretically grounded. In addition to offering similarities between distinct values of the same attribute, it even offers similarities between different attributes. For example, we can measure the similarity between a *Car Type* and a *Gender*. *REP* may be computed efficiently using a relaxation algorithm that scales very well with data base size. Most importantly, *REP* shows improved performance on real tasks using real data.

The remainder of this paper is organized as follows. Section 2 describes the related work. Section 3, provides background material on random walks and electrical circuits and proposes our two graph-construction algorithms. Section 4 presents three interesting, but flawed, random walk based similarity or distance functions. Section 5 presents *REP*, correcting these flaws. We experimentally validate *REP* in Sec. 6.

2   Related Work

The specific problem of clustering categorical data has been actively studied recently [5,6,7]. Clustering algorithms are developed (not similarity functions) and is thus only peripherally related to our work.

Jeh and Widow propose a distance function between nodes of a graphs that is based on random walks [8]. The distance between nodes u and v is measured using two random web surfer, one at u and the other at v. In lock-step, the surfers randomly surf the web. The expected number of steps until they meet at a common node is computed. This method has some surprising properties. All nodes in a cycle are infinitely far apart (because random walkers walking in lock step will never meet). Such a property may be unfortunate. Moreover, it is expensive to compute. Given a graph G with n nodes, the computation is based on G^2, a graph with n^2 nodes, making it impractical for our experiments.

Das and Manila proposed the only prior work on external similarity [3]. Given a probe set P and attributes A and B, define the distance between A and B as:

$$D_{fr,P}(A, B) = \sum_{D \in P} |fr(A, D) - fr(B, D)| \tag{1}$$

where $fr(x, y)$ is the fraction of rows containing both x and y in the data base. Two comments are in order at this point. First, without loss of generality, we can remove the probe set. The probe set is really a projection of the attributes and we will always use all attributes for the probe set (except A and B). Second, $D_{fr,P}$ does not define distances recursively: *Milk*, *Bud*, and *Coors* are all equally far apart in the toy example from the Introduction.

Finally, Klein and Randic proposed one of the alternatives that we consider in Sec. 4 as the similarity between molecules: *resistance distance* [9]. This is the reciprocal of the *current similarity* that we discuss. When we evaluate the *current similarity*, we will see that it is not appropriate for our task.

3   Background and Definitions

3.1   Definitions

Categorical Table. Has n rows and m columns. The total number of (column) distinct attributes values is M. In Fig. 1, there are $n = 3$ rows, $m = 7$ columns and $M = 14$ distinct attributes (a 0 or 1 in each column).

Basic Electricity. In a circuit, the *current* (I), *voltage* (V) and *resistance* (R) are related by the equation $I = V/R$. *Conductance* (C) is the reciprocal of resistance ($C = 1/R$). Voltage is measured as a decrease in electric potential between pairs of points. We say *voltage at x* whenever the second point is obviously inferred (such as the ground). At every point in an electrical circuit the total influx of current is the same as the total outflow of current (the *Kirchoff's law* of conservation of current).

Walks. Let G be an edge weighted undirected graph with vertices V, edges E and edge weights $w : V \times V \rightarrow \mathcal{R}^+$. A (u, v)-*walk* is a sequence of vertices starting with u and ending with v, $(u = x_0 \rightarrow x1 \rightarrow \cdots \rightarrow x_n = v)$, such that (x_{i-1}, x_i) is an edge in E. We write $u \rightarrow^* v$ to refer to a (u, v) walk. A *constrained* (u, v)-*walk*, $u \rightarrow^* v/S$, is a (u, v)-walk in which none of the intermediate vertices belong to the set S. That is, $u \rightarrow^* v/\{u, v\}$ is a walk which starts from u and then never returns to u before it reaches v (for the first time). Notationally, let $u \rightarrow^* v/x_1, x_2, \cdots, x_k$ be equivalent to $u \rightarrow^* v/\{x_1, x_2, \cdots, x_k\}$.

Random walks. Let $C(u)$ be the total weight of all edge incident with u. Define the probability of stepping from u to v as $w(u, v)/C(u)$. I.e., the probability of transitioning from u to v is proportional to the weight of (u, v). Define $P(u \rightarrow^* v/S)$ as the probability that a random walk from u is a $u \rightarrow^* v/S$ walk.

Commute distance between nodes u and v in G is the expected length of a tour that begins at u and passes through v prior to returning to u.

Escape probability is the probability that a walk starting from x will reach S before it returns to x.

3.2   Electricity vs. Walks

Electricity and random walks on graphs are intimately related. Doyle and Snell provide an excellent introduction to this synergy in [4]. The circuit corresponding to a graph (and vice-versa) is defined by replacing each edge (with weight $w(u, v)$) with a resister with resistance $R(u, v) = 1/w(u, v)$ (or $C(u, v) = w(u, v)$).

Theorem 1. *Let Z be a circuit with a battery attached to u (+1 volt) and v (ground). Let G be the graph corresponding to Z and let it be connected. Let C be the total of the conductance of all resisters in Z. Let $I(u, v)$ be the current flowing from u to v. Then, $2C/I(u, v)$ is the commute distance between u and v in G.*

Proof. The unweighted version of this proof is in [2], Theorem 2.1. The weighted version follows naturally.   □

Theorem 2. *Let Z be a circuit with a battery attached to u (+1 volt) and grounded to each element in S ($u \notin S$). Let G be the graph corresponding to Z. Let $E(x)$ be the voltage drop at x. Then $E(x) = P(x \rightarrow^* u/S)$ for all x.*

Proof. (sketch) - This proof uses a uniqueness theorem for harmonic functions. Two harmonic functions are equal if they have the same boundary values. We therefore write $E(x)$ and $P(x \rightarrow^* u/S)$ as harmonic functions and then show equality of their boundary values. Consider $E(x)$ for any $x \neq u$ and $x \notin S$. The law of conservation of current states that:

$$\sum_{y \sim x} \frac{E(y) - E(x)}{R(x, y)} = 0$$
$$\Rightarrow E(x) \cdot 1/R(x) = \sum_{y \sim x} \frac{E(y)}{R(x, y)}$$
$$\Rightarrow E(x) = \sum_{y \sim x} \frac{E(y) \cdot R(x)}{R(x, y)}$$

Consider $f(x) = P(x \to^* u/S)$ for $x \neq u$ and $x \notin S$. Since x is not an end point of a candidate walk we can write:

$$f(x) = P(x \to^* u/S) = \sum_{y \sim x} P(x \to y) \cdot P(y \to^* u/S) = \sum_{y \sim x} w(x, y) \cdot f(y)$$

$E(x)$ and $f(x)$ are harmonic functions with boundary values at u and S. Since, by definition $E(u) = 1 = P(u \to^* u/S)$ and $E(s) = 0 = P(s \to^* u/S)$ for all $s \in S$, they share boundary values. By the uniqueness theorem of harmonic functions, $f(x) = E(x)$. □

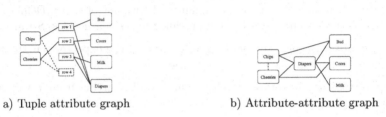

a) Tuple attribute graph   b) Attribute-attribute graph

Fig. 2. Two graph representations for the table in Fig. 1a)

3.3 Converting Categorical Data to a Graph

We propose two methods for converting a table of categorical values into a weighted graph (circuit) for recursive similarity computations. The first method preserves the tuples while the second method produces a more compact graph. To illustrate the constructions, we will use the toy shopping basket data from Fig. 1a) where the non-existent values are treated as NULLs.

Tuple-Attribute Graph. Create a node, r_i, corresponding to each of the n rows. Create a node, a_j, corresponding to each of the M attribute values. Place an edge (weight 1) between r_i and a_j iff row i contains the attribute j. This construction results in the bipartite graph in Fig. 2a).

Attribute-Attribute Graph. Create a node, a_j, corresponding to each of the M attribute values. The edge set is implied by the weight function $w(a_i, a_j) = fr(a_i, a_j)$, the fraction of rows that contain both attributes i and j. This procedure generates the graph in Fig. 2b).

There is obviously a strong relationship between these two graphs. In this paper we concentrate on the smaller attribute-attribute graph. The use of the *Tuple-Attribute graph* is an area that we are exploring.

4   Electric Similarity Functions

Using the attribute-attribute graph of a table of categorical data, we can imagine several natural and intuitive similarity functions. In this section, we consider three such functions and find that all are flawed. We then use these flaws to motivate our proposed similarity measure, *REP*.

4.1   Flawed Similarity/Distance Functions

Electrical Current Similarity. Define $I(x, y)$ to be the current flowing between x and y in the electrical circuit with a 1 volt battery across x and y. $I(x, y)$ has two excellent properties: $I(x, y)$ is larger if there are more walks between x and y, and $I(x, y)$ is larger if there are shorter walks between x and y. Unfortunately, current also has serious scale issues. Take for example the simple table in Fig. 3a) in which x and y occur with a every time they appear and p and q appear with b every time they appear. We expect the similarity between x and y to be the same as the similarity between p and q because their appearances of equivalent, except for scale. Unfortunately, $I(x, y) = 2 \cdot I(p, q)$. This is a serious practical issue, causing pairs of very frequent attributes to be very similar and making $I(x, y)$ useless for measuring the similarity between attribute values.

Commute Distance. To correct the scale problem, we considered the commute distance (expected length of a commute starting from x, reaching y and then returning to x). By Theorem 1, the commute distance is simply $2C/I(x, y)$ and is a normalized form of the electrical current similarity. This corrects the scale issue of the first example and the flaw here is more subtle. For data that is relatively uniform (the out degree distribution of the graph does not follow a skewed distribution), the commute distance actually appears useful. However, for realistic data where the degree distribution follows a Zipf or power-law relationship, the commute distance displays a bias toward high degree nodes. This is due to the fact that high degree nodes will have a much higher stationary probability (probability that a random walk will be at the high degree node at any given time) and consequently all the distances are skewed toward the largest nodes. This was discovered when we began the experiments discussed in Sec. 6. When clustering attribute values, the highest degree node was invariably the focal point of the clustering and distances were not particularly useful.

Escape Probability. The escape probability is the probability of a non-trivial walk starting at x will return to x before reaching y. This has a natural definition in terms of circuits. Place a $+1$ volt battery at x and grounded at y, and then measure the effective conductance, C, between x and y ($C = I(x, y)$ since we have a 1 volt drop) and let $C(x)$ be the conductance of x (for a proof see [4], page 42). We can then define

$$Pesc(x, y) = I(x, y)/C(x) = I(x, y) \cdot R(x) \tag{2}$$

$$S_{esc}(x, y) = \frac{Pesc(x, y) + Pesc(y, x)}{2} \tag{3}$$

where S_{esc} is defined to make it symmetric. Since $C(x)$ is the number of rows containing x, the escape probability helps correct the commute distance problems by normalizing the distance in terms of the degree. However, S_{esc} is also flawed because it does not account for the length of a (x, y) walk, only its existence. For example, see Fig. 3b). Here there is a direct relationship between x and y and only an indirect relationship (through x and y!) between $z1$ and $z6$. However, the escape probability similarity makes $z1$ and $z6$ more similar than x and y: $S_{esc}(x, y) = .125$ and $S_{esc}(z1, z6) = .25$.

X or Y	A or B
x	a
x	a
y	a
y	a
p	b
q	b

X or Y	Z
x	z1
x	z2
x	z3
x	z0
y	z0
y	z4
y	z5
y	z6

a) Electrical current similarity   b) Escape probability

Fig. 3. Examples with poor behaviour for the proposed similarity/distance functions

4.2   Proposal: Refined Escape Probability (REP)

We now combine the positive points of each method discussed above to define our proposed similarity function, the *Refined Escape Probability* (REP) similarity. Commute distance was appealing because it accounts for the length of (x, y) walks. Escape probability normalized according to frequency of an attribute but became walk-length agnostic. To correct this problem with the escape probability, we will use the concept of a *sink* node, s, that we attach to all nodes in the network. We assign resistances such that the probability of stepping from any node to the sink is *sink_p*. We then can measure the probability of (x, y) walks that do not pass through either x, y or s. The sink de-weights longer walks correcting the problem with the escape probability. Additionally, it still addresses the normalization problem with the commute distance. *REP* is thus:

$$S_{REP}(x, y) = R(y) \cdot P(x \to^* y/x, y, \text{ or } s) \qquad (4)$$

In the next section we convert this to two electrical circuit computations and show that it is symmetric (which explains $R(y)$ as a normalization factor).

5   *REP* Algorithm

Given the high level description of *REP*, we now turn out attention to making the procedure concrete and considering the efficiency of implementation. Recall that the similarity between x and y is the probability of a walk from x to y that does not pass through x, y or the new "sink" node. The sink node provides a bias toward short walks and the normalization factor makes this symmetric. We now complete the algorithm by formalizing the addition of the *sink* node, by showing that S_{REP} is symmetric, by converting the probability definition of S_{REP} to an electricity problem and then finally by using a relaxation algorithm (based on Kirchhoff's laws) to solve the circuit problem.

5.1 Adding *sink* Nodes to a Graph

From a table of categorical attributes (or some other source), we have a graph G_0. The problem is to construct a graph, G, from G_0 by adding a new sink node, s, and adding an edge from every node, x, of G_0 to s such that the probability of a random step from x to s is some constant, $sink\_p$. To do this, we note that $P(x \rightarrow s) = w(x, s)/C_G(x)$ where $C_G(x)$ is the total conductance of x in the graph G and $w(x, s)$ is the weight of the edge. Now, since $C_G(x) = C_{G_0}(x) + w(x, s)$ we just solve the equation to add the required weighted edges to the graph:

$$\frac{w(x,s)}{C_{G_0}(x)+w(x,s)} = sink\_p$$
$$\Rightarrow w(x, s) = C_{G_0}(x)\frac{1}{1/sink\_p-1}$$

5.2 S_{REP} Is Symmetric

Theorem 3. $S_{REP}(x, y) = S_{REP}(y, x)$.

Proof. Let $W = (x = u_0 \rightarrow u_1 \rightarrow \cdots \rightarrow u_k = y)$ be any (x, y) walk where $u_i \notin \{x, y, s\}$ for $1 \leq i < k$. From W we can also define the walk $\bar{W} = (y = u_k \rightarrow u_{k-1} \rightarrow \cdots \rightarrow u_0 = x)$. There is a 1-1 correspondence between the (x, y) and (y, x) walks. Thus to prove the result, we must show that $R(y) \cdot P(W) = R(x) \cdot P(\bar{W})$. Using $C(u_i) = 1/R(u_1)$ and $P(u_i \rightarrow u_{i+1}) = w(u, u_{i+1})/C(u_i)$ we can write:

$$R(y) \cdot P(W) = P(W) \cdot R(y)$$
$$= P(u_0 \rightarrow u_1) \cdot P(u_1 \rightarrow u_2) \cdots P(u_{i-1} \rightarrow u_k) \cdot R(y)$$
$$= \frac{w(x,u_1)}{C(x)} \cdot \frac{w(u_1,u_2)}{C(u_1)} \cdots \frac{w(u_{k-2},u_{k-1})}{C(u_{k-2})} \cdot \frac{w(u_{k-1},y)}{C(u_{k-1})} \cdot \frac{1}{C(y)}$$

which we can rewrite by simple "shifting" the denominators to the left, substituting $w(u_i, u_{i+1}) = w(u_{i+1}, u_i)$ (by definition) and reordering the terms:

$$R(y) \cdot P(W) = \frac{1}{C(x)} \cdot \frac{w(x,u_1)}{C(u_1)} \cdot \frac{w(u_1,u_2)}{C(u_2)} \cdots \frac{w(u_{k-2},u_{k-1})}{C(u_{k-1})} \cdot \frac{w(u_{k-1},y)}{C(y)}$$
$$= \frac{1}{C(x)} \cdot \frac{w(u_1,x)}{C(u_1)} \cdot \frac{w(u_2,u_1)}{C(u_2)} \cdots \frac{w(u_{k-1},u_{k-2})}{C(u_{k-1})} \cdot \frac{w(y,u_{k-1})}{C(y)}$$
$$= \frac{1}{C(x)} \cdot \frac{w(y,u_{k-1})}{C(y)} \cdot \frac{w(u_{k-1},u_{k-2})}{C(u_{k-1})} \cdots \frac{w(u_2,u_1)}{C(u_2)} \cdot \frac{w(u_1,x)}{C(u_1)}$$
$$= R(x) \cdot P(\bar{W})$$

$\square$

5.3 S_{REP} as Electrical Currents

We proposed a similarity function based on the quantity $P(x \rightarrow^* y/x, y, s)$ which we now convert to an equivalent electrical current problem. A slightly simpler form can be easily handled by Theorem 2: $P(x \rightarrow^* y/y, s)$ is the same as the voltage drop at x when a battery is attached across y and s. For the required result, we must address the walks beginning at x and returning to x before reaching y or s. We can write:

$$P(x \to^* y/y, s) = \left(\sum_{i=0}^{\infty} P(x \to^* x/x, y, s)^i\right) \cdot P(x \to^* y/x, y, s)$$
$$\Rightarrow P(x \to^* y/x, y, s) = (1 - P(x \to^* x/x, y, s)) \cdot P(x \to^* y/y, s)$$

Let $E_{y,s}(x)$ be the voltage drop at x with the battery across y and s. Let $E_{y:\{x,s\}}(u)$ be the voltage drop at u when the battery is attached to y and grounded at $\{x, s\}$. Then, we can define the probability of a loop from x to x by considering a single step and then using Theorem 2 again:

$$P(x \to^* x/x, y, s) = \sum_{u \sim x} P(x \to u) \cdot P(u \to^* x/x, y, s)$$
$$= \sum_{u \sim x} \frac{w(x,u)}{C(x)} \cdot E_{x:\{y,s\}}[u]$$

That is, we can define

$$P(x \to^* y/x, y, s) = E_{y:s}[x] \cdot \sum_{u \sim x} \frac{w(x, u)}{C(x)} \cdot E_{x:\{y,s\}}[u] \tag{5}$$

5.4   Kirchhoff Relaxation Algorithm for Voltages

We need to compute $E_{S_1:S_2}$ which is the set of voltage drops between all nodes in the circuit when the nodes in set S_1 are fixed at 1 volt and the nodes in set S_2 are fixed at 0 volts (ground). We assume that S_1 and S_2 are disjoint. For brevity of notation, let $V_i[u]$ be the ith approximation to $E_{S_1:S_2}[u]$. We initialize the relaxation algorithm by setting $V_0[u] = 1$ iff $u \in S_1$ and then at each step of the relaxation algorithm we update

$$\begin{cases} V_i[u] = 1 & \text{if } u \in S_1 \\ V_i[u] = 0 & \text{if } u \in S_2 \\ V_i[u] = \sum_{v \sim u} \frac{w(u,v)}{C(u)} \cdot V_0[v] & \text{otherwise} \end{cases}$$

where the last sub-equation is simply the basic identify $I = V/R$ subject to Kirchhoff's law (conservation of current):

$$\sum_{v \sim u} \frac{V[u] - V[v]}{R(u,v)} = 0$$
$$\Rightarrow \left(V[u] \cdot \sum_{v \sim u} 1/R(u, v)\right) - \sum_{v \sim u} \frac{V[v]}{R(u,v)} = 0$$
$$\Rightarrow V[u] \cdot C[u] = \sum_{v \sim u} \frac{V[v]}{R(u,v)}$$
$$\Rightarrow V[u] = \sum_{v \sim u} \frac{C(u,v)}{C(u)} \cdot V[v]$$

It is easy to verify that these approximation follow a monotone increasing relationship $0 \le V_0[u] < \cdots < V_i[u] < \cdots < V_l[u] = E_{S_1:S_2}[u]$ (exercise 1.2.5 of [4]) and then converge to the true value. We truncate the approximation sequence when a suitably accurate result is found.

5.5   Running Time of *REP*

There are two phases to our algorithm. First, a graph is constructed from a table of categorical values. This requires the addition of $O(n \cdot m^2)$ edges (n rows and m columns) and can be done $O(n \cdot m^2)$ time.

Computing S_{REP} requires that we use the relaxation algorithm described in the previous section. In that algorithm, the running time to compute V_0 is $O(n)$ where n is the number of nodes. Each improvement step computes V_i from V_{i-1} and requires $O(\#\ edges)$ time. Since a graph from a categorical table has one node for each distinct attribute value (and the sink) it has $M + 1$ nodes. The worst case bound on the number of edges is $O(M^2)$, which is rarely reached in practice. Thus, the time to compute S_{REP} is the time to do two calls to the relaxation algorithm and is thus $O(M^2)$ in the worst case. It is very interesting is that the time to compute the similarities is independent of the number of rows in the table and that the pre-computation grows linearly with the number of rows in the input table. We will explore this in the next section.

6   Experiments

Since the $D_{fr,P}$ algorithm represents the "state of the art" in computing distances for categorical data, the following experiments attempt to compare *REP* to the $D_{fr,P}$ algorithm. To do so, we define a distance function between attributes. Additionally, we will perform nearest neighbour classification and require a distance functions between tuples. These two distance functions are defined as:

$$d_{REP}(x, y) = 1/S_{REP}(x, y) \qquad (6)$$

$$d_{REP}(< x_1, \cdots, x_k >, y) = || < d_{REP}(x_1, y), \cdots, d_{REP}(x_k, y) > || \qquad (7)$$

6.1   Clustering

The purpose of the clustering experiments is to visualize the distance function over different attributes and different data sets. We will find that *REP* provides similarity functions that match our expectations better than the distance functions computed by $D_{fr,P}$. Since $D_{fr,P}$ has been previously evaluated using single link hierarchical clustering, we will do the same here [3]. The three data sets we evaluated are:

Adult. A selection of fields from the 1994 census data collected in the United States [1]. There are 32,561 training examples and 16,281 test examples with 6 numeric fields and 8 categorical fields. In the experiments that follow, we treat the numeric fields as categorical fields (each value is a category). Similar results were obtained by simply ignoring the numeric fields.

Autos. (Imports-85) a data base of automobile specifications and insurance risk measures [1]. There are 205 instances with 16 numeric fields and 10 categorical fields. Most of the numeric fields are actually drawn from a small ranges and it is appropriate to treat them as categorical fields.

Reuters. We extracted the subject keywords from the standard *Reuters-21578* text collection and used each keyword as a binary attribute. There are 19,716 instances with 445 boolean fields. This is the same data used in [3].

Figure 4 shows four different clusterings. The left column is always *REP* and the right column is always $D_{fr,P}$. Overall, *REP* appears to match our understand of the data better than $D_{fr,P}$. In particular, in parts a) and b), we see that *REP* creates clusters for the Latin American countries which are not well represented by $D_{fr,P}$. In parts c) and d) where we have clustered by maximum level of education attained, our *REP* results are basically perfect (represents the real hierarchy of education levels) while $D_{fr,P}$ is reasonably good but failed to identify the post-high school degree vs. high school degree split seen in part c). In parts e) and f) where we have clustered by a person's occupation type, we see that *REP* creates three clusters: manual labour, lower level occupations and senior occupations. Conversely, $D_{fr,P}$ has left *Private house servant* as an outlier, combined *Clerical* with *Other service* and combined *Sales* and *Technical support*. The final pair of clusterings in parts g) and h) show the makes of cars in the Auto data set. The comparison here is more subtle, but the *REP* clustering has a more natural looking structure and three very distinct clusters for the luxury cars, the family cars and the imports. $D_{fr,P}$ on the other hand has combined *Mercury* with the *Alfa Romeo* and the *Porsche* which is somewhat surprising.

Overall, we see that even with very different data sets and different choices of the attribute on which to cluster, the *REP* distance function appears both more natural and more "correct" than the $D_{fr,P}$ distance function.

6.2   Classification

REP appears to be producing excellent distance measures and we now attempt to quantitatively evaluate its performance. Using the vector definition (Eq. 7) we can perform nearest neighbour classification. For comparison, we also run C4.5 [10], an excellent benchmark of quality and a NN algorithm using the hamming distance between instances. We do not report results for $D_{fr,P}$ as it is only defined for single attributes, not for tuples. We ran this classification task for 5 data sets available from the UCI collection [1] and the results are summarized in Table 1 which reports the percent errors.

REP offers performance similar to C4.5. It has lower error for two data sets, nearly identical error for two data sets and higher error for only one data set. Using nearest neighbour classification with hamming distance is actually surprisingly good on some of the simpler tasks but is quite poor for the adult data set. Thus we see that NN classification with *REP* is on par with C4.5 and better than NN with the hamming distance.

6.3   Sensitivity to *sink_p*

REP has one parameter, *sink_p*, which is the transition probability from each node to the terminal sink node. Larger values of *sink_p*, de-weight longer paths.

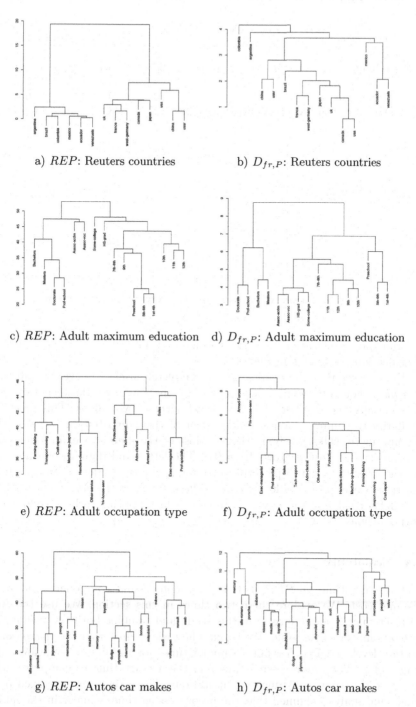

a) *REP*: Reuters countries

b) $D_{fr,P}$: Reuters countries

c) *REP*: Adult maximum education

d) $D_{fr,P}$: Adult maximum education

e) *REP*: Adult occupation type

f) $D_{fr,P}$: Adult occupation type

g) *REP*: Autos car makes

h) $D_{fr,P}$: Autos car makes

Fig. 4. Comparison of *REP* and $D_{fr,P}$ by using clustered output

Table 1. Error rates: *REP* is similar to C4.5 and better than NN (hamming)

Data set name	*REP*	C4.5	NN-hamming
Adult	15.1 %	13.6 %	21.5 %
Audiology	23.1 %	15.4 %	30.8 %
Letter recognition	28.2 %	32.4 %	30.5 %
Mushrooms	0.8 %	0.2 %	0 %
Optical recognition of handwritten digits	12.2 %	43.2 %	14.4 %

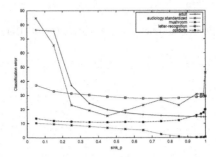

Fig. 5. Classification results are not very sensitive to *sink_p*

We ran the classification task for a range of values of *sink_p* and report the classification error rates in Fig. 5.

Here we see that the results can be completely useless for very small values of *sink_p*. For very small values of *sink_p*, the similarity between two values is essentially the number of walks between these two values. This gives high similarity to common values, independent of their distribution (because there are just more walks involving them). Most of the data sets exhibit increasing performance as *sink_p* grows from 0.5. The letter and digit recognition tasks are the exceptions which perform worse for very large values of *sink_p*. Overall the results are quite stable for many different values of *sink_p*. We used 0.85 as a default value of the *sink_p* parameter which appears to be a good general purpose choice.

6.4   Scalability

The running time of *REP* includes a preprocessing component and a per-distance cost. We explore these two times in this section. We use the Adult data set since it is the largest in this study and combine both the training and test sets into a single table. We repeat the task of clustering the *maximum education* level and Fig. 6 reports running times for the first x rows for various values of x. The preprocessing time and the average time to compute each of the 16×16 required distances computations. These times are averaged over 3 runs. Our analysis claimed that the preprocessing time is linear in the number of edges in the graph and in terms of the number of input rows. We see this behaviour in our experiment and further see that it is the cost of the edges that are dominating in our experiment. The time becomes almost constant at the point where all pairs of attributes that will appear in the same tuple have appeared in the same tuple. The average time to compute a distance array element varies from .03 seconds to .13 and scales excellently with the input size!

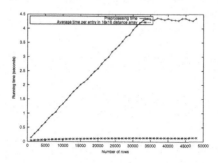

Fig. 6. Running time scales at most linearly with data set size

7 Conclusion

In this paper we presented a node similarity function for graphs. We used this node similarity function to provide an external similarity function called *REP*. Experimentally we found *REP* superior to the best existing external distance function. *REP* has a theoretical foundation, something that was missing from the prior research. Moreover, *REP* can handle cross-attribute similarity computations such as comparing a *Gender* and a *Car Type* in an insurance data base. *REP* also provided a natural definition for tuple similarity that let us build a very effective nearest neighbour classifier. Finally, our implementation was evaluated and we found that it provides excellent scalability with input size, and is not sensitive to its parameter.

References

1. C.L. Blake and C.J. Merz. UCI repository of machine learning databases, 1998.
2. A. K. Chandra, P. Raghavan, W. L. Ruzzo, and R. Smolensky. The electrical resistance of a graph captures its commute and cover times. In *Proceedings of the Twenty First Annual ACM Symposium on Theory of Computing*, 1989.
3. G. Das, H. Mannila, and P. Ronkainen. Similarity of attributes by external probes. In *Knowledge Discovery and Data Mining*, pages 23–29, 1998.
4. P. G. Doyle and J. L. Snell. Random Walks and Electric Networks.
5. V. Ganti, J. Gehrke, and R. Ramakrishnan. CACTUS - clustering categorical data using summaries. In *Knowledge Discovery and Data Mining*, pages 73–83, 1999.
6. D. Gibson, J. M. Kleinberg, and P. Raghavan. Clustering categorical data: An approach based on dynamical systems. *VLDB Journal*, 8(3–4):222–236, 2000.

7. S. Guha, R. Rastogi, and K. Shim. ROCK – a robust clusering algorith for categorical attributes. In *Proceedings of IEEE International Conference on Data Engineering*, 1999.

8. G. Jeh and J. Widom. Simrank: A measure of structural-context similarity. In *Eigth ACM SIGKDD Internation Conference on Knowledge Discovery and Data Mining*, 2002.

9. D. J. Klein and M. Randic. Resistance distance. *J. of Math. Chemistry*, 1993.

10. R. Quinlan. C4.5 decision tree generator.

Weighted Proportional k-Interval Discretization for Naive-Bayes Classifiers

Ying Yang and Geoffrey I. Webb

School of Computer Science and Software Engineering
Monash University
Melbourne, VIC 3800, Australia
{yyang, geoff.webb}@csse.monash.edu.au

Abstract. The use of different discretization techniques can be expected to affect the classification bias and variance of naive-Bayes classifiers. We call such an effect *discretization bias* and *variance*. Proportional k-interval discretization (PKID) tunes discretization bias and variance by adjusting discretized interval size and number proportional to the number of training instances. Theoretical analysis suggests that this is desirable for naive-Bayes classifiers. However PKID is sub-optimal when learning from training data of small size. We argue that this is because PKID equally weighs bias reduction and variance reduction. But for small data, variance reduction can contribute more to lower learning error and thus should be given greater weight than bias reduction. Accordingly we propose weighted proportional k-interval discretization (WPKID), which establishes a more suitable bias and variance trade-off for small data while allowing additional training data to be used to reduce both bias and variance. Our experiments demonstrate that for naive-Bayes classifiers, WPKID improves upon PKID for smaller datasets[1] with significant frequency; and WPKID delivers lower classification error significantly more often than not in comparison to three other leading alternative discretization techniques studied.

1 Introduction

Numeric attributes are often discretized for naive-Bayes classifiers [5,9]. The use of different discretization techniques can be expected to affect the naive-Bayes classification bias and variance. We call such an effect *discretization bias* and *variance*. A number of previous authors have linked the number of discretized intervals to classification error. Pazzani [17] and Mora et al. [16] have mentioned that the interval number has a major effect on the naive-Bayes classification error. If it is too small, important distinctions are missed; if it is too big, the data are over-discretized and the probability estimation may become unreliable. The best interval number depends upon the size of the training data. Torgo and

---

[1] 'Small' is a relative rather than an absolute term. Of necessity, we here utilize an arbitrary definition, deeming datasets with size no larger than 1000 as 'smaller' datasets, otherwise as 'larger' datasets.

K.-Y. Whang, J. Jeon, K. Shim, J. Srivatava (Eds.): PAKDD 2003, LNAI 2637, pp. 501–512, 2003.

Gama [21] have noticed that an interesting effect of increasing the interval number is that after some threshold the learning algorithm's performance decreases. They suggested that it might be caused by the decrease of the number of training instances per class per interval leading to unreliable estimates due to overfitting the data. Gama et al. [8] have suggested that discretization with fewer number of intervals tends to have greater utility. By minimizing the number of intervals, the dependence of the set of intervals on the training data will be reduced. This fact will have positive influence on the variance of the generated classifiers. In contrast, if there are a large number of intervals, high variance tends to be produced since small variation on the training data will be propagated on to the set of intervals. Hussain et al. [10] have proposed that there is a trade-off between the interval number and its effect on the accuracy of classification tasks. A lower number can improve understanding of an attribute but lower learning accuracy. A higher number can degrade understanding but increase learning accuracy. Hsu et al. [9] have observed that as the interval number increases, the classification accuracy will improve and reach a plateau. When the interval number becomes very large, the accuracy will drop gradually. How fast the accuracy drops will depend on the size of the training data. The smaller the training data size, the earlier and faster the accuracy drops.

Our previous research [24] has proposed proportional k-interval discretization (PKID). To the best of our knowledge, PKID is the first discretization technique that adjusts discretization bias and variance by tuning interval size and interval number. We have argued on theoretical grounds that PKID suits naive-Bayes classifiers. Our experiments have demonstrated that when learning from large data, naive-Bayes classifiers trained by PKID can achieve lower classification error than those trained by alternative discretization methods. This is particularly desirable since large datasets with high dimensional attribute spaces and huge numbers of instances are increasingly used in real-world applications; and naive-Bayes classifiers are widely deployed for these applications because of their time and space efficiency. However, we have detected a serious limitation of PKID which reduces learning accuracy on small data. In this paper, we analyze the reasons of PKID's disadvantage. This analysis leads to *weighted proportional k-interval discretization* (WPKID), a more elegant approach to managing discretization bias and variance. We expect WPKID to achieve better performance than PKID on small data while retaining competitive performance on large data. Improving naive-Bayes classifiers' performance on small data is of particular importance as they have consistently demonstrated strong classification accuracy for small data. Thus, improving performance in this context is improving the performance of one of the methods-of-choice for this context.

The rest of this paper is organized as follows. Section 2 introduces naive-Bayes classifiers. Section 3 discusses discretization, bias and variance. Section 4 reviews PKID and three other leading discretization methods for naive-Bayes classifiers. Section 5 proposes WPKID. Section 6 empirically evaluates WPKID against previous methods. Section 7 provides a conclusion.

2   Naive-Bayes Classifiers

In classification learning, each instance is described by a vector of attribute values and its class can take any value from some predefined set of values. Training data, a set of instances with their classes, are provided. A test instance is presented. The learner is asked to predict the class of the test instance according to the evidence provided by the training data. We define:

- C as a random variable denoting the class of an instance,
- $X < X_1, X_2, \cdots, X_k >$ as a vector of random variables denoting observed attribute values (an instance),
- c as a particular class,
- $x < x_1, x_2, \cdots, x_k >$ as a particular observed attribute value vector (a particular instance),
- $X = x$ as shorthand for $X_1 = x_1 \land X_2 = x_2 \land \cdots \land X_k = x_k$.

Expected classification error can be minimized by choosing $argmax_c(p(C = c \,|\, X = x))$ for each x. Bayes' theorem can be used to calculate the probability:

$$p(C = c \,|\, X = x) = p(C = c)\, p(X = x \,|\, C = c)\, /\, p(X = x). \qquad (1)$$

Since the denominator in (1) is invariant across classes, it does not affect the final choice and can be dropped, thus:

$$p(C = c \,|\, X = x) \propto p(C = c)\, p(X = x \,|\, C = c). \qquad (2)$$

Probabilities $p(C = c)$ and $p(X = x \,|\, C = c)$ need to be estimated from the training data. Unfortunately, since x is usually an unseen instance which does not appear in the training data, it may not be possible to directly estimate $p(X = x \,|\, C = c)$. So a simplification is made: if attributes $X_1, X_2, \cdots, X_k$ are conditionally independent of each other given the class, then

$$p(X = x \,|\, C = c) = p(\land X_i = x_i \,|\, C = c) = \prod p(X_i = x_i \,|\, C = c). \qquad (3)$$

Combining (2) and (3), one can further estimate the probability by:

$$p(C = c \,|\, X = x) \propto p(C = c) \prod p(X_i = x_i \,|\, C = c). \qquad (4)$$

Classifiers using (4) are called naive-Bayes classifiers.

Naive-Bayes classifiers are simple, efficient, effective and robust to noisy data. One limitation, however, is that naive-Bayes classifiers utilize the attributes independence assumption embodied in (3) which is often violated in the real world. Domingos and Pazzani [4] suggest that this limitation is ameliorated by the fact that classification estimation under zero-one loss is only a function of the sign of the probability estimation. In consequence, the classification accuracy can remain high even while the probability estimation is poor.

3   Discretization, Bias, and Variance

We here describe how discretization works in naive-Bayes learning, and introduce discretization bias and variance.

3.1   Discretization

An attribute is either categorical or numeric. Values of a categorical attribute are discrete. Values of a numeric attribute are either discrete or continuous [11].

A categorical attribute often takes a small number of values. So does the class label. Accordingly $p(C = c)$ and $p(X_i = x_i | C = c)$ can be estimated with reasonable accuracy from corresponding frequencies in the training data. Typically, the Laplace-estimate [3] is used to estimate $p(C = c)$: $\frac{n_c+k}{N+n\times k}$, where n_c is the number of instances satisfying $C = c$, N is the number of training instances, n is the number of classes and $k = 1$; and the M-estimate [3] is used to estimate $p(X_i = x_i | C = c)$: $\frac{n_{ci}+m\times p}{n_c+m}$, where n_{ci} is the number of instances satisfying $X_i = x_i \wedge C = c$, n_c is the number of instances satisfying $C = c$, p is the prior probability $p(X_i = x_i)$ (estimated by the Laplace-estimate) and $m = 2$.

A numeric attribute usually has a large or even an infinite number of values, thus for any particular value x_i, $p(X_i = x_i | C = c)$ might be arbitrarily close to 0. Suppose S_i is the value space of X_i within the class c, the probability distribution of $X_i | C = c$ is completely determined by a probability density function f which satisfies [19]:

1. $f(X_i = x_i | C = c) \geq 0, \forall x_i \in S_i$;
2. $\int_{S_i} f(X_i | C = c)\mathrm{d}X_i = 1$;
3. $\int_{a_i}^{b_i} f(X_i | C = c)\mathrm{d}X_i = p(a_i < X_i \leq b_i | C = c), \forall (a_i, b_i] \in S_i$.

Specifying f gives a description of the distribution of $X_i | C = c$, and allows associated probabilities to be found [20]. Unfortunately, f is usually unknown for real-world data. Thus it is often advisable to aggregate a range of values into a single value for the purpose of estimating probabilities [5,9]. Under discretization, a categorical attribute X_i^* is formed for X_i. Each value x_i^* of X_i^* corresponds to an interval $(a_i, b_i]$ of X_i. X_i^* instead of X_i is employed for training classifiers. Since $p(X_i^* = x_i^* | C = c)$ is estimated as for categorical attributes, there is no need to assume the format of f. But the difference between X_i and X_i^* may cause information loss.

3.2   Bias and Variance

Error of a machine learning algorithm can be partitioned into a *bias* term, a *variance* term and an *irreducible* term [7,12,13,22]. Bias describes the component of error that results from systematic error of the learning algorithm. Variance describes the component of error that results from random variation in the training data and from random behavior in the learning algorithm, and thus measures how sensitive an algorithm is to the changes in the training data. As the algorithm becomes more sensitive, the variance increases. Irreducible error describes the error of an optimal algorithm (the level of noise in the data). Consider a classification learning algorithm A applied to a set S of training instances to produce a classifier to classify an instance x. Suppose we could draw a sequence of training sets $S_1, S_2, ..., S_l$, each of size m, and apply A to construct classifiers, the average error of A at x can be defined as: $Error(A, m, x) = Bias(A, m, x) + Variance(A, m, x) + Irreducible(A, m, x)$. There is often a 'bias and variance trade-off' [12]. As one modifies some aspect of the learning algorithm, it will have opposite effects on bias and variance. For example, usually as one increases the number of degrees of freedom in the algorithm, the bias decreases but the variance increases. The optimal number of degrees of freedom (as far as the expected loss is concerned) is the number that optimizes this trade-off between bias and variance.

When discretization is employed to process numeric attributes in naive-Bayes learning, the use of different discretization techniques can be expected to affect the classification bias and variance. We call such an effect *discretization bias* and *variance*. Discretization bias and variance relate to *interval size* (the number of training instances in each interval) and *interval number* (the number of intervals formed). The larger the interval $(a_i, b_i]$ formed for a particular numeric value x_i, the more training instances in it, the lower the discretization variance, and thus the lower the probability estimation variance by substituting $(a_i, b_i]$ for x_i. However, the larger the interval, the less distinguishing information is obtained about x_i, the higher the discretization bias, and hence the higher the probability estimation bias. Low learning error can be achieved by tuning the interval size and interval number to find a good trade-off between the discretization bias and variance.

4   Rival Discretization Methods

We here review four discretization methods, each of which is either designed especially for, or is in practice often used by naive-Bayes classifiers. We believe that it is illuminating to analyze them in terms of discretization bias and variance.

4.1   Fixed k-Interval Discretization (FKID)

FKID [5] divides sorted values of a numeric attribute into k intervals, where (given n observed instances) each interval contains n/k instances. Since k is determined without reference to the properties of the training data, this method potentially suffers much attribute information loss. But although it may be deemed simplistic, FKID works surprisingly well for naive-Bayes classifiers. One reason suggested is that discretization approaches usually assume that discretized attributes have Dirichlet priors, and 'Perfect Aggregation' of Dirichlets can ensure that naive-Bayes with discretization appropriately approximates the distribution of a numeric attribute [9].

4.2   Fuzzy Discretization (FD)

FD [14,15][2] initially discretizes the value range of X_i into k equal-width intervals $(a_i, b_i]$ $(1 \leq i \leq k)$, and then estimates $p(a_i < X_i \leq b_i \,|\, C = c)$ from all training instances rather than from instances that have values of X_i in $(a_i, b_i]$. The influence of a training instance with value v of X_i on $(a_i, b_i]$ is assumed to be normally distributed with the mean value equal to v and is proportional to $P(v, \sigma, i) = \int_{a_i}^{b_i} \frac{1}{\sigma\sqrt{2\pi}} e^{-\frac{1}{2}(\frac{x-v}{\sigma})^2} \mathrm{d}x$. σ is a parameter to the algorithm and is used to control the 'fuzziness' of the interval bounds. σ is set equal to $0.7 \times \frac{max-min}{k}$ where max and min are the maximum value and minimum value of X_i respectively[3]. Suppose there are N training instances and there are N_c training instances with known value for X_i and with class c, each with influence $P(v_j, \sigma, i)$ on $(a_i, b_i]$ $(j = 1, \cdots, N_c)$:
$p(a_i < X_i \leq b_i \,|\, C = c) = \frac{p(a_i < X_i \leq b_i \wedge C = c)}{p(C = c)} \approx \frac{\sum_{j=1}^{N_c} P(v_j, \sigma, i)}{N \times p(C = c)}$. The idea behind FD is that small variation of the value of a numeric attribute should have small effects on the attribute's probabilities, whereas under non-fuzzy discretization, a slight difference between two values, one above and on below the cut point can have drastic effects on the estimated probabilities. But when the training instances' influence on each interval does not follow the normal distribution, FD's performance can degrade.

Both FKID and FD fix the number of intervals to be produced (decided by the parameter k). When the training data are very small, intervals will have small size and tend to incur high variance. When the training data are very large, intervals will have large size and tend to incur high bias. Thus they control well neither discretization bias nor discretization variance.

4.3   Fayyad & Irani's Entropy Minimization Discretization (FID)

FID [6] evaluates as a candidate cut point the midpoint between each successive pair of the sorted values for a numeric attribute. For each evaluation of a candidate cut point, the data are discretized into two intervals and the resulting class information entropy is calculated. A binary discretization is determined by selecting the cut point for which the entropy is minimal amongst all candidate cut points. The binary discretization is applied recursively, always selecting the best cut point. A minimum description length criterion (MDL) is applied to decide when to stop discretization. FID was developed in the particular context of top-down induction of decision trees. It uses MDL as the termination condition. This has an effect to tend to minimize the number of resulting intervals, which

---

[2] There are three versions of fuzzy discretization proposed by Kononenko for naive-Bayes classifiers. They differ in how the estimation of $p(a_i < X_i \leq b_i \,|\, C = c)$ is obtained. Because of space limits, we present here only the version that, according to our experiments, best reduces the classification error.

[3] This setting of σ is chosen because it achieved the best performance in Kononenko's experiments [14].

is desirable for avoiding the fragmentation problem in the decision tree learning [18]. As a result, FID always focuses on reducing discretization variance, but does not control bias. This might work well for training data of small size, for which it is credible that variance reduction can contribute more to lower naive-Bayes learning error than bias reduction [7]. However, when training data size is large, it is very possible that the loss through bias increase will soon overshadow the gain through variance reduction, resulting in inferior learning performance.

4.4   Proportional k-Interval Discretization (PKID)

PKID [24] adjusts discretization bias and variance by tuning the interval size and number, and further adjusts the naive-Bayes' probability estimation bias and variance to achieve lower classification error. As described in Section 3.2, increasing interval size (decreasing interval number) will decrease variance but increase bias. Conversely, decreasing interval size (increasing interval number) will decrease bias but increase variance. PKID aims to resolve this conflict by setting the interval size and number proportional to the number of training instances. With the number of training instances increasing, both discretization bias and variance tend to decrease. Bias can decrease because the interval number increases. Variance can decrease because the interval size increases. This means that PKID has greater capacity to take advantage of the additional information inherent in large volumes of training data than previous methods. Given a numeric attribute, supposing there are N training instances with known values for this attribute, the desired interval size is s and the desired interval number is t, PKID employs (5) to calculate s and t:

$$s \times t = N$$
$$s = t. \tag{5}$$

PKID discretizes the ascendingly sorted values into intervals with size s. Experiments have shown that although it significantly reduced classification error in comparison to previous methods on larger datasets, PKID was sub-optimal on smaller datasets. We here suggest the reason. Naive-Bayes learning is probabilistic learning. It estimates probabilities from the training data. According to (5), PKID gives equal weight to discretization bias reduction and variance reduction by setting the interval size equal to the interval number. When N is small, PKID tends to produce a number of intervals with small size. In particular, small interval size should result in high variance. Thus fewer intervals each containing more instances would be of greater utility.

5   Weighted Proportional k-Interval Discretization

The above analysis leads to *weighted proportional k-interval discretization* (WP-KID). This new discretization techniques sets a minimum interval size **m**. As the training data increase, both the interval size *above* the minimum and the interval number increase. Given the same definitions of N, s and t as in (5), we calculate s and t by:

$$s \times t = N$$
$$s - \mathbf{m} = t$$
$$\mathbf{m} = 30. \tag{6}$$

We set $\mathbf{m}$ as 30 since it is commonly held to be the minimum sample from which one should draw statistical inferences [23]. WPKID should establish a more suitable bias and variance trade-off for training data of small size. By introducing $\mathbf{m} = 30$, we ensure that in general each interval has enough instances for reliable probability estimation for naive-Bayes classifiers. Thus WPKID can be expected to improve upon PKID by preventing intervals of high variance. For example, with 100 training instances, WPKID will produce 3 intervals containing approximately 33 instances each, while PKID will produce 10 intervals containing only 10 instances each. At the same time, WPKID still allows additional training data to be used to reduce both bias and variance as PKID does.

6   Experiments

We want to evaluate whether WPKID can better reduce classification errors of naive-Bayes classifiers compared with FKID, FD, FID and PKID.

6.1   Experimental Design

We run experiments on 35 natural datasets from UCI machine learning repository [2] and KDD archive [1]. This experimental suit comprises two parts. One is all the 29 datasets used by PKID [24]. The other is an addition of 6 datasets[4] with size smaller than 1000, since WPKID is expected to improve upon PKID for small data. Table 1 describes each dataset, including the number of instances (Size), numeric attributes (Num.), categorical attributes (Cat.) and classes (Class). Datasets are ascendingly ordered by their sizes and broken down to smaller and larger ones. 'Small' is a relative rather than an absolute term. Of necessity, we here utilize an arbitrary definition, deeming datasets with size no larger than 1000 as 'smaller' datasets, otherwise as 'larger' datasets. For each dataset, we implement naive-Bayes learning by conducting a 10-trial, 3-fold cross validation. For each fold, the training data is separately discretized by FKID ($k = 10$), FD ($k = 10$), FID, PKID and WPKID. The intervals so formed are separately applied to the test data. The experimental results are recorded in Table 1 as: **classification error** is the percentage of incorrect predictions of naive-Bayes classifiers in the test averaged across all folds in the cross validation; and **classification bias and variance** are defined and calculated using the method of Webb [22].

---

[4] They are Pittsburgh-Bridges-Material, Flag-Landmass, Haberman, Ecoli, Dermatology and Vowel-Context.

Table 1. Experimental datasets; and classification error, bias and variance (%)

Dataset	Size	Num.	Cat.	Class	Error					Bias		Variance	
					WPKID	PKID	FID	FD	FKID	WPKID	PKID	WPKID	PKID
Labor-Negotiations	57	8	8	2	8.6	7.2	9.5	12.8	8.9	4.6	5.3	4.0	1.9
Echocardiogram	74	5	1	2	25.7	25.3	23.8	27.7	29.2	19.7	20.8	5.9	4.5
Pittsburgh-Bridges-Material	106	3	8	3	11.9	13.0	12.6	10.5	12.1	9.9	10.2	2.0	2.8
Iris	150	4	0	3	6.9	7.5	6.8	5.3	7.5	4.7	12.7	2.1	1.9
Hepatitis	155	6	13	2	15.9	14.6	14.5	13.4	14.7	14.7	1.1	1.2	1.9
Wine-Recognition	178	13	0	3	2.0	2.2	2.6	3.3	2.1	1.3	2.1	0.7	1.1
Flag-Landmass	194	10	18	6	29.0	30.7	29.9	32.0	30.5	21.5	21.5	7.5	9.2
Sonar	208	60	0	2	23.7	25.7	26.3	26.8	25.2	19.3	19.6	4.4	6.2
Glass-Identification	214	9	0	3	38.4	33.6	34.9	40.7	34.1	29.7	21.3	8.7	12.3
Heart-Disease(Cleveland)	270	7	6	2	16.7	17.5	17.5	16.3	17.1	14.9	15.6	1.8	2.0
Haberman	306	3	0	2	25.8	27.7	26.5	25.1	27.1	22.3	23.3	3.5	4.4
Ecoli	336	5	0	8	17.6	19.0	17.9	16.0	19.0	12.3	11.5	5.2	7.4
Liver-Disorders	345	6	0	2	35.5	38.0	37.4	37.9	37.1	26.3	28.8	9.2	9.2
Ionosphere	351	34	0	2	10.3	10.6	11.1	8.5	10.2	8.5	9.8	1.8	0.8
Dermatology	366	1	33	6	1.9	2.2	2.0	1.9	2.2	1.5	1.5	0.4	0.7
Horse-Colic	368	7	14	2	20.7	20.9	20.7	20.7	20.9	18.9	19.0	1.7	1.8
Credit-Screening(Australia)	690	6	9	2	14.3	14.2	14.5	15.2	14.5	12.8	12.1	1.5	2.1
Breast-Cancer(Wisconsin)	699	9	0	2	2.7	2.7	2.7	2.8	2.6	2.6	2.6	0.1	0.1
Pima-Indians-Diabetes	768	8	0	2	25.5	26.3	26.0	24.8	25.9	22.0	21.8	3.5	4.5
Vehicle	846	18	0	4	38.2	38.2	38.9	42.4	40.5	31.5	31.4	6.7	6.8
Annealing	898	6	32	6	2.2	2.2	1.9	3.9	2.3	1.9	1.6	0.2	0.5
Vowel-Context	990	10	1	11	39.2	43.0	41.4	38.0	38.4	20.1	19.2	19.1	23.9
German	1000	7	13	2	25.4	25.5	25.1	25.2	25.4	22.0	21.7	3.4	3.7
ME	-	-	-	-	19.0	19.5	19.3	19.6	19.5	14.9	14.7	4.1	4.8
GM	-	-	-	-	1.00	1.02	1.02	1.05	1.03	1.00	0.98	1.00	1.16
Multiple-Features	2000	3	3	10	31.4	31.5	32.6	30.8	31.9	27.6	27.3	3.8	4.2
Hypothyroid	3163	7	18	2	2.1	1.8	1.7	2.6	2.8	1.8	1.6	0.3	0.3
Satimage	6435	36	0	6	17.7	17.8	18.1	20.1	18.9	16.9	17.0	0.8	0.7
Musk	6598	166	0	2	8.5	8.3	9.4	21.2	19.2	7.9	7.6	0.6	0.7
Pioneer-MobileRobot	9150	29	7	57	1.8	1.7	14.8	18.2	10.8	0.9	0.8	0.9	0.9
Handwritten-Digits	10992	16	0	10	12.2	12.0	13.5	13.2	13.2	10.7	10.7	1.5	1.4
Australian-Sign-Language	12546	8	0	3	36.0	35.8	36.5	38.7	38.2	34.2	34.1	1.9	1.8
Letter-Recognition	20000	16	0	26	25.7	25.8	30.4	34.7	30.7	22.4	22.5	3.2	3.2
Adult	48842	6	8	2	17.0	17.1	17.2	18.5	19.2	16.4	16.6	0.6	0.5
Ipums-la-99	88443	20	40	13	19.9	19.9	20.1	32.0	20.5	15.3	15.3	4.6	4.6
Census-Income	299285	8	33	2	23.3	23.3	23.6	24.7	24.5	23.1	23.1	0.2	0.2
Forest-Covertype	581012	10	44	7	31.7	31.7	32.1	32.2	32.9	30.3	30.3	1.4	1.4
ME	-	-	-	-	18.9	18.9	20.8	23.9	21.9	17.3	17.2	1.7	1.7
GM	-	-	-	-	1.00	0.98	1.22	1.47	1.34	1.00	0.98	1.00	0.98

6.2 Experimental Statistics

Three statistics are employed to evaluate the experimental results.

Mean error is the arithmetic mean of errors across all datasets, listed in 'ME' rows of Table 1. It provides a gross indication of relative performance. It is debatable whether errors in different datasets are commensurable, and hence whether averaging errors across datasets is very meaningful. Nonetheless, a low average error is indicative of a tendency toward low errors for individual datasets.

Geometric mean error ratio has been explained by Webb [22]. It allows for the relative difficulty of error reduction in different datasets and can be more reliable than the mean ratio of errors across datasets. The 'GE' rows of Table 1 lists the results of alternative methods against WPKID.

Win/lose/tie record comprises three values that are respectively the number of datasets for which WPKID obtains lower, higher or equal classification error, compared with alternative algorithms. Table 2 shows the results of WP-KID compared with alternatives on smaller datasets, larger datasets and all datasets respectively. A one-tailed sign test can be applied to each record. If the test result is significantly low (here we use the 0.05 critical level), it is reasonable to conclude that the outcome is unlikely to be obtained by chance and hence the record of wins to losses represents a systematic underlying advantage to WPKID with respect to the type of datasets studied.

Table 2. Win/Lose/Tie Records of WPKID against Alternatives

Datasets	Smaller				Larger				All			
WPKID	Win	Lose	Tie	Sign Test	Win	Lose	Tie	Sign Test	Win	Lose	Tie	Sign Test
PKID	15	5	3	0.02	4	5	3	0.50	19	10	6	0.07
FID	15	6	2	0.04	11	1	0	< 0.01	26	7	2	< 0.01
FD	11	10	2	0.50	11	1	0	< 0.01	22	11	2	0.04
FKID	17	5	1	< 0.01	12	0	0	< 0.01	29	5	1	< 0.01

6.3   Experimental Results Analysis

WPKID is devised to overcome PKID's disadvantage for small data while retaining PKID's advantage for large data. It is expected that naive-Bayes classifiers trained on data preprocessed by WPKID are able to achieve lower classification error, compared with those trained on data preprocessed by FKID, FD, FID or PKID. The experimental results support this expectation.

1. For Smaller Datasets
   - WPKID achieves the lowest mean error among all the methods.
   - The geometric mean error ratios of the alternatives against WPKID are all larger than 1. This suggests that WPKID enjoys an advantage in terms of error reduction on smaller datasets.
   - With respect to the win/lose/tie records, WPKID achieves lower classification error than FKID, FID and PKID with frequency significant at 0.05 level. WPKID and FD have competitive performance.
2. For Larger Datasets
   - WPKID achieves mean error the same as PKID and lower than FKID, FD and FID.
   - The geometric mean error ratio of PKID against WPKID is close to 1, while those of other methods are all larger than 1. This suggests that WPKID retains PKID's desirable performance on larger datasets.
   - With respect to the win/lose/tie records, WPKID achieves lower classification error than FKID, FD and FID with frequency significant at 0.05 level.

- WPKID achieves higher classification error than PKID for only one dataset more than the reverse.
3. For All Datasets
   - The win/lose/tie records across all datasets favor WPKID over FKID, FD and FID with frequency significant at 0.05 level. WPKID also achieves lower error more often than not compared with PKID.
   - It seems possible to attribute WPKID's improvement upon PKID primarily to variance reduction. WPKID has lower variance than PKID for 19 datasets but higher variance for only 8. This win/lose record is significant at 0.05 level (sign test = 0.03). In contrast, WPKID has lower bias than PKID for 13 datasets while higher bias for 15.

7   Conclusion

We have previously argued that discretization for naive-Bayes classifiers can tune classification bias and variance by adjusting the interval size and number proportional to the number of training instances, an approach called proportional k-interval discretization (PKID). However, PKID allocates equal weight to bias reduction and variance reduction. We argue that this is inappropriate for learning from small data and propose weighted proportional k-interval discretization (WPKID), which more elegantly manages discretization bias and variance by assigning a minimum interval size and then adjusting the interval size and number as more training data allow. This strategy is expected to improve on PKID by preventing high discretization variance. Our experiments demonstrate that compared to previous discretization techniques FKID, FD and FID, WPKID reduces the naive-Bayes classification error with significant frequency. Our experiments also demonstrate that when learning from small data, WPKID significantly improves on PKID by achieving lower variance, as predicted.

References

1. BAY, S. D. The UCI KDD archive [http://kdd.ics.uci.edu], 1999. Department of Information and Computer Science, University of California, Irvine.
2. BLAKE, C. L., AND MERZ, C. J. UCI repository of machine learning databases [http://www.ics.uci.edu/~mlearn/mlrepository.html], 1998. Department of Information and Computer Science, University of California, Irvine.
3. CESTNIK, B. Estimating probabilities: A crucial task in machine learning. In *Proc. of the European Conf. on Artificial Intelligence* (1990), pp. 147–149.
4. DOMINGOS, P., AND PAZZANI, M. On the optimality of the simple Bayesian classifier under zero-one loss. *Machine Learning 29* (1997), 103–130.
5. DOUGHERTY, J., KOHAVI, R., AND SAHAMI, M. Supervised and unsupervised discretization of continuous features. In *Proc. of the Twelfth International Conf. on Machine Learning* (1995), pp. 194–202.
6. FAYYAD, U. M., AND IRANI, K. B. Multi-interval discretization of continuous-valued attributes for classification learning. In *Proc. of the Thirteenth International Joint Conf. on Artificial Intelligence* (1993), pp. 1022–1027.

7. FRIEDMAN, J. H. On bias, variance, 0/1-loss, and the curse-of-dimensionality. *Data Mining and Knowledge Discovery 1*, 1 (1997), 55–77.

8. GAMA, J., TORGO, L., AND SOARES, C. Dynamic discretization of continuous attributes. In *Proc. of the Sixth Ibero-American Conf. on AI* (1998), pp. 160–169.

9. HSU, C. N., HUANG, H. J., AND WONG, T. T. Why discretization works for naive Bayesian classifiers. In *Proc. of the Seventeenth International Conf. on Machine Learning* (2000), pp. 309–406.

10. HUSSAIN, F., LIU, H., TAN, C. L., AND DASH, M. Discretization: An enabling technique, 1999. Technical Report, TRC6/99, School of Computing, National University of Singapore.

11. JOHNSON, R., AND BHATTACHARYYA, G. *Statistics: Principles and Methods*. John Wiley & Sons Publisher, 1985.

12. KOHAVI, R., AND WOLPERT, D. Bias plus variance decomposition for zero-one loss functions. In *Proc. of the Thirteenth International Conf. on Machine Learning* (1996), pp. 275–283.

13. KONG, E. B., AND DIETTERICH, T. G. Error-correcting output coding corrects bias and variance. In *Proc. of the Twelfth International Conf. on Machine Learning* (1995), pp. 313–321.

14. KONONENKO, I. Naive Bayesian classifier and continuous attributes. *Informatica 16*, 1 (1992), 1–8.

15. KONONENKO, I. Inductive and Bayesian learning in medical diagnosis. *Applied Artificial Intelligence 7* (1993), 317–337.

16. MORA, L., FORTES, I., MORALES, R., AND TRIGUERO, F. Dynamic discretization of continuous values from time series. In *Proc. of the Eleventh European Conf. on Machine Learning* (2000), pp. 280–291.

17. PAZZANI, M. J. An iterative improvement approach for the discretization of numeric attributes in Bayesian classifiers. In *Proc. of the First International Conf. on Knowledge Discovery and Data Mining* (1995).

18. QUINLAN, J. R. *C4.5: Programs for Machine Learning*. Morgan Kaufmann Publishers, 1993.

19. SCHEAFFER, R. L., AND MCCLAVE, J. T. *Probability and Statistics for Engineers*, fourth ed. Duxbury Press, 1995.

20. SILVERMAN, B. *Density Estimation for Statistics and Data Analysis*. Chapman and Hall Ltd., 1986.

21. TORGO, L., AND GAMA, J. Search-based class discretization. In *Proc. of the Ninth European Conf. on Machine Learning* (1997), pp. 266–273.

22. WEBB, G. I. Multiboosting: A technique for combining boosting and wagging. *Machine Learning 40*, 2 (2000), 159–196.

23. WEISS, N. A. *Introductory Statistics*, vol. Sixth Edition. Greg Tobin, 2002.

24. YANG, Y., AND WEBB, G. I. Proportional k-interval discretization for naive-Bayes classifiers. In *Proc. of the Twelfth European Conf. on Machine Learning* (2001), pp. 564–575.

Dealing with Relative Similarity in Clustering: An Indiscernibility Based Approach

Shoji Hirano and Shusaku Tsumoto

Department of Medical Informatics, Shimane Medical University, School of Medicine
89-1 Enya-cho, Izumo, Shimane 693-8501, Japan
hirano@ieee.org, tsumoto@computer.org

Abstract. In this paper we propose a clustering method that works on relative proximity. The key process of this method is iterative refinement of N binary classifications, where N denotes the number of objects. First, for each of N objects, an equivalence relation that classifies all the other objects into two classes, similar and dissimilar, is assigned by refering to their relative proximity. Next, for each pair objects, we count the number of binary classifications in which the pair is included in the same class. We call this number as indiscernibility degree. If indiscernibility degree of the pair is larger than a user-defined threshold value, we modify the equivalence relations so that all of them commonly classify the pair into the same class. This process is repeated until clusters become stable. Consequently we get the clusters that follows granularity of the given threshold without using geometric measures.

1 Introduction

Relative proximity is a class of proximity measures that can be used to represent subjective similarity or dissimilarity such as human judgment about likeness of persons. Unlike distance-based proximity measures, relative proximity is not necessarily required to satisfy triangular inequality, e.g., for given objects A, B and C, $d(B, C)$ may be larger than $d(A, C) + d(A, B)$. However, this property makes most of the widely used clustering methods [1] hard to work because centroids the clusters are determined inappropriately.

This paper presents a new clustering method based on indiscernibility degree of objects. It gives good partition to objects even when their proximities are represented as relative ones. The key ideas of this method are the use of the indiscernibility degree and iterative refinement of equivalence relations, both of which are defined in the context of rough set theory [2]. First, each object independently groups up similar and dissimilar objects according to the relative dissimilarity from itself to other objects. An equivalence relation is assigned to each object to represent the classification rule. Second, global similarity of any two objects is derived as the indiscernibility degree. An indiscernibility degree corresponds to the ratio of objects that give the same classification to the two objects. When the degree is larger than a predefined threshold value, the two objects are considered to be indiscernible, and all the equivalence relations are

K.-Y. Whang, J. Jeon, K. Shim, J. Srivatava (Eds.): PAKDD 2003, LNAI 2637, pp. 513–518, 2003.

refined to classify them into the same cluster. The refinement process is iterated several times, because the new candidates to be merged could appear on the refined set of equivalence relations. As a result, globally similar objects are merged into the same cluster and coarse clusters are obtained. The usefulness of this method is demonstrated on the numerical data sets. The results show that good clusters are obtained without using distance-oriented parameters such as cluster centers or deviances.

2   Clustering Method

2.1   Assignment of Initial Equivalence Relations

When the dissimilarity is defined relatively, the only information available for object x_i is dissimilarity of x_i to other objects, for example $d(x_i, x_j)$. This is because those for other pairs of objects, namely $d(x_j, x_k)$, $x_j, x_k \neq x_i$, are determined independently of x_i. Therefore, we independently assign an initial equivalence relation to each object and evaluate relative dissimilarity observed from the corresponding object.

Let $U = \{x_1, x_2, ..., x_N\}$ be the set of objects we are interested in. An equivalence relation R_i for object x_i is defined by

$$U/R_i = \{P_i,\ U - P_i\},$$

where

$$P_i = \{x_j|\ d(x_i, x_j) \leq Th_{di}\},\quad \forall x_j \in U.$$

$d(x_i, x_j)$ denotes dissimilarity between objects x_i and x_j, and Th_{di} denotes an upper threshold value of dissimilarity for object x_i. Equivalence relation R_i classifies U into two categories: P_i containing objects similar to x_i and $U - P_i$ containing objects dissimilar to x_i. When $d(x_i, x_j)$ is smaller than Th_{di}, object x_j is considered to be indiscernible to x_i. U/R_i can be alternatively written as $U/R_i = \{\{[x_i]_{R_i}\}, \{\overline{[x_i]_{R_i}}\}\}$, where $[x_i]_{R_i} \cap \overline{[x_i]_{R_i}} = \phi$ and $[x_i]_{R_i} \cup \overline{[x_i]_{R_i}} = U$ hold.

Definition of dissimilarity measure $d(x_i, x_j)$ is arbitrary. If all the attribute values are numerical, ordered and independent of each other, conventional Minkowski distance

$$d(x_i, x_j) = \left(\sum_{a=1}^{N_a} |x_{ia} - x_{ja}|^p \right)^{\frac{1}{p}},$$

where N_a denotes the number of attributes, x_{ia} denotes the a-th attribute of object x_i, and p denotes a positive integer, is a reasonable measure. More generally, any type of dissimilarity measure can be used regardless of whether or not triangular inequality is satisfied among objects.

Threshold of dissimilarity Th_{di} for object x_i is automatically determined based on the denseness of the objects. The procedure is summarized as follows.

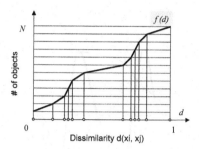

Fig. 1. An example of function $f(d)$ generated by $d(x_i, x_s)$.

1. Sort $d(x_i, x_j)$ in the increasing order. For simplicity we denote the sorted dissimilarity using the same representation $d(x_i, x_s)$, $1 \leq s \leq N$.
2. Generate a function $f(d)$. For a given dissimilarity d, function f returns the number of objects whose dissimilarity to x_i is smaller than d. Figure 1 shows an example. Function $f(d)$ can be generated by linearly interpolating $f(d(x_i, x_s)) = n$, where n corresponds to the index of x_s in the sorted dissimilarity list.
3. Calculate the second-order derivative of $f(d)$ by convoluting $f(d)$ and the second-order derivative of Gaussian function as follows.

$$f''(d) = \int_{-\infty}^{\infty} f(d) \frac{-(d-u)}{\sigma^3 \sqrt{2\pi}} e^{-(d-u)^2/2\sigma^2} du,$$

where $f(d) = 1$ and $f(d) = N$ are used for $d < 0$ and $d > 1$ respectively. The first-order derivative of $f(d)$ represents denseness of objects because it returns increase or decrease velocity of the objects induced by the change of dissimilarity. Therefore, by calculating its further derivative as $f''(d)$, we find the sparse region between two dense regions. Figure 2 illustrates relations between $f(d)$ and its derivatives. The most sparse point d^* should take a local minimum of the denseness where the followings conditions are satisfied.

$$f''(d^* - \Delta d) < 0 \text{ and } f''(d^* + \Delta d) > 0.$$

Usually, there are some d^*s in $f(d)$ because $f(d)$ has some local minima. The value of σ in the above Gaussian function can be adjusted to eliminate meaninglessly small minima.
4. Choose the smallest d^* and the object x_{j^*} whose dissimilarity is the closest to but not larger than d^*. Finally, dissimilarity threshold Th_{di} is obtained as $Th_{di} = d(x_i, x_{j^*})$.

2.2   Iterative Refinement of Initial Equivalence Relations

In the second stage, we perform refinement of the initial equivalence relations. Generally, objects should be classified into the same category when most of the

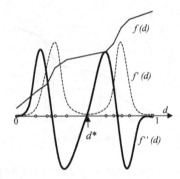

Fig. 2. Relations between $f(d)$, $f'(d)$ and $f''(d)$.

equivalence relations commonly regard them as indiscernible. However, these similar objects will be classified into different categories if there exists at least one equivalence relation that has ability to discern the objects. In such a case, unpreferable clustering result containing small and fine categories will be obtained. Let us give one example consisting 9 objects below.

$$U/R_1, U/R_2, U/R_3, U/R_4 = \{\{x_1, x_2, x_3, x_4, x_5\}, \{x_6, x_7, x_8, x_9\}\},$$
$$U/R_5 = \{\{x_4, x_5\}, \{x_1, x_2, x_3, x_6, x_7, x_8, x_9\}\},$$
$$U/R_6, U/R_7, U/R_8, U/R_9 = \{\{x_6, x_7, x_8, x_9\}, \{x_1, x_2, x_3, x_4, x_5\}\},$$
$$U/\mathbf{R} = \{\{x_1, x_2, x_3\}, \{x_4, x_5\}, \{x_7, x_8, x_9\}\}.$$

Here, an equivalence relation, R_5, makes slightly different classification compared to the others. R_5 classifies objects x_4 and x_5 into an independent category whereas all other relations regard x_5 as an indiscernible object to x_1, x_2, x_3 and x_4. Consequently, three small categories are obtained. The purpose of this stage is to refine such equivalence relations so that the resultant indiscernibility relation gives adequately coarse classification to the object.

First, we define an *indiscernibility degree*, $\gamma(x_i, x_j)$, of two objects x_i and x_j as follows.

$$\gamma(x_i, x_j) = \frac{1}{|U|} \sum_{k=1}^{|U|} \delta_k(x_i, x_j),$$

where

$$\delta_k(x_i, x_j) = \begin{cases} 1, \text{ if } x_i \in [x_k]_{R_k} \wedge x_j \in [x_k]_{R_k} \\ 0, \text{ otherwise.} \end{cases}$$

The higher $\gamma(x_i, x_j)$ represents that x_i and x_j are commonly regarded as indiscernible objects by large number of the equivalence relations. Therefore, if an equivalence relation discerns the objects that have high γ value, we consider that it gives excessively fine classification and refine it according to the procedure given below.

Let $R_i, R_j \in \mathbf{R}$ be initial equivalence relations and let $R_i', R_j' \in \mathbf{R}'$ be equivalence relations after refinement. For an initial equivalence relation R_i, a refined equivalence relation R_i' is defined as

$$U/R_i' = \{P_i', \ U - P_i'\}$$

where P_i' denotes a subset of objects represented by

$$P_i' = \{x_j | \gamma(x_i, x_j) \geq T_h\}, \quad \forall x_j \in U.$$

The value T_h denotes the lower threshold value of indiscernibility degree to regard x_i and x_j as the indiscernible objects.

Because refinement of one equivalence relation may change indiscernibility degree of other objects, we iterate the refinement process using the same T_h until the clusters become stable. Note that each refinement process is performed using the previously 'refined' set of equivalence relations.

3   Experimental Results

We applied the proposed method to some artificial numerical data sets and evaluated its clustering ability. Note that we used numerical data, but clustered them without using any types of geometric measures.

We first examined effect of refinement of initial equivalence relations. We created two-dimensional numerical data that follow normal distribution. The number of clusters was set to 5. Each of the clusters contained approximately 100 objects, and total 491 objects were included in the data. We evaluated validity of the clustering result based on the following validity measure: $v_{\mathbf{R}}(C) = \min(\frac{|X_{\mathbf{R}} \cap C|}{|X_{\mathbf{R}}|}, \frac{|X_{\mathbf{R}} \cap C|}{|C|})$ where $X_{\mathbf{R}}$ and C denote the cluster obtained by the proposed method and its expected cluster, respectively. The threshold value for refinement was set to $T_h = 0.2$, meaning that if two objects were commonly regarded as indiscernible by 20% of objects in the data, all the equivalence relations were modified to regard them as indiscernible objects.

Without refinement, the method produced 461 small clusters. Validity of the result was 0.011, which was the smallest of all. This was because small size of clusters produced very low coverage, namely, overlaps between the generated clusters and their corresponding expected clusters were very small compared with the size of the expected clusters.

By performing refinement one time, the number of clusters was reduced to 429, improving validity to 0.013. As the refinement proceeds, the small clusters became merged and validity of the results kept increasing. Finally, 10 clusters were formed after 6th refinement as shown in Figure 3. Validity of the clusters was 0.927. One can observe that a few tiny clusters, for example clusters 5 and 6, were formed between the large clusters. These objects were classified into independent clusters as the result of competition of the large clusters containing almost the same populations. Except for that, the results showed that the proposed method automatically produced good clusters that have high correspondence to the original ones.

Due to the constraint on page space, we omit the results on the further analysis and instead summarize observations below.

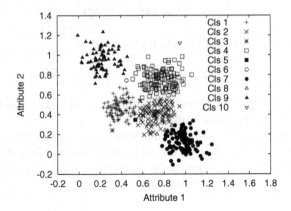

Fig. 3. Clusters after 6th refinement.

- For any values of T_h, the number of clusters decreased following the number of refinement and finally saturated after several times of refinement.
- Validity increased following the times of refinement, and finally saturated after several times of refinement.

Both characteristics enable us to stop refinement when the clusters become stable, regardless of the selection of T_h.

4   Conclusions

In this paper, we have presented an indiscernibility-based clustering method that can deal with relative proximity and empirically validated its usefulness by clustering a numerical dataset using relative proximity. It remains as future work to reduce computational complexity of the method and to apply the method to large and complex databases.

Acknowledgments. This work was supported in part by the Grant-in-Aid for Scientific Research on Priority Area (B)(No.759) "Implementation of Active Mining in the Era of Information Flood" by the Ministry of Education, Culture, Science and Technology of Japan.

References

1. P. Berkhin (2002): Survey of Clustering Data Mining Techniques. Accrue Software Research Paper. URL: http://www.accrue.com/products/researchpapers.html.
2. Z. Pawlak (1991): Rough Sets, Theoretical Aspects of Reasoning about Data. Kluwer Academic Publishers, Dordrecht.

Considering Correlation between Variables to Improve Spatiotemporal Forecasting*

Zhigang Li[1], Liangang Liu[2], and Margaret H. Dunham[1]

[1] Department of Computer Science and Engineering, Southern Methodist University
Dallas, TX 75275-0122, USA
{zgli,mhd}@engr.smu.edu
[2] Department of Statistical Science, Southern Methodist University
Dallas, TX 75275-0122, USA
{lliu}@mail.smu.edu

Abstract. The importance of forecasting cannot be overemphasized in modern environment surveillance applications, including flood control, rainfall analysis, pollution study, nuclear leakage prevention and so on. That is why we proposed STIFF (SpatioTemporal Integrated Forecasting Framework) in previous work [11], trying to answer such a challenging problem of doing forecasting in natural environment with both spatial and temporal characteristics involved. However, despite its promising performance on univariate-based data, STIFF is not sophisticated enough for more complicated environmental data derived from multiple correlated variables. Therefore in this paper we add multivariate analysis to the solution, take the correlation between different variables into account and further extend STIFF to address spatiotemporal forecasting involving multiple variables. Our experiments show that this introduction and integration of multivariate correlation not only has a more reasonable and rational interpretation of the data itself, but also produces a higher accuracy and slightly more balanced behavior.

1 Introduction

As one of data mining techniques, forecasting is widely used to predict the unknown future based upon the patterns hidden in the current and past data [6]. Spatiotemporal forecasting, termed after the simultaneous consideration of both spatial and temporal (or even more) factors in order to achieve a better forecasting result, has proven to be very helpful in environment surveillance [4,9, 3] and other fields, like fishery management [16] and market analysis [12]. Generally speaking, the application of spatiotemporal forecasting is quite difficult, since there are so many factors to be taken into account. For example, the flood forecasting at a certain location along a river, not only depends on the rainfall in the past days at this location, but also on the rainfall at its upper stream, plus the saturation of the soil, the forest coverage percentage, the cross-section

---

* Supported by the National Science Foundation under Grant No. IIS-9820841.

K.-Y. Whang, J. Jeon, K. Shim, J. Srivatava (Eds.): PAKDD 2003, LNAI 2637, pp. 519–531, 2003.
© Springer-Verlag Berlin Heidelberg 2003

area of the river, and possibly even more factors that are not thoroughly understood by people or simply hard to be measured and assessed precisely. As a first step to address this challenging problem, a univariate-based forecasting framework STIFF (SpatioTemporal Integrated Forecasting Framework) was proposed in our previous work [11]. Targeted to the flood forecasting problem, STIFF took a *divide-and-conquer* approach, by 1) first constructing an ARIMA time series model to obtain temporal forecasting at each location, 2) then creating a neural network model, in line with the spatial characteristic of all the locations in the catchment, to make the spatial forecasting, 3) and finally generating the overall spatiotemporal forecasting using a fusion of the individual predictions obtained from the time series model and the neural network model based upon statistical regression methods.

However, as our research has continued, STIFF's limitation that it can only handle one variable at a time has become obvious when working under more complicated situations, where the forecasting accuracy is more dependent on data correlation between different variables. A typical example is demonstrated below regarding Tropical Atmosphere Ocean project [10], abbreviated as TAO, which, with a number of sensors moored at different locations, aims at collecting hourly real time oceanographic and meteorological data that can be used to help monitor and forecast the climate swings associated with El Niño and La Niña.

Example 1. Suppose in TAO we pick up seven sensors at different locations (denoted by a pair of latitude and longitude), including 0N165E, 2S180W, 5N156E, 5N165E, 5S156E, 5S165E, 8N180W, and the first one is the target location where forecasting is going to be conducted. Moreover at each location, we choose five main representative variables – SST (Sea Surface Temperature), RH (Relative Humidity), AIRT (Air Temperature), HWND (Horizontal Wind Velocity) and VWND (Vertical Wind Velocity), among which SST is the principal variable deciding if El Niño or La Niña has been present while others contribute to SST's fluctuation more or less somehow. The problem now is, how can we generate an optimal forecasting regarding SST at the target location 0N165E with the full usage of all the variables at all locations.

Directly applying STIFF is at odds with the nature of TAO data. If we consider SST only, we will miss knowledge from other influential variables and the forecasting accuracy is not very good (shown later on). If STIFF is repeatedly carried out on all different variables one at a time, the correlation between variables will be lost. In addition, since there is no inherent structure of the ocean data (unlike that found in a river catchment), how to effectively create and train the neural network is not clear. These problems have motivated us to extend STIFF to utilize multiple variables simultaneously without losing correlation between them.

In this paper we present an improved forecasting approach based upon STIFF. First we introduce multivariate analysis concepts into the forecasting problem, then incorporate the inherent correlation in model construction and optimization by temporarily turning correlated variables into uncorrelated combinations based upon principal component analysis, and finally come up with a multivariate spatiotemporal forecasting method that has a higher accuracy and more balanced behaviors[1]. The contribution of our work is that it does enrich the spatiotemporal forecasting literature from a multivariate perspective.

1.1   Related Work

There has been a lot of work addressing spatiotemporal forecasting. But, much of it only targets at specific problems (like flood forecasting) rather than creating a general methodology. In addition, even less pays enough attention to the multivariate aspect of the problem.

In [9] HMM (Hidden Markov Model) was used to simulate rainfall within a certain time period, which was then further disaggregated unevenly across a whole area. Cressie et al. [3] took a different approach in spatiotemporal modelling. They built their model based upon spatial statistics analysis in which the temporal characteristics was condensed into a "three-day area of influence". Furthermore, in a contrary manner to [3], Deutsch et al. in [4] based their work mainly upon time series analysis, with spatial influence recorded as 1 or 0 in a simple neighboring distance matrix. As far as multivariate analysis concerned, the work in [13] was carried out to study joint influence of several kinds of fertilizers on agricultural produce, based upon a not-well-justified assumption that a current event was only affected by its immediate temporal predecessor. In [18], for its powerful discriminating capability involving several variables, multivariate analysis was used in biology research to identify motifs in a DNA sequence from human genes.

1.2   Paper Organization

The remaining of the paper is organized as follows. In section 2 we first define the multivariate spatiotemporal forecasting problem and then present the new algorithm, M-STIFF, which is targeted to solve the problem. Details regarding multivariate analysis integration and model construction is demonstrated in Section 3. Forecasting experiments are shown in Section 4, which ends with some concluding remarks.

2   Problem Definition and Algorithm Outline

Compared to our previous work STIFF [11], which was based upon analyzing the single variable, the biggest improvement of this new approach is the simultaneous consideration of multiple variables, which, although complicates the model construction further, does provide us more information for a better forecasting, as stated in [8,14], through considering correlation between different variables. First of all, frequently-used symbols are listed in the following Table 1.

---

[1] Here we mean the number of over-estimation tends to be equal to the number of under-estimation when talking about "balanced".

Table 1. Symbols summary

symbols	meaning
A	set of all locations
$A_i(i = 0, 1, ..., n)$	the (i)th location
V	set of all variables being considered
$V_i(i = 0, 1, ..., m)$	the ith variable
T	set of all time series data
$T_i(i = 0, ..., n)$	time series data in the ith location
$T_{ij}(i = 0, ..., n)(j = 0, ..., m)$	the jth variable time series in the ith location
$l\ (=|T_{ij}|)$	length of available time series data
$T_{ijk}(i = 0, ..., n)(j = 0, ..., m)(k = 1, ..., l)$	the k observation of time series data T_{ij}
s	steps to forecast ahead

We further explain these symbols by re-examining Example 1. Here $A = \{$0N165E, 2S180W, 5N156E, 5N165E, 5S156E, 5S165E, 8N180W$\}$. The variable set is $V = \{$SST, RH, AIRT, HWND, VWND$\}$. T includes all 35 time series data scattered at each location for each variable. By following the similar definition in [11], we extend the multivariate spatiotemporal forecasting problem as shown in Definition 1.

Definition 1. *Given a set of locations A, composed of $n+1$ locations $A_0, ..., A_n$. Suppose for each location $A_i \in A$ there are $m + 1$ associated variables, denoted as $V_0, ..., V_m(\in V)$, and each of them has time series data $T_{ij}(i = 0, ..., n)(i = 0, ..., m)$. Also suppose the forecast-ahead steps s is known. The multivariate spatiotemporal forecasting problem is to find a mapping relationship f, defined as*

$$f : \{A, V, T, s\} \vdash \{\hat{T}_{00(l+1)}, \hat{T}_{00(l+2)}, ..., \hat{T}_{00(l+s)}\} \tag{1}$$

that is as good as possible.

Please note in the definition, without loss of generality, we assume A_0, the target location, is the only location where forecasting will be carried out, and V_0 is the principal variable to be forecasted with the consideration of correlation from other variables.

We outline the new M-STIFF (Multivariate STIFF) algorithm as follows:

Step 1. To ensure the applicability of M-STIFF, the forecasting problem may need to be restated in terms of the above definition. If needed, clustering or classification techniques, like K-$Means$, PAM, etc. [6], might be employed to separate locations.

Step 2. For each location $A_i \in A$ build a time series model TS_i that provides the necessary temporal forecasting capability at the concerned location. The model construction cannot be based upon only one variable (even for the principal one), due to the presence of correlation between variables. In reality, uncorrelated linear combinations of variables based upon PCA (Principal Components Analysis) are used instead, which are then transposed back to the original variables. Necessary data preprocessing and transformation are also performed here.

Step 3. Based upon the spatial correlation of all locations $A_i(i \neq 0)$ against A_0, an appropriate artificial neural network is built to capture the spatial influence over the principal variable V_0 (at A_0). All forecasting of V_0 from time series models $TS_i(i \neq 0)$ is fed into the network and its output is taken as spatially-influenced forecasting of V_0 (at A_0).

Step 4. For the two forecasting values, one from TS_0 and the other from the neural network, both regarding V_0 in the target location A_0, a statistical regression mechanism combines them together, which will be our final multivariate spatiotemporal forecasting value.

Notice that, instead of aiming only at a very specific multivariate spatiotemporal forecasting problem, M-STIFF is actually a general procedure that combines existing techniques (clustering, ARIMA, PCA, neural networks) in a particular manner thus creating a unified process to solve the spatiotemporal forecasting problem.

To facilitate the comparison of the new approach with other models, we employ the same error measurement NARE defined in [11]

$$NARE = \frac{\sum_{t=1}^{s} |\hat{T}_{00(l+t)} - T_{00(l+t)}|}{\sum_{t=1}^{s} |T_{00(l+t)}|}. \tag{2}$$

NARE is an easy-interpretation but very useful error measurement from an empirical viewpoint. When studying Nile river flood control with the help of neural networks, Shaheen et al. [15] took a very similar approach, in which the absolute operation was replaced with square root for the authors' preference.

3   Build Multivariate Spatiotemporal Forecasting Model

Steps 1 and 4 of the M-STIFF model are almost the same as those in the STIFF approach [11]. Therefore, we only focus on the explanation of step 2 and 3, for they have become quite different or more complicated.

3.1   A More Parsimonious Methodology to Remove Hidden Periodicity

Normally, data transformation and differencing operations are needed to stabilize the data and remove hidden periodic components before any time series models can be fitted [1,17]. The way to do that is to inspect the original time series plot and related autocorrelation and spectrum density graphs. Take the TAO data [10] discussed above as an example. The SST data that recorded sea surface temperature hourly was extracted from location 0N165E. Its original plot as well as related autocorrelation and spectrum density are shown in Figures 1 and 2 respectively.

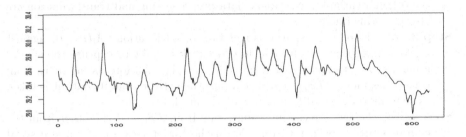

Fig. 1. Time series plot of SST from location 0N165E

Looking at Figure 1 does not reveal any obvious abnormality regarding either mean or covariance, which implies no data transformation is needed at this time. But, it turns out that the data has a disturbed periodic behavior with an indication of periodicity between 20 to 30 hours. Moreover, autocorrelation displays a damped sinusoidal pattern with periodicity of 24 hours, which strongly hints to us that the data is non-stationary with periodic components contained. As a further proof, the second highest peak in spectrum density tells us a frequency around 0.04. Considering the physical nature of SST data that was recorded hourly, we can be quite confident about the presence of periodicity.

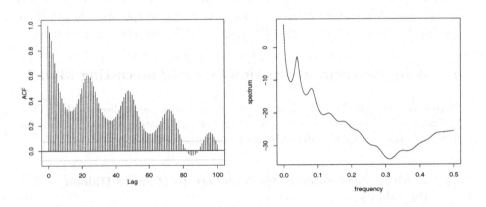

Fig. 2. Autocorrelation (left) and Spectrum Density (right) plots

A common practice to remove the hidden periodicity is to take a 24 differencing operation, namely $(1 - B^{24})$, suggested by Box et al [1]. But this kind of approach has two pitfalls. Firstly, it means we have to sacrifice 24 observations for the sake of differencing between each pair separated 24 hours apart. More important, the implication behind the brute-forth 24 differencing, is that in the corresponding *characteristic equation* there are at least 24 roots lying on the unit circle [1,17]. But no matter how hard we over-fit the data with higher autoregression degrees, in a hope to check if there are indeed at least 24 roots lying on the unit circle, only three stand up, one real and two conjugated complex, which pronouncedly and uniformly approach the unit circle with the moduli almost equal to 1 (difference less than 0.03) and are regarded as on the unit circle.

So here we suggest to follow a more parsimonious methodology for periodicity removal from a spectrum perspective. For the real root, which means a very long periodicity corresponding to the highest peak around 0 frequency in spectrum density, we use a random linear trend differencing $(1 - B)$ to describe it. For the two conjugated complex roots, $(1 - \phi_1 B - \phi_2 B^2)$ is employed, where ϕ_1 and ϕ_2 satisfy the following equation [1]

$$\cos(2\pi f_0) = \frac{\phi_1}{2\sqrt{-\phi_2}}. \tag{3}$$

By setting $\phi_2 = -1$ (so that the moduli of the complex roots are both equal to 1) we have the following answer

$$\phi_1 = 2\cos(2\pi f_0) \tag{4}$$

where $f_0 = \frac{1}{24}$, shown by the second highest peak around 0.04 in spectrum density plot of Figure 2 .

Combining them together de facto gives us a three-degree differencing $(1 - B)(1 - \phi_1 B - \phi_2 B^2)$, which not only greatly cuts down the number of lost observations, but also successfully dismiss the main periodicity (via frequency) from the data. What's more important is that this parsimonious differencing better interprets the data by focusing on its inherent spectral characteristics, and does not impose any extra root, which is not supposed to lie on the unit circle, on the data. We will take this differencing approach in the follow-up work whenever needed.

3.2   Multivariate Time Series Models Based upon PCA

Time series analysis is employed here to capture the temporal relationship at each location A_i. Detailed introduction about time series analysis can be found in [1,17], which, unfortunately, only addresses the univariate situation and lacks a merging mechanism for correlation between different variables.

Take a look at Table 2 that records the correlation between each pair of variables at location 0N165E. The strongest one, between AIRT and RH, and the weakest one, between VWND and RH, are further charted in Figure 3. Obviously there is more or less correlation between each pair of variables, which should be utilized to contribute to the forecasting.

We therefore introduce PCA (Principal Components Analysis) to help capture the hidden correlation in model construction process. Generally speaking, PCA originated as a useful technique aiming mainly at data reduction and interpretation [14,8]. Algebraically, for n variables $X_1, X_2, ..., X_n$, principal components are linear combinations of all these variables. Geometrically, these linear

Table 2. Correlation between variables

	HWND	VWND	AIRT	SST	RH
HWND	1.0000000				
VWND	0.1999811	1.00000000			
AIRT	-0.1372548	-0.06538648	1.00000000		
SST	-0.4262996	-0.29556301	0.40866301	1.0000000	
RH	0.1310432	0.00794552	-0.85613661	-0.3496958	1.00000000

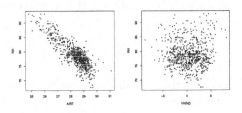

Fig. 3. The strongest and weakest correlation

combinations represent the selection of a new coordinate system obtained by rotating the original one. What is most important is that it does have a pretty good characteristic – principal components are uncorrelated with each other. Therefore each principal component can be used individually without considering correlation any more. For a comparison to Table 2, we list the correlation between five principal components in Table 3, which shows how greatly PCA can minimize correlation.

Table 3. Correlation between principal components

	1st PC	2nd PC	3rd PC	4th PC	5th PC
1st PC	1.0000000				
2nd PC	9.005625e-016	1.00000000			
3rd PC	9.228538e-017	2.214611e-016	1.00000000		
4th PC	-2.841340e-016	8.810815e-016	1.303865e-015	1.0000000	
5th PC	3.776149e-015	6.091742e-015	1.370957e-015	1.682662e-014	1.00000000

Obtaining these principal components is quite straightforward and has been well studied in the literature [14,8]. The main steps are outlined as follows. Suppose there is a random vector of $X^T = [X_1, X_2, ..., X_n]$. First, compute Σ, the covariance matrix of the random vector. Then, calculate Σ's singular decomposition VLV^T, where $L = diag(l_1, l_2, ..., l_n)$ is a diagonal matrix of eigenvalues of Σ with $l_1 \geq l_2 \geq ... \geq l_n \geq 0$, and $V = [\vec{v_1}, \vec{v_2}, ..., \vec{v_n}]$ is the matrix containing the corresponding eigenvectors. Finally obtain the principal component as $Y_i = \vec{v_i}^T X$.

Based upon PCA, we summarize how to build the multivariate time series model for each location A_i:

Step 1. PCA is used to disassemble the correlation between the original variables $X_1, X_2, ..., X_n$ into several uncorrelated principal components $Y_1, Y_2, ..., Y_n$.

Step 2. For each uncorrelated principal component Y_i build a time series model and get the corresponding forecasting of Y_i.

Step 3. Transpose the forecasted Y_i back to X_i with the following equation.

$$[X_1, X_2, ..., X_n] = [Y_1, Y_2, ..., Y_n]V^{-1} \tag{5}$$

Table 4. SST forecasting based upon multivariate and univariate time series model

	Multivariate			Univariate		
	NARE	Over No.	Under No.	NARE	Over No.	Under No.
24	0.00469	24	0	0.0067	15	9
48	0.0058	46	2	0.00704	25	23
72	0.00791	67	5	0.00804	35	37

In Table 4 we compare the forecasting performance from multivariate time series model, which incorporates correlation as described above, and the univariate time series model for 24, 48 and 72 hours ahead at location 0N165E. Although univariate model has a relatively more balanced behavior, it lags behind multivariate model in term of the forecasting accuracy.

3.3   Artificial Neural Network Training with Time Lag Delay and Location Correlation

Artificial neural network, also known as connectionist models or parallel distributed processing, is introduced to acquire the spatial influence of all the non-target location A_i over A_0 on the principal variable V_0. Neural network's powerful capability to find hidden patterns, especially in "situations where the underlying physical relationships are not fully understood" [5], has been well justified in the literature, such as [15,2].

As demonstrated in [11], the complete network construction depends upon a variety of factors, some of which include how to determine the appropriate number of the neurons in different layers, how to carry out the pruning based upon the special topology of all locations A, how to set the starting parameters to be trained, and so on. Instead, in this paper for the space limitation, we just focus on how to derive the useful information from the data itself, which can be used to guide the construction and training process and deserves more attention when multiple variables are concerned.

Following Example 1 presented in the introduction section about TAO project [10], here we construct a neural network for spatial correlation derivation regarding SST data. With the same approach employed in [11] a fully-connected feedforward neural network with 6 neurons, corresponding to the 6 non-target locations respectively, in the input layer is put up for training.

However, in order to maximize the correlation between time series data coming from different locations, a shifting operation is first introduced to find out the possible time lag (against the one in the target location). Intuitively speaking, an appropriate time lag is picked up when it minimizes the following difference sum

$$\sum_{k=1,\ldots,l} (|T_{i0(k+shift)} - T_{00k}|). \tag{6}$$

In order not to lose too many observations that may undermine the neural network training, we further limit the shifting with a maximal value of 50. Moreover, for the sake of simplicity, the time series in the target location is fixed, while data from other locations can only be moved in one direction, arbitrarily set towards the past.

Table 5 summarizes the shifting result. Except the last one at 8N180W all correlation rises to some extent. Especially the one at 5N165E, jumps from an original correlation of 0.5 with more than 36% up. It also has a minimal difference sum compared with its neighboring locations, which shows data from this location, 5N165E, shares the most similarity with data from the target location, 0N165E. In network training phase, the input data is shifted with the corresponding time lag, and the concerned correlation serves as the weights for links and is used for adjusting the learning process accordingly.

Table 5. Time lag and its corresponding correlation against time series data at 0N165E

	2S180W	5N156E	5N165E	5S156E	5S165E	8N180W
minimal sum	188.12	181.09	95.44	218.51	115.01	484.98
shift steps	47	1	48	2	24	50
correlation	0.227	0.43	0.683	0.182	0.268	-0.039

4   Experiments and Concluding Remarks

We will work on the problem raised in Example 1 of the introduction section, and present the experiment result and performance evaluation here, to show how this multivariate spatiotemporal forecasting method works. Our lookahead time windows are 24, 48 and 72 hours respectively, and the new approach, M-STIFF, is compared against our previous model STIFF, and a common time series model base upon SST only.

Figure 4 depicts the forecasting from three different models and the real observation. Related accuracy measurement and over & under forecasting behaviors are summarized in Table 6.

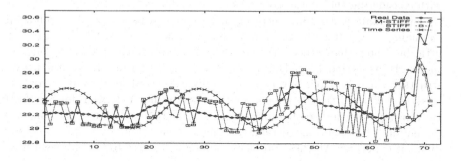

Fig. 4. Forecasting from different models

In fact, all the three models have pretty high forecasting accuracy, partially due to the fact that data itself is quite stable and does not fluctuate drastically. Forecasting from time series model almost follows a sinusoidal wave, which right falls into our expectation, since, as stated in [7], time series model with only two conjugated complex roots on the unit circle will exactly follow a *sin* function. However, it is right the sinusoidal pattern that prevents time series model from forecasting extreme values that are not present in the training data, as illustrated at the end of Figure 4 where there comes a big leap in SST. Another drawback for time series model is it lacks an efficient way to update itself with the latest data, which partially contributes its failure to capture the unprecedented observations.

Table 6. Comparison between different models

Hours	M-STIFF			STIFF			TS		
	NARE	Over #	Under #	NARE	Over #	Under #	NARE	Over #	Under #
24	0.00428	13	11	0.00534	11	13	0.0067	15	9
48	0.00504	23	25	0.00624	27	21	0.00704	25	23
72	0.00751	34	38	0.0082	39	33	0.00804	35	37

However, the problem has almost been solved in the hybrid model M-STIFF where artificial neural network can dynamically modify itself through link weights change once the forecasted data is available, which somehow enables the model to follow the extreme values that are usually hinting us exceptional phenomena are approaching. At the same time M-STIFF bears a higher forecasting accuracy compared with other models, which confirms the introduction of correlation between variables into model construction is beneficial. Furthermore, M-STIFF tends to have a more balanced forecasting behavior, especially in the short period of 24 hours. Compared with Table 4 where only multivariate time series model was constructed, the improvement in term of forecasting balance with considering the spatial correlation is obvious.

Unfortunately, carefully checking in Table 6 discloses a point somehow discouraging to us. It looks the forecasting accuracy from M-STIFF degrades a little bit faster than others. Due to the time limitation, we cannot conduct a thorough inspection here. But we believe it might mean the neural network structure (in this paper a very simple fully-connected feedforward network is used) is not sophisticated enough or the training has not been done perfectly. By the way, another possible drawback is M-STIFF calls for much human intervention (in model construction, training, optimization, etc.) and is subject to different expertise. All these problems remain in the list to be addressed in our future work.

References

1. George E. P. Box, Gwilym M. Jenkins, and Gregory C. Reinsel. *Time Series Analysis : Forecasting and Control*. Prentice-Hall, 1994.
2. Sung-Bae Cho. Neural-network classifiers for recognizing totally unconstrained handwritten numerals. *IEEE Transactions on Neural Networks*, 8(1):43–53, 1997.
3. N. Cressie and J. J. Majure. Spatio-temporal statistical modeling of livestock waste in streams. *Journal of Agricultural, Biological and Environmental Statistics*, 1997.
4. S. J. Deutsch and J. A. Ramos. Space-time modeling of vector hydrologic sequences. *Water Resource Bulletin*, 22:967–980, 1986.
5. M. Dougherty, Linda See, S. Corne, and S. Openshaw. Some initial experiments with neural network models of flood forecasting on the river ouse. In *Proceedings of the 2nd Annual Conference of GeoComputation*, 1997.
6. Margaret H. Dunham. *Data Mining : Introductory and Advanced Topics*. Prentice Hall, 2003.
7. Henry L. Gray. *Theory and Applications of ARMA Models*. Springer-Verlag, 2002.
8. R. A. Johnson and D. W. Wichern. *Applied Multivariate Statistical Analysis*. Prentice Hall, 2002.
9. C. Jothityangkoon, M. Sivapalan, and N. R. Viney. Tests of a space-time model of daily rainfall in southwestern australia based on nonhomogeneous random cascades. *Water Resource Research*, 2000.
10. Pacific Marine Environmental Laboratory. Tropic atmosphere ocean project. http://www.pmel.noaa.gov/tao/.
11. Zhigang Li and Margaret H. Dunham. Stiff : A forecasting framework for spatio-temporal data. In *Workshop on Knowledge Discovery in Multimedia and Complex Data, PAKDD 02*, 2002.
12. Phillip E. Pfeifer and Stuart J. Deutsch. A test of space-time arma modelling and forecasting of hotel data. *Journal of Forecasting*, 9:225–272, 1990.
13. D. Pokrajac and Z. Obradovic. Improved spatial-temporal forecasting through modeling of spatial residuals in recent history. In *First SIAM International Conference on Data Mining (SDM'2001)*, 2001.

14. A. C. Rencher. *Methods of Multivariate Analysis*. John Wiley & Sons, 2002.
15. S. I. Shaheen, A. F. Atiya, S. M. El-Shoura, and M. S. El-Sherif. A comparison between neural network forecasting techniques. *IEEE Transactions on Neural Networks*, 10(2):402–409, 1999.
16. D. S. Stofer. Estimation and identification of space-time armax models in the presence of missing data. *Journal of American Statistical Association*, 81:762–772, 1986.
17. William W. S. Wei. *Time Series Analysis : Univariate and Multivariant Methods*. Addison Wesley, 1990.
18. Michael Q. Zhang. Discriminant analysis and its application in dna sequence motif recognition. *Briefings in Bioinformatics*, 1(4):1–12, 2000.

Correlation Analysis of Spatial Time Series Datasets: A Filter-and-Refine Approach

Pusheng Zhang*, Yan Huang, Shashi Shekhar**, and Vipin Kumar**

Computer Science & Engineering Department, University of Minnesota,
200 Union Street SE, Minneapolis, MN 55455, U.S.A.
[pusheng|huangyan|shekhar|kumar]@cs.umn.edu

Abstract. A spatial time series dataset is a collection of time series, each referencing a location in a common spatial framework. Correlation analysis is often used to identify pairs of potentially interacting elements from the cross product of two spatial time series datasets. However, the computational cost of correlation analysis is very high when the dimension of the time series and the number of locations in the spatial frameworks are large. The key contribution of this paper is the use of spatial autocorrelation among spatial neighboring time series to reduce computational cost. A filter-and-refine algorithm based on coning, i.e. grouping of locations, is proposed to reduce the cost of correlation analysis over a pair of spatial time series datasets. Cone-level correlation computation can be used to eliminate (filter out) a large number of element pairs whose correlation is clearly below (or above) a given threshold. Element pair correlation needs to be computed for remaining pairs. Using experimental studies with Earth science datasets, we show that the filter-and-refine approach can save a large fraction of the computational cost, particularly when the minimal correlation threshold is high.

1   Introduction

Spatio-temporal data mining [14,16,15,17,13,7] is important in many application domains such as epidemiology, ecology, climatology, or census statistics, where datasets which are spatio-temporal in nature are routinely collected. The development of efficient tools [1,4,8,10,11] to explore these datasets, the focus of this work, is crucial to organizations which make decisions based on large spatio-temporal datasets. A spatial framework [19] consists of a collection of locations and a neighbor relationship. A time series is a sequence of observations taken sequentially in time [2]. A spatial time series dataset is a collection of time series, each referencing a location in a common spatial framework. For example,

---

* The contact author. pusheng@cs.umn.edu. Tel: 1-612-626-7515

** This work was partially supported by NASA grant No. NCC 2 1231 and by Army High Performance Computing Research Center contract number DAAD19-01-2-0014. The content of this work does not necessarily reflect the position or policy of the government and no official endorsement should be inferred. AHPCRC and Minnesota Supercomputer Institute provided access to computing facilities.

K.-Y. Whang, J. Jeon, K. Shim, J. Srivatava (Eds.): PAKDD 2003, LNAI 2637, pp. 532–544, 2003.
© Springer-Verlag Berlin Heidelberg 2003

the collection of global daily temperature measurements for the last 10 years is a spatial time series dataset over a degree-by-degree latitude-longitude grid spatial framework on the surface of the Earth.

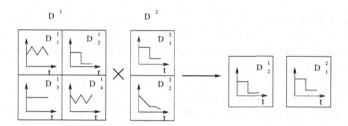

Fig. 1. An Illustration of the Correlation Analysis of Two Spatial Time Series Datasets

Correlation analysis is important to identify potentially interacting pairs of time series across two spatial time series datasets. A strongly correlated pair of time series indicates potential movement in one series when the other time series moves. For example, El Nino, the anomalous warming of the eastern tropical region of the Pacific, has been linked to climate phenomena such as droughts in Australia and heavy rainfall along the Eastern coast of South America [18]. Fig. 1 illustrates the correlation analysis of two spatial time series datasets D^1 and D^2. D^1 has 4 spatial locations and D^2 has 2 spatial locations. The cross product of D^1 and D^2 has 8 pairs of locations. A highly correlated pair, i.e. (D^1_2, D^2_1), is identified from the correlation analysis of the cross product of the two datasets.

However, a correlation analysis across two spatial time series datasets is computationally expensive when the dimension of the time series and number of locations in the spaces are large. The computational cost can be reduced by reducing time series dimensionality or reducing the number of time series pairs to be tested, or both. Time series dimensionality reduction techniques include discrete Fourier transformation [1], discrete wavelet transformation [4], and singular vector decomposition [6].

The number of pairs of time series can be reduced by a cone-based filter-and-refine approach which groups together similar time series within each dataset. A filter-and-refine approach has two logical phases. The filtering phase groups similar time series as cones in each dataset and calculates the centroids and boundaries of each cone. These cone parameters allow computation of the upper and lower bounds of the correlations between the time series pairs across cones. Many All-True and All-False time series pairs can be eliminated at the cone level to reduce the set of time series pairs to be tested by the refinement phase. We propose to exploit an interesting property of spatial time series datasets, namely spatial auto-correlation [5], which provides a computationally efficient method to determine cones. Experiments with Earth science data [12] show that the filter-and-refine approach can save a large fraction of computational cost, especially when the minimal correlation threshold is high. To the best of our knowledge, this is the first paper exploiting spatial auto-correlation among time series at nearby locations to reduce the computational cost of correlation analysis over a pair of spatial time series datasets.

Scope and Outline: In this paper, the computation saving methods focus on reduction of the time series pairs to be tested. Methods based on non-spatial properties (e.g. time-series power spectrum [1,4,6]) are beyond the scope of the paper and will be addressed in future work.

The rest of the paper is organized as follows. In Section 2, the basic concepts and lemmas related to cone boundaries are provided. Section 3 proposes our filter-and-refine algorithm, and the experimental design and results are presented in Section 4. We summarize our work and discuss future directions in Section 5.

2   Basic Concepts

In this section, we introduce the basic concepts of correlation calculation and the multi-dimensional unit sphere formed by normalized time series. We define the cone concept in the multi-dimensional unit sphere and prove two lemmas to bound the correlation of pairs of time series from two cones.

2.1   Correlation and Test of Significance of Correlation

Let $x = \langle x_1, x_2, \ldots, x_m \rangle$ and $y = \langle y_1, y_2, \ldots, y_m \rangle$ be two time series of length m. The correlation coefficient [3] of the two time series is defined as:

$$corr(x, y) = \frac{1}{m-1} \sum_{i=1}^{m} \left(\frac{x_i - \overline{x}}{\sigma_x} \right) \cdot \left(\frac{y_i - \overline{y}}{\sigma_y} \right) = \widehat{x} \cdot \widehat{y}$$

where $\overline{x} = \frac{\sum_{i=1}^{m} x_i}{m}$, $\overline{y} = \frac{\sum_{i=1}^{m} y_i}{m}$, $\sigma_x = \sqrt{\frac{\sum_{i=1}^{m}(x_i - \overline{x})^2}{m-1}}$, $\sigma_y = \sqrt{\frac{\sum_{i=1}^{m}(y_i - \overline{x})^2}{m-1}}$, $\widehat{x}_i = \frac{1}{\sqrt{m-1}} \frac{x_i - \overline{x}}{\sigma_x}$, $\widehat{y}_i = \frac{1}{\sqrt{m-1}} \frac{y_i - \overline{y}}{\sigma_y}$, $\widehat{x} = \langle \widehat{x}_1, \widehat{x}_2, \ldots, \widehat{x}_m \rangle$, and $\widehat{y} = \langle \widehat{y}_1, \widehat{y}_2, \ldots, \widehat{y}_m \rangle$. A simple method to test the null hypothesis that the product moment correlation coefficient is zero can be obtained using a Student's t-test [3] on the t statistic as follows: $t = \sqrt{m-2} \frac{r}{\sqrt{1-r^2}}$, where r is the correlation coefficient between the two time series. The freedom degree of the above test is $m - 2$. Using this we can find a $p - value$ or find the critical value for a test at a specified level of significance. For a dataset with larger length m, we can adopt Fisher's Z-test [3] as follows: $Z = \frac{1}{2} \log \frac{1+r}{1-r}$, where r is the correlation coefficient between the two time series. The correlation threshold can be determined for a given time series length and confidence level

2.2   Multi-dimensional Sphere Structure

In this subsection, we discuss the multi-dimensional unit sphere representation of time series. The correlation of a pair of time series is related to the cosine measure between their unit vector representations in the unit sphere.

Fact 1 (Multi-dimensional Unit Sphere Representation.) *Let*
$x = \langle x_1, x_2, \ldots, x_m \rangle$ *and* $y = \langle y_1, y_2, \ldots, y_m \rangle$ *be two time series of length*
m. *Let* $\widehat{x}_i = \frac{1}{\sqrt{m-1}} \frac{x_i - \overline{x}}{\sigma_x}$, $\widehat{y}_i = \frac{1}{\sqrt{m-1}} \frac{y_i - \overline{y}}{\sigma_y}$, $\widehat{x} = \langle \widehat{x}_1, \widehat{x}_2, \ldots, \widehat{x}_m \rangle$, *and* $\widehat{y} = $
$\langle \widehat{y}_1, \widehat{y}_2, \ldots, \widehat{y}_m \rangle$. *Then* $\widehat{x}$ *and* $\widehat{y}$ *are located on the surface of a multi-dimensional*
unit sphere and $corr(x, y) = \widehat{x} \cdot \widehat{y} = \cos(\angle(\widehat{x}, \widehat{y}))$ *where* $\angle(\widehat{x}, \widehat{y})$ *is the angle of* $\widehat{x}$
and $\widehat{y}$ *in* $[0, 180°]$ *in the multi-dimensional unit sphere .*

Because the sum of the $\widehat{x}_i^2$ is equal to 1: $\sum_{i=1}^{m} \widehat{x}_i^2 = \sum_{i=1}^{m} \left(\frac{1}{\sqrt{m-1}} \frac{x_i - \overline{x}}{\sqrt{\frac{\sum_{i=1}^{m} (x_i - \overline{x})^2}{m-1}}} \right)^2$

$= 1$, $\widehat{x}$ is located in the multi-dimensional unit sphere. Similarly, $\widehat{y}$ is also located
in the multi-dimensional unit sphere. Based on the definition of $corr(x, y)$, we
have $corr(x, y) = \widehat{x} \cdot \widehat{y} = \cos(\angle(\widehat{x}, \widehat{y}))$.

Fact 2 (Correlation and Cosine.) *Given two time series x and y and a user*
specified minimal correlation threshold θ where $0 < \theta \leq 1$, $|corr(x, y)| = $
$|\cos(\angle(\widehat{x}, \widehat{y}))| \geq \theta$ if and only if $0 \leq \angle(\widehat{x}, \widehat{y}) \leq \theta_a$ or $180° - \theta_a \leq \angle(\widehat{x}, \widehat{y}) \leq 180°$,
where $\theta_a = \arccos(\theta)$.

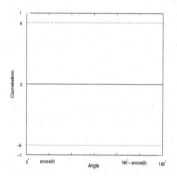

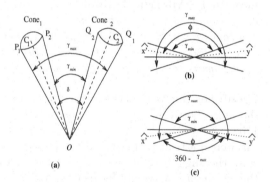

Fig. 2. Cosine Value vs. Central Angle

Fig. 3. Angle of Time Series in Two Spherical Cones

Fig. 2 shows that $|corr(x, y)| = |\cos(\angle(\widehat{x}, \widehat{y}))|$ falls in the range of $[\theta, 1]$
or $[-1, -\theta]$ if and only if $\angle(\widehat{x}, \widehat{y})$ falls in the range of $[0, arccos(\theta)]$ or $[180° - arccos(\theta), 180°]$.

The correlation of two time series is directly related to the angle between the
two time series in the multi-dimensional unit sphere. Finding pairs of time series
with an absolute value of correlation above the user given minimal correlation
threshold θ is equivalent to finding pairs of time series $\widehat{x}$ and $\widehat{y}$ on the unit multi-
dimensional sphere with an angle in the range of $[0, \theta_a]$ or $[180° - \theta_a, 180°]$.

2.3   Cone and Correlation between a Pair of Cones

This subsection formally defines the concept of cone and proves two lemmas to
bound the correlations of pairs of time series from two cones. The user specified

minimal correlation threshold is denoted by θ ($0 < \theta \leq 1$), and $\arccos(\theta)$ is denoted by θ_a accordingly.

Definition 1 (Cone). *A cone is a set of time series in a multi-dimensional unit sphere and is characterized by two parameters, the center and the span of the cone. The center of the cone is the mean of all the time series in the cone. The span τ of the cone is the maximal angle between any time series in the cone and the cone center.*

We now investigate the relationship of two time series from two cones in a multi-dimensional unit sphere as illustrated in Fig. 3 (a). The largest angle($\angle P_1OQ_1$) between two cones C_1 and C_2 is denoted as γ_{max} and the smallest angle ($\angle P_2OQ_2$) is denoted as γ_{min}. We prove the following lemmas to show that if γ_{max} and γ_{min} are in specific ranges, the absolute value of the correlation of any pair of time series from the two cones are all above θ (or below θ). Thus all pairs of time series between the two cones satisfy (or dissatisfy) the minimal correlation threshold.

Lemma 1 (All-True Lemma). *Let C_1 and C_2 be two cones from the multi-dimensional unit sphere structure. Let $\hat{x}$ and $\hat{y}$ be any two time series from the two cones respectively. If $0 \leq \gamma_{max} \leq \theta_a$, then $0 \leq \angle(\hat{x}, \hat{y}) \leq \theta_a$. If $180° - \theta_a \leq \gamma_{min} \leq 180°$, then $180° - \theta_a \leq \angle(\hat{x}, \hat{y}) \leq 180°$. If either of the above two conditions is satisfied, $\{C_1, C_2\}$ is called an All-True cone pair.*

Proof: For the first case, it is easy to see from Fig. 3 that if $\gamma_{max} \leq \theta_a$, then the angle between $\hat{x}$ and $\hat{y}$ is less or equal to θ_a. For the second case, when $180° - \theta_a \leq \gamma_{min} \leq 180°$, we need to show that $180° - \theta_a \leq \angle(\hat{x}, \hat{y}) \leq 180°$. If this were not true, there exist $\hat{x}' \in C_1$ and $\hat{y}' \in C_2$ where $0 \leq \angle(\hat{x}', \hat{y}') < 180° - \theta_a$ since the angle between any pairs of time series is chosen from 0 to 180°. From this inequality, we would have either $\gamma_{min} \leq \phi = \angle(\hat{x}', \hat{y}') < 180° - \theta_a$ as shown in Fig. 3 (b) or $360° - \gamma_{max} \leq \phi = \angle(\hat{x}', \hat{y}') < 180° - \theta_a$ as shown in Fig. 3 (c). The first condition contradicts our assumption that $180° - \theta_a \leq \gamma_{min} \leq 180°$. The second condition implies that $360° - \gamma_{max} < \gamma_{min}$ since $180° - \theta_a \leq \gamma_{min}$. This contradicts our choice of γ_{min} as the minimal angle of the two cones.   □

Lemma 1 shows that when two cones are close enough, any pair of time series from the two cones is highly positively correlated; and when two cones are far enough apart, any pair of time series from the two cones are highly negatively correlated.

Lemma 2 (All-False Lemma). *Let C_1 and C_2 be two cones from the multi-dimensional unit sphere; let $\hat{x}$ and $\hat{y}$ be any two time series from the two cones respectively. If $\theta_a \leq \gamma_{min} \leq 180°$ and $\gamma_{min} \leq \gamma_{max} \leq 180° - \theta_a$, then $\theta_a \leq \angle(\hat{x}, \hat{y}) \leq 180° - \theta_a$ and $\{C_1, C_2\}$ is called an All-False cone pair.*

Proof: The proof is straightforward from the inequalities.   □

Lemma 2 shows that if two cones are in a moderate range, any pair of time series from the two cones is weakly correlated.

3   Cone-Based Filter-and-Refine Algorithm

Our algorithm consists of four steps as shown in Algorithm 1: Pre-processing (line 1), Cone Formation (line 2), Filtering i.e. Cone-level Join (line 4), and Refinement i.e. Instance-level Join (lines 7–11). The first step is to pre-process

Input: 1) $S^1 = \{s_1^1, s_2^1, \ldots, s_n^1\}$: n_1 spatial referenced time series
where each instance references a spatial framework SF_1;
2) $S^2 = \{s_1^2, s_2^2, \ldots, s_n^2\}$: n_2 spatial referenced time series
where each instance references a spatial framework SF_2;
3) a user defined correlation threshold θ;
Output: all pairs of time series each from S^1 and S^2 with correlations
above θ;
Method:

$$
\begin{array}{ll}
\texttt{Pre-processing}(S^1);\ \texttt{Pre-processing}(S^2); & (1) \\
CN_1 = \texttt{Cone\_Formation}(S^1, SF_1);\ CN_2 = \texttt{Cone\_Formation}(S^2, SF_2); & (2) \\
\texttt{for all pair } c_1 \texttt{ and } c_2 \texttt{ each from } CN_1 \texttt{ and } CN_2 \texttt{ do } \{ & (3) \\
\quad Filter\_Flag = \texttt{Cone-level\_Join}(c_1,\ c_2,\ \theta); & (4) \\
\quad \texttt{if } (Filter\_Flag == \texttt{ALL\_TRUE}) & (5) \\
\quad\quad \texttt{output all pairs in the two cones} & (6) \\
\quad \texttt{else if } (Filter\_Flag\ != \texttt{ALL\_FALSE}) \{ & (7) \\
\quad\quad \texttt{for all pair } s_1 \texttt{ and } s_2 \texttt{ each from } c_1 \texttt{ and } c_2 \texttt{ do } \{ & (8) \\
\quad\quad\quad High\_Corr\_Flag = \texttt{Instance-level\_Join}(s_1, s_2,\ \theta); & (9) \\
\quad\quad\quad \texttt{if } (High\_Corr\_Flag) \texttt{ output } s_1 \texttt{ and } s_2; & (10) \\
\quad\quad \} & (11) \\
\} & (12)
\end{array}
$$

Algorithm 1: Correlation Finder

the raw data to the multi-dimensional unit sphere representation. The second step, cone formation, involves grouping similar time series into cones in spatial time series datasets. Clustering the time series is an intuitive approach. However, clustering on time-series datasets may be expensive and sensitive to the clustering method and its objective function. For example, K-means approaches [9] find globular clusters while density-based clustering approaches [9] find arbitrary shaped clusters with user-given density thresholds. Spatial indexes, such as R^* trees, which are built after time series dimensionality reduction [1,4] could be another approach to group similar time series together. In this paper, we explore spatial auto-correlation for the cone formation. First the space is divided into disjoint cells. The cells can come from domain experts, such as the El Nino region, or could be as simple as uniform grids. By scanning the dataset once, we map each time-series into its corresponding cell. Each cell contains similar time series and represents a cone in the multi-dimensional unit sphere representation. The center and span are calculated to characterize each cone.

Example 1 (Spatial Cone Formation). Fig. 4 illustrates the spatial cone formation for two datasets, namely land and ocean. Both land and ocean frameworks consist of 16 locations. The time series of length m in a location s is denoted as $F(s) = F_1(s), F_2(s), \ldots, F_i(s), \ldots F_m(s)$. Fig. 4 only depicts a time series for $m = 2$. Each arrow in a location s of ocean or land represents the vector $< F_1(s), F_2(s) >$ normalized to the two dimensional unit sphere. Since the dimension of the time series is two, the multi-dimensional unit sphere reduces to a unit circle, as shown in Fig. 4 (b). By grouping the time series in each dataset into 4 disjoint cells according to their spatial proximity, we have 4 cells each for ocean and land. The ocean is partitioned to $L_1 - L_4$ and the land is partitioned to $O_0 - O_4$, as shown in Fig. 4 (a). Each cell represents a cone in the multi-dimensional unit sphere. For example, the patch L_2 in Fig. 4 (a) matches L_2 in the circle in Fig. 4 (b).

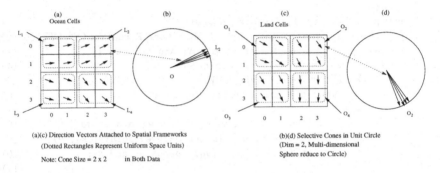

(a)(c) Direction Vectors Attached to Spatial Frameworks
(Dotted Rectangles Represent Uniform Space Units)

Note: Cone Size = 2 x 2   in Both Data

(b)(d) Selective Cones in Unit Circle
(Dim = 2, Multi-dimensional
Sphere reduce to Circle)

Fig. 4. An Illustrative Example for Spatial Cone Formation

After the cone formation, a cone-based join is applied between the two datasets. The calculation of the angle between each pair of cone centers is carried out, and the minimum and maximum bounds of the angles between the two cones are derived based on the spans of the two cones. The All-False cone pairs or All-True cone pairs are filtered out based on the lemmas. Finally, the candidates which cannot be filtered are explored in the refinement step.

Example 2 (Filter-and-refine). The join operations between the cones in Fig. 4 (a) are applied as shown in Table 1. The number of correlation computations is used in this paper as the basic unit to measure computation costs. Many All-False cone pairs and All-True cone pairs are detected in the filtering step and the number of candidates explored in the refinement step are reduced substantially. The cost of the filtering phase is 16. Only pairs (O_1, L_1), (O_3, L_4), and (O_4, L_4) cannot be filtered and need to be explored in the refinement step. The cost of the refinement step is 3×16 since there are 4 time series in both the ocean and land cone for all 3 pairs. The total cost of filter-and-refine adds up to 64. The number of correlation calculations using the simple nested loop is 256, which is greater than the number of correlation calculations in the filter-and-refine approach. Thus when the cost of the cone formation phase is less than 192 units, the filter-and-refine approach is more efficient.

Completeness and Correctness. Based on the lemmas in Section 2, All-True cone pairs and All-False cone pairs are filtered out so that a superset of results is obtained after the filtering step. There are no false dismissals for this filter-and-refine algorithm. All pairs found by the algorithm satisfy the given minimal correlation threshold.

Table 1. Cone-based Join in Example Data

Ocean-Land	Filtering	Refinement	Ocean-Land	Filtering	Refinement
$O_1 - L_1$	No	16	$O_3 - L_1$	All-True	
$O_1 - L_2$	All-False		$O_3 - L_2$	All-True	
$O_1 - L_3$	All-False		$O_3 - L_3$	All-True	
$O_1 - L_4$	All-False		$O_3 - L_4$	No	16
$O_2 - L_1$	All-False		$O_4 - L_1$	All-True	
$O_2 - L_2$	All-False		$O_4 - L_2$	All-True	
$O_2 - L_3$	All-False		$O_4 - L_3$	All-True	
$O_2 - L_4$	All-False		$O_4 - L_4$	No	16

4   Performance Evaluation

We wanted to answer two questions: (1)How does the spatial auto-correlation based inexpensive grouping algorithm affect filtering efficiency? In particular, how do we identify the proper cone size to achieve better overall savings? (2) How does the minimal correlation threshold influence the filtering efficiency? These questions can be answered in two ways: algebraically, as discussed in section 4.1 and experimentally, as discussed in section 4.2. Fig. 5 describes the experimental setup to evaluate the impact of parameters on the performance of the algorithm. We evaluated the performance of the algorithm with a dataset from NASA Earth science data [12]. In this experiment, a correlation analysis between the East Pacific Ocean region (80W - 180W, 15N - 15S) and the United States was investigated. The time series from 2901 land cells of the United States and 11556 ocean cells of the East Pacific Ocean were obtained under a 0.5 degree by 0.5 degree resolution.

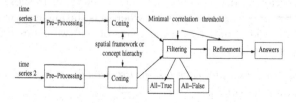

Fig. 5. Experiment Design

Net Primary Production (NPP) was the attribute for the land cells, and Sea Surface Temperature (SST) was the attribute for the ocean cells. NPP is the net photo-synthetic accumulation of carbon by plants. Keeping track of NPP is important because NPP includes the food source of humans and all other organisms and thus, sudden changes in the NPP of a region can have a direct impact on the regional ecology. The records of NPP and SST were monthly data from 1982 to 1993.

4.1   Parameter Selections

In this section we investigate the selective range of the cone spans to improve filtering efficiency. Both All-False and All-True filtering can be applied in the filtering step. Thus we investigate the appropriate range of the cone spans in each of these filtering categories. Here we define the fraction of time series pairs reduced in the filtering step as FAR, i.e. $FAR = \frac{N_{time\ series\ pairs-filtered}}{|D_1| \times |D_2|}$. Thus FAR in the cone level is represented as FAR_{cone}.

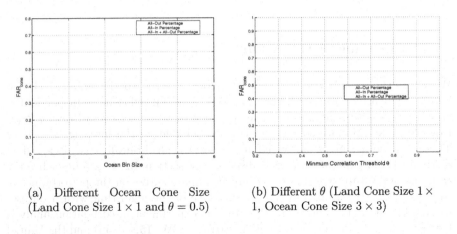

(a) Different Ocean Cone Size (Land Cone Size 1×1 and $\theta = 0.5$)

(b) Different θ (Land Cone Size 1×1, Ocean Cone Size 3×3)

Fig. 6. All-True and All-False Filtering Percentages for Different Parameters

Fig. 6 demonstrates the FAR_{cone} is related to the cone size and minimal correlation threshold. The proper cone size and larger minimal correlation threshold improve the filtering ability. Given a minimal correlation threshold θ ($0 < \theta < 1$), $\gamma_{max} = \delta + \tau_1 + \tau_2$ and $\gamma_{min} = \delta - \tau_1 - \tau_2$, where δ is the angle between the centers of two cones, and the τ_1 and τ_2 are the spans of the two cones respectively. For simplicity, suppose $\tau_1 \simeq \tau_2 = \tau$.

Lemma 3. *Given a minimal correlation threshold θ, if a pair of cones both with span τ is an All-True cone pair, then $\tau < \frac{\arccos(\theta)}{2}$.*

Proof: Assume that a cone pair satisfies the All-True Lemma, i.e., either $\gamma_{max} < \arccos(\theta)$ or $\gamma_{min} > 180° - \arccos(\theta)$ is satisfied. In the former scenario, the angle

δ is very small, and we get $\delta + 2\tau < \arccos(\theta)$, i.e., $\tau < \frac{\arccos(\theta) - \delta}{2}$. In the latter scenario, the angle δ is very large, and we get $\delta - 2\tau > 180° - \arccos(\theta)$, i.e., $\tau < \frac{\arccos(\theta) + \delta - 180°}{2}$. The τ is less than $\frac{\arccos(\theta)}{2}$ in either scenario since $\tau < 180°$.

$\square$

Lemma 4. *Given a minimal correlation threshold θ, if a pair of cones both with span τ is an All-False cone pair, then $\tau \geq \frac{180°}{4} - \frac{\arccos(\theta)}{2}$.*

Proof: Assume that a cone pair satisfies the All-False Lemma, i.e., the conditions $\gamma_{min} > \arccos(\theta)$ and $\gamma_{max} < 180° - \arccos(\theta)$ hold. Based on the two inequations above, $\gamma_{max} - \gamma_{min} < 180° - 2\arccos(\theta)$ and $\gamma_{max} - \gamma_{min} = 4\tau < 180° - 2\arccos(\theta)$ are true. Thus when the All-False lemma is satisfied, $\tau < \frac{180°}{4} - \frac{\arccos(\theta)}{2}$. $\square$

The range of τ is related to the minimal correlation thresholds. In this application domain, the pairs with absolute correlations over 0.3 are interesting to the domain experts. As shown in Fig. 6, All-False filtering provides stronger filtering than All-True filtering for almost all values of cone sizes and correlation thresholds. Thus we choose the cone span τ for maximizing All-False filtering conditions. The value of $\arccos(\theta)$ is less than 72.5° for $\theta \in (0.3, 1]$, so the cone span τ should not be greater than $\frac{180°}{4} - \frac{\arccos(\theta)}{2} = 8.75°$.

4.2 Experimental Results

Experiment 1: Effect of Coning. The purpose of the first experiment was to evaluate under what coning sizes the savings from filtering outweighs the overhead. When the cone is small, the time series in the cone are relatively homogeneous, resulting in a small cone span τ. Although it may result in more All-False and All-True pairs of cones, such cone formation incurs more filtering overhead because the number of cones is substantially increased and the number of filtered instances in each All-False or All-True pair is small. When the cone is large, the value of the cone span τ is large, resulting in a decrease in the number of All-False and All-True pairs. The effects of the All-False and All-True filtering in the given data are investigated.

Experiment 2: Effect of Minimal Correlation Thresholds. In this experiment, we evaluated the performance of the filtering algorithm when the minimal correlation threshold is changed. Various minimal correlation thresholds were tested and the trends of filtering efficiency were identified with the change of minimal correlation thresholds.

Effect of Coning. This section describes a group of experiments carried out to show the net savings of the algorithm for different cone sizes. For simplicity, we only changed the cone size for one dataset. According to the analysis of the previous section, the land cone size is fixed at 1×1. We carried out a series of experiments using the fixed minimal correlation threshold, the fixed land cone size, and various ocean cone sizes. The minimal correlation threshold θ was fixed at 0.5. Fig. 7 (a) shows the net savings as a percentage of the computational cost of the nested loop join algorithm for different ocean cone sizes. The x-axis represents the different cone sizes ranging from 1×1 to 6×6, and the y-axis represents the net savings in computational cost as a percentage of the costs using the simple nested loop join algorithm. The net savings range from 40 percent to 62 percent.

Effect of Minimal Correlation Thresholds. In this experiment, we investigated the effects of minimal correlation threshold θ on the savings in computation cost for correlation analysis. The land and ocean cone sizes were fixed at 1×1 and 3×3 respectively, and a series of experiments was carried out for different θs. Fig. 7 (b) shows the total savings as a percentage of the computational cost of the nested loop join algorithm for different θs. The x-axis represents the different cone sizes ranging from 1×1 to 6×6, and the y-axis represents the total savings as a percentage of the computational cost of the nested loop join algorithm. The net savings percentages range from 44 percent to 88 percent with the higher savings at higher values of correlation thresholds. Thus when other parameters are fixed, the filtering algorithm generally achieves better performance as the minimal correlation threshold is increased.

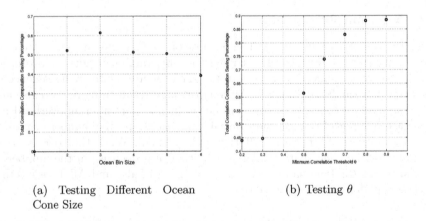

(a) Testing Different Ocean
Cone Size

(b) Testing θ

Fig. 7. Testing Different Cone Sizes and Minimal Correlation Threshold θ

5   Conclusion and Future Work

In this paper, a filter-and-refine correlation analysis algorithm for a pair of spatial time series datasets is proposed. Experimental evaluations using a NASA Earth science dataset are presented. The total savings of correlation analysis computation range from 40 percent to 88 percent. In future work, we would like to explore other coning methods, such as clustering and time series dimensionality reduction and indexing methods [1,4,6]. Clustering and spatial methods using other schemes may provide higher filtering capabilities but possibly with higher overheads. Time series dimensionality reduction and indexing methods will also be explored to determine the tradeoff between filtering efficiency and overhead.

Acknowledgment. This work was partially supported by NASA grant No. NCC 2 1231 and by Army High Performance Computing Research Center(AHPCRC) under the auspices of the Department of the Army, Army Research Laboratory(ARL) under contract number DAAD19-01-2-0014. The content of this work does not necessarily reflect the position or policy of the government and no official endorsement should be inferred. AHPCRC and Minnesota Supercomputer Institute provided access to computing facilities. We are particularly grateful to NASA Ames Research Center collaborators C. Potter and S. Klooster for their helpful comments and valuable discussions. We would also like to express our thanks to Kim Koffolt for improving the readability of this paper.

References

1. R. Agrawal, C. Faloutsos, and A. Swami. Efficient Similarity Search In Sequence Databases. In *Proc. of the 4th Int'l Conference of Foundations of Data Organization and Algorithms*, 1993.
2. G. Box, G. Jenkins, and G. Reinsel. *Time Series Analysis: Forecasting and Control.* Prentice Hall, 1994.
3. B.W. Lindgren. *Statistical Theory (Fourth Edition).* Chapman-Hall, 1998.
4. K. Chan and A. W. Fu. Efficient Time Series Matching by Wavelets. In *Proc. of the 15th ICDE*, 1999.
5. N. Cressie. *Statistics for Spatial Data.* John Wiley and Sons, 1991.
6. Christos Faloutsos. *Searching Multimedia Databases By Content.* Kluwer Academic Publishers, 1996.
7. R. Grossman, C. Kamath, P. Kegelmeyer, V. Kumar, and R. Namburu, editors. *Data Mining for Scientific and Engineering Applications.* Kluwer Academic Publishers, 2001.
8. D. Gunopulos and G. Das. Time Series Similarity Measures and Time Series Indexing. *SIGMOD Record*, 30(2):624–624, 2001.
9. J. Han and M. Kamber. *Data Mining: Concepts and Techniques.* Morgan Kaufmann Publishers, 2000.
10. E. Keogh and M. Pazzani. An Indexing Scheme for Fast Similarity Search in Large Time Series Databases. In *Proc. of 11th Int'l Conference on Scientific and Statistical Database Management*, 1999.
11. Y. Moon, K. Whang, and W. Han. A Subsequence Matching Method in Time-Series Databases Based on Generalized Windows. In *Proc. of ACM SIGMOD*, Madison, WI, 2002.
12. C. Potter, S. Klooster, and V. Brooks. Inter-annual Variability in Terrestrial Net Primary Production: Exploration of Trends and Controls on Regional to Global Scales. *Ecosystems*, 2(1):36–48, 1999.
13. J. Roddick and K. Hornsby. Temporal, Spatial, and Spatio-Temporal Data Mining. In *First Int'l Workshop on Temporal, Spatial and Spatio-Temporal Data Mining*, 2000.
14. S. Shekhar and S. Chawla. *Spatial Databases: A Tour.* Prentice Hall, 2002.

15. S. Shekhar, S. Chawla, S. Ravada, A. Fetterer, X. Liu, and C.T. Lu. Spatial databases: Accomplishments and research needs. *IEEE Transactions on Knowledge and Data Engineering*, 11(1):45–55, 1999.

16. M. Steinbach, P. Tan, V. Kumar, C. Potter, S. Klooster, and A. Torregrosa. Data Mining for the Discovery of Ocean Climate Indices. In *Proc of the Fifth Workshop on Scientific Data Mining*, 2002.

17. P. Tan, M. Steinbach, V. Kumar, C. Potter, S. Klooster, and A. Torregrosa. Finding Spatio-Temporal Patterns in Earth Science Data. In *KDD 2001 Workshop on Temporal Data Mining*, 2001.

18. G. H. Taylor. Impacts of the El Niño/Southern Oscillation on the Pacific Northwest. http://www.ocs.orst.edu/reports/enso_pnw.html.

19. Michael F. Worboys. *GIS – A Computing Perspective*. Taylor and Francis, 1995.

When to Update the Sequential Patterns of Stream Data?

Qingguo Zheng, Ke Xu, and Shilong Ma

National Lab of Software Development Environment
Department of Computer Science and Engineering
Beijing University of Aeronautics and Astronautics, Beijing 100083
{zqg, kexu, slma }@nlsde.buaa.edu.cn

Abstract. In this paper, we first define a difference measure between the old and new sequential patterns of stream data, which is proved to be a distance. Then we propose an experimental method, called TPD (Tradeoff between Performance and Difference), to decide when to update the sequential patterns of stream data by making a tradeoff between the performance of increasingly updating algorithms and the difference of sequential patterns. The experiments for the increasingly updating algorithm IUS on the alarm data show that generally, as the size of incremental windows grows, the values of the speedup and the values of the difference will decrease and increase respectively. It is also shown experimentally that the incremental ratio determined by the TPD method does not monotonically increase or decrease but changes in a range between 20 and 30 percentage for the IUS algorithm.

1   Introduction

To enhance the performance of algorithms of data mining, many researchers [1,2,3,4,5,6,7] have focused on increasingly updating association rules and sequential patterns. But if we update association rules or sequential patterns too frequently, the cost of computation will increase significantly. For the problem above, Lee and Cheung [8] studied the problem "Maintenance of Discovered Association Rules: When to update?" and proposed an algorithm called DELL to deal with it. The important problem about "When to update" is to find a suitable distance measure between the old and new association rules. In [8], the symmetric difference was used to measure the difference of association rules. But Lee and Cheung only considered the difference of association rules, and did not consider that the performance of increasingly updating algorithms will change with the size of added transactions. Ganti et al. [9,10] focused on the incremental stream data mining model maintenance and change detection under block evolution. However, they also didn't consider the performance of incremental data mining algorithms for the evolving data.

Obviously, with the increment of the size of incremental windows, the performance of increasingly updating algorithms will decrease. If the difference between the new and old sequential patterns is too high, the size of increasingly window will become too large, therefore the performance of increasingly updating algorithm will be reduced greatly. On the other hand, if the difference is too small, the

K.-Y. Whang, J. Jeon, K. Shim, J. Srivatava (Eds.): PAKDD 2003, LNAI 2637, pp. 545–550, 2003.

increasingly updating algorithm will update the old sequential patterns very frequently, which will also consume too many computing resource. In all, we must make a tradeoff between the performance of the updating algorithms and the difference of the sequential patterns. In this paper, we use the speedup as a measure of increasingly updating algorithms and define a metric distance as the difference measure to detect the change of the sequential patterns of stream data. Based on those measures, we propose an experimental method, called TPD (Tradeoff between Performance and Difference), to estimate the suitable range of the incremental ratio of stream data by making a tradeoff between the performance of the increasingly updating algorithm and the difference of sequential patterns.

By the TPD method, we estimate the suitable range of incremental ratio of stream data for the increasingly updating algorithm IUS [12]. The experiments on the alarm data in Section 4 show that generally, as the size of incremental windows grows, the values of the speedup and the values of the difference will decrease and increase respectively. By the experiments, we can discover that as the size of original windows increases, the incremental ratio determined by TPD method does not monotonically increase or decrease but changes in a range between 18 and 30 percentage for the IUS algorithm.

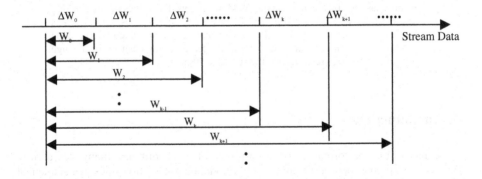

Fig. 1. Sliding stream viewing windows on the stream queue

2   Problem Definition

1. Definitions of initial window, incremental window and incremental ratio

Given stream view window W_i (i=0,1,2,3, …), ΔW_i (i=0,1,2,3, …) is called **incremental window**, iff i=0, ΔW_0 is called **initial window**, where $W_{i+1}=W_i+\Delta W_{i+1}$ (i=0,1,2,3, …). The ratio of the size of incremental window to that of the initial window, i.e. $|\Delta W_1|/|W_0|$, is called **incremental ratio** of stream data.

2. Definition of difference measure between the old and new sequential patterns

Before updating the frequent sequences L^{W_K} in W_k, we must estimate the difference between L^{W_K} and $L^{W_{K+1}}$. If the difference is very large, then we should update the sequential patterns as soon as possible. But if the difference is very small, then we do not need to update the sequential patterns. In order to measure the difference between the old and new frequent sequence sets, we can define a measure as follows

$$d(L^{W_k}, L^{W_{k+1}}) = \frac{|L^{W_k} \Delta L^{W_{k+1}}|}{|L^{W_k} \bigcup L^{W_{k+1}}|} \quad \text{if} \quad L^{W_k} \neq \Phi \text{ or } \quad L^{W_{k+1}} \neq \Phi, \quad \text{otherwise} \tag{1}$$

$d(L^{W_k}, L^{W_{k+1}}) = 0$, where $L^{W_k} \Delta L^{W_{k+1}}$ is the symmetric difference between L^{W_k} and $L^{W_{k+1}}$.

We know that a measure is not necessarily a metric. A metric distance must satisfy the triangle inequality which means that for three objects A, B and C in the metric space, the distance between A and C is greater than the distance between A and B plus the distance between B and C. For the measure defined above, we can prove that it is also a metric (please see Appendix A in [13]).

3 The TPD (Tradeoff between Performance and Difference) Method of Deciding When to Update Sequential Patterns

Lee and Cheung [8] only considered the difference between the old and new association rules, but that they didn't consider the change of the performance of increasingly updating algorithms. As mentioned before, too large difference between the new and old sequential patters will result in poor performance of increasingly updating algorithms, while too small difference will increase the computations lose significantly. Therefore, we must make a tradeoff between the difference of sequential patterns and the performance of increasingly updating algorithms and find the suitable range of the incremental ratio of stream data.

We propose an experimental method, called TPD (Tradeoff between Performance and Difference), to find the suitable range of incremental ratio of the initial window for deciding when to update sequential patterns of stream data. The TPD method uses the speedups as the measurement of increasingly updating algorithms and adopts the measure defined in Section 3 as the difference between the new and old sequential patterns of stream data. With the increment of the size of incremental window, the speedup of the algorithm will decrease, while the difference of the old and new sequential patterns will increase. According to two main factors of the increasingly updating algorithms, the TPD method maps the curve of the speedup and the difference changing with the size of incremental windows into the same graph under the same scale, and the points of intersection of the two curves are the suitable range of the incremental ratio of the initial windows for the increasingly updating algorithm.

In this paper, by the experiments in Section 4, we study the suitable range of incremental ratio of the initial windows for the increasingly updating algorithm: IUS [12] by TPD (Tradeoff between Performance and Difference) method. In the experiments, the speedup ratio of the IUS algorithm is defined as **speedup=the execution time of Robust_search / the execution time of IUS**, where Robust_search is an algorithm to discover sequential patterns from stream data [11] and IUS is an increasingly updating sequential patterns algorithm based on Robust_search algorithm [12]. We use the distance i.e. $d(L^{W_k}, L^{W_{k+1}})$ defined above as the difference measure between the old frequent sequences L^{W_k} and the new frequent sequences $L^{W_{k+1}}$.

The experiments of Section 4 show that generally, as the size of incremental windows grows, the values of the speedup and the values of the difference will decrease and increase respectively. By making data transform, called Min-max normalization [14] for the values of the speedup and the difference, we can map the speedup and the difference with the increment of the size of incremental windows into the same graph under the same scale, and then from the graph we can discover the intersection point of two lines, obviously, by which we can compute the suitable range of incremental ratio of the initial window to update the sequential patterns.

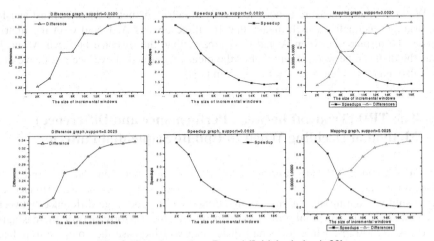

Fig. 2. Experiment 1 on Data_1 |Initial window|=20k

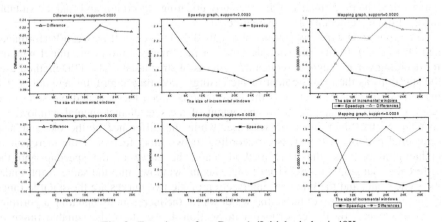

Fig. 3. Experiment 2 on Data_1 |Initial window|=40K

4   Experiments

The experiments were on the DELL PC Server with 2 CPU Pentium II, CPU MHz 397.952211, Memory 512M, SCSI Disk 16G. The Operating system on the server is

Red Hat Linux version 6.0. The data_1 in experiments are the alarms in GSM Networks, which contain 194 alarm types and 100k alarm events. The time of alarm events in the data_1 range from 2001-08-11-18 to 2001-08-13-17.

The Speedup graph in Figure 2,3 is the speedup of IUS Algorithm [11] to the Robust_search Algorithm [12] with the size of incremental windows. The Difference graph in Figure 2,3 is the difference measure of frequent sequences between the initial window and the incremental windows with the size of that increasing. In the experiments of this paper, we map the value of difference and speedup into the same range [0,1] by let $new\_min_A=0$ and $new\_max_A=1$ [14]. Then, we map the broken lines of the difference and speedup into the same graph i.e. Mapping Graph under the same scale.

In the experiment 1 on data_1, we choose the initial window $|W_0|=20K$, and update the initial sequential patterns by the incremental size of 2K, 4K, 6K, 8K, 10K, 12K, 14K, 16K, and 18K, i.e. the size of incremental window ΔW_i. The results of experiment 1 are illustrated in Figure 2. In order to find the suitable size of incremental windows, we map the graphs of speedup and difference into the same graph by the data transform above, by which we can locate the intersection point of the two lines. The intersection point is a tradeoff between the speedup and the difference, which is a suitable point to update sequential patterns. In the mapping graphs of Figure 2, the intersection point is about 6K, so the suitable range of incremental ratio of initial window is about 30 percent of initial windows W_0.

In the experiment 2 on data_1, we choose the initial window $|W_0|=40K$, and update the initial sequential patterns by the incremental size of 4K, 8K, 16K, 20K, 24K, 28K, and 32K. The results of experiment 2 are illustrated in Figure 3. In the mapping graphs of Figure 3, the intersection point is between 9K and 10K, so the suitable range is about 22.5 to 25 percent of initial windows W_0.

In all, by the experiments above, in general, as the size of incremental windows grows, the values of the speedup and the values of the difference will decrease and increase respectively. Based on the TPD method we proposed, it is shown experimentally that the suitable range of incremental ratio of initial windows to update is about 20 to 30 percent of the size of initial windows for the IUS algorithm.

5   Conclusion

In this paper, we first proposed a metric distance as the difference measure between the sequential patterns. Then we present an experimental method, called TPD (Tradeoff between Performance and Difference), to decide when to update sequential patterns of stream data. The TPD method can determine a reasonable ratio of the size of incremental window to that of original window for increasingly updating algorithms, which may depend on the concrete data. We also do two experiments of IUS algorithm [11] to verify the TPD method. From the experiments, we can see that as the size of original windows increases, the incremental ratio determined by the TPD method does not monotonically increase or decrease but changes in a range between 20 and 30 percentage. So in practice, by use of the TPD method, we can do some initial experiments to discover a suitable incremental ratio for this kind of data and then use this ratio to decide when to update sequential patterns is better.

Acknowledgement. Thanks Professor Jiawei Han's Summer Course in Beijing 2002. Thanks Professor Wei Li for the choice of subject and guidance of methodology, Professor YueFei Sui and the anonymous Reviewers' suggestions for this paper. This research was supported by National 973 Project of No.G1999032701 and No.G1999032709.

References

1. D. W. Cheung, J. Han, V. T. Ng, and C. Y. Wong, "Maintenance of discovered association rules in large databases: An incremental update technique," *In Proceedings of 12th ICDE'9)*, pages 106–114, Louisiana, USA, February 1996.
2. D. W. Cheung, S. D. Lee, and B. Kao, "A General Incremental Technique for Maintaining Discovered Association Rules," *In Proceedings of the 5th DASFAA'97*, pages 185–194, Melbourne, Australia, April 1997.
3. F. Masseglia, P. Poncelet and M.Teisseire, "Incremental Mining of Sequential Patterns in Large Databases (PS)," *BDA'00*, Blois, France, October 2000.
4. S. Parthasarathy, M. J. Zaki, M. Ogihara, and S. Dwarkadas, "Incremental and Interactive Sequence Mining," *In Proceedings of the 8th CIKM'99*, pages 251–258, USA, November 1999.
5. Necip Fazil Ayan, Abdullah Uz Tansel, and Erol Arkun, "An efficient algorithm to update large itemsets with early pruning," *Proceedings of the fifth KDD'99*, August 15-18, 1999, San Diego, CA USA pp. 287–291.
6. S. Thomas, S. Bodagala, K. Alsabti, and S. Ranka, "An efficient algorithm for the incremental updating of association rules in large database," *In Proceedings of the 3rd KDD'97*, pages 263–266, California, USA, August 1997.
7. Ahmed Ayad, Nagwa El-Makky and Yousry Taha, "Incremental Mining of Constrained Association Rules," *ICDM'2001 MINING*, April 5–7, 2001, USA
8. S.D. Lee and D.W. Cheung, "Maintenance of Discovered Association Rules: When to update?," *Proc. 1997 Workshop on Data Mining and Knowledge Discovery (DMKD'97)* with ACM-SIGMOD'97,Tucson, Arizona, May 11, 1997.
9. V. Ganti, J. Gehrke, R. Ramakrishnan, and W.-Y. Loh, "A frame work for measuring changes in data characteristics," In Proceedings of the 18th Symposium on Principles of Database Systems, 1999.
10. V. Ganti, J. Gehrke and R. Ramakrishnan, "Mining Data Streams under Block Evolution," SIGKDD Explorations, Volume 3, Issue 2, pp. 1–11, January, 2002.
11. Q. Zheng, K. Xu, W. Lv, S. Ma, "Intelligent Search of Correlated Alarms from Database Containing Noise Data," Proc. of the 8th International IFIP/IEEE NOMS 2002, Florence, Italy, April, 2002, http://arXiv.org/abs/cs.NI/0109042
12. Q. Zheng, K. Xu, W. Lv, S. Ma, "The Algorithms of Updating Sequential Patterns" The Second SIAM Data mining'2002: workshop HPDM , Washington, USA, April 2002. http://arXiv.org/abs/cs.DB/0203027
13. Q. Zheng, K. Xu, S. Ma, "When to Update the sequential patterns of stream data", Tecnical Report, NLSDE, China, 2002, http://arXiv.org/abs/cs.DB/0203028
14. J. Han and M. Kambr, "DATA MINING Concepts and Techniques", p. 115, Morgan Kaufmann Publisher, 2000

A New Clustering Algorithm for Transaction Data via Caucus

Jinmei Xu[1], Hui Xiong[2], Sam Yuan Sung[1], and Vipin Kumar[2]

[1] Department of Computer Science, National University of Singapore
Kent Ridge 117543, Singapore
{xujinmei, ssung}@comp.nus.edu.sg
[2] Department of Computer Science, University of Minnesota-Twin Cities,
Minneapolis MN 55455, USA
{huix, kumar}@cs.umn.edu

Abstract. The fast-growing large point of sale databases in stores and companies sets a pressing need for extracting high-level knowledge. Transaction clustering arises to receive attentions in recent years. However, traditional clustering techniques are not useful to solve this problem. Transaction data sets are different from the traditional data sets in their high dimensionality, sparsity and a large number of outliers. In this paper we present and experimentally evaluate a new efficient transaction clustering technique based on cluster of buyers called *caucus* that can be effectively used for identification of center of cluster. Experiments on real and synthetic data sets indicate that compare to prior work, caucus-based method can derive clusters of better quality as well as reduce the execution time considerably.

1 Introduction

A core issue in data mining is to automatically summarize huge amount of information into simpler and more understandable categories. One of the most common and well-suited ways of categorizing is done by partitioning the data objects into disparate groups called *clusters*. Transaction clustering is an important application of clustering and has received increasing attentions in recent developments of data mining [9,11,14,15]. Transaction clustering is to partition transaction database into clusters such that transactions of the same cluster are as similar as possible, and transactions from different clusters are as dissimilar as possible. Clustering techniques have been extensively studied in a wide range of application domains, but only in recent years have new clustering algorithms been created by the data mining domain. A recent review of these methods is given in [11]. However, most traditional clustering approaches are not suitable to solve the transaction clustering problem. This is because most existing clustering methods adopt a similarity that is measured on the basis of pair-wise comparisons among transactions. The comparisons are performed either based on distance or probability. These approaches can effectively cluster transactions when the dimensionality of the space (i.e., the number of different items) is low

K.-Y. Whang, J. Jeon, K. Shim, J. Srivatava (Eds.): PAKDD 2003, LNAI 2637, pp. 551–562, 2003.

and most of the items are presented [2], but are not suitable to a data model where items are large and each transaction is short.

A new direction of approaches is to measure similarity using the concept of "clustering of items" or "large items" [14,15]. An item in a cluster is called a large item if the support of that item exceeds a pre-specified *minimum support threshold* (i.e., an item that appeared in a sufficient number of transactions) [8]. Otherwise, it is defined as a small item. The criterion of a good clustering in [14] is that there are many large items within a cluster and there is little overlapping of such items across clusters. In [9], items are first partitioned into clusters of items, and each transaction is then assigned to the cluster, which has the highest percentage of intersection with the respective transaction. However, there are some problems that cannot be well handled using these approaches. We will discuss the details in next section.

In order to group transactions into meaningful clusters, we define an underlying model of the buyer's behavior, in which a transactional dataset is partitioned into three main parts: *a background table, a foreground table* and *a relationship table*. The background table is a collection of background attributes related to the customer profile such as *age, gender, employment, work hours, income level,* etc. The attributes of the foreground table are purchase transactions of a supermarket basket type.

In this paper we use *caucus*, the cluster of buyers, for clustering transactions. We define caucus as a set of background attributes together with the associated values/value ranges that cause specific type of buying behavior. Ideally, a caucus has two characteristics. One is that the customers in each caucus behave in a highly similar way. The other is that customers in different caucuses behave differently. Suppose that we have a database of customers on insurance policy options. Each policy is designed purposefully for a particular kind of customers. The database describes attributes of the customers, such as their name, age, income, residence area, occupation and credit rating. The customers can be classified according to the attributes as to form different segments. Caucus differs from the segments for it represents a group of buyers exhibit similar behavior and similar profile.

The basic idea of introducing caucus to cluster transactions is primarily motivated by the fact that the initial choice of centers of clusters is sensitive for later clustering. With the goal to find a "good" cluster center close to prototype, caucus is used as an integrated part of our clustering scheme. It turns out that caucus is particularly useful in the initial analysis of clustering transaction data. Since caucuses differ in the purchase characteristics, large items in each caucus may be distinct from other caucuses. Thus, we can view the frequently purchased set of items in a caucus as center of this caucus.

One feature of our clustering procedure is we group on customers first then transactions, rather than directly group on transactions. In the latter, the mistakes of assigning transactions into wrong clusters may result in serious problems as reflected in the quality of clustering. Researchers have found that it is not easy to correct these errors. The new approach tries to improve the clustering's accuracy level by decomposing related customer population into distinct segments based on real transaction data and customer profile. Clusters are then built on the bases of these segments. Another advantage of the caucus method is its low sensitivity for the order of input data. As a result, small insertion and deletion of records will not affect the clustering negatively. In addition, the caucus-based method can achieve better interpretability and usefulness of the resulting clusters rather than offering raw clusters. In this way, the efficiency of a number of data mining tasks like profile association rules mining and customer predicting can be enhanced dramatically.

The remainder of this paper is organized as follows. Section 2 provides a review on the related works. In section 3, we describe how to cluster transactions with caucuses. Section 4 shows experimental results. Finally, in section 5, we conclude the paper.

2 Related Work

The increasingly growing sales data demand cost-effective transaction clustering approaches. Approach in [14] uses large item as similarity measure of a cluster when clustering transactions. A global cost function which is based on a degree of similarity is as follows. For a clustering $C = \{C_1, ..., C_k\}$, the cost of C is

$$Cost(C) = \omega \times Intra(C) + Inter(C)$$

where ω is a weight with default value of 1. $Intra(C)$ is the intra-cluster cost and is defined as: $Intra(C) = |\bigcup_{i=1}^{k} Small_i|$, where $Small_i$ is the set of small items in cluster C_i. $Inter(C)$ is the inter-cluster cost and is defined as: $Inter(C) = \sum_{i=1}^{k} |Large_i| - |\bigcup_{i=1}^{k} Large_i|$, where $Large_i$ is the set of large items in cluster C_i. Given a collection of transactions and a minimum support (for determining large items), objective of this approach is to find a clustering C such that $Cost(C)$ is minimum.

Consider the sample transaction dataset shown in Figure 1. The rows are transaction records and the columns are items purchased by customers. The darker the region, the more items were purchased. Suppose A and B are sets of large items purchased of category S_1 and S_2, respectively. In category S_3, we represent the set of small items as C, which we can ignore in our discussion.

For large item based approach, when clustering with two clusters, the choices are {A, B}, {A, AB}, or {B, AB}, as shown in Figure 2. The best 2-clusters clustering is Figure 2(a) or 2(c), which has the minimum cost, $|B|$, where $|B|$ represents the number of large items in set of B. Unfortunately, this is obviously not a good clustering. The best clustering should be a clustering of 3-clusters, {A, B, AB}, as shown in Figure 2(d). However, this clustering has a higher cost of $|A| + |B|$, and therefore cannot be produced as the optimal clustering using large item based approach. An improved method in [15], which we call SLR method, attempts to speedup large item based method by introducing a new metric called *small large ratio*. However, since it inherits the same cost function from [14], method in [15] inevitably suffers from the problem mentioned above.

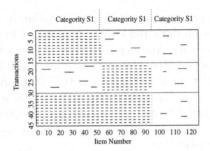

Fig. 1. A sample transaction dataset.

In [9], items are first grouped based on association rules into disjoint "clusters of items". Clustering of transactions is then performed by assigning each transaction to exactly one of the "clusters of items" discovered in the first phase. Thus, there is the same number of "clusters of transactions" as the number of "clusters of items". Therefore, overlapping of items cannot occur between different clusters of transactions. In the above example, using this clustering of items approach, the choice is either a 1-cluster {AB} (if only one cluster of item), or a 2-clusters {A, B} (if there are two clusters of items). The 3-clusters clustering of Figure 2(d) can never occur. we observe that in transaction clustering,

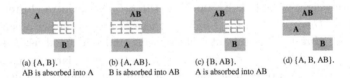

(a) {A, B}.
AB is absorbed into A

(b) {A, AB}.
B is absorbed into AB

(c) {B, AB}.
A is absorbed into AB

(d) {A, B, AB}.

Fig. 2. Clustering using 2-clusters or 3-clusters

the difficulty arises when clusters have overlap. The approach in [9] does not provide a mechanism to handle the overlapping case. All clusters found are non-overlapping. The approach in [14] discourages splitting the large items between clusters by increasing inter-cost. This adversely forces transactions containing the same large items stick together, and therefore cannot cope with the overlapping case. We propose a new method to handle this problem.

3   Caucus Construction and Transaction Clustering

The problem of finding cluster centers has been extensively investigated both in the database and computer geometry domains. A well-known general method is k-medoids method [10], which uses points in the original data set to serve as surrogate centers for clusters. Approaches in computational geometry try to devise sophisticated data structures to speed up the algorithm. Since the initial choice of cluster's center is critical for later clustering, we are motivated to find valid set of cluster centers by combining customer profiles and behavioral information which can be achieved through *caucus*.

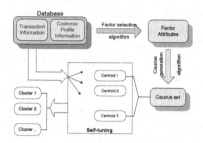

Fig. 3. The architecture of caucus-based clustering algorithm

Following the above consideration, our caucus-based clustering method proceeds in three phases: *factor extraction, caucus construction* and *clustering trans- actions with caucus.* The first phase removes redundant background attributes and keeps the most relevant attributes (factors) that cause the purchase differ- ence of customers. Since one major focus in this paper will be on generating *caucus,* which we can derive using the factors. Phase 2 gives methodology of caucus construction. In phase 3, clusters are formed on the basis of caucuses constructed. Figure 3 presents an overview of caucus-based clustering algorithm.

3.1   Background Attributes

The background table contains a set of background attributes. Background at- tribute refers to customer profiles such as marketing or socio-economic attributes. With the availability of web tools and information technology, collection of cus- tomer background information becomes simpler. In general, transactions of par- ticular members can be stored up and tagged in the sale database. Investigating the transaction data and customer background information, useful data could be extracted for further decision.

It is observed that in our model, there is a presence of interesting correlation between background attributes and buying behavior. Although different buy- ers purchase different items in different departments and in different amounts, customers with similar profiles tend to have similar purchase behavior. Some evidence that this is the case is provided e.g. by Aggarwal. C. C. *et.al* [3] who investigated the highly related patterns between customer profile and behavioral information. For example, ({marital status = married, income level = low} ⇒ {fast food}). In the preceding example, it can be seen that *marital status* and *income level* are critical in determining the purchase behavior of such customers. We define such causative background attribute as *factor.*

3.2 Phase 1: Identify Potentially Important Background Attributes

Potentially important background attributes are those that affect the purchase behavior of customers. We call such attributes factors. Specifically, background attribute A_i is factor if A_i gives rise to bigger impact on the purchase behavior.

Our work of factor extraction is most similar to feature subset selection or attribute relevance analysis, where the term feature represents an object or data instance from the original data. Feature selection may be performed using methods such as C4.5, genetic algorithm, or neural network. C4.5 runs over the training data set and the features that appear in the pruned decision tree are selected. This method has higher efficiency on some data but poor accuracy of features extracted [12]. Genetic algorithm and neural network method select features with a higher level of accuracy [13], whereas, the algorithms time complexity are expensive.

For our purpose, we propose a simple yet efficient method to extract factor attributes. Our method adopts entropy [6] as a measure of the goodness of the background attributes with the caveat that the best entropy of an attribute is obtained when customers with the same attribute value having the same purchase. Thus, factors can be found by ranking their entropies. Table 1 shows the notations used throughout this paper.

Table 1. Notations

A	Set of background attributes. $A_1, A_2, ..., A_w$	C	Set of caucuses
n	Number of items	I	Set of items $I_1, I_2, ..., I_n$
d	Number of factors selected	F	Set of factors. $F_1, F_2, ..., F_d$
t	Number of transactions in database	T	Set of transactions
m_i	Number of values in attribute A_i	K	Number of final caucuses

Let background attribute A_i ($A_i \in A$) has N value sets, $V_{A_i} = \{v_1, v_2, ..., v_N\}$. Then each value set consists of one value or multiple values. If each v_r has only one value, then $N = m_i$. We write V_{A_i} as V if A_i is understood. For each value set v in V, there is an n-dimensional vector f_{v,A_i}. The kth dimension in f_v is defined as $f_{vk} = |T_{A_i=v}|$, where $|T_{A_i=v}|$ is the number of transactions in the database having item I_k bought by people whose attribute A_i's value is in the value set v. For example, assume A_i is age, which ranges from 1 to 90. For each value set v in age, there is a corresponding vector $f_{v,age}$ whose dimension is n (number of items in database). The kth dimension of $f_{v,age}$ is then defined as the number of transactions in the database having item I_k bought by people whose age fall in the range of the value set v. In this way, for any two value sets v_r and v_s, we define the entropy between them as:

$$E_{(v_r,v_s),A_i} = -\left(\frac{1 - S_{rs}}{2} log \frac{1 - S_{rs}}{2} + \frac{1 + S_{rs}}{2} log \frac{1 + S_{rs}}{2}\right) \quad (1)$$

where S_{rs} is the similarity value normalized to [0, 1], between f_{v_r} and f_{v_s}. The similarity value $S_{rs} = e^{-\alpha \times D_{rs}}$, where D_{rs} is the distance between f_{v_r} and f_{v_s}, and α is a parameter. Specifically, in [4] the decay constant α is assigned to be 0.5 since it is robust for all kinds of data sets. Euclidean distance measure is used to calculate the distance D_{rs}. It is defined as: $D_{rs} = [\sum_{k=1}^{n}(f_{v_r k} - f_{v_s k})^2]^{\frac{1}{2}}$.

For attribute A_i, if it is a categorical attribute, then for each value v_r of A_i, we define $E_{r,A_i} = E_{(v_r, \bar{v}_r), A_i}$, where $\bar{v}_r$ represents the value set $V_{A_i} - v_r$. If A_i is a continuous attribute, then for each v_r, we define $E_{r,A_i} = E_{(v_{r'}, v_{r''}), A_i}$, where $v_{r'}$ represents the value set $\{v | v \leq v_r\}$ and $v_{r''}$ represents the value set $\{v | v > v_r\}$. With the entropy for each value that partitions the whole values into two value sets, we can calculate the average entropy of all values, which is given as: $E_{V,A_i} = \frac{1}{N} \sum_{r=1}^{N} E_{r,A_i}$.

3.3 Phase 2: Caucus Construction

Since we have d factors, we then have d-dimensional factor space of caucuses. To achieve the goal of minimizing the difference of customers within the same caucus as well as maximizing the variance between caucuses, we propose the following approach.

```
Input: C = {C₁, C₂, ..., Cₗ}
Output: Eₛ // Eₛ keeps the largest entropies for each caucus
1. Eₛ = LargestEntropySet(C)
2. While( |C| ≤ K ) {
3.    Let E_{Vₗ,Fₚ,Cₖ} = max(Eₛ)
4.    Eₛ = Eₛ − E_{Vₗ,Fₚ,Cₖ}
5.    Divide Cₖ into two new caucuses with Fₚ and vₗ as
Cₖ₁ = Cₖ ∧ {Fₚ < vₗ}, Cₖ₂ = Cₖ ∧ {Fₚ ≥ vₗ}, C = C − Cₖ + Cₖ₁ + Cₖ₂
6.    Let C' = {Cₖ₁, Cₖ₂}
7.    Eₛ = Eₛ ∪ LargestEntropySet(C') }\*while*\

LargestEntropySet(C) {
1.    For each Caucus Cᵢ in C {
2.       For each Fⱼ (Fⱼ ∈ F) {
3.          For each value v of Fⱼ, and {Fⱼ = v} in Cᵢ {
4.             Divide Fⱼ into two value sets in Cᵢ: Vᵥ = {{Fⱼ < v}, {Fⱼ ≥ v}}
5.             Compute E_{Vᵥ,Fⱼ,Cᵢ} and Let E_{Vₗ,Fₚ,Cᵢ} be the maximal entropy for the caucus Cᵢ
6.          Let E = {E_{Vₗ,Fₚ,Cᵢ}|Cᵢ ∈ C} }}
7. return E }
```

Fig. 4. The caucus construction algorithm.

We can first construct a base caucus set by combining values of categorical factors and quantitative factors having a few values. Assuming that we get l base caucuses, we have $C = \{C_1, C_2, ..., C_l\}$. Suppose quantitative factor F_i, like *age* or *yearly income*, $V = \{v_1, v_2, ..., v_N\}$. For each value set v in V and the caucus C_j, an n-dimensional vector f_{v,F_i,C_j} can be defined. For the kth dimension in f_v, $f_{vk} = \frac{|T_{C_j \wedge F_i = v}|}{|T_{C_j}|}$, where $|T_{C_j}|$ is the number of transactions in T having item I_k bought by people in caucus C_j and $|T_{C_j \wedge F_i = v}|$ is the number of transaction in T having item I_k bought by people in caucus C_j whose attribute F_i's value is in the value set v. We can calculate entropy of each possible split using the entropy function (1).

The procedure $LargestEntropySet(C)$ computes entropy values for the caucus set C and returns the set of the largest entropy for each caucus. Line 1 gets the largest entropy values for the base caucuses set C. Line 7 obtains the largest entropies for the two newly-generated caucuses. Furthermore, E_s is updated by uniting with these two caucuses.

Recall that entropy is a measure of the unpredictable nature of a set of possible elements in a training data set [6,7]. The higher the level of variation within the training set, the higher the entropy. Line 5 divides one caucus into two new caucuses with maximal entropy for this division, which help find two new caucuses having the most different buying behavior.

We examine the worst case complexity of caucus construction algorithm. In $LargestEntropySet(C)$, step 6 executes $O(tn)$ times. This is because that the database is scanned for calculation of f_{vr} and f_{vs}. Thus the overall time of procedure $LargestEntropySet(C)$ is $O(KdNtn)$, where $N = \sum_{j=1}^{d} N_j$. In the main program, since the number of caucuses in C ranges from l to K, the "while" loop will execute $K - l + 1$ times. Step 2 to Step 7 take $O(l(K - l + 1))$ time. Thus, the overall computational complexity of caucus generation algorithm is $O(l(K - l + 1)) + O(KdNtn)$, where $N = \sum_{i=1}^{d} N_i$. Since the value of d, K and l are expected to be far lower than the dimensionality of database, we effectively maintain linear time complexity. The total time complexity is $O(Ntn)$.

3.4   Phase 3: Caucuses for Transaction Clustering

In this section, we discuss how to generate transaction clusters. With given K caucuses, a direct k-clusters clustering is computed as follows. Initially, each transaction is assigned to a cluster according to the caucus it belongs to. So, exactly K clusters are generated and each cluster corresponds to one caucus.

For each transaction t, we calculate the similarity between t and centers of existing clusters. The similarity between t and cluster C_i is computed using the Hamming distance, which is defined as $Hamming(x, y) = \sum_{j=1}^{n} \chi(x_j, y_j)$, where $\chi(x_j, y_j)$ equals 1 if x_j and y_j are different, and equals to 0 otherwise. Transaction t is then assigned to the most similar cluster. Suppose we have K initial clusters, the caucus-based clustering algorithm is given in Figure 5.

The procedure $ModifyLargeItem(C_w)$ in line 10 modifies the large items of cluster C_w, namely, updates the center vector of cluster C_w. In the implementation, the update of clusters' centroids is performed only when clusters have new members. The refinement is an iterated process. The iteration will stop when clustering becomes stable. The time complexity of clustering algorithm above is $O(K^2 t)$, where K is the number of caucuses.

4   Experiments

This section includes experimental evaluation of our algorithm. We carried out experiments on various data sets. Experiments are performed on a Dell work-

/*Cluster Allocation Phase*/

1. relocate = **false**

2. **While** not the end of the database file {

3.   Read the next transaction $< t,$ cluster_number$>$

4.   **For** each cluster C_i {

5.   Compute the distance between t and cluster C_i

6.   Search for C_w that minimize the distances among all clusters }

7.   **If** t.cluster_number $\neq w$ {

8.   write $< t, C_w >$, relocate = **true**;

9. t.cluster_number $= w$;

10. $ModifyLargeItem(C_w);$} }*while*\

/*Cluster Refinement Phase*/

11. **While** relocate = **true**; {

12.   eliminate empty clusters; relocate = **false**;

13. **Repeat** Line 2 to 10}

Fig. 5. Caucus-based clustering algorithm

station with a 1.6 GHz Pentium 4 processor and a memory of 512 MB. We test two performance functions with our algorithm:

(1) **Accuracy:** The accuracy of caucus-based method is determined by two aspects: the optimal cluster number and the quality of clusters. For our evaluation purpose, we apply a relative cluster validity index, the Davies-Bouldin validity index [5], which is given as: $DB = \frac{1}{n}\sum_{i=1}^{n} \max_{i \neq j}\{\frac{Q_i + Q_j}{M(i,j)}\}$, where n is the number of clusters, Q_i is the dispersion (e.g. average distance of all objects from the cluster to their cluster center) of cluster C_i and C_j, and $M(i, j)$ is the distance between clusters centers. This index tries to measure the ratio of the sum of within-cluster scatter to between-cluster separation. We also apply this index to find the optimal cluster number. Since the number of clusters correlates closely with initial caucus number, by tuning the number of final caucuses, we try to find one minimizing the DB index.

(2) **Scalability:** We test the performance of the algorithm as the number of transactions and the dimensionality (number of items in database) increase.

4.1 Real Dataset

It is possible but difficult to find transaction data sets containing customer background information. We take the real data set in [16], which consists of 9822 tuples of insurance policy transactions. Each transaction contains 43 statistically socio-demographic attributes. All insurance polices in it is treated as items. There are 21 distinct items involved in this transaction database. Five attributes are extracted as factors for caucus construction, namely, *age, car number, customer type, education level* and *marital status* with their entropy values are 0.17, 0.11, 0.14, 0.11 and 0.13 respectively. Thus, a base caucus set of 18 caucuses can be yielded through combining values of factor *car number* (0, 1 and 2), *education level*(low, middle and high) and *marital status* (married and single).

We compare caucus-based method with two alternative approaches introduced in [14] and [15], which we call LargeItem and SLR methods in the following sections. The experimental result is shown in Table 2.

The experiments show that LargeItem, SLR and caucus-based method are able to detect outlier transactions. We can see that the clusters produced by three methods differ. Note that small values of DB occur for a solution with small variance within segments and high variance between segments. The smaller DB value is, the better quality of clusters. Here better clusters means that clusters are internally compact and well separated. From Table 2, we can see that caucus-based method outperforms largeItem and SLR methods with regard to the quality of clusters produced.

Table 2. DB index of LargeItem, SLR and caucus method on real data

Caucus Method		LargeItem Method		SLR Method	
Cluster No	DB index	Cluster No	DB index	Cluster No	DB index
8	0.1227	15	0.2153	9	0.1630

We also perform tests of sensitivity on the order of input data for caucus method. We randomly rearrange the sequence of transactions in the input file. The results have a slight difference from the original clustering result with a best DB value reaching 0.1074 for 8 clusters and a worst DB value reaching 0.1320 for 8 clusters. The reason is that in caucus-based method cluster's center is identified by caucus. Although later refinement process moves transactions between clusters, centroids of clusters are not changed much. Thus, the resulting set of clusters is not significantly affected by the order of input data. These results show that, compared to caucus-based method, LargeItem and SLR methods are more sensitive to the order of input data.

4.2 Synthetic Data

In order to have better control over the impact of several data characteristics, we use transaction data files created by the well-known synthetic data generator [8]. We modify the method to include multiple caucuses in a single data set. The generation of caucuses involves the generation and assignment of different items to different caucuses. Additionally, we generate large itemsets for each caucus from its items. These two together result in the difference between the large itemsets of the different caucuses. We also generate a customer profile database. For each record, it is generated with 9 fields: *gender, marital status, race, income level, education level, work hours,work place, occupation* and *age*. The value of each field is generated following different distributions such as *age* follows normal distribution and *gender* follows uniform distribution. The support value is set to be 0.5%. For the transactions of a caucus, they are generated using the itemsets for that particular caucus. We set a metric c for corruption level associated with each itemset. An item is dropped from an itemset being added to a transaction as long as a uniformly distributed random number between 0 and 1 is less than this metric c. The mean and standard deviation of c are 0.5 and 0.1 respectively. We generate three data sets T1, T2 and T3 with the number of items are 500, 1000 and 5000 respectively and the number of tuples in database are 5000, 10000 and 50000. The average size of an itemset is set to be 4 and the average size of transaction is set to be 10.

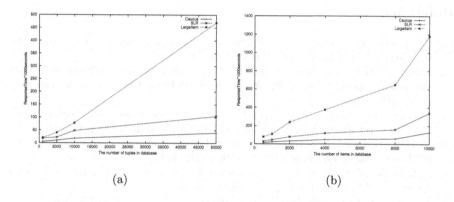

(a)   (b)

Fig. 6. Performance test: a) CPU time vs. Database Size b) CPU time vs. Item number.

Figure 6 shows the scalability test of largeItem, SLR and caucus-based methods. As the database size and item number increase, the overall running time of LargeItem increases rapidly while that of SLR and caucus-based algorithm increases linearly. This is very useful because it is very common to have very large databases in practice. In addition, the caucus-based method is radically faster than LargeItem, being able to cluster 10000 transactions in the same CPU time that LargeItem takes for only slightly more than 1200 transactions. The efficiency of caucus-based method is better compared to SLR method.

5 Conclusion

This paper described a clustering algorithm for transactional data. Fundamentally different from most clustering algorithms, this approach attempts to group customers with similar behavior, which we call caucus, first and then transactions. In this approach, one first determines a set of background attributes that would be considered significant. A set of caucuses is then constructed to identify the initial centers of clusters. Experiments of the proposed algorithm are performed against large synthetic and real data sets. The experimental results show that the proposed method can improve the clustering quality and has a good scale-up ability for large sale database.

Acknowledgments. This work was partially supported by NSF grant # ACI-9982274, DOE contract # DOE/LLNL W-7045-ENG-48 and by Army High Performance Computing Research Center contract number DAAD19-01-2-0014. The content of this work does not necessarily reflect the position or policy of the government and no official endorsement should be inferred.

References

1. R. Agrawal, R. Srikant: Fast algorithms for mining association rules. Proc. 1994 Int. Conf. on Very Large Data Bases. (1994) 487–499.
2. Charu C. Aggarwal, Philip S. Yu: Finding generalized projetedclusters in high dimensional spaces. Proc. 2000 ACM SIGMOD Inter. Conf. on Management of Data, (2000) 70–81.
3. Charu C. Aggarwal, Zheng S., Philip S. Yu: Finding Profile Association Rules. Proc. 1998 Int. conf. on Information and Knowledge Management, (1998).
4. M. Dash, H. Liu, J. Yao: Dimensionality reduction of unsupervised data. Proc. 9th IEEE Inter. Conf. on Tools with AI, (1997) 532–539.
5. D. L. Davies, S. W. Bouldin: A cluster separation measure. IEEE Trans. on Pattern Analysis and Machine Intelligence, (1979) PAMI-1(2) 224–227.
6. J. D. Fast: Entropy: the significance of the concept of entropy and its applications in science and technology. Chapter of The Statistical Significance of the Entropy Concept. Eindhoven: Philips Technical Library, (1962).
7. M. Gluck, J. Corter: Information, uncertainty, and the utility of categories. Proc. 7th Conf. Cognitive Science Society, (1985) 83–87.
8. R. Agrawal, R. Srikant: Fast algorithms for mining association rules. Proc. of the 20th Int'l Conf. on Very Large Database (VLDB), 1994 487–499.
9. E-H(Sam) Han, G. Karypis, B. Mobashad: Hypergraph Based Clustering In High-dimensional Data Sets: A Summary of Results. In Bulletin of the Technical Committee on Data Engineering, (1998).
10. L. Kaufman, P. J. Rousseeuw: Finding Groups in Data: An Introduction to Cluster Analysis. New York: John Wiley & Sons, (1990).
11. B. Pavel: Survey of Clustering Data Mining Techniques. (2002) http://www.accrue.com/products/rp_cluster_review.ps
12. J. R. Quinlan: C4.5: programme for machine learning. Morgan Kaufmann (1993).
13. R. Setiona, H. Liu: Neural Network Feature Selector IEEE Trans. Neural Networks, Aug, (1997) 654–662.
14. K. Wang, C. Xu, B. Liu: Clustering Transactions Using Large Items.Proc. 1999 Int. Conf on Information and Knowledge Management, (1999) 483–490.
15. Ch. H. Yun, K. T. Chuang, M. S. Chen: An Efficient Clustering Algorithm for Market Basket Data Based on Small Large Ratio. Proc. 25th Annual International Computer Software and Applications Conference. (2001).
16. The Coil data set can be found at: http://www.liacs.nl/putten/library/cc2000/problem.html

DBRS: A Density-Based Spatial Clustering Method with Random Sampling

Xin Wang and Howard J. Hamilton

Department of Computer Science
University of Regina
Regina, SK, Canada S4S 0A2
{wangx, hamilton}@cs.uregina.ca

Abstract. In this paper, we propose a novel density-based spatial clustering method called DBRS. The algorithm can identify clusters of widely varying shapes, clusters of varying densities, clusters which depend on non-spatial attributes, and approximate clusters in very large databases. DBRS achieves these results by repeatedly picking an unclassified point at random and examining its neighborhood. A theoretical comparison of DBRS and DBSCAN, a well-known density-based algorithm, is also given in the paper.

1 Introduction

A *spatial database system* is a database system for the management of spatial data. Rapid growth is occurring in the number and the size of spatial databases for applications such as geo-marketing, traffic control, and environmental studies [1]. *Spatial data mining*, or *knowledge discovery in spatial databases*, refers to the extraction from spatial databases of implicit knowledge, spatial relations, or other patterns that are not explicitly stored [2]. Previous research has investigated generalization-based knowledge discovery, spatial clustering, spatial associations, and image database mining [3]. Finding clusters in spatial data is an active research area, with recent research on effectiveness and scalability of algorithms [4, 5, 6, 7].

In this paper, we are interested in clustering in spatial datasets with the following four features. First, clusters may be of widely varying shapes. For example, in the dataset containing all addresses for current and historical students at our university, clusters of addresses may match the bounds of groups of apartment buildings (roughly rectangular) or may be strung along major bus routes (long thin lines).

Secondly, clusters may be of varying densities. For example, suppose a supermarket company is investigating the distribution of its customers across the whole country. For metropolitan areas, clusters of customers may be very dense, because the population is large and people live very close together. On the other hand, for towns and the countryside, the population is smaller and customers are more sparsely distributed.

Thirdly, the spatial dataset may have significant non-spatial attributes. Spatial clustering has previously been based on only the topological features of the data. However, one or more non-spatial attributes may have a significant influence on the results of the clustering process. For example, in image processing, the general procedure for region-based segmentation compares a pixel with its immediate

K.-Y. Whang, J. Jeon, K. Shim, J. Srivatava (Eds.): PAKDD 2003, LNAI 2637, pp. 563–575, 2003.

surrounding neighbors. Region growing is heavily based on not only the location of pixels but also the non-spatial attributes or 'a priori' knowledge [9].

Fourthly, we are interested in very large datasets, including those of at least 100 000 points. (We refer to each spatial data object in a dataset as a *point*.) In this paper, we introduce the problem of *α-approximate density based clustering*. The goal is to find clusters until the probability that any cluster of at least *MinPts* points has not been found is at most $α$. In practice, not every application requires complete and accurate results. For example, if a construction company wants to distribute its advertisements to clusters of similar residences according to the residence type (non-spatial attribute) and location (spatial attribute) of residences in a city's title-registration database, then an *α*-approximate clustering may be satisfactory. Although the clustering literature is extensive, we were unable to identify any previous research on the problem of approximate clustering with a controlled degree of inaccuracy.

In this paper, we propose a new density-based spatial clustering method called ***Density-Based clustering with Random Sampling (DBRS)***. DBRS can discover density-based clusters with noise (Figure 1 (a)) and follow clusters of many shapes (Figure 1 (b)). In addition, when density varies in the dataset, DBRS scales well on clusters with very dense neighborhoods (Figure 1 (c)). It also handles non-spatial attributes along with spatial attributes by paying attention to the purity (or consistency) of a neighborhood. It can avoid creating clusters of points with different values for non-spatial attributes even though they are close to each other according to the spatial attributes (Figure 1(d)).

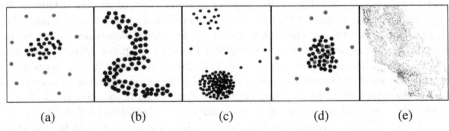

Fig. 1. DBRS (a) discovers density-based clusters with noise; (b) follows snaky clusters; (c) scales well on very dense-neighborhood clusters; (d) pays attention to purity; (e) can operate heuristically on very large datasets.

DBRS repeatedly picks an unclassified point at random and examines its neighborhood. If the neighborhood is sparsely populated or the purity of the points in the neighborhood is too low, the point is classified as noise. Otherwise, if any point in the neighborhood is part of a known cluster, this neighborhood is joined to that cluster. If neither of these two possibilities applies, a new cluster is begun with this neighborhood. With this approach, DBRS discovers multiple clusters concurrently. For large datasets, where reducing the run time is crucial, we propose one heuristic which saves significant amounts of time, but which somewhat reduces the accuracy and precision of the results. With the heuristic, DBRS can perform α-approximate density based clustering on very large datasets (Figure1 (e)).

The remainder of the paper is organized as follows. In Section 2, we briefly discuss existing spatial clustering algorithms, including DBSCAN and CLARANS, and we define the concepts of reachability and connectivity, which are needed to explain the expansion process of DBRS. In Section 3, the DBRS algorithm is presented with its complexity analysis. A heuristic is introduced to reduce run time by sacrificing a probabilistically controlled amount of accuracy. In Section 4, we compare DBRS and DBSCAN from a theoretical view with respect to the neighborhood graph. Section 5 presents an empirical evaluation of the effectiveness of DBRS on the type of datasets described above. We compare the results of DBRS and DBSCAN on this type of dataset. We also examine the costs and benefits of incorporating the heuristic. Section 6 concludes with a summary and some directions for future work.

2   Related Work

In this section, we briefly survey two spatial clustering algorithms, CLARANS [7] and DBSCAN [6]. To find a clustering, CLARANS finds a medoid for each of k clusters. Informally, a medoid is the center of a cluster. The clustering process can be described as searching a graph where every node is a potential solution. In this graph, each node is represented by a set of k medoids. Two nodes are neighbors if their sets differ by only one object. CLARANS processes in the following steps: first, it selects an arbitrary possible clustering node *current*. Next, the algorithm continues to randomly pick a neighbor of *current*. After that, CLARANS compares the quality of clusterings at *current* and the neighbor node. A swap between *current* and a neighbor is made as long as such a swap would result in an improvement of the clustering quality. The number of neighbors to be randomly tried is restricted by a parameter called *maxneighbor*. If the swap happens, CLARANS moves to the neighbor's node and the process is started again, otherwise the current clustering produces a local optimum.   If the local optimum is found, CLARANS starts with new randomly selected node in search for a new local optimum. The number of local optima to be searched is also bounded by a parameter called *numlocal*.

CLARANS suffers from some problems [6]. First, it assumes that the objects to be clustered are all stored in main memory. Secondly, the run time of the algorithm is prohibitive on large databases. Without considering the parameter *numlocal*, the computational complexity of CLARANS is $\Omega(kn^2)$ where n is the size of the dataset.

DBSCAN was the first density-based spatial clustering method proposed [6]. The key idea is that to define a new cluster or extend an existing cluster, a neighborhood around a point of a given radius (*Eps*) must contain at least a minimum number of points (*MinPts*), i.e., the density in the neighborhood is determined by the choice of a distance function for two points p and q, denoted by $dist(p,q)$. DBSCAN uses an efficient spatial access data structure, called an R*-tree, to retrieve the neighborhood from the database. The average case time complexity of DBSCAN is $O(n \log n)$. The greatest advantages of DBSCAN are that it can follow the shape of the clusters and that it requires only a distance function and two input parameters [6].

Given a dataset D, a distance function *dist*, and parameters *Eps* and *MinPts*, the following definitions are used to define DBSCAN (adapted from [6]).

Definition 1. The ***Eps-neighborhood*** of a point p, denoted by $N_{Eps}(p)$, is defined by $N_{Eps}(p) = \{ q \in D \mid dist(p,q) \leq Eps \}$.

Definition 2. A point p is ***directly density-reachable*** from a point q with respect to Eps and $MinPts$ if (1) $p \in N_{Eps}(q)$ and (2) $\mid N_{Eps}(q) \mid \geq MinPts$.

Definition 3. A point p is ***density-reachable*** from a point q with respect to Eps and $MinPts$ if there is a chain of points $p_1,...,p_n$, $p_1=q$, $p_n=p$ such that p_{i+1} is directly density-reachable from p_i.

Definition 4. A point p is ***density-connected*** to a point q with respect to Eps and $MinPts$ if there is a point o such that both p and q are density-reachable from o with respect to Eps and $MinPts$.

Definition 5. Let D be a set of data points. A ***density-based cluster*** C with respect to Eps and $MinPts$ is a non-empty subset of D satisfying the following conditions: 1) $\forall p$, q: if $p \in C$ and q is density-reachable from p with respect to Eps and $MinPts$, then $q \in C$; 2) $\forall p$, $q \in C$: p is density-connected to q with respect to Eps and $MinPts$.

Once the two parameters Eps and $MinPts$ are defined, DBSCAN starts to group data points from an arbitrary point q. It begins by finding the ***neighborhood*** of point q, i.e., all points that are directly density reachable from point q. To find this neighborhood requires performing a ***region query***, i.e., a look up in the R*-tree. If the neighborhood is sparsely populated, i.e., it contains fewer than MinPts points, then point q is labeled as noise. Otherwise, a cluster is created and all points in q's neighborhood are placed in this cluster. Then the neighborhoods of <u>all</u> q's neighbors are examined to see if they can be added to the cluster. If a cluster cannot be expanded from this point with respect to the density reachable and density-connected definitions, DBSCAN chooses another arbitrary ungrouped point and repeats the process. This procedure is iterated until all data points in the dataset have been placed in clusters or labeled as noise. For a dataset containing n points, n region queries are required.

Although DBSCAN gives extremely good results and is efficient in many datasets, it may not suitable for the following cases. First, if a dataset has clusters of widely varying densities, DBSCAN is not able to handle it efficiently. On such a dataset, the density of the least dense cluster of dataset must be applied to the whole dataset, regardless of the density of the other clusters in the dataset. Since all neighbors are checked, much time may be spent in dense clusters examining the neighborhoods of all points. Nonetheless, it has been argued that with density-based algorithms, random sampling cannot be applied in the typical manner to reduce the input size [5]. Unless sample sizes are large, the density of points within clusters can vary substantially in the random sample.

Secondly, if non-spatial attributes play a role in determining the desired clustering result, DBSCAN is not appropriate, because it does not take into account any non-spatial attributes in the database.

Thirdly, DBSCAN is not suitable for finding approximate clusters in very large datasets. DBSCAN starts to create and expand a cluster from a randomly picked point. It works very thoroughly and completely accurately on this cluster until all points in the cluster have been found. Then another point outside the cluster is randomly selected and the procedure is repeated. This method is not suited to stopping early with an approximate identification of the clusters.

3   Density-Based Clustering with Random Sampling

In this section, we propose a novel spatial clustering algorithm called Density-Based Clustering with Random Sampling (DBRS).

Given a dataset D, a symmetric distance function *dist*, parameters *Eps* and *MinPts*, and a property *prop* defined with respect to one or more non-spatial attributes, the following definitions are used to define DBRS.

Definition 6. The ***matching neighborhood*** of a point p, denoted by $N'_{Eps}(p)$, is defined as $N'_{Eps}(p) = \{q \in D|\ dist(p,q) \leq Eps$ and $p.prop = q.prop\}$.

DBRS is to handle non-spatial attributes in the neighbor finding function and to use a minimum purity threshold, called *MinPur*, to control the purity (or consistency) of the neighborhood. A ***core point*** is a point whose matching neighborhood is dense enough, i.e., it has at least *MinPts* and over *MinPur* percentage of matching neighbors. A ***border point*** is a neighbor of a core point, that is not a core point itself. Points other than core points and border points are ***noise***.

Definition 7. A point p and a point q are ***directly purity-density-reachable*** with respect to *Eps*, *MinPts* and *MinPur* from each other if (1) $p \in N'_{Eps}(q)$ and $q \in N'_{Eps}(p)$; (2) $|\ N'_{Eps}(q)| \geq MinPts$ or $|\ N'_{Eps}(p)| \geq MinPts$; and (3) $|\ N'_{Eps}(q)|\ /\ |\ N_{Eps}(q)| \geq MinPur$ or $|\ N'_{Eps}(p)|\ /\ |\ N_{Eps}(p)| \geq MinPur$.

Directly purity-density-reachable is symmetric for core points as well as one core point and one border point. But it is not symmetric for two border points.

Definition 8. A point p and a point q are ***purity-density-reachable(PD-reachable)*** with respect to *Eps*, *MinPts* and *MinPur* from each other, denoted by $PD(p, q)$, if there is a chain of points $p_1,...,p_n$, $p_1=q$, $p_n=p$ such that p_{i+1} is directly purity-density-reachable from p_i.

Definition 9. Let D be a database of points. A ***purity-density-based cluster*** C is a non-empty subset of D satisfying the following condition: $\forall p,\ q \in D$: if $p \in C$ and $PD(p, q)$ holds, then $q \in C$.

It is obvious that for $\forall p,\ q \in C$, $PD(p, q)$ holds.

The intuition behind DBRS is that a cluster can be viewed as a minimal number of core points (called *skeletal points*) and their neighborhoods. To find a cluster, it is sufficient to perform region queries on the skeletal points. However, it is not possible to identify skeletal points before examining the dataset. Instead, we can randomly select sample points, find their neighborhoods, and merge their neighborhoods if they intersect. The sample points may not be the skeletal points, but the number of region queries can be significantly reduced for datasets with widely varying densities.

Based on the above definitions, in many cases, DBRS will find the same clusters as DBSCAN. However, when two groups have a common border point, since the points from two groups are PD-reachable from each other through the common point, DBRS will identify the two groups as one cluster. DBSCAN will separate the two groups into two clusters and the common point will be assigned to the cluster discovered first.

In Figure 2, we present the DBRS algorithm. D is the dataset, i.e., a set of points. *Eps* and *MinPts* are global density parameters, determined either manually or according to the heuristics presented in [6]. *MinPur* is the minimum fraction of points in a neighborhood with property *prop* required to define a cluster.

DBRS starts with an arbitrary point q and finds its matching neighborhood, i.e., all points that are purity-density-reachable from q. In the algorithm, the region query D.matchingNeighbors(q,Eps,prop) in line 4 finds this matching neighborhood, which is called qseeds.

If the number of matching neighbors of q is at least *MinPts* and the purity is at least *MinPur*, then q is a core point; otherwise q is noise or a border point, but it is tentatively classified as noise in line 6. If q is a core point, the algorithm checks whether its neighborhood intersects any known cluster. The clusters are organized in a list called ClusterList. If qseeds intersects with a single existing cluster, DBRS merges qseeds into this cluster. If qseeds intersects with two or more existing clusters, the algorithm merges qseeds and those clusters together, as shown in lines 11-18. Otherwise, a new cluster is formed from qseeds in lines 19–21.

Algorithm DBRS(D, Eps, MinPts, MinPur, prop)

```
1   ClusterList = Empty;
2   while (!D.isClassified( ))
3      { Select one unclassified point q from D;
4        qseeds = D.matchingNeighbors(q, Eps, prop);
5        if ((|qseeds| < MinPts) or (qseed.pur < MinPur))
6            q.clusterID = -1; /*q is noise or a border point */
7        else
8            { isFirstMerge = True;
9                Cᵢ = ClusterList. firstCluster;
                 /* compare qseeds to all existing clusters */
10             while (Cᵢ != Empty)
11                { if ( hasIntersection(qseeds, Cᵢ) )
12                    if (isFirstMerge)
13                       { newCᵢ = Cᵢ.merge(qseeds);
14                          isFirstMerge = False; }
15                   else
16                      { newCᵢ = newCᵢ.merge(Cᵢ);
17                         ClusterList.deleteCluster(Ci);}
18                 Cᵢ = ClusterList. nextCluster;
                } // while != Empty
         /*No intersection with any existing cluster */
19            if (isFirstMerge)
20            { Create a new cluster Cⱼ from qseeds;
21               ClusterList = ClusterList.
addCluster(Cⱼ);
              }   //if isFirstMerge
            }  //else
       }  // while !D.isClassified
```

Fig. 2. DBRS Algorithm

After examining the neighborhood of one point, the algorithm selects another arbitrary, unclassified point and repeats the above procedure. The procedure is iterated until every data point in the dataset is clustered or is labeled as noise.

The crucial difference from DBSCAN is that once DBRS has labeled q's neighbors as part of a cluster, it does not examine the neighborhood for each of these neighbors. For dense clusters, where every point is a neighbor of many other points, this difference can lead to a significant time savings.

As previously mentioned, the region query D.matchingNeighbors(q,Eps,prop) in line 4 is the most time-consuming part of the algorithm. A region query can be answered in O(log n) time using spatial access methods such as R*-trees [8]. In the worst case, where all n points in a dataset are noise, DBRS performs n region queries and the time complexity of DBRS is O(n log n); however, if any clusters are found, it will perform fewer region queries.

With the addition of a heuristic stopping condition, DBRS can be used for α-approximate density-based clustering to handle very large datasets more efficiently. With density-based algorithms, using ordinary random sampling to reduce the input size may not be feasible. Unless sample sizes are large, the density of points could vary substantially within each cluster in the random sample.

DBRS works in a different way. It picks a point randomly. This arbitrary point might be noise or it might be in any of the clusters remaining to be discovered. In particular, if most unclassified points are in clusters, it is likely to choose such a point. Large clusters and particularly dense neighborhoods are likely found early because they have more points to be randomly selected. After randomly selecting points for a certain number of iterations, the increase in the number of newly clustered points may fall to zero, if only noise points are being selected.

The heuristic is based on the following observation. Given a dataset D with n points, the chance of missing a particular cluster of $MinPts$ points by choosing the first single random point and finding its neighborhood is at most $\frac{n - MinPts}{n}$. Let n_i be the number of points removed after examining the ith randomly selected point. For noise, n_i is 1. For core points, n_i is at least one; and if the core point is in a newly found cluster, it is at least $MinPts$. Let $N_i = \sum_{i=1}^{j-1} n_i$ be the cumulative number of points removed after j-1 points have been examined. After examining k points, the chance of not finding a cluster of at least $MinPts$ is $\alpha_k = \prod_{i=1}^{k} \left(\frac{n - N_i - MinPts}{n - N_i} \right)$. The value of α_k becomes small as k increases, and each time a new cluster is found, it decreases more rapidly. Although determining α_k analytically is not possible, it is easily computed while DBRS is running.

The heuristic value of α-approximate confidence is calculated as follows. The number of points left unclassified at step i is $L_i = n - N_i$. Initially, we let $\alpha_0 = 1$ and $L_0 = n$. At step i, for $i > 0$, we compute: $\alpha_i = \alpha_{i-1} \cdot \left(\frac{L_{i-1} - MinPts}{L_{i-1}} \right)$, $L_i = L_{i-1} - n_i$.

DBRS is stopped if $\alpha_i \leq \alpha_{max}$ or if all points are clustered or labeled as noise. Although the user must choose a value for α_{max}, it can be thought of as similar to statistical significance. We set α_{max} as 0.01 in our experiments, so the algorithm stops when it is 99% confident that no cluster of at least $MinPts$ points remains.

4   Theoretical Comparison of DBRS and DBSCAN

In this section, we compare DBS and DBSCAN from the theoretical view with respect to the neighborhood graphs they generate during the clustering process.

Definition 10. The ***neighborhood graph*** for a spatial relation *neighbor* is a graph $G = (V, E)$ with the set of vertices V and the set of directed edges E such that each vertex corresponds to an object of the database and two vertices $v1$ and $v2$ are connected iff *neighbor(v1, v2)* holds. Given different *neighbor* relations, a neighborhood graph can be directed or undirected.

Definition 11. A neighborhood graph (or a neighborhood subgraph) is **connected** iff for any pair of vertices in the graph (or subgraph) there is an undirected path joining the vertices.

Definition 12. A directed neighborhood graph (or a neighborhood subgraph) $G = (V, E)$ is **strongly connected** iff for any two nodes p, q with *neighbor(p, q)* holding, there is a directed path from p to q.

Lemma 1. If the density-reachable relation is the *neighbor* relation, a connected neighborhood (sub-)graph represents a cluster generated by DBSCAN.

Proof: We must prove that: (1) a cluster defined by DBSCAN in Definition 5 is represented in a connected neighborhood (sub-)graph; (2) only the points belonging to the same cluster represented in the same connected neighborhood graph.

Let p, $q \in D$ and in the following v_p and v_q are used to represent the corresponding vertices of p and q in the neighborhood graph.

(1) Since the cluster that generated by DBSCAN defined in Definition 5 is determined by the two conditions, we only need to prove that the cluster defined by the two conditions is in a connected neighborhood graph.

(1.1) If q is density-reachable from p, then *neighbor(p, q)* holds. By the definition of the neighborhood graph, if *neighbor* relation holds for two objects, the vertices in the neighborhood graph are connected. So in the neighborhood graph, v_p and v_q are connected. That is, v_q belongs to the same subgraph in which v_p exists.

(1.2) For the condition 2, $\forall p$, $q \in C$, p is density-connected to q. According to the definition of the density-connected relation, there exists an object $o \in C$, p and q are density-reachable from o. So *neighbor(o, p)* and *neighbor(o, q)* hold. It indicates that in the neighborhood graph, there are paths connecting v_o and v_p, and v_o and v_q, respectively. So v_p and v_q are connected by an undirected path through v_o.

From (1.1) and (1.2), we can conclude that a cluster defined by DBSCAN is represented in a connected neighborhood graph.

(2) Assume p is not density-reachable from q, but v_p and v_q are represented in a same connected neighborhood graph. Since p is not density-reachable from q, so *neighbor(v_p, v_q)* does not hold. By the definition of the neighborhood graph, two vertices are connected only if *neighbor* relation holds for them. So there is no path connecting them. The assumption is wrong; the proof by contradiction is complete. ◆

From Lemma 1, given n points, the clustering process can be viewed abstractly as constructing neighborhood graphs. Each time a core point is found, the algorithm finds the directly density-reachable relation between the core point and its neighbors. The directly density-reachable relation holding for the two objects corresponds to the directed edge between the two corresponding vertices in the neighborhood graph.

Each cluster in the dataset is constructed as a connected neighborhood subgraph. If the dataset has k clusters, its neighborhood graph has k connected subgraphs.

Lemma 2. If the density-reachable relation is the *neighbor* relation, DBSCAN's clustering process corresponds to constructing the strongly connected neighborhood graph.

Proof: From Lemma 1, we know each cluster generated by DBSCAN is represented as a connected neighborhood subgraph. To prove the connected neighborhood graph is strongly connected, we need to prove that there is a directed path connecting two vertices p and q iff *neighbor(p, q)* holds. The proof requires two parts: (1) if *neighbor(p, q)* holds, there is a directed path connecting v_p and v_q; (2) only if *neighbor(p, q)* holds, there is a directed path from v_p to v_q.

(1): If *neighbor(p, q)* holds, then q is density-reachable from p. By the definition of the density-reachable relation, there exists an ordered list of objects $[p_1, p_2, ...p_n]$ from p to q, $p_1 = p$ and $p_n = q$. Among the list, p_{i+1} is directly density-reachable from p_i. For DBSCAN the directly density-reachable relation holding for the two objects can be viewed as the directed edge between the two corresponding vertices in the neighborhood graph. So if p_{i+1} is directly density-reachable from p_i, then there is a directed edge from v_{pi} to v_{pi+1}. Thus, there is a directed path from v_{p1} to v_{pn}, i.e., there is a directed path from v_p to v_q.

(2): Assume *neighbor(p, q)* does not hold, but there is a directed path from v_p to v_q. Suppose the path from v_p to v_q is composed of $[v_{p1}, v_{p2}, ...,v_{pn}]$, $v_{p1} = v_p$ and $v_{pn} = v_q$. Because there is an edge from v_{pi} to v_{pi+1}, by the definition of neighborhood graph, *neighbor(p_i, p_{i+1})* holds. So p is density-reachable from q. That is, *neighbor(p, q)* holds. Thus, the assumption is wrong, and the proof by contradiction is complete. ♦

In the following lemma, we prove that a neighborhood graph is also connected by applying PD-reachable as the neighbor relation.

Lemma 3. If the PD-reachable relation is the *neighbor* relation, the neighborhood graph generated by DBRS is connected.

Proof: We must prove two parts: (1) a cluster defined by DBRS in Definition 9 is represented in a connected neighborhood graph; (2) only the points belonging to the same cluster are represented in the same connected neighborhood graph.

(1) If q and p are PD-reachable from each other, then *neighbor(p, q)* holds. Since *neighbor(p, q)* holds, then by the definition of the neighborhood graph, there is a path connecting v_p and v_q. So v_p and v_q belong to same subgraph.

(2) Assume that p and q are not undirected PD-reachable from each other, but v_p and v_q belong to the same connected subgraph. Since p and q are not PD-reachable from each other, *neighbor(p, q)* does not hold. If the *neighbor(p, q)* does not hold, there is no path connecting v_p and v_q according to the definition of the neighborhood graph. Thus, the assumption is wrong, and the proof by contradiction is complete. ♦

Unless all points are noise, constructing a strongly connected neighborhood graph is more expensive than constructing a connected neighborhood graph. When constructing an undirected connected graph, fewer points are checked for their neighborhood. In DBRS, for any two PD-reachable points, there is at least one undirected path connecting them. In DBSCAN, for any two directly density-reachable core points, there are always two directed paths (edges) connecting them. Regardless of whether the connectivity is directed or undirected, all connected points should belong to the same cluster. It is irrelevant whether a point is density reachable via a directed neighborhood path or via an undirected path. In the worst case, i.e., all points are noise, the costs of constructing the two neighborhood graphs are the same because no directed or undirected edges are generated.

5   Performance Evaluation

We use both the synthetic and real datasets to compare DBRS with DBSCAN. Table 1 shows the result of our experiments on the synthetic data sets. Each dataset includes x, y coordinators and one non-spatial property with 2-10 clusters. For preliminary testing, we implemented DBRS and re-implemented DBSCAN without using R*-trees. All experiments were run on a 500MHz PC with 256M memory. In the Table 1, the second and third rows are the results when the radius is 7, MinPts is 17, and MinCon is assigned to 0.75 and 0.98, respectively. The run times, in seconds, are shown for our version of DBSCAN in the fourth row. The performance of both algorithms will be greatly improved by using R*-trees, but the number of region queries will be unchanged in both cases.

Table 1. Run Time in Seconds

Number of Points (with 10% noise)	1k	2k	3k	4k	5k	6k	7k	8k	9k	10k	15k	25k	50k
Time (sec) (7-17-0.75)	1	1	3	4	6	8	11	15	18	23	31	165	663
Time (sec) (7-17-0.98)	1	2	3	6	10	12	19	26	31	36	49	173	829
DBSCAN* Time (sec)	2	7	14	24	37	54	75	95	120	151	370	927	3778

Figure 3(a) shows the clusters found by DBRS in a dataset with 1000 points with a neighborhood consistency over 98%. The shapes at the left-bottom corner and right-top corner are both clustered as two clusters because of a difference in a non-spatial property. As shown in Figure 3(b), DBSCAN does not separate the clusters in the pairs because it ignores non-spatial attributes.

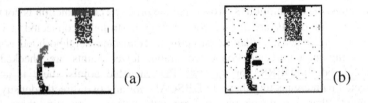

(a)   (b)

Fig. 3. Clusters Discovered (a) by DBRS, (b) by DBSCAN

Figure 4 shows the number of region queries for varying amounts of noise in datasets ranging from 1k to 100k. For every dataset, if the percentage of noise is 100%, DBRS has the same number of region queries (equal to total number of points) as DBSCAN. As noise decreases, far fewer region queries are needed for DBRS.

Figure 5 shows the number of region queries required for a 10k dataset with clusters of varying densities. With DBSCAN, the number of region queries does not change as the radius *Eps* increases. With DBRS, the number of the region queries decreases as the radius of the neighborhood (query region) increases. For our data, increasing the radius is equivalent to reducing the density of the overall cluster. Thus, for higher-density clusters, DBRS can achieve better performance than DBSCAN.

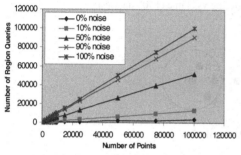 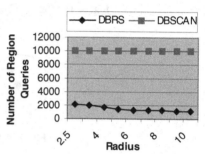

Fig. 4. The Number of Region Queries for Datasets with Different Percentages of

Fig. 5. Radius Vs. Number of Region Queries (10k Dataset)

We ran a series of experiments to determine the rate at which points are placed in clusters. We observed that after a certain amount of time, most remaining points are noise. Figure 6(a) shows the size of the increase (delta) in the number of clustered points, with respect to time, for datasets of 100k points with 10% noise. Figure 6(b) shows the percentage for 100k points with 10% noise with respect to time. As shown in Figure 6(b), after 580 seconds, most points (over 85% of the points) are clustered.

Table 2 lists the run time performance, the points clustered, percentage of the points clustered and the percentage not clustered when using the heuristic. For example, the first line shows that a dataset with 25k points, including 10 clusters and 2500 (10%) noise points, was processed in 59 seconds (original run time is 164 seconds) by using the first heuristic. In this run, 22162 points were clustered, i.e., 88.65% of the whole dataset. Of 22500 points that should have been clustered, 338 were not clustered, representing 1.35% of all points.

We applied DBRS with the heuristic set to 0.01 to all Canadian addresses of University of Regina students since 1999. The addresses were transformed to longitude and latitude using GeoPinpoint. Eps was set to 0.5 and MinPts to 40. The whole dataset includes 52216 records, and 45609 records were clustered to form 19 clusters. A domain expert considered the clusters to be reasonable. The biggest cluster is in Saskatchewan, which includes 41253 points. Other clusters are distributed in other provinces. The number of region queries is 870.

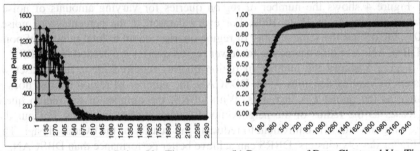

(a) Increase in Clustered Points Vs. Time   (b) Percentage of Data Clustered Vs. Time

Fig. 6. The Rate of Clustering vs. Time (100k dataset)

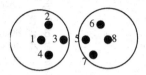

Fig. 7. A Difficult Case

6 Conclusion

Clustering spatial data has been extensively studied in the knowledge discovery literature. In this paper, we proposed a new spatial clustering method called DBRS, which aims to reduce the number of region queries for datasets with varying densities. Our preliminary experiments suggest that it scales well on the high-density clusters. As well, DBRS deals with both spatial and non-spatial attributes. It can take account of a property related to non-spatial attribute(s), by means of a purity threshold, when finding the matching neighborhood. To increase the efficiency of clustering on very large datasets, we introduced a heuristic that can reduce the run time significantly at the cost of missing a probabilistically controlled number of clusters.

Table 2. Performance Using the Heuristic

Number of Points (% noise)	Time (sec)	DBRS Original Time (sec)	DBSCAN (sec)	Points Clustered	Number of Clusters Found	Percentage of Points Clustered	Non-noise Points Not Placed in Clusters (% of All)
25k (10%)	59	164	927	22162	10/10	88.65	338 (1.35)
25k (20%)	88	274	989	19707	10/10	78.83	293 (1.17)
50k (10%)	253	668	3778	44348	10/10	88.70	652 (1.30)
50k (20%)	371	1160	3987	39410	10/10	78.82	590 (1.18)

The DBRS approach still needs improvement. First, our implementation must be improved by incorporating R*-trees or other specialized data structures for performing region queries. Secondly, the algorithm may miss joining certain clusters. For the case shown in Figure 7, Point 1 and Point 8 are close together and should be placed in the same cluster. However, if we first pick Point 1 and Point 5, all points will be clustered, and no unclustered point will remain that we could pick to merge the two subgraphs. Such cases were rare in our experiments, but to reduce their occurrence, the algorithm could pick some random border points and check their neighborhoods.

References

[1]   Ester, M., Kriegel, H., and Sander, J.: Spatial Data Mining: A Database Approach. In: Proc. 5th Int'l Symp. on Large Spatial Databases (SSD'97), Berlin (1997) 47–66.

[2]   Koperski, K. and Han, J.: Discovery of Spatial Association Rules in Geographic Information Databases. In: Proc. 4th Int'l Symp. on Large Spatial Databases (SSD95), Maine, USA (1995) 47–66.

[3]   Koperski, K., Adhikary, J. and Han, J.: Spatial Data Mining: Progress and Challenges. In: SIGMOD'96 Workshop on Research Issues on Data Mining and Knowledge Discovery (DMKD'96), Montreal, Canada (1996) 55–70.

[4]   Jain, A.K., Murty, M.N., and Flynn, P.J.: Data Clustering: A Review. ACM Computing Surveys, 31(3) (1999) 264–323.

[5]   Guha, S., Rastogi, R., and Shim, K.: CURE: An efficient clustering algorithm for large databases. In: Proc. of ACM SIGMOD, New York (1998) 73–84.

[6]   Ester, M., Kriegel, H., Sander, J., and Xu, X.: A Density-Based Algorithm for Discovering Clusters in Large Spatial Databases with Noise. In: Proc. of 2nd Int'l Conf. On Knowledge Discovery and Data Mining, Portland, OR (1996) 226–231.

[7]   Ng, R. and Han, J.: Efficient and Effective Clustering Method for Spatial Data Mining. In: Proc. of 1994 Int'l Conf. on Very Large Data Bases, Chile (1994) 144–155.

[8]   Beckermann, N., Kriegel, H., and Schneider, R.: The R *-Tree: An Efficient and Robust Access Method for Points and Rectangles. In: Proc. ACM SIGMOD Int'l Conf. on Management of Data, Atlantic City, USA (1990) 322–331.

[9]   Cramariuc B., Gabbouj M., and Astola J., Clustering Based Region Growing Algorithm for Color Image Segmentation, In: Proc. Digital Signal Processing, Stockholm, Sweden (1997) 857–860.

Optimized Clustering for Anomaly Intrusion Detection

Sang Hyun Oh and Won Suk Lee

Department of Computer Science Yonsei University
134 Seodaemoon-gu, Shinchon-dong
Seoul, 120-749, Korea
{osh, leewo}@amadeus.yonsei.ac.kr

Abstract. Although conventional clustering algorithms have been used to classify data objects in a data set into the groups of similar data objects based on data similarity, they can be employed to extract the common knowledge i.e. properties of similar data objects commonly appearing in a set of transactions. The common knowledge of the activities in the transactions of a user is represented by the occurrence frequency of similar activities by the unit of a transaction as well as the repetitive ratio of similar activities in each transaction. This paper proposes an optimized clustering method for modeling the normal pattern of a user's activities. Furthermore, it also addresses how to determine the optimal values of clustering parameters for a user as well as how to maintain identified common knowledge as a concise profile. As a result, it can be used to detect any anomalous behavior in an online transaction of the user.

1 Introduction

For anomaly detection, previous works [1][2] are concentrated on statistical techniques. The typical system of statistical analysis is NIDES [2] developed in SRI. In NIDES, for detecting an anomaly, the information of the online activities of the user is summarized into a short-term profile, and then it is compared with the long-term profile of the user. If the difference between two profiles is large enough, the online activities are considered as anomalous behavior. The strong point of statistical analysis is that it can generate a concise profile containing only a statistical summary, which can lessen the burden of computation overhead for real-time intrusion monitoring. However, since statistical analysis represents the diverse behavior of a user's normal activities as a statistical summary, it can often fail to model the normal behavior of activities accurately when their deviation is large.

In this paper, a DBSCAN [3]-like algorithm is used to model user normal behavior. While the number of data objects in a specific range of a domain is an important criterion for creating a cluster in conventional density-based clustering algorithms, the number of distinct transactions i.e., the transaction support of data objects in a specific range is an important criterion in the proposed method. Since the purpose of clustering is no longer classification, the properties of identified clusters can be summarized as a profile concisely. Besides the transaction support of a cluster, there is another type of common knowledge to be modeled in an audit data set. If there exists the common repetitive ratio of similar data objects in each transaction, it can be

K.-Y. Whang, J. Jeon, K. Shim, J. Srivatava (Eds.): PAKDD 2003, LNAI 2637, pp. 576–581, 2003.

modeled as meaningful common knowledge to find any anomaly of a new transaction. However, in order to utilize a clustering algorithm, selecting the optimal values of clustering parameters is not a trivial task. This paper proposes how to determine the optimal values of clustering parameters for clustering the transactions of a user accurately.

2   Audit Data Clustering

A transaction in an audit data set contains a set of data objects each of which lies in the n-dimensional feature space of an application domain. The value domain of each dimension can be non-categorical like real or integer values. Given a set of modeling features and a set of transactions TD containing the actual activities of a user, let TD^k denote a set of transactions each of which contains those activities that are related to the kth feature. Let $|TD^k|$ denote the total number of these transactions and let D^k denote an object data set which contains the entire activities of all transactions in TD^k. In this paper, the values of the same feature for these activities are denoted by data objects to be clustered together. Each transaction T_i of TD^k is represented by $T_i = \{o^1_i, o^2_i, ..., o^w_i\}$ and the number of data objects in a transaction T_i is denoted by $|T_i|$. The feature value of an activity o^j_i on the kth feature is denoted by $v_k(o^j_i)$.

Unlike conventional clustering algorithms [3][4], in the proposed algorithm, a cluster expands based on the proper number of distinct transactions that contain similar data objects together. A predefined clustering range λ is used as a similarity measure for data objects. The transaction support of a data group of similar data objects is the ratio of the number of distinct transactions in which the objects of the data group appear over the total number of transactions. A data group of similar data objects can become a cluster when its transaction support is greater than or equal to a predefined minimum transaction support *minsup*.

The formal description of the proposed clustering method is presented in [5]. It discovers a set of clusters in an object data set D^k of the kth feature for the given values of a minimum transaction support *minsup* and a clustering range λ. If the support of a data group is greater than or equal to *minsup*, it becomes an initial cluster. This initial cluster can be expanded by being merged with adjacent clusters. The characteristics of each cluster identified by the proposed clustering algorithm are represented by a set of the properties defined in Definition 1.

Definition 1. Properties of a cluster C^k
Given a cluster C^k of similar data objects for the kth feature, let $O_C^k \subseteq D^k$ denote those data objects that belong to the cluster C^k. The properties of a cluster C^k are denoted by a tuple *C^k(center, cdev, min, max, tcount, ratio, rdev)*.   □

In Definition 1, the range of a cluster is represented by the minimum value *min(C^k)* and the maximum value *max(C^k)* among the values of the data objects in O_C^k. The transaction count of the data group O_C^k of a cluster C^k is represented by tcount(C^k). Based on this value, the support of the cluster C^k can be calculated i.e., support(C^k) =

tcount(C^k)/|TD^k|. The central value of the cluster C^k is represented by center(C^k). Let $avg_i(C^k)$ denote the average of the value of those data objects in O_C^k that appear in a transaction $T_i \in TD^k$. The standard deviation of the center of a cluster C is denoted by cdev(C^k). Let $r_i(C^k)$ denote the individual repetitive ratio of a cluster C^k for all data objects in an individual transaction $T_i \in TD^k$. It is defined by $r_i(C^k)=|S_i|/|T_i|$ where $|S_i|$ denotes the number of data objects whose values are within the range of the cluster C^k. The repetitive ratio $ratio(C^k)$ of a cluster C^k is the average value of individual ratios $r_i(C^k)$, $1 \le i \le |TD^k|$. The standard deviation of the repetitive ratio of a cluster C^k is denoted by rdev(C^k).

3   Anomaly Detection

For each feature, a profile contains the clustering result of common knowledge extraction. It is composed of two different summaries: internal and external. The internal summary contains the properties of each cluster while the external summary represents the statistics of noise data objects i.e. the data objects in the outside of all clusters. The external summary is represented by two specialized properties. One is an external repetitive ratio *extratio* and its standard deviation *extratio_dev*. The other is an external distance *extdist* and its standard deviation *extdist_dev*. When the number of identified clusters for the kth feature is denoted by m, the value of *extratio^k* for the kth feature is represented as follows.

$$extratio^k = \frac{1}{|TD^k|} \cdot \sum_{i=1}^{|TD^k|} \left(1 - \sum_{j=1}^{m} r_i(C_j^k) \right) = 1 - \sum_{j=1}^{m} ratio(C_j^k) \cdot support(C_j^k) \qquad (1)$$

The external distance of a noise data object is calculated by its distance to the nearest cluster. Let P^k be the profile of the kth feature and $dist(P^k, o_i^j)$ denotes the distance between a noise data object o_i^j and a selected cluster C^k in the internal summary of P^k. It is defined as follows.

$$dist(P^k, o_i^j) = \begin{cases} |v_k(o_i^j) - \min(C^k)| & \text{if } v_k(o_i^j) < \min(C^k) \\ |v_k(o_i^j) - \max(C^k)| & \text{if } v_k(o_i^j) > \max(C^k) \end{cases} \qquad (2)$$

Consequently, let E_i^k be the set of noise data objects in $T_i \in TD^k$ for the kth feature and $|E_i^k|$ denote the number of noise data objects in E_i^k. The individual external distance $ex\_d(P^k, T_i)$ can be calculated as follows.

$$ex\_d(P^k, T_i) = \frac{1}{|E_i^k|} \sum_{j=1}^{|E_i^k|} dist(P^k, o_i^j) \quad \text{where } o_i^j \in E_i^k \qquad (3)$$

Therefore, the external distance *extdist^k* of a set of transactions TD^k is represented by the average of their individual external distances.

By comparing a new transaction with the profile of each feature, an anomaly in a new transaction can be identified. The result of this comparison is expressed by both internal and external abnormalities for each feature. The internal abnormality is a measure for representing the difference between the internal summary of a profile and those data objects in a new transaction that are within the ranges of clusters. Similarly, the external abnormality is a measure for representing the difference between the external summary of a profile and noise data objects in a new transaction. Each abnormality is further categorized by a distance difference and a ratio difference. Let *MS* denote a set of four measures of abnormality *MS* = {*ID, IR, ED, ER*} where *ID, IR, ED* and *ER* denote an internal distance, an internal ratio, an external distance and an external ratio respectively.

For an online transaction T_v, the internal distance difference in the cluster C^k is defined by the distance between the center of the cluster C^k and $avg_v(C^k)$. Since the deviation of the center of each cluster is different, this internal distance difference should be normalized. Consequently, the internal distance difference $ind\_diff_{ID}(C^k, T_v)$ between a cluster C^k and a transaction T_v is normalized by its standard deviation $cdev(C^k)$. Likewise, the deviation of the repetitive ratio of each cluster can also be different, so that the internal ratio difference $ind\_diff_{IR}(C^k, T_v)$ should also be normalized by its standard deviation $rdev(C^k)$.

$$\text{ind\_diff}_{ID}(C^k, T_v) = \frac{|\text{center}(C^k) - \text{avg}_v(C^k)|}{\gamma \cdot \text{cdev}(C^k)} \quad \text{ind\_diff}_{IR}(C^k, T_v) = \frac{|\text{ratio}(C^k) - r_v(C^k)|}{\gamma \cdot \text{rdev}(C^k)} \quad (4)$$

When the $cdev(C^k)$ or $rdev(C^k)$ of a cluster C^k is quite small, its corresponding difference becomes very large, so that the overall difference of a new transaction is influenced by the specific cluster. In order to avoid this, the values of these differences are set to a user-defined maximum value when they are larger than the maximum value. The normalizing factor γ is a user-defined parameter that can control the effect of an internal difference. Ultimately, when the number of identified clusters is *m*, each of the two internal differences for the kth feature is defined by the sum of its corresponding differences obtained by all the clusters of the feature respectively as follows:

$$\text{diff}_{ID}(P^k, T_v) = \sum_{i=1}^{m} \text{ind\_diff}_{ID}(C_i^k, T_v) \cdot \text{support}(C_i^k) \quad (5)$$

$$\text{diff}_{IR}(P^k, T_v) = \sum_{i=1}^{m} \text{ind\_diff}_{IR}(C_i^k, T_v) \cdot \text{support}(C_i^k) \quad (6)$$

Since the support of each cluster is different, the internal difference of a cluster should be weighted by its support to obtain the total internal difference of a transaction in Equation (5) and (6).

Similarly, the external differences between the profile P^k of the kth feature and a new transaction T_v is almost identical to the internal differences. For the kth feature, let $diff_{ED}(P^k, T_v)$ and $diff_{ER}(P^k, T_v)$ denote external distance difference and external ratio difference respectively. For the kth feature, when the number of noise data objects in a new transaction T_v is denoted by $|E_v^k|$, the two external differences can be defined as follows.

$$\text{diff}_{ED}(P^k, T_v) = \frac{|\text{extdist}^k - \text{ex\_d}(P^k, T_v)|}{\gamma \cdot \text{extdist\_dev}^k} \quad \text{diff}_{ER}(P^k, T_v) = \frac{|\text{extratio}^k - |E_v|/|T_v||}{\gamma \cdot \text{extratio\_dev}^k} \quad (7)$$

Therefore, for each measure $\mu \in MS$, when the number of features in n, the overall abnormality of a new transaction T_v is represented by the average of the individual differences of all features as follows (P= {$P^1, P^2, ..., P^n$}):

$$\text{overall}_{\mu}(P, T_v) = \frac{1}{n} \sum_{k=1}^{n} \text{diff}_{\mu}(P^k, T_v) \quad (8)$$

As a result, for each measure $\mu \in MS$, average overall abnormality Φ_μ and its standard deviation sd_μ of all past transactions can be calculated as follows.

$$\Phi_\mu(TD) = \frac{1}{|TD|} \cdot \sum_{i=1}^{|TD|} overall_\mu(P,T_i) \qquad sd_\mu = \sqrt{\frac{1}{|TD|} \cdot \sum_{i=1}^{|TD|} overall_\mu(P,T_i)^2 - \Phi_\mu(TD)^2} \qquad (9)$$

If $\Phi_\mu(TD) + sd_\mu \cdot \xi < overall_\mu(P,T_v)$, there exist an anomaly in the online transaction T_v. Let a detecting factor ξ denote a user-defined parameter that defines the boundary of an anomaly.

4   Experimental Results

In order to demonstrate the performance of the proposed algorithm, several experiments have been performed by log data collected in Solaris 2.6 for two months. The log data of users has been extracted by BSM (Basic Security Module) [6] of Solaris 2.6. The BSM has been widely used as a tool that provides a C2 level security, and it recognizes 228 kernel signals. Among these signals, 28 signals are used as basic features in the experiments. In the collected log data, a transaction is defined by a user's session containing commands from login to logout in a host computer.

Among the two clustering parameters λ and *minsup*, the clustering range λ affects the accuracy of the shape of a cluster. As it is set to be smaller, the shape of an identified cluster is more precisely determined. As a result, the value of a clustering range is set to the smallest unit $\lambda=1$ in all experiments since most of the values denote the occurrence frequency of a signal. Furthermore, a term Di/Dj denotes an experiment where the transactions of a data set Di are used as online transactions for the profile of a data set Dj.

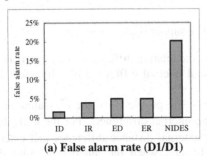

(a) False alarm rate (D1/D1)

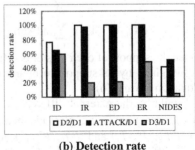

(b) Detection rate

Fig. 1. Anomaly detection ($\xi=2$)

Fig. 1 shows the result of anomaly detection for the data set D1. In this figure, D1 and D3 are the log data of two different programming users and contain the information of signals for general programming and shell activities. D2 is produced by administrative activities in the root account of a host computer. In this experiment, the minimum support is set to 1% and the detecting factor ξ is set to 2. In order to demonstrate the intrusion detection capability of the proposed method, D2 and an attack data set (ATTACK) are used as anomalous data sets. In ATTACK, 50 attack transactions each of which contains about 200 activities are generated. The data set contains the activity of a buffer overflow intrusion, the activity of password guessing and the behavior of an intruder that contains a considerable number of damaging activities such as searching directory, deleting files and copying files. In Fig.1(a), the false alarm rate of D1/D1 are less than 5% while the detection rates of D2/D1, ATTACK/D1 and D3/D1 shown in Fig. 1(b) are high enough in the proposed method. However, when the false alarm rate of NIDES is 20% shown in Fig. 1(a), its detection rates for D2/D1, ATTACK/D1 and D3/D1 are 41%, 51% and 4% respectively. These rates are much lower than the corresponding rates of the proposed method respectively. As shown in Fig. 1(b), the performance of ID is lower than the other measures. This is because most of attacks or anomalous activities are in the outside of the clusters of the target data set D1.

5 Conclusion

In this paper, a method of modeling user normal behavior is introduced. After clusters are found by the proposed algorithm for each feature, their statistical information is used to model the normal behavior of a user in terms of each cluster. For this purpose, this paper proposes two major measures: internal difference and external difference. By applying the proposed clustering algorithm, the domain of a feature is divided by two areas. One is the frequent range of a user's activities and the other is the infrequent range. The normal behavior of these two areas is further analyzed by distance difference and ratio difference. As a result, the various aspects of the normal behavior of a user can be analyzed separately. Furthermore, a method of determining the optimal values of clustering parameters is presented in order to model the normal behavior of a user more precisely.

References

1. Harold S.Javitz and Alfonso Valdes, The NIDES Statistical Component Description and Justification, Annual report, SRI International, 333 Ravenwood Avenue, Menlo Park, CA 94025, March 1994.
2. Phillip A. Porras and Peter G. Neumann, "*EMERALD: Event Monitoring Enabling Responses to Anomalous Live Disturbances*," 20th NISSC, October 1997.
3. Martin Ester, Hans-Peter Kriegel, Sander, Michael Wimmer, Xiaowei Xu, "*Incremental Clustering for Mining in a Data Warehousing Environment*", Proceedings of the 24th VLDB Conference, New York, USA, 1998.
4. Tian Zhang, Raghu Ramakrishnan, and Miron Livny, "*Birch: An Efficient data clustering method for very large databases,*" Proceedings for the ACM SIGMOD Conference on Management of Data, Montreal, Canada, June 1996.
5. Sang Hyun Oh and Won Suk Lee, "*Clustering Normal User Behavior for Anomaly Intrusion Detection,*" Proceeding of the IASTED International Conference on Applied Modeling and Simulation, Cambridge, MA, USA, November 4–6, 2002.
6. Sun Microsystems. SunShield Basic Security Module Guide.

Finding Frequent Subgraphs from Graph Structured Data with Geometric Information and Its Application to Lossless Compression

Yuko Itokawa[1], Tomoyuki Uchida[2], Takayoshi Shoudai[3],
Tetsuhiro Miyahara[2], and Yasuaki Nakamura[2]

[1] Faculty of Human and Social Environment,
Hiroshima International University, Kurose 724-0695, Japan
y-itoka@he.hirokoku-u.ac.jp
[2] Faculty of Information Sciences,
Hiroshima City University, Hiroshima 731-3194, Japan
{uchida@cs, miyahara@its, nakamura@cs}.hiroshima-cu.ac.jp
[3] Department of Informatics, Kyushu University 39, Kasuga 816-8580, Japan
shoudai@i.kyushu-u.ac.jp

Abstract. In this paper, we present an effective algorithm for extracting characteristic substructures from graph structured data with geometric information, such as CAD, map data and drawing data. Moreover, as an application of our algorithm, we give a method of lossless compression for such data. First, in order to deal with graph structured data with geometric information, we give a layout graph which has the total order on all vertices. As a knowledge representation, we define a layout term graph with structured variables. Secondly, we present an algorithm for finding frequent connected subgraphs in given data. This algorithm is based on levelwise strategies like Apriori algorithm by focusing on the total order on vertices. Next, we design a method of lossless compression of graph structured data with geometric information by introducing the notion of a substitution in logic programming. In general, analyzing large graph structured data is a time consuming process. If we can reduce the number of vertices without loss of information, we can speed up such a heavy process. Finally, in order to show an effectiveness of our method, we report several experimental results.

1 Introduction

Background:

Large amount of graph structured data such as map data, CAD and drawing data are stored in databases. Such databases become larger day by day. In analyzing graph structured data, we must often solve the subgraph isomorphism problem which is known to be NP-complete. Furthermore, in extracting characteristic patterns from graph structured data, the number of possible patterns becomes combinatorially explosive. Hence, it is very difficult to analyze graph structured data in practical time. Many graph structured data such as map data, drawing

K.-Y. Whang, J. Jeon, K. Shim, J. Srivatava (Eds.): PAKDD 2003, LNAI 2637, pp. 582–594, 2003.
© Springer-Verlag Berlin Heidelberg 2003

data and CAD have geometric information. Then, the purpose of this paper is to give an effective algorithm for extracting characteristic patterns from graph structured data with geometric information and to present a method of reducing the size of such data without loss of information.

Data model:

To achieve our purpose, first of all, we present a layout graph as a data model for graph structured data with geometric information. A layout graph is a graph having the total orders on all vertices with respect to directions. These orders are given in a layout graph by directed edges labeled with special symbols x and y, which are called x-edges and y-edges, respectively. For example, in Fig. 1, we present a layout graph G corresponding to the map Map in Fig. 1. Each vertex of G corresponds with a crossing in Map, and each edge between vertices of G corresponds with a road between crossings in Map.

Next, we define a layout term graph as a knowledge representation for graph structured data with geometric information. A layout term graph is a layout graph having structured variables. A variable can be substituted by an arbitrary layout graph.

Main Results:

In this paper, we consider the problem of extracting characteristic substructures from graph structured data with geometric information, such as CAD, map data and drawing data. In order to solve this problem, for a set S of layout graphs and a minimum support ρ, we present an algorithm for finding all frequent layout graphs G such that the ratio of layout graphs which contains G in S is more than or equal to ρ. This algorithm is based on levelwise search strategy like Apriori by focusing on the total order on vertices. Next, we give a lossless compression method of graph structured data with geometric data by introducing the notion of a substitution in logic programming. In general, analyzing large graph structured data is a time-consuming process in data mining. If we can reduce the number of vertices without loss of information, we can speed up such a heavy process. Finally, in order to show an effectiveness of our algorithm, we report some experimental results.

Related Words:

Many data mining techniques for analyzing general graph structured data are proposed [2,3,4,5,6]. In [6], we presented a knowledge discovery system KD-FGS which uses Formal Graph System (FGS) [8] as a knowledge representation and refutably inductive inference as a learning method. In [2,3], Cook et al. presented a graph-based knowledge discovery system SUBDUE based on Minimum Description Length principle. Matsuda [5] gave graph-based induction (GBI) to seek the frequent patterns for graph structured data. For general graph structured data, Inokuchi [4] presented an Apriori-based algorithm for mining frequent substructures. For graph structured data, in [7], we presented Layout Formal Graph System (LFGS) as a knowledge representation, that is a logic programming system having a layout term graph as a term. We designed a knowledge discovery

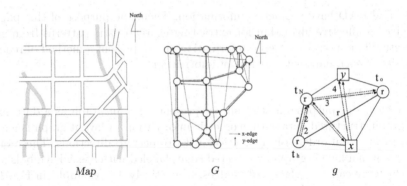

Fig. 1. Map data *Map*, a layout graph *G* which is data model of *Map*, and a layout term graph *g*. A variable is represented by a box with thin lines to its elements and its variable label is in the box. An edge is represented by a thick line.

system which uses LFGS as a knowledge representation language and refutably inductive inference as a learning method.

Organization:
This paper is organized as follows. In the next section, we define a layout graph and a layout term graph with structured variables as a data model and as a knowledge representation for graph structured data with geometric information, respectively. In Section 3, we present an effective algorithm for finding frequent substructures in given data. Furthermore, by using a concept of a substitution in logic programming, we give a lossless compression method for a given layout graph. In Section 4, we report some experimental results by implementing our compression algorithms and applying our compression algorithm to artificial data. In Section 5, we conclude this paper.

2   Knowledge Representation for Graph Structured Data with Geometric Information

In order to deal with graph structured data with geometric information such as CAD, Map data, drawing data, in this section, we introduce a layout graph, which has total order on vertices, as a data model for graph structured data with geometric information. Moreover, we formally define a layout term graph as a knowledge representation for graph structured data with geometric information.

2.1   Data Model

Let Σ and Λ be finite alphabets. An element in Σ and $\Lambda \cup \{x, y\}$ is called a *vertex label* and an *edge label*, respectively. We assume that $\Lambda \cap \{x, y\} = \emptyset$. Let V, E and F be a finite set, a subset of $V \times \Lambda \times V$, and a subset of $V \times \{x, y\} \times V$, respectively. An element in V, E and F is called a *vertex*, an *edge* and a *layout edge*, respectively. An edge having an edge label $a \in \Lambda \cup \{x, y\}$ is called *a-edge*. For E and F, we allow multiple edges and multiple layout edges but disallow self-loops. Let $\mathcal{N}$ be the set of non-negative integers and $\mathcal{N}^+ = \mathcal{N} - \{0\}$. Let $dist : F \rightarrow \mathcal{N}$ be a function which gives a distance between two vertices. A layout edge (u, x, v) (resp. (u, y, v)) means that the vertex u must be placed in the left (resp. lower) side of the vertex v, so that the distance between u and v is more than $dist((u, x, v))$ in the x-direction (resp. $dist((u, y, v))$ in the y-direction). For any edge label $s \in \{x, \dot{y}\}$, an *s-path* is a sequence of layout edges $(u_1, s, u_2), (u_2, s, u_3), ..., (u_n, s, u_{n+1})$ such that $u_i \neq u_j$ for $1 \leq i < j \leq n + 1$, where each u_i $(1 \leq i \leq n + 1)$ is a vertex. If $u_1 = u_{n+1}$, the s-path is called an *s-cycle*.

A *layout graph* is a graph $G = (V, E, F)$ satisfying the following conditions (1) and (2) as a data model for graph structured data with geometric information.

(1) For any two vertices in V, there exist an x-path and a y-path between them.
(2) G has no x-cycle and y-cycle.

By the above conditions (1) and (2), we guarantee that the layout graph G has the two total orders on vertices with respect to x-edges and y-edges.

Example 1. In Fig. 1, we give the graph $G = (V, E, F)$ for the map data *Map*. A vertex in V corresponds with a crossing, a layout edge in E shows that there exists a road between crossings, and an edge in F gives geometric information of crossings.

Let $G = (V, E, F)$ be a layout graph. From the definition of a layout graph, there exists an x-path which have all vertices in V. This x-path is called a *Hamiltonian x-path*. The occurrence order on vertices is easily shown to be unique for all Hamiltonian x-paths. Let $G = (V, E, F)$ be a layout graph. Let P^X and P^y be a longest Hamiltonian x-path and a longest Hamiltonian y-path, respectively. The *minimum layout edge set* of G is the subset F' of F such that $F' = F - \bigcup_{s \in \{x,y\}} \{(c, s, d) \in F \mid (c, s, d)$ is not in P^s and the total of distances between c and d over P^s is greater than or equal to $dist((c, s, d))\}$. Layout graphs $P = (V_P, E_P, F_P)$ and $Q = (V_Q, E_Q, F_Q)$ are *isomorphic*, which is denoted by $P \simeq Q$, if there exists a bijection $\pi : V_P \rightarrow V_Q$ satisfying the following conditions (1)-(4). Let F'_P and F'_Q be the minimum layout edge set of P and Q, respectively. For a layout graph G, let $\varphi_G : V \rightarrow \Sigma$ be a vertex labeling function of G.

(1) $\varphi_P(v) = \varphi_Q(\pi(v))$ for any $v \in V_P$.
(2) $(u, a, v) \in E_P$ if and only if $(\pi(u), a, \pi(v)) \in E_Q$.
(3) For each $s \in \{x, y\}$, $(c, s, d) \in F'_P$ if and only if $(\pi(c), s, \pi(d)) \in F'_Q$ and $dist((c, s, d))$ of P is equal to $dist((\pi(c), s, \pi(d)))$ of Q.

Since the occurrence order of vertices is unique for all Hamiltonian x-paths, the following theorem holds by using polynomial time algorithm for finding a Hamiltonian path for a directed acyclic graph in [7].

Theorem 1. *Let P and Q be layout term graphs. The problem of deciding whether or not P and Q are isomorphic is solvable in polynomial time.*

2.2   Layout Term Graph as a Knowledge Representation

In this section, we formally define a layout term graph as a knowledge representation for graph structured data with geometric information.

Let $G = (V, E, F)$ be a layout graph. Let $\mathcal{X}$ be an alphabet with $(\Sigma \cup \Lambda \cup \{x, y\}) \cap \mathcal{X} = \emptyset$ and H a multi-set of lists of distinct vertices in V. An element in $\mathcal{X}$ and H is called a *variable label* and a *variable*, respectively. For a variable h in H, we denote a set of all elements in h by $V(h)$ and $V(H)$ denotes $\bigcup_{h \in H} V(h)$. We assume two functions, called *rank* and *perm*, for the variable label set $\mathcal{X}$. The first function $rank: \mathcal{X} \to \mathcal{N}^+$ assigns a positive integer for each variable label. A positive integer $rank(x)$ is called the *rank* of x. The second function *perm* assigns a permutation over $rank(x)$ elements for each variable label $x \in \mathcal{X}$. That is, for a variable label $x \in \mathcal{X}$, $perm(x) = \begin{pmatrix} 1 & 2 & \cdots & i & \cdots & k \\ \xi(1) & \xi(2) & \cdots & \xi(i) & \cdots & \xi(k) \end{pmatrix}$ is an operation which changes the i-th element to the $\xi(i)$-th element for each $1 \le i \le k$, where $k = rank(x)$ and $\xi : \{1, \dots, k\} \to \{1, \dots, k\}$ is a permutation. Applying a permutation $perm(x)$ to a variable $h = (v_1, v_2, \dots, v_k)$ is defined as follows. $h \cdot perm(x) = (v_1, v_2, \dots, v_k) \cdot perm(x) = (v_{\xi^{-1}(1)}, v_{\xi^{-1}(2)}, \dots, v_{\xi^{-1}(k)})$. For a list or a set S, the number of elements in S is denoted by $|S|$. Each variable $h \in H$ is labeled with a variable label in $\mathcal{X}$ whose rank is $|h|$. Let I be a subset of $(V \cup H) \times \{x, y\} \times (V \cup H) - V \times \{x, y\} \times V$. An element in I is a layout edge between a variable and a vertex, or between variables. Later we define a *substitution* which replaces variables with layout graphs. In order to specify the positions of the resulting layout graphs after applying a substitution, we give relations between a vertex and a variable, or two variables, in advance, by *dist* and layout edges in I. We assume that each layout edge e in I has information necessary to update geometric information between both endpoints of e in the resulting layout graphs as its edge label. For two variables $p = (u_1, u_2, \dots, u_s)$ and $q = (v_1, v_2, \dots, v_t)$ in H, if there exists neither an x-path from u_s to v_1 nor from v_t to u_1, then we say that variables p and q *overlap* with respect to x-edge each other. Otherwise, we say that variables p and q does not overlap with respect to x-edge. Similarly, for a variable $p = (u_1, u_2, \dots, u_s)$ with a variable label x and a variable $q = (v_1, v_2, \dots, v_t)$ with a variable label y in H, if there exists neither a y-path from $u_{\xi^{-1}(s)}$ to $v_{\mu^{-1}(1)}$ nor from $v_{\mu^{-1}(t)}$ to $u_{\xi^{-1}(s)}$, then we say that variables p and q *overlap* with respect to y-edge each other, where $(u_1, u_2, \dots, u_s) \cdot perm(x) = (u_{\xi^{-1}(1)}, u_{\xi^{-1}(2)}, \dots, u_{\xi^{-1}(s)})$ and $(v_1, v_2, \dots, v_t) \cdot perm(y) = (v_{\mu^{-1}(1)}, v_{\mu^{-1}(2)}, \dots, v_{\mu^{-1}(t)})$. Otherwise, we say that variables p and q does not *overlap* with respect to y-edge.

Definition 1. A 5-tuple $g = (V, E, F, H, I)$ is called a *layout term graph* if it satisfies the following conditions.

(1) $G = (V, E, F)$ is a layout graph.
(2) Let p and q are variables. An x-edge (p, x, q) is in I if and only if p and q overlap with respect to x-edge each other. Similarly, a y-edge (p, y, q) is in I if and only if p and q overlap with respect to y-edge each other.

(3) Let $h = (u_1, u_2, \ldots, u_s)$ be a variable in H and v a vertex in $V - V(h)$. An x-edge (v, x, h) is in I if and only if there exist x-paths from u_1 to v and from v to u_s. Similarly, a y-edge (v, y, h) is in I if and only if there exist y-paths from $u_{\xi^{-1}(1)}$ to v and from v to $u_{\xi^{-1}(s)}$, where $(u_1, u_2, \ldots, u_s) \cdot perm(\lambda(h)) = (u_{\xi^{-1}(1)}, u_{\xi^{-1}(2)}, \ldots, u_{\xi^{-1}(s)})$.

(4) For any variable $h \in H$ and any vertex $v \in V(h)$, I contains no layout edge between h and v.

(5) There is no x-cycle and y-cycle in g.

(6) For any variable $h = (v_1, \ldots, v_k) \in H$ whose variable label is x, there exist an x-path from v_i to v_{i+1} and a y-path from $v_{\xi^{-1}(i)}$ to $v_{\xi^{-1}(i+1)}$ for all $1 \le i \le k - 1$, where $perm(x) = \begin{pmatrix} 1 & 2 & \cdots & k \\ \xi(1) & \xi(2) & \cdots & \xi(k) \end{pmatrix}$.

We remark that a layout term graph $g = (V, E, F, H, I)$ is a layout graph if $H = \emptyset$. If both (u, a, v) and (v, a, u) are in E, we treat the two edges as one undirected edge between u and v.

Example 2. In Fig. 1, as example, we give a layout term graph $g = (V, E, F, H, I)$, where $V = \{v_1, v_2, v_3\}$, $E = \{(v_1, a, v_2), (v_2, a, v_1), (v_2, a, v_3), (v_3, a, v_2)\}$, $F = \{(v_2, x, v_1), (v_1, x, v_3), (v_2, y, v_1), (v_1, y, v_3)\}$, $H = \{(v_2, v_3), (v_1, v_3)\}$ and $I = \{((v_2, v_3), x, (v_1, v_3)), ((v_1, x, (v_2, v_3)), ((v_2, v_3), y, (v_1, v_3)), ((v_1, v_3), y, v_1)\}$. $rank(x) = rank(y) = 2$, $perm(x) = \begin{pmatrix} 1 & 2 \\ 1 & 2 \end{pmatrix}$ and $perm(y) = \begin{pmatrix} 1 & 2 \\ 2 & 1 \end{pmatrix}$. Then, $(v_2, v_3) \cdot perm(x) = (v_2, v_3) \begin{pmatrix} 1 & 2 \\ 1 & 2 \end{pmatrix} = (v_2, v_3)$ and $(v_1, v_3) \cdot perm(y) = (v_1, v_3) \begin{pmatrix} 1 & 2 \\ 2 & 1 \end{pmatrix} = (v_3, v_1)$. $dist((v_2, x, v_1)) = 2$, $dist((v_1, x, v_3)) = 4$, $dist((v_2, y, v_3)) = 2$, and $dist((v_3, y, v_1)) = 3$.

Example 3. Let $g = (V, E, F, H, I)$ be the layout term graph in Fig. 1. Sequences of layout edges $((v_2, x, v_1), (v_1, x, v_3))$ and $((v_2, y, v_1), (v_1, y, v_3))$ are the Hamiltonian x-path and the Hamiltonian y-path of g, respectively.

Let $g = (V, E, F, H, I)$ be a layout term graph, σ be a list $(v_1, v_2, \ldots, v_k)$ of k vertices in V, x be a variable label in $\mathcal{X}$ with $perm(x) = \begin{pmatrix} 1 & 2 & \cdots & k \\ \xi(1) & \xi(2) & \cdots & \xi(k) \end{pmatrix}$. The form $x := [g, \sigma]$ is called a *binding* of x if there are x-paths from v_i to v_{i+1} of g, there are y-paths from $v_{\xi^{-1}(i)}$ to $v_{\xi^{-1}(i+1)}$ of g for all $1 \le i \le k - 1$, and rightmost vertex, leftmost vertex, uppermost vertex and lowermost vertex in g is contained in σ. For a list S of vertices, we denote by $S[m]$ the m-th element of S. A *substitution* θ is a finite collection of bindings $\{x_1 := [g_1, \sigma_1], \ldots, x_n := [g_n, \sigma_n]\}$, where x_i's are mutually distinct variable labels in $\mathcal{X}$ and each g_i $(1 \le i \le n)$ has no variable label in $\{x_1, \ldots, x_n\}$. In the same way as logic programming system, we obtain a new layout term graph f, denoted by $g\theta$, by applying a substitution $\theta = \{x_1 := [g_1, \sigma_1], \ldots, x_n := [g_n, \sigma_n]\}$ to a layout term graph $g = (V, E, F, H, I)$ in the following way. Let $N = |V|$ and $r_i = rank(x_i)$, and the number of vertices of g_i is denoted by N_i for all $1 \le i \le n$.

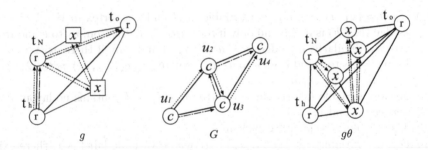

Fig. 2. A layout term graph g and layout graphs G and $g\theta$. In $g\theta$, we show only layout edges which are added by applying θ to g.

(1) First, for all $1 \leq i \leq n$, we replace all variables having the variable label x_i with the layout term graph g_i as follows. Let $h_i^1, h_i^2, \ldots, h_i^{k_i}$ be all variables which are labeled with the variable label x_i. And let C_i be the set of all layout edges incident to one of the variables $h_i^1, h_i^2, \ldots, h_i^{k_i}$. Then, we attach the k_i layout term graphs $g_i^1, g_i^2, \cdots, g_i^{k_i}$, which are copies of g_i, to g according to the k_i lists $\sigma_i^1, \sigma_i^2, \ldots, \sigma_i^{k_i}$, which are the k_i copies of σ_i in the following way. We remove all variables $h_i^1, h_i^2, \ldots, h_i^{k_i}$ from H and all layout edges in C_i from F, and identify the m-th element $h_i^j[m]$ of h_i^j and the m-th element $\sigma_i^j[m]$ of σ_i^j for all $1 \leq j \leq k_i$ and all $1 \leq m \leq r_i$. Then, the resulting graph is denoted by f_0. The vertex label of each vertex $h_i^j[m]$ ($1 \leq m \leq r_i$) is used for f_0, that is, the vertex label of $\sigma_i^j[m]$ is ignored in f_0.

(2) Next, for all $i = 1, \ldots, n$, we add new layout edges to a layout graph f_0 so that the resulting graph f become to be a layout term graph, by using information which each layout edge in C_i has. Then, the resulting graph f is obtained from f_0.

Example 4. Let g and G be a layout term graph and a layout graph in Fig. 2. We assume that layout edges $(v_1, \mathsf{x}, (v_2, v_3))$, $((v_2, v_3), \mathsf{x}, v_1)$, $((v_2, v_3), \mathsf{y}, v_1)$ and $((v_2, v_3), \mathsf{y}, (v_1, v_3))$ have information necessary to update geometric information. For example, the layout edge $(v_1, \mathsf{x}, (v_2, v_3))$ has the information that v_1 is in left side of any vertex of layout graph which is replaced in a substitution. Let $\theta = \{\mathsf{x} := [G, (u_1, u_4)]\}$ be a substitution. Then, $g\theta$ can be obtained from g and θ by applying θ to g.

3 Algorithms for Finding Frequent Substructures and for Lossless Compression of Layout Graphs

In this section, we present effective algorithms for finding frequent substructures from graph structured data with geometric data, and for lossless compression of such data.

3.1   Algorithm for Finding Frequent Substructures

Let P and Q be layout graphs. *the occurrence count* of Q in P, which is denoted by $Occ_P(Q)$, is the number of subgraphs of P which are isomorphic to Q. Let $S = \{G_1, G_2, ..., G_m\}$ be a set of layout graphs. Then, *the occurrence count* of a layout graph Q w.r.t. S, which is denoted by $Occ_S(Q)$, is the number of layout graphs in S which have a subgraph isomorphic to Q. The *frequency* of a layout graph Q w.r.t. S is denoted by $Supp_S(Q) = Occ_S(Q)/m$. Let ρ be a real number where $0 < \rho \leq 1$. A layout graph Q is ρ-frequent w.r.t. S if $Supp_S(Q) \geq \rho$. In general, the above real number ρ is given by a user and is called a *minimum support*. In natural, we define the problem of finding frequent substructures from graph structured data with geometric data as follows.

Finding Frequent Subgraphs Problem
Instance: A set S of layout graphs and a minimum support $0 < \rho \leq 1$.
Problem: Find all ρ-frequent layout graphs w.r.t. S.

In Fig.3, we present an algorithm for solving Finding Frequent Subgraphs Problem. Our algorithm is based on Apriori heuristic presented by Agrawal [1]. As input for our algorithm, given a set S of layout graphs and a minimum support $0 < \rho \leq 1$, our algorithm outputs the set F of all ρ-frequent layout graphs w.r.t. S. In order to find efficiently all ρ-frequent layout graphs from a set of layout graphs, we focus on the total order on vertices in each given layout graph with respect to x-edges. The idea of our algorithm is that we construct candidate layout graphs having $k + 1$ vertices obtained from a layout graph G having k vertices by expanding one vertex from the rightmost vertex of G.

First of all, the algorithm computes the set of all ρ-frequent layout graphs having only one vertex w.r.t. S in line 1 in Fig. 3. For a layout graph $G = (V, E, F)$, the rightmost vertex of G is denoted by rightmost_vertex(G). $G -$ rightmost_vertex(G) is a layout graph $(V - \{$rightmost_vertex(G)$\}, E', F')$ obtained from G such that $E' = E - \{e \mid e$ is incident to rightmost_vertex(G)$\}$ and $F' = F \cup \{(u, y, v) \mid (u, y,$ rightmost_vertex(G)$) \in F, ($rightmost_vertex(G)$, y, v) \in F\} - \{f \mid f$ is a layout edge incident to rightmost_vertex(G)$\}$.

The algorithm repeats the following levelwise search strategy with respect to the number of vertices until $F[k + 1] = \emptyset$ between line 3 and line 13 in Fig. 3. For any two layout graphs G_1 and G_2 in $F[k]$ such that $(G_1 -$ rightmost_vertex(G_1)$) \simeq (G_2 -$ rightmost_vertex(G_2)$)$, a set of candidate layout graphs is obtained from G_1 and G_2 by carrying out the procedure *Apriori_Gen* in line 8. *Apriori_Gen* is a procedure which products a candidate layout graph having $k + 1$ vertices from two layout graphs G_1 and G_2 by expanding the rightmost vertex of G_2 from the rightmost vertex of G_1. There exist at most 4 cases by the positional relations between the rightmost vertices of G_1 and G_2 in the x-direction and in the y-direction. For each cases, there exist at most 2 cases by the existence of an edge between them. Then, we can see that the number of candidate layout graphs is at most 8. As example, in Fig. 4, we show the set of 8

Algorithm Find_Freq_Layout_Graphs

Input: A set S of layout graphs and a minimum support $0 < \rho \leq 1$
Output: The set of F of all ρ-frequent layout graphs w.r.t. S.

1. Let $F[1]$ be all ρ-frequent layout graphs having only one vertex w.r.t. S;
2. $k = 1$;
3. **while** $F[k] \neq \emptyset$ **do**
4. $F[k + 1] = \emptyset$;
5.   **for each** pair of G_1 and G_2 in $F[k]$ such that
            $(G_1 - \text{rightmost\_vertex}(G_1)) \simeq (G_2 - \text{rightmost\_vertex}(G_2))$ **do**
6. $CG[k + 1] = Apriori\_Gen(G_1, G_2)$;
7.   **for each** $G \in CG[k + 1]$ **do**
8.   add G to $F[k + 1]$ if $\dfrac{Occ_S(G)}{|S|} \geq \rho$;
9.   **end;**
10.   **end;**
11. $k = k + 1$;
12. return $F = F[1] \cup F[2] \cup \cdots \cup F[k]$;

Fig. 3. The algorithm for finding all ρ-frequent layout graphs w.r.t. S

candidate layout graphs obtained from the procedure $Apriori\_Gen(G_1, G_2)$, for layout graphs G_1 and G_2 in Fig. 4.

In lines 9 and 10 of Fig. 3, the all ρ-frequent layout graphs obtained from G_1 and G_2 are got by deleting all candidate layout graphs whose frequency is less than ρ from $CG[k+1]$. Finally, the algorithm outputs the set F of all ρ-frequent layout graphs.

3.2 Lossless Compression Algorithm

Let G and P be layout graphs. If $Occ_G(P) \geq 2$ then a description length of G may be reduced by replacing all subgraphs of G which are isomorphic to P with structured variables with same variable label x and adding a binding $x := [P, \sigma]$ to a substitution. By replacing frequent subgraphs with structured variables so that isomorphic subgraphs have same variable labels and adding a corresponding binding to a substitution θ, we can construct a layout term graph g and a substitution θ from G such that G and $g\theta$ are isomorphic. If the total of description lengths of g and θ is less than the description length of G, then G can be compressed by g and θ without loss of information.

In this paper, we simply define the description length of a layout graph $G = (V, E, F)$ as $|G| = |V| + 2 \times |E| + 2 \times |F|$. It is called the *size* of G. In the same way, the size of a layout term graph $g = (V, E, F, H, I)$ is defined as $|g| = |V| + 2 \times |E| + 2 \times |F| + \Sigma_{k \in H} |k| + 2 \times |I|$. Furthermore, the size of a substitution θ is defined as $|\theta| = \Sigma_{x := [G, \sigma] \in \theta} (|G| + |\sigma|)$. For a set S of layout graphs, the description length of S is defined as $DL(S) = \Sigma_{t \in S} (|t|)$.

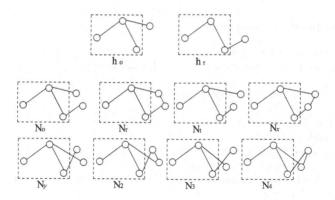

Fig. 4. The set $\{C_1, C_2, \ldots, C_8\}$ of candidate layout graphs which Apriori_Gen(G_1, G_2) outputs.

For a layout graph G, a layout term graph g and a substitution θ such that $G \simeq g\theta$, the compression ratio of G is defined as $\dfrac{|g| + |\theta|}{|G|}$. In the same way, for a set $S = \{G_1, G_2, \ldots, G_n\}$ of layout graphs, a set $D = \{g_1, g_2, \ldots, g_n\}$ of layout term graphs and a substitution θ such that for $1 \leq i \leq n$, $G_i \simeq g_i\theta$, the compression ratio of S is defined as $\dfrac{DL(D) + |\theta|}{DL(S)}$. If we can replace larger subgraphs with a variable whose size is smaller, we can get better compression of a given layout graph.

We consider Lossless Compression Problem defined as follows.

Lossless Compression Problem
Instance: A set S of layout graphs.
Problem: Find a set D of layout term graphs and a substitution θ such that

$$\text{for } 1 \leq i \leq n,\ G_i \simeq g_i\theta \text{ and } \frac{DL(D) + |\theta|}{DL(S)} \text{ is minimal.}$$

When a set of large layout graphs is given as an input, solving Lossless Compression Problem within real time is very hard. Then, in Fig. 5, we present a algorithm for producing a set $D = \{g_1, g_2, \ldots, g_n\}$ of layout term graphs and a substitution θ within real time for a given set $S = \{G_1, G_2, \ldots, G_n\}$ of layout graphs such that for $1 \leq i \leq n$, $G_i \simeq g_i\theta$ and the description length of D and the size of θ are as small as possible.

In line 1 of Fig. 5, we can compute the set F of all ρ-frequent layout graphs w.r.t. S by applying S and ρ to the algorithm Find_Freq_Layout_Graphs. From lines 4 to 20, the algorithm finds a most suitable layout graph P from F in order to compress D, products a new set of layout term graphs from D and adds a new binding to θ. The algorithm repeats the above process until $F = \emptyset$. Finally, the algorithm outputs a set D of layout term graphs and a substitution θ.

Algorithm Lossless_Compression

Input: A set S of layout graphs and a minimum support $0 < \rho \leq 1$.
Output: A set D of layout term graphs and a substitution θ.

1. $F = \text{Find\_Freq\_Layout\_Graphs}(S, \rho)$;
2. $D = S; \theta = \emptyset$;
3. **while** $F \neq \emptyset$ **do**
4.   **for each** $G \in F$ **do**
5.   **for each** layout graph $P = (V_P, E_P, F_P)$ in S having a subgraph
               isomorphic to G **do**
6.   **for each** subgraph $Q = (V_Q, E_Q, F_Q)$ in P with $Q \simeq G$ **do**
7. $h[Q] = ()$;
8.   **for each** $v \in V_Q$ **do**
9.   **if** $\exists(v, a, u) \in E_P - E_Q$ or $\exists(u, a, v) \in E_P - E_Q$ **then**
10.   Append v to $h[Q]$;
11.   **end**;
12.   **end**;
13.   **end**;
14.   Let $Ratio_D(G)$ be the compression ratio of D computed by replacing
               each subgraph Q isomorphic to G with the variable $h[Q]$;
15.   **if** $Ratio_D(G) \geq 1$ **then** $F = F - \{G\}$;
16.   **end**;
17.   Let P be a layout graph in F such that $Rate_D(P) = \min\{Rate_D(G) \mid G \in F\}$;
18.   Let x be a new variable label which is not appeared in θ;
19. $D = \left\{ g \; \middle| \; \begin{array}{l} \text{layout term graph } g \text{ obtained from } f \in D \text{ by replacing each sub-} \\ \text{graph } Q \text{ in } f \text{ isomorphic to } P \text{ with the variable } h[Q] \text{ with } x \end{array} \right\}$;
20.   Add a binding $x := [P, \sigma]$ to θ;
21. $F = F - \{P\}$;
22. **end**;

Fig. 5. The algorithm for lossless compression for a set of layout graphs

4   Experiments

We have implemented our algorithms given in Section 3, and have evaluated our algorithms for artificial data. In this section, we report several experimental results of applying our algorithms to a set of artificial layout graphs. We prepared three artificial layout graphs G_{40}, G_{60} and G_{80} whose number of vertices are 40, 60 and 80, respectively, so that each layout graph has some subgraphs whose occurrence counts are more than or equal to 2. We also prepared two sets of artificial layout graphs, which are created randomly, so that the degree of each vertex is at most 4. In order to evaluate the performance of our algorithm for compression, one is the set S_{10} of 50 layout graphs whose number of vertices is 10 such that S_{10} contains 10 same layout graphs. The other is the set S_{20} of 50 layout graphs whose number of vertices is 20 such that S_{20} contains 10 same layout graphs.

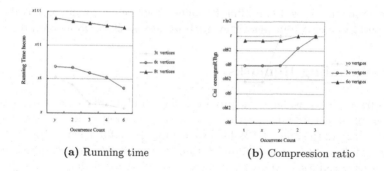

(a) Running time   **(b)** Compression ratio

Fig. 6. Experimental results of compressing a layout graph.

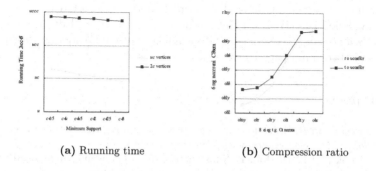

(a) Running time   **(b)** Compression ratio

Fig. 7. Experimental results of compressing a set of layout graph.

We have improved Lossless_Compression given in Fig. 5 for an integer $k > 0$ and a layout graph G, to find all connected subgraphs of G whose occurrence counts are more than or equal to k, and to compress the layout graph by replacing appropriate subgraphs with variables. We have implemented this algorithm in PC with two 2.8GHz CPUs and 2.00GB main memory. Then, we have some experiments of applying this algorithm to three artificial layout graphs G_{40}, G_{60} and G_{80} by varying the minimum occurrence count k from 2 to 6, respectively. Fig. 6 (a) and (b) show the running time and the compression ratio for each occurence count.

Next, we have some experiments of finding all connected ρ-frequent subgraphs by applying our algorithm given in Fig. 5 to S_{10} and S_{20}, respectively, by varying the minimum support ρ from 0.05 to 0.3. The machine used in these experiments is a PC with two 2.4GHz CPUs and 1.00GB main memory. Fig. 7 (a) and (b) show the running time and the compression ratio for S_{10} and S_{20}, respecitively.

From Fig. 6 (a) and Fig. 7 (a), we can see that all of the running time are slightly short when the occurrence counts are high or the minimum supports are low. These results show the effectness of levelwise search strategy which our algorithm is based on. From Fig. 6 (b) and Fig. 7 (b), we can see that the compression ratio are better when the number of founded frequent subgraphs is larger. From these results, finding frequent subgraphs as many as possible leads good performance of our algorithm with respect to the compression to us.

5   Concluding Remarks

In this paper, we considered the problems of (1) finding all frequent substructures from graph structured data with geometric information such as CAD, map data and drawing data, and of (2) lossless compression for such data. We gave a layout graph as a data model and formulated a layout term graph, which has structured variables, as a knowledge representation for such data. We presented effective algorithms for solving the above problems (1) and (2). By implementing and applying our algorithms to a set of artificial layout graphs, we showed the effectiveness of our algorithms. This work is partly supported by Grant-in-Aid for Young Scientists (B) No. 14780303 from the Ministry of Education, Culture, Sports, Science and Technology and Grant for Special Academic Research No. 2117 from Hiroshima City University.

References

1. R. Agrawal and R. Srikant. Fast algorithms for mining association rules. *Proc. of the 20th VLDB Conference*, pages 487–499, 1994.
2. D. J. Cook and L. B. Holder. Substructure discovery using minimum description length and background knowledge. *Journal of Artificial Intelligence Research*, 1:231–255, 1994.
3. D. J. Cook and L. B. Holder. Graph-based data mining. *IEEE Intelligent Systems*, 15(2):32–41, 2000.
4. A. Inokuchi, T. Washio, and H. Motoda. An Apriori-based algorithm for mining frequent substructures from graph data. *Proc. PAKDD-2000, Springer-Verlag, LNAI 1805*, pages 13–23, 2000.
5. T. Matsuda, T. Horiuchi, H. Motoda, and T. Washio. Extension of graph-based induction for general graph structured data. *Proc. PAKDD-2000, Springer-Verlag, LNAI 1805*, pages 420–431, 2000.
6. T. Miyahara, T. Uchida, T. Shoudai, T. Kuboyama, K. Takahashi, and H. Ueda. Discovering knowledge from graph structured data by using refutably inductive inference of formal graph systems. *IEICE Trans. Inf. Syst.*, E84-D(1):48–56, 2001.
7. T. Uchida, Y. Itokawa, T. Shoudai, T. Miyahara, and Y. Nakamura. A new framework for discovering knowledge from two-dimensional structured data using layout formal graph system. *Proc. ALT-00, Springer-Verlag, LNAI 1968*, pages 141–155, 2000.
8. T. Uchida, T. Shoudai, and S. Miyano. Parallel algorithm for refutation tree problem on formal graph systems. *IEICE Trans. Inf. Syst.*, E78-D(2):99–112, 1995.

Upgrading ILP Rules to First-Order Bayesian Networks

Ratthachat Chatpatanasiri and Boonserm Kijsirikul

Department of Computer Engineering, Chulalongkorn University,
Pathumwan, Bangkok, 10330, Thailand.
`ratthachat.c@student.chula.ac.th, boonserm.k@chula.ac.th`

Abstract. Inductive Logic Programming (ILP) is an efficient technique for relational data mining, but when ILP is applied in imperfect domains, the rules induced by ILP often struggle with the overfitting problem. This paper proposes a method to learn first-order Bayesian network (FOBN) which can handle imperfect data powerfully. Due to a high computation cost for directly learning FOBN, we adapt an ILP and a Bayesian network learner to construct FOBN. We propose a feature extraction algorithm to generate features from ILP rules, and use these features as the main structure of the FOBN. We also propose a propositionalisation algorithm for translating the original data into the single table format to learn the remaining parts of the FOBN structure and its conditional probability tables by a standard Bayesian network learner.

1 Introduction

Inductive Logic Programming (ILP) plays the central role to *relational* data mining, but the first-order rules induced by ILP often struggle with the *overfitting* problem (see [10]). Recently, a very nice combination between a first-order and a graphical model is proposed by Koller et al. [6] and by Kersting and De Raedt [7] in the name of *probabilistic relational model (PRM)* and *Bayesian logic program (BLP)*, respectively. In this paper, we will call both models as *first-order Bayesian network (FOBN)*. The model of FOBN is very powerful to handle noisy data by the power of probabilistic theories, and also an expressive model by the power of first-order model combining with Bayesian network. However, the algorithms to learn FOBN proposed in [6] and [7] centrally concern in discovery tasks while the algorithm to learn FOBN as a classifier has not been well discussed. Hence, this paper proposes an efficient framework to learn FOBN as classifier.

2 The Framework Overview

Botta et al. [2] have recently shown that the relational learning problem is linked to the exponential complexity, and Chickering [3] has also shown that learning a Bayesian

K.-Y. Whang, J. Jeon, K. Shim, J. Srivatava (Eds.): PAKDD 2003, LNAI 2637, pp. 595–601, 2003.

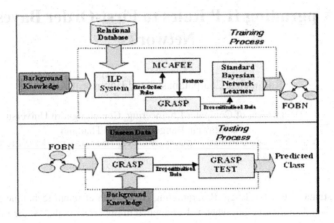

Fig. 1. All processes of the framework.

network is an NP-complete problem. Thus, directly learning FOBN which is the combination of the relational and Bayesian network models from relational data requires extremely more computational cost. This is the main reason that our framework adapts a standard Bayesian network learner and ILP to learn FOBN. All processes of constructing an FOBN classifier and predicting unseen data are shown in Figure 1. To build an FOBN classifier (the training step), our framework induces ILP rules as usual, and we use the MCAFEE algorithm described in Section 3 to extract the features from the rules. We then use these features as *the main structure* of the output FOBN. The main structure of FOBN, in our context, can be constructed by using the class attribute as the output node (node which has no child), using each feature as a parent node of the target class node, and finally, using all literals of each feature as parent nodes of that feature. The reason to do this is because we want to remain the core knowledge of the first-order rule induced by an ILP system. After that, we are ready to learn *the remaining parts of the FOBN structure* and its conditional probability tables (CPTs) by a standard Bayesian network learner. The remaining parts of the structure are additional links those fit the training data but are not explored by our feature extraction algorithm. In order to use a standard Bayesian network learner to learn the rest of the FOBN classifier, we have to translate relational data into the propositional representation, i.e., the single database table. This propositionalisation process is done by our proposed algorithm named GRASP described in Section 3. The testing process in Figure 1 will also be described in Section 3. Here, let us briefly describe our main idea to make the ILP rules more flexible by upgrading them to FOBN. When no ILP rule exactly matches an unseen data, some features of the rules may match that data. Giving these matched features, we can calculate the posterior probability to predict the most appropriate class of the data by the inference method of FOBN.

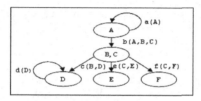

Fig. 2. Dependency graph for rule `r(A) :- a(A),b(A,B,C),c(B,D),d(D),e(C,E),f(C,F)`.

3   The Methods

In this section, we use the notion of *chain* to be a part of a first-order rule (subset of literals in the body of that rule). MCAFEE (Minimal ChAin FEature Extraction) is an algorithm extracting only *significant chains* (*features*) which are not *meaningless* or *redundant* from induced ILP rules. A meaningless chain occurs when some variables occurring in the chain are not introduced by any other literal of that chain before (see [8,9]). The main idea of MCAFEE to extract only meaningful chains is best understood by viewing each rule as a directed dependency graph. For instance, consider the rule `r(A) :- a(A),b(A,B,C),c(B,D),d(D),e(C,E),f(C,F)`. This rule has a corresponding dependency graph as shown in Figure 2. The root node of the graph is the set of variables occurring in the head of the rule. Each of the other nodes represents a set of new variables introduced by a literal in the body, and an edge to the node represents the literal. From the graph, only meaningful chains are extracted by generating *valid chain* which can be defined as follows.

Definition 1 (Valid Chain). In a directed dependency graph of each rule, any path in the graph is called *valid chain* iff that path satisfies one of the two following conditions: (1) some edge(s) of the path connects to any node which is already in the path, or (2) some edge(s) of the path connects to a *leaf* node of the graph.

In our context, a leaf node is one that has no edge leaving out of it. Since every edge connecting to a new node which contains new variable(s) is a literal that introduces the new variable(s), every valid chain guarantees to be meaningful. However, recall that all features extracted by MCAFEE are the parents of the class attribute. This makes some valid chains redundant if we select them all as features. This is because every combination of all parent nodes will be used to construct the CPT of the class attribute. For instance, from Figure 2, the valid chain "a(A), b(A,B,C),f(C,F)" is redundant because it can be constructed by the combination of two valid chains "a(A)" and "b(A,B,C),f(C,F)". MCAFEE, therefore, will extract only meaningful and non-redundant chains as features. This kind of chains is called *minimal valid chain*, and formally defined in Definition 2. Note that by the partial order property [11], every other valid chain can be constructed by the union of two or more minimal valid chains, so all extracted features now guarantee to be non-redundant and meaningful.

Definition 2 (Minimal Valid Chain). Any valid chain r is a *minimal valid chain* iff there is no any other valid chain r' such that r' ⊆ r.

Table 1. A simple example of an incorrect support network.

Example	Class	binding	Feature₁	Feature₂	Feature₃	Feature₄
E_1	+	θ_{E11}	1	0	0	1
E_2	+	θ_{E21}	1	0	0	1
E_3	+	θ_{E31}	0	0	1	1
E_4	+	θ_{E41}	0	0	0	1
E_5	-	θ_{E51}	1	0	1	0
E_6	-	θ_{E61}	1	0	1	0
E_7	-	θ_{E71}	0	0	0	1
		θ_{E72}	0	0	0	1
		θ_{E73}	0	0	0	1
		θ_{E74}	0	0	0	1

As described in Section 2, we adapt a standard Bayesian network learner which receives the flat table as its input to construct the remaining structure and the CPTs of the output FOBN. However, some relational problems cannot be simply translated or solved by propositional learners especially for relational problems containing *non-determinate literals* [9] since a number of columns (attributes) and rows (data cases/ examples) of the output table can be exponential growth. GRASP (GRound substitution AS example Propositionalisation), therefore, is proposed for translating the original data into the single table format efficiently. To prevent the size problem, the number of columns in the output table is *limited* by GRASP. All generated columns are the features extracted by MCAFEE as well as all parents of the features. To generate the rows in the output table, there are many factors to concern. Due to non-determinate literals, each training example may have many different variable bindings so that it may correspond to more than one row in the translated table. Following [5], they regard each *binding (ground substitution) θ as a new example* instead of the original example. Using each binding as example has two advantages when constructing FOBN. First, obviously, with the new definition of the data cases, we can learn FOBN from a standard Bayesian network learner *directly*. Second, when dealing with the non-determinate problem, some methods try to select only one best binding of each original example to be a new example (e.g. [8]). However, when taking a closer look at constructing process of FOBN, the rules obtained by using all bindings as examples are more correct. This property is called *strong completeness* in [5]. Nonetheless, the use of this method still causes two problems. The first problem is about the huge number of the rows in translated table, and the other problem is that, in some cases, a *support network* [7] of each attribute can be incorrect. To illustrate an incorrect support network, consider Table 1. By our method (every binding as example) we get a posterior probability P(+ | Feature₄) = 0.5 and a prior probability P(+) = 0.57, so, by our method, the Bayesian network learner would not choose Feature₄ as a parent node of the class attribute. This is a wrong result because, actually, only one negative example contradicts four positive examples (in fact, P(+ | Feature₄) = 0.8). The wrong result in this case is because of duplicated bindings in the same original negative example. Conversely, duplicated bindings in the same original positive example also lead to an incorrect Bayesian network. Fortunately, both the size and incorrect support network problems can be solved concurrently by selecting only *non-duplicated maximal spe-*

cific bindings as new examples. First of all, we have to give a definition of *maximal specific binding* as follows.

Definition 3 (Maximal Specific Binding). Given a set of all original examples E, for each $e \in E$, a maximal specific binding of e is any variable binding θ such that there is no other variable binding α of e that $\theta \subseteq \alpha$.

Using only maximal specific bindings can prune the redundant binding (see [1]) efficiently, and thus we can decrease the size problem described above. However, the incorrect support network still occurs because of the duplicated bindings. Therefore, GRASP will select only non-duplicated maximal specific bindings in each original example as new examples. By the technique described above, both size and incorrect support network problems are solved by GRASP. Moreover, the important information of the original data still remains in the translated table (see [1]). Since we use all non-duplicated maximal specific bindings as the examples in the training process (constructing FOBN), it is reasonable to use this kind of bindings in the predicting process of unseen data also. As shown in Figure 1, the data will also be transformed by GRASP to get all of its non-duplicated maximal specific bindings. However, there are many ways to predict the class of the unseen data which is now extracted into many new data cases by GRASP. For example, we can use a simple prediction method such that we first calculate all posterior probability for each new data case, and then use only the maximum one to predict the most appropriate class. Hence, in our framework, we use GRASP-TEST which allows the user to specify the best prediction method for the problem.

Table 2. Compared accuracy of our framework with PROGOL on the mutagenesis dataset.

Noise Level in Dataset	PROGOL 0% noise setting	PROGOL 5% noise setting	PROGOL 10% noise setting	PROGOL 15% noise setting	PROGOL +FOBN
0%	**84.58**	82.99	77.14	77.14	84.34
15%	60.56	59.02	61.54	65.31	**74.33**

4 Preliminary Experiments

We have evaluated our framework by performing experiments on the Mutagenesis dataset. In the experiments, we selected PROGOL (CProgol4.2) [11], for learning first-order rules, and WinMine [4] as a standard Bayesian network learner. In the following classification experiments, we used three-fold cross-validation for all testing processes, and for the prediction method in GRASP-TEST, we simply used the maximum posterior probability of all bindings in each original example to predict the most appropriate class as described in Section 3. The average results over three-fold data of all systems are shown in Table 2. "x% noise setting" in the table indicates that noise was set to x% as an option of PROGOL. From the table, when noise was added into the dataset, the accuracy of our framework ("PROGOL + FOBN" in the table) was much higher than PROGOL. There are two main reasons that the FOBN rules induced

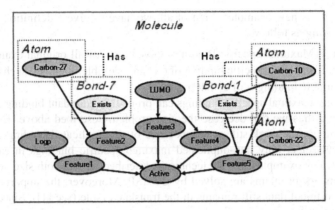

Fig. 3. The example of PRM learned from the mutagenesis problem.

by our framework are more robust against noise than original rules. The first reason is due to the partial matching technique as shown by Kijsirikul et al. [8] that even using the partial matching technique with the extracted features alone, the prediction performance still outperforms using ILP rules directly. The second reason is because of the power of the probabilistic theory which makes the use of the features more flexible by assigning weight into each feature. More details about the efficiency of "ground substitutions as examples" technique, we refer to [5]. As the discovery task, a sample of the learned FOBN classifier of the mutagenesis problem can be demonstrated in PRM representations as shown in Figure 3.

References

1. E. Alphonse and C. Rouveirol. Lazy Propositionalisation for Relational Learning, In Horn W., editor, Proc. of *14th ECAI, Berlin, Allemagne* , IOS Press, 2000.
2. M. Botta and A. Giordana and L. Saitta and M. Sebag. Relational learning: Hard Problems and Phase transition. Selected papers from AIIA'99, Springer-Verlag, 2000.
3. D. M. Chickering. Learning Bayesian Networks is NP-Complete. In D. Fisher and H. J. Lenz, editors, *Learning from Data: Artificial Intelligence and Statistics* V, 1996.
4. D. M. Chickering. WinMine Toolkit. *Technical Report MSR-TR-2002-103*, Microsoft, 2002
5. D. Fensel, M.Zickwolff, and M. Weise. Are substitutions the better examples? In L. De Raedt, editor, *Proc. of the 5th International Workshop on ILP*, 1995.
6. L. Getoor, N. Friedman, D. Koller, and A. Pfeffer. Learning Probabilistic Relational Models. In S. Dzeroski and N. Lavrac, editors, *Relational Data Mining*, 2001
7. K. Kersting, L. De Raedt. Basic Principles of Learning Bayesian Logic Programs. *Technical Report No. 174*, Institute for Computer Science, University of Freiburg, Germany, June 2002
8. B. Kijsirikul, S. Sinthupinyo, and K. Chongkasemwongse. Approximate Match of Rules Using Backpropagation Neural Networks. *Machine Learning Journal*, Volume 44, September, 2001

9. S. Kramer, N. Lavrac and P. Flach. Propositionalization Approaches to Relational Data Mining, in: Dzeroski S., Lavrac N, editors, *Relational Data Mining*, 2001.
10. N. Lavrac and S. Dzeroski. *Inductive Logic Programming: Techniques and Applications.* Ellis Horwood, New York, 1994
11. S. Muggleton. Inverse entailment and Progol. *New Generation Computing*, Special issue on Inductive Logic Programming, 13(3-4):245–286, 1995.

A Clustering Validity Assessment Index

Youngok Kim and Soowon Lee

School of Computing, Soongsil University
1-1 Sang-Do Dong, Dong-Jak Gu, Seoul 156-743 Korea
{yokim, swlee}@computing.ssu.ac.kr
http://mining.ssu.ac.kr/

Abstract. Clustering is a method for grouping objects with similar patterns and finding meaningful clusters in a data set. There exist a large number of clustering algorithms in the literature, and the results of clustering even in a particular algorithm vary according to its input parameters such as the number of clusters, field weights, similarity measures, the number of passes, etc. Thus, it is important to effectively evaluate the clustering results a priori, so that the generated clusters are more close to the real partition. In this paper, an improved clustering validity assessment index is proposed based on a new density function for inter-cluster similarity and a new scatter function for intra-cluster similarity. Experimental results show the effectiveness of the proposed index on the data sets under consideration regardless of the choice of a clustering algorithm.

1 Introduction

Clustering is a method for grouping objects with similar patterns and finding meaningful clusters in a data set. Clustering is important in many fields including data mining, information retrieval, pattern recognition, etc [1,2,3]. A wide variety of algorithms have been proposed in the literature for different applications and types of data set [1,2,4]. These algorithms can be classified according to the clustering methods used, such as partitional vs. hierarchical, monothetic vs. polythetic, and so on [1]. However, even a particular clustering algorithm behave in a different way depending on its features of the data set, field weights, similarity measure, etc [1,2,4]. A more critical issue is to decide the optimal number of clusters that best fits the underlying data set. Thus, it is important to effectively evaluate the clustering results a priori, so that the generated clusters are more close to the real partition.

The goal of this paper is to propose a new clustering validity index, S_Dbw^*, which takes advantage of the concept underlying S_Dbw index [3], while resolving problems in S_Dbw index. S_Dbw^* index can deal with inter-cluster similarity and intra-cluster similarity more robustly and enables the selection of optimal number of clusters more effectively for a clustering algorithm. The rest of this paper is organized as follows. Section 2 presents the related work and Section 3 defines a new validity index, S_Dbw^*. Then, in Section 4, we show the experimental results of S_Dbw^* index com

K.-Y. Whang, J. Jeon, K. Shim, J. Srivatava (Eds.): PAKDD 2003, LNAI 2637, pp. 602–608, 2003.
© Springer-Verlag Berlin Heidelberg 2003

pared with the results of *S_Dbw* index. Finally, in Section 5, we conclude by briefly presenting our contributions and provide directions for further research.

2 Related Work

There exist many indices for clustering validity assessment including *Dunn and Dunn-like* indices [5], *DB*(the Davies-Bouldin) index [6], *RMSSDT*(Root-mean-square standard deviation of the new cluster) index [7], *SPR*(Semi-partial R-squared) index [7], *RS*(R-squared) index [7], *CD*(Distance between two clusters) index [7], *S_Dbw* (Scatter and Density between clusters) index [3], etc. Recently research shows that *S_Dbw* index has better performance and effectiveness than other indices [2,3].

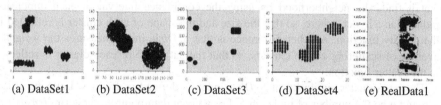

| (a) DataSet1 | (b) DataSet2 | (c) DataSet3 | (d) DataSet4 | (e) RealData1 |

Fig. 1. Sample synthetic datasets (a,b,c,d) and a real dataset(e) [3].

Fig. 1 depicts five data sets used in [3]. Intuitively, DataSet1 has four clusters, DataSet2 has two or three clusters, DataSet3 has seven clusters, DataSet4 has three clusters, and RealData1 has three clusters, respectively. Table 1 shows the optimal number of clusters proposed by *RS, RMSSTD, DB, SD, S_Dbw* indices for the data sets given in Fig. 1, indicating that *S_Dbw* index can find optimal clusters for all given data sets [3].

Table 1. Optimal number of clusters proposed by different validity indices [3].

Index	DataSet1	DataSet2	Dataset3	DataSet4	RealData1
	Optimal number of clusters				
RS, RMSSTD	3	2	5	4	3
DB	6	3	7	4	3
SD	4	3	6	3	3
S_Dbw	4	2	7	3	3

3 *S_Dbw** Index

The definition of *S_Dbw* index indicates that both criteria of inter-cluster and intra-cluster similarity are properly combined, enabling reliable evaluation of clustering results [3]. Though the definition of *S_Dbw* index is proper for both criteria of inter-cluster and intra-cluster similarity, *S_Dbw* index has some problems. First, *Dens_bw* of *S_Dbw* index for inter-cluster similarity finds the neighborhood simply within the boundary of *stdev*. Thus, the detected neighborhood does not reflect the inter-cluster similarity precisely in the case of non-circular clusters. Second, *Scat* of *S_Dbw* index for intra-cluster similarity measures the scattering by calculating relative ratio between the average variance of clusters and the variance of the entire data set. The problem here is *Scat* does not consider the number of tuples in each cluster. If two clusters' variances are almost equal, their variances have the same effect on *Scat* value, regardless of the number of tuples in the clusters. Moreover, since *Scat* value is calculated by averaging the variances of clusters, *S_Dbw* tends to prefer small divided clustering to large combined clustering. These properties of *S_Dbw* index may effect negatively on finding good partition.

To overcome the problems of *S_Dbw*, a new index *S_Dbw** is defined in this paper. *S_Dbw** is based on the same clustering evaluation concepts as *S_Dbw* (inter-, intra-cluster similarity). However, *S_Dbw** index is different from *S_Dbw* index in the sense that it finds the neighborhood by using the confidence interval of each dimension rather than a hyper-sphere, so that the size and the shape of each cluster is reflected in the density function. Another difference is that *S_Dbw** index measures the weighted average of scattering within clusters, so that the data ratio of each cluster is reflected in the scatter function.

The inter-cluster similarity of *S_Dbw** index is defined as follows. Let S be a data set, and n be the number of tuples in S. For a given data point m, the term *density**(m) is defined similarly as in equation (1):

$$density^*(m) = \sum_{l=1}^{n} f^*(x_l, m) .$$ (1)

Here, the function $f^*(x,m)$ denotes whether the distance between two data points x and m is less or equal to the confidence interval for each dimension. Let k be the number of dimensions for the data set, then $f^*(x,m)$ is defined as:

$$f^*(x,m) = \begin{cases} 1 : CI^p \le d(x^p, m^p) \le CI^p, \ (\forall p, \ 1 \le p \le k) \\ 0 : otherwise \end{cases}$$ (2)

where CI^p denotes the confidence intervals of p-th dimension. With the confidence interval of 95%, CI^p is defined as follows:

$$CI^p = u^p \pm (1.96 \times \frac{\sigma^p}{\sqrt{n}})$$ (3)

where u^p and σ^p represent the average and the standard deviation of p-th dimension, respectively. Let m_{ij} be the middle point of the line segment defined by the centers of clusters c_i, c_j. Then, terms in equation (3) are substituted with u_{ij}^p, σ_{ij}^p, n_{ij}, where

$$u_{ij}^p = \frac{u_i^p + u_j^p}{2}, \sigma_{ij}^p = \frac{\sigma_i^p + \sigma_j^p}{2}, n_{ij} = n_i + n_j$$ (4)

From the above definitions, *Dens_bw**(c) is defined as in equation (5):

$$Dens\_bw^*(c) = \frac{1}{c(c-1)} \sum_{i=1}^{c} \left[\sum_{\substack{j=1 \\ i \neq j}}^{c} \frac{density^*(m_{ij})}{\max\{density^*(v_i), density^*(v_j)\}} \right] \quad (5)$$

The basic idea of $Dens\_bw^*$ is that it detects the neighborhood by using the confidence interval for each dimension instead of *stdev* boundary, resulting the different boundary for each dimension according to the cluster's shape. This implies that $Dens\_bw^*$ is more flexible than $Dens\_bw$ of S_Dbw index in detecting neighborhood in case of non-circular or lined up clusters.

For the intra-cluster similarity of S_Dbw^* index, $Scat^*(c)$ is defined as:

$$Scat^*(c) = \frac{1}{c} \sum_{i=1}^{c} \frac{n-n_i}{n} \left\| \sigma(v_i)^2 \right\| \Big/ \left\| \sigma(S)^2 \right\|, \quad where \ \|x\| = (xx^T)^{\frac{1}{2}} \quad (6)$$

where n is the total number of tuples in S and n_i is the number of tuples in cluster c_i.

If two clusters have the same variance but different numbers of tuples, the variance of cluster with more tuples has more effect on the scatter function than the one with less tuples. By minimizing the influence of small clusters or noisy clusters, $Scat^*$ can be more robust than $Scat$ of S_Dbw index.

From equation (5) and (6), we have $S\_Dbw^*(c)$ index, which is defined as follows:

$$S\_Dbw^*(c) = Dens\_bw^*(c) + Scat^*(c). \quad (7)$$

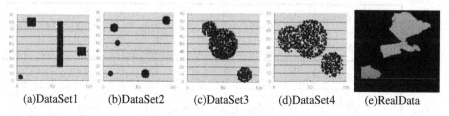

| (a)DataSet1 | (b)DataSet2 | (c)DataSet3 | (d)DataSet4 | (e)RealData |

Fig. 2. Experimental data sets((a)-(d)) and a real data set(e).

4   Experiment

In this section, we experimentally evaluate the proposed validity index, S_Dbw^*, in comparison with S_Dbw index. Fig. 2 shows the experimental data sets consisting of four artificial data sets and one real data set. Intuitively, DataSet1 has four clusters, but one cluster consists of lined up points. DataSet2 has five circular clusters. DataSet3 has three clusters, but two clusters' boundaries are very closely connected. DataSet4 has roughly two clusters, but all points spread broadly. RealData is the North East data set which has 50,000 postal addresses(points) representing three metropolitan areas(New York, Philadelphia and Boston), hence three clusters are in the form of uniformly distributed rural areas and smaller population centers [8].

S_Dbw^* index is evaluated in comparison with S_Dbw index experimentally by applying three representative clustering algorithms from different categories. Clustering algorithms used in the experiments are K-Means(partition-based), EM(statistical model-based), and SOM(Kohonen network-based). For each algorithm, the given number of clusters varies from two to ten. We applied EM and SOM algorithms to DataSet1 and DataSet4, since K-Means algorithm is not appropriate in this case because of its weakness on non-circular and sparse data sets. EM and K-Means algorithms are applied to DataSet2 and DataSet3 to have circular clusters. RealData is evaluated by EM algorithm.

Table 2. Optimal number of clusters found by S_Dbw and S_Dbw^* for each algorithm and data.

Data	Algorithm	Index	Number of Clusters							Not Optimal	
			Optimal								
Data Set1	EM	S_Dbw	4	5	8	6	7	9	10	3	2
			0.125	0.155	0.167	0.178	0.181	0.240	0.312	0.503	0.740
		S_Dbw^*	4	5	6	8	3	2	10	7	9
			0.089	0.109	0.111	0.364	0.409	0.455	0.512	0.520	0.538
	SOM	S_Dbw	10	6	7	9	4	5	8	3	2
			0.133	0.148	0.160	0.163	0.167	0.168	0.175	0.597	0.740
		S_Dbw^*	4	5	8	7	10	9	6	3	2
			0.123	0.228	0.238	0.252	0.307	0.326	0.370	0.428	0.455
Data Set2	EM	S_Dbw	8	4	6	7	9	3	5	10	2
			0.003	0.019	0.134	0.135	0.139	0.145	0.161	0.380	0.716
		S_Dbw^*	8	9	5	7	4	10	6	3	2
			0.003	0.122	0.156	0.182	0.273	0.292	0.301	0.331	2.353
	K-Means	S_Dbw	5	4	6	2	7	8	9	10	3
			0.004	0.019	0.077	0.087	0.099	0.112	0.138	0.142	0.149
		S_Dbw^*	5	6	9	7	8	10	4	3	2
			0.080	0.019	0.077	0.087	0.099	0.112	0.138	0.142	0.149
Data Set3	EM	S_Dbw	10	9	8	6	7	4	5	2	3
			0.259	0.281	0.317	0.358	0.376	0.409	0.413	0.473	0.514
		S_Dbw^*	3	4	10	6	2	9	8	7	5
			0.272	0.306	0.384	0.400	0.442	0.473	0.488	0.509	0.556
	K-Means	S_Dbw	10	9	5	6	2	3	4	8	7
			0.306	0.310	0.440	0.472	0.473	0.482	0.491	0.505	0.530
		S_Dbw^*	3	9	2	4	8	7	10	6	5
			0.281	0.436	0.442	0.442	0.514	0.514	0.550	0.553	0.558
Data Set4	EM	S_Dbw	10	9	8	6	7	5	4	3	2
			0.309	0.317	0.355	0.398	0.420	0.429	0.528	0.722	1.02
		S_Dbw^*	9	2	10	8	5	7	3	6	4
			0.503	0.558	0.561	0.598	0.611	0.648	0.659	0.703	0.907
	SOM	S_Dbw	9	10	8	6	7	5	4	3	2
			0.328	0.329	0.350	0.390	0.445	0.608	0.703	0.934	0.989
		S_Dbw^*	2	9	8	10	5	7	4	3	6
			0.324	0.482	0.487	0.500	0.616	0.670	0.683	0.707	0.712
Real Data	EM	S_Dbw	4	7	9	2	8	10	5	6	3
			1.21	1.31	1.33	1.34	1.34	1.35	1.57	1.62	1.66
		S_Dbw^*	3	4	9	8	5	2	7	10	6
			0.250	0.296	0.348	0.353	0.406	0.421	0.422	0.459	0.468

Table 2 compares the ranks and index values evaluated by S_Dbw and S_Dbw^*. For DataSet1, 3, 4 and RealData, two indices show the remarkable different results. For Dataset1, S_Dbw index evaluates ten clusters as the best one in SOM, while S_Dbw^* index evaluates four clusters as the best one. For DataSet3, S_Dbw index evaluates ten clusters as the best one both in EM and K-Means (and evaluates three clusters as the worst one in EM and the sixth one in SOM), while S_Dbw^* evaluated three clusters as the best one in both algorithms. For Dataset4, S_Dbw index evaluated nine clusters as the best one and two clusters as the worst one in SOM, while S_Dbw^* finds two clusters correctly. Also, for RealData, S_Dbw index evaluated four clusters as the best one, while S_Dbw^* finds three clusters exactly.

In DataSet2, EM algorithm finds five clusters when the number of clusters is given as eight. EM is known to be an appropriate optimization algorithm for constructing proper statistical models on the data. It assigns the density probability of each data to be included in a cluster according to the constructed model. Though EM algorithm calculates the density probability for eight clusters, data points are assigned to only five clusters in this case.

5 Conclusions

In this paper we addressed the important issue of assessing the validity of clustering algorithms' results. We defined a new validity index, S_Dbw^*, for assessing the results of clustering algorithms. For inter-cluster similarity, S_Dbw^* uses the confidence interval for each dimension and for each cluster, while for intra-cluster similarity, S_Dbw^* uses weighted average variance between clusters. We evaluated the flexibility of S_Dbw^* index experimentally on five data sets. The results implies that S_Dbw^* index outperforms S_Dbw index for all clustering algorithm used in the experiments, especially when the real partition is noncircular or sparse.

The essence of clustering is not a totally resolved issue, and depending on the application domain, we may consider different aspects as more significant ones. Having an indication that a good partitioning can be proposed by the cluster validity index, the domain experts may analyze further the validation procedure.

References

1. Jain, A. K., Murty, M.N., Flynn, P.J.: Data Clustering: A Review. *ACM Computing Surveys*, Vol. 31. No. 3 (1999) 264–323
2. Halkidi, M., Batistakis, I., Varzirgiannis, M.: On Clustering Validation Techniques, *Journal of Intelligent Information Systems*, Vol. 17. No. 2-3 (2001) 107–145
3. Halkidi, M., Varzirgiannis, M.: Clustering Validity Assesment: Finding the Optimal Partitioning of a Data Set. *ICDM* (2001) 187–194
4. Fasulo, D.: An Analysis of Recent Work on Clustering Algorithms. *Technical Report*, University of Washington (1999)

5. Dunn, J. C.: Well Separated Clusters and Optimal Fuzzy Partitions. J. Cybern. Vol. 4 (1974) 95–104
6. Davies, DL., Douldin, D. W.: A Cluster Separation Measure. *IEEE Transactions on Pattern Analysis and Machine Intelligence*. Vol. 1. No. 2 (1979)
7. Sharma, S.C.: Applied Multivariate Techniques. John Willy & Sons (1996)
8. Theodoridis, Y.: Spatial Datasets – an unofficial collection. (1996) http://dias.cti.gr/~ytheod/research/datasets/spatial.html

Author Index